Lipids: Biochemistry and Health

Lipids: Biochemistry and Health

Editor: Theo Stewart

R CALLISTO REFERENCE

www.callistoreference.com

Callisto Reference,
118-35 Queens Blvd., Suite 400,
Forest Hills, NY 11375, USA

Visit us on the World Wide Web at:
www.callistoreference.com

ISBN: 978-1-64116-260-9 (Hardback)

Cataloging-in-Publication Data

Lipids : biochemistry and health / edited by Theo Stewart.
    p. cm.
Includes bibliographical references and index.
ISBN 978-1-64116-260-9
1. Lipids. 2. Steroids. 3. Biomolecules. 4. Biochemistry. 5. Health. I. Stewart, Theo.
QP751 .L57 2020
572.57--dc23

Table of Contents

Preface

Over the recent decade, advancements and applications have progressed exponentially. This has led to the increased interest in this field and projects are being conducted to enhance knowledge. The main objective of this book is to present some of the critical challenges and provide insights into possible solutions. This book will answer the varied questions that arise in the field and also provide an increased scope for furthering studies.

A biomolecule that is soluble in nonpolar solvents is called lipid. They are sometimes defined as amphiphilic or hydrophobic small molecules. Hydrocarbons that are used to dissolve other naturally occurring hydrocarbon lipid molecules, which do not dissolve in water such as waxes sterols, triglycerides, fatty acids, etc., are called non-polar solvents. Lipids perform various biological functions that include signaling, acting as structural components of the cell membrane and storing energy. It encompasses molecules such as fatty acids and their derivatives as well as other sterol-containing metabolites such as cholesterol. Their applications are present in nanotechnology as well as in the cosmetic and food industries. Lipids are divided into various categories such as prenol lipids, glycerolipids, sphingolipids, sterol lipids, etc. Different approaches, evaluations, methodologies and advanced studies on lipids have been included in this book. It presents this complex subject in the most comprehensible and easy to understand language. This book is a resource guide for experts as well as students.

I hope that this book, with its visionary approach, will be a valuable addition and will promote interest among readers. Each of the authors has provided their extraordinary competence in their specific fields by providing different perspectives as they come from diverse nations and regions. I thank them for their contributions.

Editor

Fatty Acid Profiles of Stipe and Blade from the Norwegian Brown Macroalgae *Laminaria hyperborea* with Special Reference to Acyl Glycerides, Polar Lipids, and Free Fatty Acids

**Lena Foseid, Hanne Devle, Yngve Stenstrøm,
Carl Fredrik Naess-Andresen, and Dag Ekeberg**

Faculty of Chemistry, Biotechnology and Food Science, Norwegian University of Life Sciences, P.O. Box 5003, 1432 Ås, Norway

Correspondence should be addressed to Hanne Devle; hanne.devle@nmbu.no

Academic Editor: Gerhard M. Kostner

A thorough analysis of the fatty acid profiles of stipe and blade from the kelp species *Laminaria hyperborea* is presented. Lipid extracts were fractionated into neutral lipids, free fatty acids, and polar lipids, prior to derivatization and GC-MS analysis. A total of 42 fatty acids were identified and quantified, including the *n*-3 fatty acids α-linolenic acid, stearidonic acid, and eicosapentaenoic acid. The fatty acid amounts are higher in blade than in stipe (7.42 mg/g dry weight and 2.57 mg/g dry weight, resp.). The highest amounts of *n*-3 fatty acids are found within the neutral lipid fractions with 590.6 ug/g dry weight and 100.9 ug/g dry weight for blade and stipe, respectively. The amounts of polyunsaturated fatty acids are 3.4 times higher in blade than stipe. The blade had the highest PUFA/SFA ratio compared to stipe (1.02 versus 0.76) and the lowest *n*-6/*n*-3 ratio (0.8 versus 3.5). This study highlights the compositional differences between the lipid fractions of stipe and blade from *L. hyperborea*. The amount of polyunsaturated fatty acids compared to saturated- and monounsaturated fatty acids is known to influence human health. In the pharmaceutical, food, and feed industries, this can be of importance for production of different health products.

1. Introduction

The increase in world population and lack of sufficient food beg for new sources of food and feed. As much as 60% of the world food energy intake is provided by the cereals wheat, rice, and corn [1]. These cereals, while high in metabolizable energy and carbohydrates, have small amounts of important nutrients such as proteins, minerals, vitamins (especially A and C), and fatty acids, especially long chained polyunsaturated *n*-3 fatty acids [2–4]. A promising supplement for food and feed is a better utilization of marine resources. World production and harvesting of micro- and macroalgae have doubled from 2004 to 2014 [5]. Still, 97% of the production and harvesting is found in Asia [5]; thus, there is a large potential for expansion in other parts of the world. Macroalgae are a diverse group of marine plants, informally divided into three groups: Rhodophyta (red algae), Chlorophyta (green algae), and Phaeophyta (brown algae). Use of seaweed as feed, food, and fertilizers in times of food shortage was common

in northern Europe from around the 10th century and up until about middle of the 18th century [6]. At present time in Scandinavia and other Western countries, the utilization of seaweeds is limited to industrial products, such as alginate, agar, carrageenan, and thickeners, and only scarcely used in food and feed industries. Biomarine processing industries have great potential in coastal regions. Norway is particularly privileged due to the long coastline combined with the presence of nutritious ocean currents (North Atlantic Drift and Norwegian Coastal Current), which ensure a good climate for growth of marine flora and fauna.

Seaweed has for many years been thought to have positive effect on human health, and consumption of these marine plants has been linked to a lower incidence of cancer, hyperlipidemia, and coronary heart disease [7]. They are also reported to possess antimicrobial, antiviral, anti-inflammatory, and immunotropic properties [8]. Many of the reported medicinal effects of marine algae have not been confirmed, but Brown et al. and Stein and Borden have published comprehensive reviews on this topic [7, 9]. Lipid profiles and compositions can be of importance for human health and for commercial application [10–12]. Eukaryotic algae contain a diverse composition of acyl lipids and their fatty acids, albeit the number of algae which have been comprehensively studied is relatively few [11]. Lipids in macroalgae can be divided into neutral lipids that include monoglycerides, diglycerides, triglycerides, sterols, and polar lipids that include glycolipids, phospholipids, and betaine lipids [10, 11, 13]. An important nutritional benefit of marine macroalgae is attributed to the high level of polyunsaturated fatty acids (PUFAs), especially n-3 and n-6 fatty acids [14]. A diet with a low n-6/n-3 ratio is reported to have suppressive effects on cardiovascular diseases, cancer, and inflammatory and autoimmune diseases [15]. With a global decline in fish stocks, seaweed may be a good option for an alternative and sustainable source of n-3 PUFAs [7]. Previous studies regarding fatty acids in brown macroalgae have had a nutritional or pharmaceutical focus. Fatty acid content has only been determined as one among several parameters, resulting in limited fatty acid profiles [16–20]. More extensive FA profiling has been done on certain species of brown macroalgae [8, 21–23]. Little is however known about fatty acid compositions of individual lipid classes of marine macrophytes in general [8]. The brown macroalgae *Laminaria hyperborea*, a common species of kelp found in the northern Atlantic, has only previously been characterized with regard to fatty acids in three studies [6, 12, 24], all presenting limited profiles identifying no more than nine fatty acids. Within-plant fatty acid distribution for this species has previously only been reported by Schmid and Stengel [24] with *Laminaria hyperborea* harvested on the west coast of Ireland. Variations in distribution of fatty acids between different parts of the seaweed depend on the morphology of the species as well as its physiological and ecological circumstances [25].

Before considering the marine algae as a food source, it is important to assess its nutritional value. In this context, the aim of this study has been to provide a thorough analysis of the fatty acid profiles in stipe and blade from the macroalgae *L. hyperborea*. In this study, the lipids were fractionated into free fatty acids (FFAs), acyl glycerides (neutral lipids, NLs), and polar lipids (PLs) by using solid phase extraction. The fatty acids from each class were identified and quantified by GC-MS.

2. Material and Methods

Chloroform and hexane were of Chromasolv® quality, heptane, diethyl ether, methanol, and NaCl puriss pa. quality, all from Sigma-Aldrich (St. Louis, MO, USA). The acetic acid was from Honeywell Riedel-de Haen (Seelze, Germany).

2.1. Standards. A fatty acid methyl ester (FAME) mix with 37 components (Food Industry FAME MIX, Restek, Bellefonte, PA, USA) was used for identification of the FAMEs. A 21-component FAME mix (Qualmix PUFA Fish M, Methyl Esters (Menhaden Oil), Larodan AB, Solna, Sweden) was used for identification of all-*cis*-6,9,12,15-octadecatetraenoic acid methyl ester, all-*cis*-8,11,14,17-eicosatetraenoic acid

methyl ester, and all-*cis*-7,10,13,16,19-docosapentaenoic acid methyl ester. In addition, the following individual FAME standards were used: nonanoic acid methyl ester (Sigma-Aldrich, St. Louis, MO, USA), 13-methyltetradecanoic acid methyl ester, *trans*-9-tetradecenoic acid methyl ester, *cis*-9-heptadecenoic acid methyl ester, *cis*-13-octadecenoic acid methyl ester, *cis*-9-eicosenoic acid methyl ester, and hexacosanoic acid methyl ester (all from Larodan AB, Solna, Sweden). Three internal standards were used (10 mg/mL, dissolved in CHCl$_3$), one for each lipid fraction; 1,2-dinonadecanoyl-sn-glycero-3-phosphatidylcholine for the PL fractions, nonadecanoic acid for the FFA fractions, and trinonadecanoin for the NL fractions (all from Larodan AB, Solna, Sweden). Nonadecanoic acid methyl ester (Larodan AB, Solna, Sweden) was added to the 37 components FAME mix for retention time identification, since C19:0 was used as internal standard in the samples.

2.2. Pretreatment of L. hyperborea. *L. hyperborea* was provided and identified by FMC BioPolymer AS. It was harvested off the west coast of Norway, outside Sør-Trøndelag County Municipality in October 2015. On board the trawler *L. hyperborea* was rinsed, crude-cut, and preserved with formalin manually. The holdfast was discarded. Once off the trawler stipe and blade were vacuum packed separately. When received at the university, stipe and blade were rinsed with water to eliminate contaminants, frozen with liquid N$_2$ (99.999%, AGA, the Linde Group, Munich, Germany), and freeze-dried (Alpha 2–4 LD plus, Martin Christ Gefriertrocknungsanlagen GmbH, Osterode am Harz, Germany). The freeze-dried material was crushed in a QMM Micromixer and pulverized in a Laboratory Mixer 3100 (Danfoss) by G. A. Lund at Pharmatech AS, Fredrikstad, Norway. The water content of fresh seaweed was measured according to ISO 11465:1993/Cor1:1994.

2.3. Lipid Extraction. Four sample replicates of both stipe and blade were used and they were all treated separately during the sample preparation stages. The lipids were extracted with a modified Folch's method [26]. In brief, 5–10 g alga powder was extracted in a separatory funnel with 10 times its volume CHCl$_3$: MeOH (2/1), and 50 μL of each internal standard was added with a Hamilton® syringe. To induce phase separation, 0.9% NaCl was added after mixing (0.2 times the volume of CHCl$_3$: MeOH). After approximately 20 min the organic phase was transferred to a test tube (Duran® 20 × 150 mm, Mainz, Germany). The polar phase was reextracted with CHCl$_3$, 30–60 mL depending on amount of alga powder. The organic phases of each sample were combined and evaporated with a vacuum evaporator (Q-101, Buchi Labortechnik AG, Flawil, Switzerland) at 40°C, redissolved in 1.00 mL chloroform, and transferred to vials for SPE.

A liquid-handling robot (Gilson, GX-271, ASPEC, Middleton, WI, USA) was used to carry out the SPE procedure. The method used was based on previous work by Pinkart et al. 1998 and Ruiz et al. 2004 [27, 28]. Aminopropyl-modified silica phase SPE columns, 500 mg, 3 mL (Chromabond, Macherey-Nagel, Düren, Germany) were conditioned with 7.5 mL hexane before 500 μL of sample was applied. The

NLs (mono-, di-, and triglycerides) were eluted with 5 mL chloroform, then the FFAs with 5 mL diethyl ether:acetic acid (98 : 2 v/v), and lastly the PLs with 5 mL methanol. The possibility of cross contaminations between any of the three classes was checked by performing tests with standards for each lipid class. The recovery was 90% or higher. The eluates were transferred to culture tubes (Duran 12×100 mm, Mainz, Germany) and evaporated under N_2 (g) at $40°C$.

2.4. Formation of FAMEs. For formation of FAMEs the NL and PL fractions were redissolved in 2 mL of heptane, before addition of 1.5 mL of 3.3 mg/mL sodium methoxide. The sodium methoxide solution was made by dissolving metallic sodium (Purum, Merck, Darmstadt, Germany) in methanol to a concentration of 3.3 mg/mL. The culture tubes were then shaken horizontally for 30 min at 350 rpm (Biosan Ltd., PSU-10i, Riga, Latvia) and left to settle vertically for 10 min before the heptane phases were transferred to vials for storage at $-20°C$ prior to GC-MS analysis. The FFA fractions were redissolved in 1 mL BF_3-MeOH (14%, Sigma-Aldrich, St. Louis, MO, USA). The samples were heated for 5 min at $70°C$ in a water bath. After heating, 1 mL heptane was added to each sample tube before mixing on a vortex mixer. The heptane phases were transferred to vials and stored at $-20°C$ prior to analysis by GC-MS.

2.5. Analysis of FAMEs by GC-MS. The analysis of the FAMEs was based on a previously published method [29]. Shortly, the analyses were carried out on an Agilent 6890 Series gas chromatograph (GC; Agilent Technology, Wilmington, DE, USA) in combination with an Autospec Ultima mass spectrometer (MS; Micromass Ltd., Manchester, England) using an EI ion source. The GC was equipped with a CTC PAL Autosampler (CTC Analytics, AG, Zwingen, Switzerland). Separation was carried out on a 60 m Restek column (Rtx®-2330) with 0.25 mm ID and a $0.2 \mu m$ film thickness of fused silica 90% biscyanopropyl/10% phenylcyanopropyl polysiloxane stationary phase (Restek Corporation, Bellefonte, PA, USA). For carrier gas, helium (99.99990%, from Yara, Rjukan, Norway) was used at 1 mL/min constant flow. The EI ion source was used in positive mode, producing 70 eV electrons at $250°C$. The MS was scanned in the range 40–600 m/z with 0.3 s scan time, 0.2 s interscan delay, and 0.5 s cycle time. The transfer line temperature was set at $270°C$. The resolution was 1000.

A split ratio of 1/10 was used with injections of $1.0 \mu L$ sample. Two injections parallels were used for each sample replicate. Identification of fatty acids was performed by comparing retention times with standards as well as MS library searches. MassLynx version 4.0 (Waters, Milford, MA, USA) and NIST 2014 Mass Spectral Library (Gaithersburg, MD, USA) were used. Relative response factors previously determined by Devle et al. [29] were employed for quantitative determination. The resulting amounts are given in $\mu g/g$ dry weight (DW). The GC oven had a start temperature of $65°C$, held for 3 min, the temperature was then raised to $150°C$ ($40°C$/min), held for 13 min, before being increased to $151°C$ ($2°C$/min), held for 20 min, with a slow increase to $230°C$ ($2°C$/min), and held for 10 min, before a final increase

to $240°C$ ($50°C$/min), and the end temperature was held for 3.7 min.

3. Results and Discussion

We have identified and quantified 42 different fatty acids in *L. hyperborea*, as shown in Table 1. This is a significantly higher number than previously reported by others for this species [6, 12, 24]. Seaweeds usually contain a lipid level of <1–5% [19, 21, 30]. The portions of the total lipids that contain molecules with fatty acids depend significantly on species, geographical location, and seasonal changes [12, 18, 24, 31, 32]. In our study, the total FA (TFA) content relative to dry weight in blade and stipe was 0.74% and 0.26%, respectively. This 3 : 1 ratio between blade and stipe is consistent with Schmid and Stengels [24] findings of within-plant variations for the same species, even though they had twice the TFA content (0.5% and 1.5% in stipe and blade, resp.). The water content was found to be 83.3% ± 0.5 and 85.6% ± 0.8 in blade and stipe, respectively. The fatty acid profile was determined for the NL, FFA, and PL fractions in stipe and blade separately. For the individual lipid fractions, the %TFA was the highest in NLs with 42.9% and 54.5% in stipe and blade, respectively. The PL fraction for the stipe consisted of 48.5% TFA versus 31.5% TFA in blades. The %TFA for the FFAs ranged from 8.6% in stipe to 13.9% in blade. While up to 41 different fatty acids were detected within a lipid fraction, the same 10 fatty acids predominated in all fractions in both stipe and blade. In this group of 10, three FAs were saturated (SFA, C14:0, C16:0, and C18:0), one was monounsaturated (MUFA, C18:1 *cis*-9) and the remaining five were polyunsaturated (PUFA, C18:2 *cis*-9,12, C18:3 *cis*-9,12,15, C18:4 *cis*-6,9,12,15, C20:4 *cis*-5,8,11,14, and C20:5 *cis*-5,8,11,14,17).

The predominating fatty acids constitute more than 90% of the total fatty acid content in all the fractions, as shown in Figure 1. They are found in amounts varying from 0.65 to 1200 $\mu g/g$ DW (Table 1). A fatty acid was classified as predominating if it was above 2% of the total fatty acid content in at least one of the lipid fractions in either stipe or blade. All these ten fatty acids corresponded to those identified by others [6, 12, 24]. Schmid and Stengel [24] identified C18:3*n*-6, at 1.2% in stipe and 5.5% in blade, which differs from our results, where C18:3*n*-6 is not above 2% in any of the blade lipid fractions. This could be due to geographical and/or seasonal variations. The same FA was not detected at all by Van Ginneken et al. [12] and also not reported by Mæhre et al. [6], who both studied *L. hyperborea* as one of several macroalgal species. Only a maximum of two *trans* fatty acids, C14:1 *trans*-9 and C16:2 *cis* or *trans*-7,10, were identified in the samples, both in relatively low amounts (≤2.53 $\mu g/g$ DW). Among the predominating fatty acids are important dietary *n*-3 fatty acids such as α-linolenic acid (ALA, C18:3*n*-3), stearidonic acid (SDA, C18:4*n*-3), and eicosapentaenoic acid (EPA, C20:5*n*-3) as well as two *n*-6 fatty acids, linoleic acid (LA, C18:2*n*-6),and arachidonic acid (AA, C20:4*n*-6). How favorable *L. hyperborea* is for the human diet (and thus also in animal feed) depends on several factors, for example, content of essential FAs (LA and ALA), other important nutritional

TABLE 1: Fatty acid content in the lipid fractions (mean ± SD, $\mu g/g$ DW) of stipe and blade from *L. hyperborea* (n = 4, two injection parallels for each of these four sampling events).

Fatty acid		Stipe NL	Stipe FFA	Stipe PL	Blade NL	Blade FFA	Blade PL
C7:0 5 methyl[a]		1.32 ± 0.09	n.d.	n.d.	0.49 ± 0.03	n.d.	n.d.
C8:0		1.47 ± 0.11	0.70 ± 0.04	n.d.	5.36 ± 0.63	0.65 ± 0.03	n.d.
C9:0		n.d.	0.41 ± 0.06	n.d.	n.d.	0.64 ± 0.03	n.d.
C10:0		0.76 ± 0.05	n.d.	n.d.	4.98 ± 0.63	0.39 ± 0.06	n.d.
C12:0		0.53 ± 0.09	1.66 ± 0.22	0.13 ± 0.01	0.73 ± 0.03	2.87 ± 0.12	0.09 ± 0.01
C13:0		n.d.	0.19 ± 0.01	n.d.	0.17 ± 0.01	0.96 ± 0.03	0.07 ± 0.01
C14:0		101.28 ± 1.80	21.94 ± 0.78	272.39 ± 9.92	223.80 ± 4.65	94.20 ± 4.02	312.39 ± 5.60
C14:0 13 methyl		n.d.	0.33 ± 0.05	n.d.	n.d.	0.39 ± 0.02	n.d.
C14:1 *trans*-9		n.d.	0.20 ± 0.02	n.d.	n.d.	1.86 ± 0.12	n.d.
C14:1 *cis*-9		0.89 ± 0.04	0.13 ± 0.02	0.27 ± 0.03	2.45 ± 0.03	0.46 ± 0.02	0.38 ± 0.01
C15:0		0.95 ± 0.05	1.00 ± 0.05	2.52 ± 0.03	8.73 ± 0.18	6.40 ± 0.18	10.43 ± 0.14
C16:0		163.36 ± 5.08	82.19 ± 3.08	276.64 ± 6.91	900.21 ± 17.85	298.88 ± 12.12	417.29 ± 5.39
C16:1[b]		n.d.	n.d.	n.d.	0.77 ± 0.10	1.31 ± 0.03	0.65 ± 0.02
C16:1[b]		n.d.	0.53 ± 0.04	0.37 ± 0.05	2.24 ± 0.09	2.75 ± 0.09	1.05 ± 0.02
C16:1 *cis*-9		74.99 ± 2.56	14.38 ± 0.21	54.60 ± 1.02	179.65 ± 1.94	74.12 ± 2.11	104.78 ± 1.14
C17:0		0.32 ± 0.02	0.52 ± 0.06	0.57 ± 0.04	6.62 ± 0.24	1.73 ± 0.07	1.83 ± 0.05
C16:2 *cis* or *trans*-7,10	(*n*-6)	0.62 ± 0.04	0.13 ± 0.02	0.49 ± 0.07	2.00 ± 0.07	1.72 ± 0.05	2.53 ± 0.05
C17:1 *cis*-9		0.33 ± 0.03	0.10 ± 0.04	0.35 ± 0.3	4.14 ± 0.24	2.02 ± 0.05	1.96 ± 0.06
C18:0		5.95 ± 0.28	17.35 ± 0.71	2.54 ± 0.09	84.00 ± 2.01	21.04 ± 0.70	4.42 ± 0.06
C18:1 *cis*-9		234.12 ± 5.87	47.46 ± 0.43	418.07 ± 10.33	1200.51 ± 29.95	219.90 ± 5.17	546.84 ± 11.59
C18:1 *cis*-11		3.37 ± 0.31	2.27 ± 0.04	4.85 ± 0.22	11.39 ± 0.21	6.64 ± 0.14	7.68 ± 0.14
C18:2 all-*cis*-9,12 (LA)[c]	(*n*-6)	44.67 ± 1.76	2.70 ± 0.09	48.39 ± 101	183.66 ± 4.19	27.79 ± 0.67	88.57 ± 1.19
C18:3 all-*cis*-6,9,12	(*n*-6)	3.23 ± 0.22	n.d.	3.59 ± 0.13	14.02 ± 0.18	3.02 ± 0.06	12.29 ± 0.19
C20:0		6.67 ± 0.54	1.35 ± 0.07	1.52 ± 0.10	49.59 ± 1.19	9.80 ± 0.25	4.79 ± 0.09
C18:3 all-*cis*-9,12,15 (ALA)[d]	(*n*-3)	7.31 ± 0.31	0.65 ± 0.03	2.37 ± 0.15	83.56 ± 1.38	21.99 ± 0.53	25.22 ± 0.68
C20:1 *cis*-9	(*n*-3)	n.d.	0.39 ± 0.08	3.34 ± 0.18	7.87 ± 0.40	4.17 ± 0.05	9.75 ± 0.19
C20:1 *cis*-11	(*n*-6)	n.d.	0.16 ± 0.04	0.39 ± 0.05	0.89 ± 0.05	0.39 ± 0.02	0.95 ± 0.03
C18:4 all-*cis*-6,9,12,15 (SDA)[e]	(*n*-3)	5.6 ± 0.3	0.93 ± 0.03	10.74 ± 0.24	68.90 ± 1.29	64.62 ± 1.33	259.58 ± 2.77
C20:2 all-*cis*-11,14	(*n*-6)	3.73 ± 0.35	0.27 ± 0.05	4.88 ± 0.18	10.35 ± 0.33	2.87 ± 0.12	8.57 ± 0.17
C20:3[b]		1.93 ± 0.16	0.12 ± 0.02	0.90 ± 0.07	1.91 ± 0.05	0.50 ± 0.03	1.08 ± 0.03
C20:3[b]		1.49 ± 0.14	0.07 ± 0.01	1.04 ± 0.06	4.50 ± 0.59	0.72 ± 0.04	2.71 ± 0.05
C20:3[b]		n.d.	0.04 ± 0.01	0.43 ± 0.03	1.20 ± 0.63	0.28 ± 0.02	0.79 ± 0.03
C20:3 all-*cis*-8,11,14	(*n*-6)	3.63 ± 0.35	0.16 ± 0.02	5.19 ± 0.20	18.23 ± 1.29	0.80 ± 0.02	4.46 ± 0.08
C22:0		n.d.	0.06 ± 0.01	n.d.	0.93 ± 0.13	0.41 ± 0.02	n.d.
C20:4 all-*cis*-5,8,11,14 (AA)[f]	(*n*-6)	344.56 ± 11.46	17.39 ± 0.39	82.89 ± 1.49	516.97 ± 11.52	66.32 ± 2.11	139.90 ± 1.25
C20:4[b]		n.d.	n.d.	0.44 ± 0.04	2.80 ± 0.14	0.92 ± 0.04	3.21 ± 0.07
C20:4 all-*cis*-8,11,14,17	(*n*-3)	3.29 ± 0.26	n.d.	2.39 ± 0.13	13.52 ± 0.33	3.76 ± 0.17	10.12 ± 0.17
C20:5 all-*cis*-5,8,11,14,17 (EPA)[g]	(*n*-3)	81.72 ± 3.28	2.91 ± 0.07	40.62 ± 0.87	409.27 ± 9.51	81.58 ± 1.90	339.09 ± 3.60

TABLE 1: Continued.

Fatty acid		Stipe			Blade		
		NL	FFA	PL	NL	FFA	PL
C24:0		n.d.	0.29 ± 0.02	n.d.	0.64 ± 0.08	1.45 ± 0.06	n.d.
C22:5 all-*cis*-7,10,13,16,19	(*n*-3)	3.01 ± 0.28	0.18 ± 0.01	1.56 ± 0.09	15.37 ± 0.50	4.97 ± 0.12	11.29 ± 0.26
C22:6 all-*cis*-4,7,10,13,16,19 (DHA)[h]	(*n*-3)	n.d.	0.08 ± 0.01	0.33 ± 0.06	n.d.	1.67 ± 0.06	0.22 ± 0.02
C26:0		n.d.	0.27 ± 0.02	n.d.	n.d.	1.08 ± 0.07	n.d.
Total		1101.14	219.51	1244.77	4042.45	1033.02	2334.99

[a]Fatty acids are identified by NIST library search only; [b]unknown isomer of fatty acid, identified by NIST library search only; [c]LA: linoleic acid; [d]ALA: alpha linolenic acid; [e]SDA: stearidonic acid; [f]AA: arachidonic acid; [g]EPA: eicosapentaenoic acid; [h]DHA: docosahexaenoic acid; n.d. = not detected. NL, neutral lipid; FFA, free fatty acid; PL, polar lipid.

Table 2: Sum of SFAs, MUFAs, and PUFAs, as well as *n*-6 and *n*-3 in *L. hyperborea* given in µg/g DW, (*n* = 4, two injection parallels for each sample replicate).

	Stipe			Blade		
	NL	FFA	PL	NL	FFA	PL
$\sum$ SFA	282.62 ± 5.04	128.26 ± 3.10	556.30 ± 10.02	1286.26 ± 18.12	440.89 ± 12.01	751.31 ± 6.15
$\sum$ MUFA	313.71 ± 6.01	65.72 ± 0.40	482.21 ± 10.12	1409.90 ± 29.37	313.60 ± 5.47	674.05 ± 12.35
$\sum$ PUFA	504.81 ± 11.14	25.63 ± 0.41	206.27 ± 1.18	1346.68 ± 11.22	278.53 ± 2.12	910.63 ± 4.22
Total	1101.14	219.51	1244.77	4042.45	1033.02	2334.99
$\sum$*n*-3	100.92 ± 3.11	4.75 ± 0.07	58.02 ± 0.91	590.62± 10.14	178.58 ± 2.23	645.52 ± 4.34
$\sum$*n*-6	402.10 ± 11.01	20.65 ± 0.40	145.44 ± 1.09	745.24 ± 11.36	97.53 ± 2.18	256.51 ± 1.11
n-6/*n*-3	3.97	4.35	2.51	1.26	0.54	0.40
PUFA/SFA	1.79	0.20	0.37	1.05	0.63	1.21

The standard deviations are the highest standard deviation among the summarized values. NL, neutral lipid; FFA, free fatty acid; PL, polar lipid; SFA, saturated fatty acid; MUFA, monounsaturated fatty acid; PUFA, polyunsaturated fatty acid.

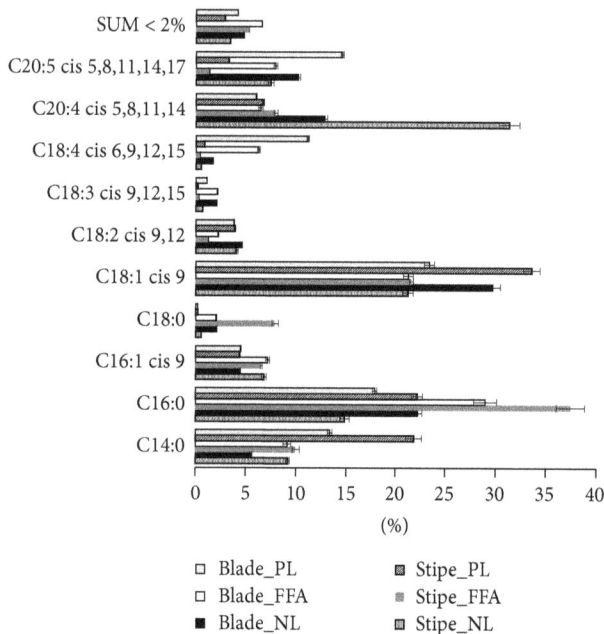

Figure 1: Fatty acid profile for fatty acids contributing more than 2% of total fatty acid content, in at least one lipid fraction. SUM < 2% is the summarized contribution of the remaining fatty acids (*n* = 4, two injection parallels for each of these four sample replicates, error bars = ±SD). NL, neutral lipid; FFA, free fatty acid; PL, polar lipid.

FAs (SDA, AA, and EPA), and the ratios between PUFA/SFA and *n*-6/*n*-3 fatty acids.

It is known that brown seaweeds grown in temperate or subarctic areas can accumulate *n*-3 and *n*-6 PUFAs [33]. In both stipe and blade the lowest amounts of *n*-3 and *n*-6 fatty acids are found in the FFA fraction. The highest amounts are found in the NL fraction, with the exception of *n*-3 in blade where the amount in the PL fraction is higher than in the NL fraction (646 and 591 µg/g DW, resp.). The highest amounts of the essential FAs LA and ALA were found in the NLs for the blade (183.7 µg/g DW and 83.6 µg/g DW, resp.). However, for stipe, the highest amount of LA was found in the PL fraction

(48.4 µg/mL) and highest amount of ALA was found in the NL fraction (7.3 µg/g DW). There was a lower proportion of the *n*-6 FA arachidonic acid in blade versus stipe (9.8% and 17.3%, resp.). This corresponds to what was reported by Schmid and Stengel [24].

The World Health Organization (WHO) recommends a daily intake of 0.25 g EPA+ C22:6 *n*-3 (DHA) as part of a healthy diet [34]. Even though seaweed can have high levels of EPA, the fatty acid DHA is generally absent or only found in small amounts in different phaeophytes [18]. The highest amount of DHA in our study was found in the blade FFA fraction (1.7 µg/g DW). With a total content of EPA+DHA in blade at 832 µg/g DW, achieving the recommended amount by consumption of seaweed alone is highly unlikely, as the daily intake would have to be approximately 300 g dried seaweed (or 1500 g fresh seaweed). Although *n*-3 and *n*-6 PUFAs are usually easily oxidized, studies have shown that these PUFAs have exhibited high oxidative stability in seaweed lipids in dried seaweed products [35]. The reason for this might be due to a protective effect of galactosyl and sulfo-quinovosyl moieties on PUFAs bonded to glycoglycerolipids (the main membrane lipids) [33].

Stipe and blade differ not only in content of the individual FAs but also in the amounts of SFAs, MUFAs, and PUFAs (Table 2). The stipe had the highest distribution of SFA with 37.7% of TFA, whereas MUFA and PUFA for the stipe were at 33.6% and 28.7%, respectively. The blades, however, had a higher distribution of PUFAs with 34.2% of TFA and a lower distribution of SFA and MUFA with 33.4% and 32.4%, respectively. Stengel 2015 also reported a higher distribution of PUFAs in blade versus stipe with 52.0% and 32.2%, respectively. The FA distribution differs significantly with geographical and seasonal variations, most likely due to nutrition, light conditions, and other biological factors. Van Ginneken et al. [12] found a PUFA distribution of 53% of TFA in *L. hyperborea* (fronds) harvested in France with 25% SFA and 22% MUFA. Mæhre et al. [6] harvested the same species (whole plant) from the Norwegian coast and reported 34.2% PUFA, 33.7% SFA, and 26.5% MUFA. These results are very similar to our findings, even though there was a significant distance in time of harvest (May/June 2010 versus October

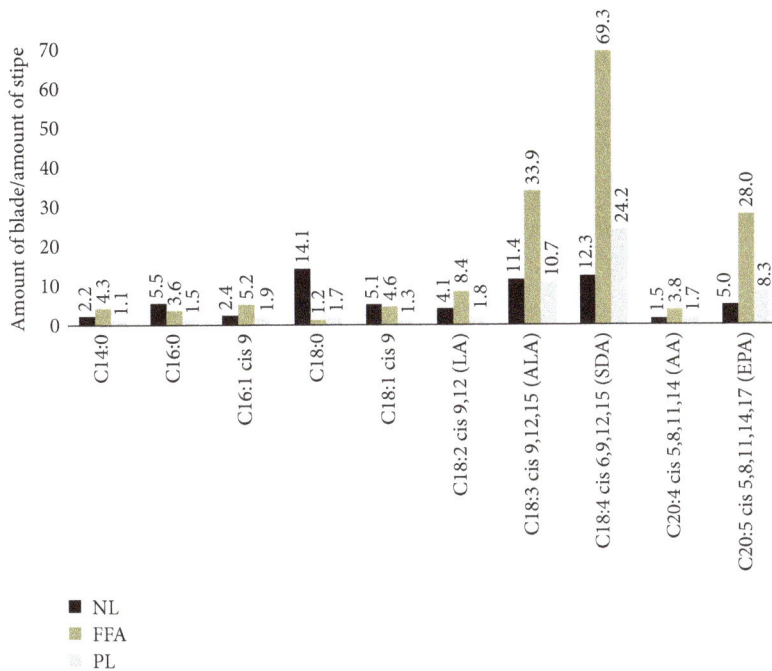

FIGURE 2: Ratio between average blade and stipe amounts in the predominating fatty acids. NL, neutral lipid; FFA, free fatty acid; PL, polar lipid.

2015). The total amount of PUFAs in blade is 3.4 times higher than in stipe, and the total amounts of SFAs and MUFAs are 2.6 and 2.8 times higher, respectively. PUFAs are preferred over SFAs from a dietary perspective, and replacing SFAs with PUFAs in the diet decreases the risk of coronary heart disease [34]. The NL fraction in stipe and the PL fraction in blade have the highest PUFA/SFA ratio of 1.79 and 1.21, respectively. When combining the lipid fractions the blade FAs had a higher PUFA/SFA ratio compared to the stipe FAs (1.02 versus 0.76).

L. hyperborea has an *n*-6/*n*-3 ratio of 0.8/1 in blade and 3.5/1 in stipe. The ratios vary between the lipid fractions, as seen in Table 2, but are higher in stipe than in blade. In Western diet the *n*-6/*n*-3 ratio is 15–20/1 and this is significantly higher than ~1, which was normal during human evolution [15, 36]. However, for health benefits lowering this ratio is considered to be beneficial and associated with prevention of inflammatory, cardiovascular, and neural disorders [12]. *n*-6/*n*-3 ratios of 2–5/1 are reported to have suppressive effects on cardiovascular, inflammatory, and autoimmune diseases [15, 37, 38]. Since there is a significant difference between the *n*-6/*n*-3 ratios in stipe and blade, using only blade could be considered if a very low *n*-6/*n*-3 ratio is desired. Though, in this context it should also be mentioned that FAO in their 2010 report [34] do not consider this ratio to be important and give no specific recommendations of such.

While the same fatty acids predominate, the amounts in blade are consistently higher than those in stipe, as seen in Figure 2. This is consistent with results found by Schmid and Stengel [24]. At minimum, the amounts in blade are 1.1 times higher than in stipe for myristic acid (C14:0) in the PL fraction, and at maximum, 69.3 times higher than in stipe for SDA in the FFA fraction. For the FFA and PL fractions, the largest differences are found in the fatty acids ALA, SDA, and EPA. In the NL fraction, the largest difference between stipe and blade amounts is found in stearic acid (C18:0), while blade and stipe amounts are almost equal in the FFA and PL fractions for the same fatty acid.

4. Conclusions

A total of 42 different fatty acids are identified and quantified in the stipe and blade of *L. hyperborea*, with maximum two fatty acids having *trans* configuration. Some fatty acids are found in either stipe or blade, while others are only present in certain lipid fractions (NL, FFA, and PL) within stipe and blade. Among the predominating fatty acids are the *n*-3 fatty acids ALA (10.4 and 131 μg/g DW), SDA (17.2 and 394 μg/g DW), and EPA (126 and 830 μg/g DW), as well as two *n*-6 fatty acids: LA (96 and 296 μg/g DW) and AA (444 and 723 μg/g DW); the values in parentheses are for stipe and blade, respectively. The ratios between *n*-6 and *n*-3 fatty acids are ≤4.4/1 in all lipid fractions but especially low (≤1.3/1) in blade. The blades also presented the highest PUFA/SFA ratio. Regarding the potential of commercialization in respect of nutritional applications of *L. hyperborea*, blade is found to represent the most suitable material, due to higher levels of PUFAs and a low *n*-6/*n*-3 ratio.

Acknowledgments

The authors would like to thank FMC Biopolymer for sampling the *L. hyperborea* and Pharmatech AS for milling all samples. The authors would also like to thank the Institute of Marine Research for allowing them to use their picture of a kelp forest in the graphical abstract. This work was supported by the Norwegian University of Life Sciences.

References

[1] "FAO: Staple foods: What do people eat?" in *Dimensions of Need: An atlas of Food and Agriculture*, 1995.

[2] O. K. Chung and Ohm J.-B., "Cereal Lipids," in *Handbook of Cereal Science and Technology*, K. Kulp and J. G. Ponte, Eds., Marcel Dekker, Inc., 2000.

[3] "FAO Food and Nutrition Series no 29 by M. C. Lathman: Chapter 26. Cereals, starchy roots and other mainly carbohydrate foods, in: Human nutrition in the developing world 1997".

[4] L. Yu, G. Li, M. Li, F. Xu, T. Beta, and J. Bao, "Genotypic variation in phenolic acids, vitamin E and fatty acids in whole grain rice," *Food Chemistry*, vol. 197, pp. 776–782, 2016.

[5] FAO, *Fishery Statistical Collections, Global Production*, Aquatic Plants, 2014.

[6] H. K. Mæhre, M. K. Malde, K.-E. Eilertsen, and E. O. Elvevoll, "Characterization of protein, lipid and mineral contents in common Norwegian seaweeds and evaluation of their potential as food and feed," *Journal of the Science of Food and Agriculture*, vol. 94, no. 15, pp. 3281–3290, 2014.

[7] E. M. Brown, P. J. Allsopp, P. J. Magee et al., "Seaweed and human health," *Nutrition Reviews*, vol. 72, no. 3, pp. 205–216, 2014.

[8] N. M. Sanina, S. N. Goncharova, and E. Y. Kostetsky, "Fatty acid composition of individual polar lipid classes from marine macrophytes," *Phytochemistry*, vol. 65, no. 6, pp. 721–730, 2004.

[9] J. R. Stein and C. A. Borden, "Causative and beneficial algae in human disease conditions: a review," *Phycologia*, vol. 23, no. 4, pp. 485–501, 1984.

[10] I. A. Guschina and J. L. Harwood, "Lipids and lipid metabolism in eukaryotic algae," *Progress in Lipid Research*, vol. 45, no. 2, pp. 160–186, 2006.

[11] J. L. Harwood and I. A. Guschina, "The versatility of algae and their lipid metabolism," *Biochimie*, vol. 91, no. 6, pp. 679–684, 2009.

[12] V. J. T. Van Ginneken, J. P. F. G. Helsper, W. De Visser, H. Van Keulen, and W. A. Brandenburg, "Polyunsaturated fatty acids in various macroalgal species from north Atlantic and tropical seas," *Lipids in Health and Disease*, vol. 10, article no. 104, 2011.

[13] A. J. Harborn, *Phytochemical Methods A Guide to Modern Techniques of Plant Analysis*, Springer, 1998.

[14] L. Mišurcová, "Chemical Composition of Seaweeds," in *Handbook of Marine Macroalgae, Biotechnology and Applied Phycology*, S.-K. Kim, Ed., Wiley-Blackwell, 2012.

[15] A. P. Simopoulos, "The importance of the ratio of omega-6/omega-3 essential fatty acids," *Biomedicine and Pharmacotherapy*, vol. 56, no. 8, pp. 365–379, 2002.

[16] G. A. Fariman, S. J. Shastan, and M. M. Zahedi, "Seasonal variation of total lipid, fatty acids, fucoxanthin content, and antioxidant properties of two tropical brown algae (Nizamuddinia zanardinii and Cystoseira indica) from Iran," *Journal of Applied Phycology*, vol. 28, no. 2, pp. 1323–1331, 2016.

[17] R. F. Patarra, J. Leite, R. Pereira, J. Baptista, and A. I. Neto, "Fatty acid composition of selected macrophytes," *Natural Product Research*, vol. 27, no. 7, pp. 665–669, 2013.

[18] H. Pereira, L. Barreira, F. Figueiredo et al., "Polyunsaturated fatty acids of marine macroalgae: potential for nutritional and pharmaceutical applications," *Marine Drugs*, vol. 10, no. 9, pp. 1920–1935, 2012.

[19] D. I. Sánchez-Machado, J. López-Cervantes, J. López-Hernández, and P. Paseiro-Losada, "Fatty acids, total lipid, protein and ash contents of processed edible seaweeds," *Food Chemistry*, vol. 85, no. 3, pp. 439–444, 2004.

[20] G. Silva, R. B. Pereira, P. Valentão, P. B. Andrade, and C. Sousa, "Distinct fatty acid profile of ten brown macroalgae," *Brazilian Journal of Pharmacognosy*, vol. 23, no. 4, pp. 608–613, 2013.

[21] C. Dawczynski, R. Schubert, and G. Jahreis, "Amino acids, fatty acids, and dietary fibre in edible seaweed products," *Food Chemistry*, vol. 103, no. 3, pp. 891–899, 2007.

[22] D. Rodrigues, A. C. Freitas, L. Pereira et al., "Chemical composition of red, brown and green macroalgae from Buarcos bay in Central West Coast of Portugal," *Food Chemistry*, vol. 183, Article ID 17304, pp. 197–207, 2015.

[23] S. V. Khotimchenko, "Fatty acids of brown algae from the Russian Far East," *Phytochemistry*, vol. 49, no. 8, pp. 2363–2369, 1998.

[24] M. Schmid and D. B. Stengel, "Intra-thallus differentiation of fatty acid and pigment profiles in some temperate Fucales and Laminariales," *Journal of Phycology*, vol. 51, no. 1, pp. 25–36, 2015.

[25] B. J. Gosch, N. A. Paul, R. de Nys, and M. Magnusson, "Seasonal and within-plant variation in fatty acid content and composition in the brown seaweed Spatoglossum macrodontum (Dictyotales, Phaeophyceae)," *Journal of Applied Phycology*, vol. 27, no. 1, pp. 387–398, 2014.

[26] J. Folch, M. Lees, and G. H. Sloane Stanley, "A simple method for the isolation and purification of total lipides from animal tissues," *The Journal of Biological Chemistry*, vol. 226, no. 1, pp. 497–509, 1957.

[27] H. C. Pinkart, R. Devereux, and P. J. Chapman, "Rapid separation of microbial lipids using solid phase extraction columns," *Journal of Microbiological Methods*, vol. 34, no. 1, pp. 9–15, 1998.

[28] J. Ruiz, T. Antequera, A. I. Andres, M. J. Petron, and E. Muriel, "Improvement of a solid phase extraction method for analysis of lipid fractions in muscle foods," *Analytica Chimica Acta*, vol. 520, no. 1-2, pp. 201–205, 2004.

[29] H. Devle, E.-O. Rukke, C. F. Naess-Andresen, and D. Ekeberg, "A GC - Magnetic sector MS method for identification and quantification of fatty acids in ewe milk by different acquisition modes," *Journal of Separation Science*, vol. 32, no. 21, pp. 3738–3745, 2009.

[30] D. B. Stengel, S. Connan, and Z. A. Popper, "Algal chemodiversity and bioactivity: Sources of natural variability and implications for commercial application," *Biotechnology Advances*, vol. 29, no. 5, pp. 483–501, 2011.

[31] J. V. Vilg, G. M. Nylund, T. Werner et al., "Seasonal and spatial variation in biochemical composition of Saccharina latissima during a potential harvesting season for Western Sweden," *Botanica Marina*, vol. 58, no. 6, pp. 435–447, 2015.

[32] B. J. Gosch, M. Magnusson, N. A. Paul, and R. de Nys, "Total lipid and fatty acid composition of seaweeds for the selection of species for oil-based biofuel and bioproducts," *GCB Bioenergy*, vol. 4, no. 6, pp. 919–930, 2012.

[33] K. Miyashita, N. Mikami, and M. Hosokawa, "Chemical and nutritional characteristics of brown seaweed lipids: A review," *Journal of Functional Foods*, vol. 5, no. 4, pp. 1507–1517, 2013.

[34] FAO Food and Nutrition Paper 91: Fats and fatty acids in human nutrition - Report of an expert consultation, 2010.

[35] R. Sugimura, M. Suda, A. Sho et al., "Stability of fucoxanthin in dried undaria pinnatifida (wakame) and baked products (scones) containing wakame powder," *Food Science and Technology Research*, vol. 18, no. 5, pp. 687–693, 2012.

[36] A. P. Simopoulos, "An increase in the Omega-6/Omega-3 fatty acid ratio increases the risk for obesity," *Nutrients*, vol. 8, no. 3, article no. 128, 2016.

[37] G. L. Russo, "Dietary n - 6 and n - 3 polyunsaturated fatty acids: From biochemistry to clinical implications in cardiovascular prevention," *Biochemical Pharmacology*, vol. 77, no. 6, pp. 937–946, 2009.

[38] P. M. Kris-Etherton, D. S. Taylor, S. Yu-Poth et al., "Polyunsaturated fatty acids in the food chain in the United States," *American Journal of Clinical Nutrition*, pp. 179S–188S, 2000.

The Prevalence of Dyslipidemia among Jordanians

Mousa Abujbara,[1] **Anwar Batieha** ⓘ**,**[2] **Yousef Khader** ⓘ**,**[3] **Hashem Jaddou,**[4]
Mohammed El-Khateeb,[1] **and Kamel Ajlouni** ⓘ[1]

[1]*The National Center (Institute) for Diabetes, Endocrinology and Genetics (NCDEG)/The University of Jordan, Jordan*
[2]*Department of Community Medicine, Jordan University of Science and Technology, Irbid, Jordan*
[3]*Jordan University of Science and Technology (JUST), Irbid, Jordan*
[4]*The Jordan University of Science and Technology (JUST), Irbid, Jordan*

Correspondence should be addressed to Kamel Ajlouni; ajlouni@ju.edu.jo

Academic Editor: Gerd Schmitz

Background. Dyslipidemia is one of the major modifiable risk factors for the development of cardiovascular disease and type two diabetes mellitus. Knowing the current prevalence of dyslipidemia is an important step for increasing awareness of the problem and for proper planning of health programs for prevention of its negative clinical effects. *Methods*. A national population based household sample was selected from north, middle, and south regions of Jordan in 2017. A total sample of 4,056 aged between 18 and 90 years were included. Selected individuals were interviewed using a structured questionnaire and fasting blood samples were collected for biochemical measurements. *Results*. The prevalence rates of hypercholesterolemia, hypertriglyceridemia, high LDL, and low HDL were 44.3%, 41.9% 75.9%, and 59.5%, respectively. Hypercholesterolemia in Jordan almost doubled from 23.0% in 1994 to 44.3% in 2017, and hypertriglyceridemia increased from 23.8% in 1994 to 41.9% in 2017. All lipid abnormalities decreased after the age of 60 years. Hypertension, diabetes, and obesity were all independently associated with hypercholesterolemia and hypertriglyceridemia. *Conclusions*. Results of this study show that dyslipidemia is a widely prevalent health problem among adult Jordanian population and that the problem has substantially increased since 1994. Encouraging healthy lifestyle and healthy diet are the basis for prevention of dyslipidemia.

1. Introduction

Noncommunicable diseases (NCDs) are considered the leading cause of death in Jordan, with 38% of deaths attributed to cardiovascular diseases (CVDs) [1]. With the rapid lifestyle changes and socioeconomic development, the prevalence of NCDs has markedly increased over the past years in Jordan [2]. It was estimated that between 1 and 3 million persons in Jordan will have hypertension, high blood cholesterol, or diabetes by 2050 [3]. Dyslipidemia is one of the major modifiable risk factors for the development of cardiovascular diseases, atherosclerosis, and type two diabetes mellitus [4]. Guidelines for screening and treating dyslipidemia indicate that proper management of lipids is both lifesaving and cost-effective. Furthermore, recent research shows that the association between blood cholesterol level and atherosclerotic cardiovascular diseases is independent of other risk factors

and causal associations [5]. Knowing the current prevalence of dyslipidemia is an important step for increasing awareness of the problem and for proper planning of health programs for prevention of the problem and its negative clinical effects. In 1994, a national community-based survey was conducted to estimate the prevalence of dyslipidemia and other cardiovascular risk factors in Jordan [6]. The present study is a follow-up survey using generally a similar methodology, to assess the current prevalence of dyslipidemia in Jordan and to identify its distribution among subgroups of the population.

2. Materials and Methods

A cross-sectional community-based survey targeting adults, ≥18 years of age in 2017, was conducted by the National Center for Diabetes, Endocrinology and Genetics (NCDEG), Jordan, Ministry of Health (MoH), and Jordan University of Science and Technology (JUST). A multistage sampling technique was used to represent rural and urban population in the three geographic regions of Jordan (north, middle, and south). As the survey procedures have to take place in health facilities (anthropometric measures, drawing blood samples...etc.). General Health Directors of each of the 12 governorates of Jordan were asked to select one to three health centers that could represent the rest of health centers in each governorate. The selected health centers had to be large enough to provide the needed space for the survey team (25 persons) to perform the required procedures. A team of two field researchers (a male and a female) went door to door to invite a systematic sample of households in the catchment area of the selected center.

The team explained the study protocol and responded to all questions. If they agreed to participate, they were asked to report to the health center at 8:00 am, fasting for ten hours, and to bring all of their medications with them to be checked by the survey team. They were also asked not to take their medications on the day of their visit to the health center. All household members ≥ 18 years of age were eligible to participate in the study. However, only subjects ≥ 25 years of age were included in this analysis. The reason for including only this age group in the analysis is to allow for comparison with the previous national survey and with results from other countries. To encourage participation in the study, visit dates to health centers were negotiated with eligible participants. The study team worked all weekdays (except Fridays) to obtain the highest possible response rate. If a person refused to participate, basic data were collected when possible such as age, gender, education, whether the person was diagnosed with any NCDs, and the reason for nonparticipation. Total response rate was 40% for males and 94% for females. A total sample of 4,056 subjects completed the survey and were included in the analysis.

The survey protocol was approved by the Ethics Committee of the National Center for Diabetes, Endocrinology, and Genetics, Amman, Jordan. A letter of support was obtained from the Minister of Health (MoH) to ensure the cooperation of the staff of the health centers. An informed consent was obtained from all study participants. Study participants were allowed to decline participation in any part of the study. All study procedures were carried out free of charge for the participants. Data were treated with high confidentiality and used only for scientific purposes.

All field researchers attended a two-day training workshop, prior to data collection. Training included explaining the study purpose and procedures, proper interview skills, and standard methods of anthropometric measurements. Health workers responsible for obtaining blood samples were trained on the standard aseptic technique for obtaining blood samples.

Selected individuals were interviewed using a structured questionnaire that inquired about socio-demographic characteristics, smoking and other variables related to this study. Height was measured without footwear using a portable stadiometer and the reading was recorded to the nearest 1 cm. Weight was measured by a standard weighing scale in light clothes and without footwear with a precision of 100 gm.

Body mass index (BMI) was calculated as weight divided by height squared (kg/m^2). Waist circumference was measured by a nonstretchable tape at the narrowest horizontal plane midway between the iliac crest and lower costal margin with a 1 cm precision. Hip circumference was measured at the widest horizontal plane at the hip level with a precision of 1 cm. Waist to height ratio was calculated as waist circumference divided by height. Three blood samples were drawn from a cannula inserted into the antecubital vein using vacutainer tubes. Collected blood samples were centrifuged within one hour in the field and transported on daily basis in cold boxes from the field to the Central Laboratory of the National Center for Diabetes, Endocrinology, and Genetics in Amman, Jordan, where all laboratory measurements were performed. All biochemical measurements were done by the same team of laboratory technicians using the same method throughout the survey period. Lipids measurements were performed according to the manufacturers' instructions, using COBAS autoanalyzer (Roche Diagnostics, Basel, Switzerland).

3. Definitions of Variables

Lipid variables were defined according to the Third Report of the National Cholesterol Education Program (NCEP-Adult Treatment Panel III). Hypercholesterolemia was defined as a serum cholesterol level of ≥ 5.17 mmol/l (200 mg/dl) or on lipid-lowering medication; high low-density lipoprotein (LDL-C) was defined as LDL-C level of ≥ 2.59 mmol/l (100 mg/dl) or on lipid-lowering medication; low high-density lipoprotein (HDL-C) was defined as HDL-C level of ≤ 1.03 mmol/l (40 mg/dl) in men, and ≤ 1.29 mmol/l (50 mg/dl) in women or on lipid-lowering medication. Hypertriglyceridemia was defined as a serum triglyceride (TG) ≥ 1.70 mmoi/l (150 mg/dl) or on lipid-lowering medication. Total cholesterol to HDL-C ratio (TC/HDL ratio) was considered high if the ratio was ≥ 5. Hypertension was defined as systolic blood pressure ≥130 mmHg or diastolic blood pressure >85 mmHg or if the patient was on antihypertensive drugs. Diabetes mellitus was defined as a fasting plasma glucose ≥ 7.0 mmol/l (126 mg/dl) or a HbA1C ≥ 6.5%, or on treatment for diabetes. Smoking was classified into nonsmoker and current smoker (smoked cigarettes daily or occasionally). Participants with BMI between 25 and 29.9 were considered overweight and with BMI of 30 or more were considered obese. Waist circumference was considered abnormal at ≥ 94 cm in men and ≥ 80 in women. Waist-to-height ratio was considered elevated if it was > 0.5.

4. Statistical Analysis

Survey data were analyzed using the Statistical Package for Social Sciences (SPSS Version 23). Range and logical checks were used to detect errors in data entry. Detected errors were corrected by returning to the original data forms. A P-value of ≤ 0.05 was considered statistically significant. Comparisons were made between participants with dyslipidemia and the rest of the sample to identify potential associations that were tested for statistical significance using the chi square test for categorical variables. Multivariate logistic regression was used to determine the effect of a given variable while simultaneously controlling for potential confounders. The magnitude of the associations was assessed by the adjusted odds ratio, and the statistical significance was assessed by the Wald chi square.

5. Results

This study included 4,056 subjects (70.6% females and 29.4% males) aged between 18 and 90 years with a mean age (SD) of 43.7 (14.2) years. The majority of the subjects were either overweight or obese (74.5%) and 14.5% of them were current smokers. The prevalence rates of hypercholesterolemia, hypertriglyceridemia, high LDL and low HDL were 44.3%, 41.9% 75.9%, and 59.5%, respectively.

Table 1 shows the prevalence of dyslipidemia according to subjects' socio-demographic characteristics. Prevalence of hypercholesterolemia (≥5.17 mmol/l) in females was 44.8% and 43.0% in males. Hypertriglyceridemia (≥ 1.70 mmol/l) was 36.5% in females significantly lower than in males (54.6%). Study participants showed no statistical difference in prevalence of low HDL-C and high LDL-C between male and female participates. On the other hand, the prevalence of TC/HDL ratio >5 in males was double that in females and the difference was statistically significant. The prevalence of hypercholesterolemia and hypertriglyceridemia reached its peak at the age group of 50-59 years then decreased again, while the prevalence of low HDL-C increased until the age of 39 years and it plateaued at 40 to 69 years to decrease at age 70. The prevalence of low LDL-C reached its peak between 40 to 49 years then decreased gradually among the older age groups. Similar significant change in the prevalence of TC/HDL ratio >5 was noticed where the prevalence increased until the age group of 50 to 59 years then decreased in the older age groups. Prevalence of hypercholesterolemia and hypertriglyceridemia was significantly higher among obese and overweight participants than normal BMI participants where the prevalence of hypercholesterolemia and TG was 50.6% and 52.7%, respectively, in obese participants. Also, the highest prevalence of low HDL-C and high LDL-C and TC/HDL ratio >5 was noticed among obese participants. Likewise, participants with waist-to-height ratio >0.5 had hypercholesterolemia, hypertriglyceridemia, low HDL-C, high LDL-C, and TC/HDL ratio >5 prevalence significantly higher than participants with waist-to-height ratio ≤ 0.5. The difference in prevalence of dyslipidemias was not statistically significant in different regions in Jordan except for hypertriglyceridemia which was lower in participants living in the middle region in comparison with participants living in other regions. Study results showed no significant difference in hypercholesterolemia between smokers and nonsmokers, while the prevalence of hypertriglyceridemia in smokers (53.9%) was significantly higher than in nonsmokers (39.7%). Low HDL-C was significantly higher in smokers than nonsmokers, while high LDL-C was not significantly different between smokers and nonsmokers. TC/HDL ratio >5 was significantly more prevalent in smokers than in nonsmokers. On bivariate analysis, the prevalence of hypercholesterolemia and high LDL-C was almost the same in diabetic and

nondiabetic subjects and in hypertensive and normotensive subjects. While, hypertriglyceridemia, low HDL-C, and TC/HDL ratio >5 prevalence were all significantly higher in diabetic subjects when compared with nondiabetics and in hypertensive subjects when compared with normotensive subjects.

Tables 2–5 shows multivariate regression findings for hypercholesterolemia, hypertriglyceridemia, low HDL-C, and high LDL-C, respectively. Table 2 shows that hypercholesterolemia was significantly associated with hypertension (Odds Ratio (OR)= 1.37, 95% Confidence Interval (CI)= 1.15-1.63) and diabetes (OR=1.25, 95% CI=1.04-1.50). Age, overweight, and obesity were all significantly associated with increased risk of hypercholesterolemia. Table 3 shows that hypertriglyceridemia was significantly associated with hypertension (OR=1.36, 95% CI=1.13-1.62), diabetes (OR=1.68, 95% CI=1.39-2.04), and cigarette smoking (OR=1.53, 95% CI=1.24-1.89). Similarly, overweight subjects were more likely to have hypertriglyceridemia (OR=2.01, 95% CI=1.61-2.50) and obese subjects were also more likely to have hypertriglyceridemia (OR=2.89, 95% CI=2.33-3.58) compared with subjects with normal BMI. Study subjects in 40-49, 50-59, and 60-69 age groups were more likely to have hypertriglyceridemia compared to subjects less than 40 years old, whereas female gender (OR=0.49, 95% CI=0.42 - 0.57), being single and living in the middle or south region were all independent protective factors from hypertriglyceridemia. Table 4 shows that low HDL-C was significantly associated with hypertension (OR=1.31, 95% CI=1.10 – 1.55), diabetes (OR=1.47, 95% CI=1.21 – 1.77), waist circumference >94 cm in men or >80 cm in women (OR=1.48, 95% CI=1.21-1.81), and obesity (OR=1.75, 95% CI=1.41-2.22), while female gender was an independent protective factor from low HDL-C (OR=0.83, 95% CI=0.71 – 0.96). Table 5 shows that high LDL-C was significantly associated with hypertension (OR=1.35, 95% CI=1.09-1.67), diabetes (OR=1.66, 95% CI=1.33-2.07), and abnormal waist circumference (OR=1.63, 95% CI=1.35-1.96). In addition, 40-49-, 50-59-, 60-69-, and a 70-year age groups were significantly more likely to have high LDL-C compared to subjects less than 40 years old. Also, overweight and obese subjects were more likely to have high LDL-C compared to subjects with normal BMI.

6. Discussion

Results of this study show that dyslipidemia is a widely prevalent health problem affecting adult Jordanian population and putting them at an enormous risk for CVD [7, 8]. To the best of our knowledge, this study is one of a few community-based studies conducted in Jordan over the past 23 years. The prevalence of hypercholesterolemia in Jordan increased from 23.0% in 1994 to 44.3% in 2017, and the prevalence of hypertriglyceridemia increased from 23.8% in 1994 to 41.9% in 2017 [6]. This marked increase could be explained by lifestyle changes and by changes in food consumption pattern. However, better survival of subjects with dyslipidemia may have contributed to this increase.

A remarkable finding of this study was that all types of lipid abnormalities decrease after the age of 60 years. Streja et

TABLE 1: Prevalence of dyslipidemia by selected variables, Jordan, 2017.

Variables	Hypercholesterolemia ≥5.17 mmol/l n (%)	Hypertriglyceridemia ≥1.70 mmol/l n (%)	Low HDL-C F ≤1.29 mmol, M ≤1.03 mmol/l n (%)	High LDL-C ≥2.59 mmol/l n (%)	TC/HDL ratio >5 n (%)
Overall	1780 (44.3)	1683 (41.9)	2391 (59.5)	3049 (75.9)	1335 (33.2)
Gender					
Male	509 (43.0)	647 (54.6)	732 (61.8)	883 (74.6)	619 (52.3)
Female	1271 (44.8)	1036 (36.5)	1659 (58.5)	2166 (76.4)	716 (25.3)
P-value	0.274	0.000	0.052	0.218	0.000
Age (years)					
25-29	83 (29.5)	58 (20.6)	120 (42.7)	179 (63.7)	44 (15.7)
30-39	324 (40.5)	281 (35.1)	449 (56.1)	611 (76.4)	219 (27.4)
40-49	562 (52.7)	516 (48.4)	665 (62.4)	907 (85.2)	419 (39.3)
50-59	467 (55.7)	457 (54.5)	550 (65.6)	696 (83.1)	377 (45.0)
60-69	208 (51.9)	245 (61.1)	269 (67.1)	314 (78.3)	178 (44.4)
≥ 70	75 (41.9)	78 (43.6)	104 (58.1)	132 (73.7)	62 (34.6)
P-value	0.000	0.000	0.000	0.000	0.000
BMI (kg/m^2)					
Normal	249 (31.0)	166 (20.7)	364 (45.3)	499 (62.1)	142 (17.7)
Overweight	602 (46.4)	548 (42.3)	735 (56.7)	1030 (79.4)	446 (34.4)
Obese	916 (50.6)	954 (52.7)	1243 (68.7)	1467 (81.1)	737 (40.7)
P-value	0.000	0.000	0.000	0.000	0.000
Waist to height ratio					
Normal	954 (40.2)	787 (33.1)	1246 (52.5)	1714 (72.2)	620 (26.1)
Abnormal	821 (50.4)	890 (54.6)	1135 (69.7)	1323 (81.2)	711 (43.6)
P-value	0.000	0.000	0.000	0.000	0.000
Geographical region					
North	579 (44.7)	602 (46.5)	785 (60.6)	991 (76.5)	441 (34.1)
Middle	784 (44.4)	648 (36.7)	1030 (58.4)	1316 (74.6)	559 (31.7)
South	417 (43.4)	433 (45.1)	576 (60.0)	742 (77.3)	335 (34.9)
P-value	0.809	0.000	0.433	0.251	0.175
Marital status					
Married	1452 (47.2)	1422 (46.2)	187 (61.0)	2438 (79.2)	1131 (36.8)
Single	141 (23.7)	88 (14.8)	298 (50.0)	331 (55.5)	74 (12.4)
Divorced or widow	180 (53.9)	164 (49.1)	206 (61.7)	271 (81.1)	122 (36.5)
P-value	0.000	0.000	0.000	0.000	0.000
Smoking					
Smoker	258 (44.4)	313 (53.9)	420 (72.4)	444 (76.6)	316 (54.5)
Non-smoker	1513 (44.3)	1356 (39.7)	1957 (57.2)	2590 (75.8)	1009 (29.5)
P-value	0.945	0.000	0.000	0.678	0.000
Hypertension					
Yes	438 (46.3)	563 (59.5)	664 (70.2)	729 (77.1)	414 (43.8)
No	1329 (43.7)	1103 (36.3)	1704 (56.0)	2296 (75.5)	908 (29.9)
P-value	0.160	0.000	0.000	0.000	0.000
Diabetes mellitus					
Yes	353 (46.3)	485 (63.6)	547 (71.9)	564 (74.1)	341 (44.8)
No	1420 (43.8)	1186 (36.6)	1832 (56.5)	2471 (76.3)	985 (30.4)
P-value	0.212	0.000	0.000	0.212	0.000

TABLE 2: Adjusted odds ratios and their 95% Cl for Hypercholesterolemia using multivariate logistic regression, Jordan 2017.

Variable	Adjusted OR	95% CI
Gender		
Male	1	
Female	1.17	1.00-1.35
Age (years)		
< 40	1	
40-49	2.37	1.99-2.82
50-59	2.87	2.36-3.49
60-69	2.73	2.11-3.53
≥ 70	1.92	3.62-6.67
BMI		
Normal	1	
Overweight	1.63	1.34-1.97
Obese	1.77	1.46-2.14
Hypertension		
No	1	
Yes	1.37	1.15-1.63
Diabetes		
No	1	
Yes	1.25	1.04-1.50

TABLE 3: Adjusted odds ratios and their 95% Cl for Hypertriglyceridemiausing multivariate logistic regression, Jordan, 2017.

Variable	Adjusted OR	95% CI
Gender		
Male	1	
Female	0.49	0.42 - 0.57
Age (years)		
< 40	1	
40-49	1.65	1.37-1.98
50-59	1.64	1.33-2.02
60-69	1.84	1.40-2.42
≥ 70	0.44	0.60-1.24
BMI		
Normal	1	
Overweight	2.01	1.61-2.50
Obese	2.89	2.33-3.58
Hypertension		
No	1	
Yes	1.36	1.13-1.62
Diabetes		
No	1	
Yes	1.68	1.39-2.04
Marital status		
Married	1	
Single	0.41	0.31-0.53
Divorced or widow	1.06	0.82-1.37
Geographical region		
North	1	
Middle	0.68	0.58-0.80
South	0.77	0.64-0.93
Smoking		
Non-smoker	1	
Smoker	1.53	1.24-1.89

al. reported similar findings in a chapter about management of dyslipidemia in the elderly [9]. A possible explanation for this finding is that many people with dyslipidemia might die before the age of 60, leaving only a few surviving to this age. Attrition or survival bias can decrease associations between harmful exposures and illnesses of aging [9].

Previous studies from our region showed variable results. In a study of Kuwaiti adults, 33.7% of men and 30.6% of women were reported to have hypercholesterolemia. Another study in Oman reported hypercholesterolemia prevalence of 33.6% [10, 11]. On the other hand, a higher prevalence of hypercholesterolemia was reported from Saudi Arabia (54%) [12] while an Iranian meta-analysis reported a hypercholesterolemia prevalence of 41.6% which is very close to our figure of 44.3% [13]. Differences in study design, dietary habits, lifestyles, and accessibility to health care may account for much of the variations in the prevalence of hypercholesterolemia.

Hypertriglyceridemia prevalence in the current study population (41.9%) was close to that reported from Saudi Arabia (40.3%) [12], and Iran (46.0%) [13].

The present study showed the independent risk factors of the different types of dyslipidemia by multivariate analysis. Diabetes mellitus and hypertension were independent risk factors for hypercholesterolemia. A study from Saudi Arabia also reported same association between diabetes mellitus, hypertension, and hypercholesterolemia [14]. Similarly, obesity was an independent risk factor for hypercholesterolemia in the current study, in an Ethiopian national survey and in a study from Beijing China [15, 16]. Additionally, hypertriglyceridemia was independently associated with diabetes

mellitus. A similar association between hypertriglyceridemia and HbA1c was reported in a Russian study [17].

An interesting finding was that living in northern region of Jordan was an independent risk factor for hypertriglyceridemia. We have no clear explanation for this finding which may need further investigation.

The prevalence of low HDL-C and prevalence of high LDL-C were positively and independently associated with obesity, increased waist circumference, diabetes, and hypertension. This is consistent with a recent study in Turkey [18]. There have been a lot of changes regarding the optimal management of dyslipidemia in the last few years. The American Association of Clinical Endocrinologists and American College of Endocrinology 2017 guidelines (AACE 2017) for management of dyslipidemia and prevention of cardiovascular disease encourage physicians to provide personalized goals of lipid levels to their patients and states that the 10-year risk of a coronary event need to be assessed in order to provide such personalized lipid management [5].

TABLE 4: Adjusted odds ratios and their 95% Cl for low HDL-C using multivariate logistic regression, Jordan 2017.

Variable	Adjusted OR	95% CI
Gender		
Male	1	
Female	0.83	0.71 – 0.96
BMI		
Normal	1	
Overweight	1.18	0.96- 1.46
Obese	1.75	1.41-2.22
Waist circumference		
Normal	1	
Increased	1.48	1.21-1.81
Hypertension		
No	1	
Yes	1.31	1.10 – 1.55
Diabetes		
No	1	
Yes	1.47	1.21 – 1.77

TABLE 5: Adjusted odds ratios and their 95% Cl for high LDL-C using multivariate logistic regression, Jordan 2017.

Variable	Adjusted OR	95% CI
Gender		
Male	1	
Female	1.19	1.00 – 1.43
Age (years)		
< 40	1	
40-49	2.90	2.33-3.60
50-59	2.76	2.16- 3.52
60-69	2.29	1.68-3.12
≥ 70	1.95	1.32 – 2.87
BMI		
Normal	1	
Overweight	2.07	1.69-2.54
Obese	2.11	1.73-2.58
Waist circumference		
Normal	1	
Increased	1.63	1.35-1.96
Hypertension		
No	1	
Yes	1.35	1.09-1.67
Diabetes		
No	1	
Yes	1.66	1.33-2.07

This study had two main limitations: firstly, it was a cross-sectional survey, in which the risk factors and dyslipidemia were assessed simultaneously. Consequently, it is not possible to judge whether dyslipidemia was present before or after the proposed risk factors; i.e., the temporal sequence from cause to effect could not be established. Secondly, our associations are based on prevalence rather incidence. Prevalence is affected by duration of the disease which may in turn be related to health care; thus differences in prevalence may merely reflect differences in survival rather differences in incidence.

In conclusion, it was clear that dyslipidemia is high in Jordan which may put a high burden on the already strained health services and leads to substantial increase in health related costs. Prevention of dyslipidemia is critical to achieve the goal of halting the rise in cardiovascular diseases. At the community level, encouraging healthy lifestyle, healthy dietary habits, and physical exercise are the cornerstone for prevention of dyslipidemia. Screening of high risk groups such as obese people is recommended.

Authors' Contributions

Mousa Abujbara wrote the manuscript, approved the protocol, analyzed the data, and approved the results. Anwar Batieha is the main author of the study protocol and a Professor of epidemiology and public health and he is responsible for the statistical part of the study. Yousef Khader shared with Dr. Batieha the responsibility for statistics and supervised the fieldwork. Hashem Jaddou is the co-author of the structured questionnaire and supervised the execution of the protocol and data collection. Mohammed El-Khateeb is responsible for all the lab procedures, samples handling, transport, and analysis. Kamel Ajlouni is the Chairman of the National Strategic Committee, the main investigator, the principle author of the study, and the guarantor, who shared in planning executing and shared with the analysis of the data.

References

[1] A. Mokdad, "Global non-communicable disease prevention: Building on success by addressing an emerging health need in developing countries," *Journal of Health Specialties*, vol. 4, no. 2, p. 92, 2016.

[2] Ministry of Health, *The national strategy and plan of action against diabetes, hypertension, dyslipidemia and obesity in Jordan*, 2011, http://www.moh.gov.jo.

[3] D. W. Brown, A. H. Mokdad, H. Walke et al., "Projected burden of chronic, noncommunicable diseases in Jordan," *Preventing Chronic Disease*, vol. 6, no. 2, article no. A78, 2009.

[4] L. Qi, X. Ding, W. Tang, Q. Li, D. Mao, and Y. Wang, "Prevalence and risk factors associated with dyslipidemia in Chongqing, China," *International Journal of Environmental Research and Public Health*, vol. 12, no. 10, pp. 13455–13465, 2015.

[5] P. S. Jellinger, Y. Handelsman, P. D. Rosenblit et al., "American Association of Clinical Endocrinologists and American College of Endocrinology guidelines for management of dyslipidemia and prevention of cardiovascular disease," *Endocrine Practice*, vol. 23, no. 4, pp. 1–87, 2017.

[6] A. Batieha, H. Y. Jaddou, and K. M. Ajlouni, "Hyperlipidemia in Jordan: A community-based survey," *Saudi Medical Journal*, vol. 18, no. 3, pp. 279–285, 1997.

[7] R. H. Nelson, "Hyperlipidemia as a Risk Factor for Cardiovascular Disease," *Primary Care: Clinics in Office Practice*, vol. 40, no. 1, pp. 195–211, 2013.

[8] H. O. Steinberg, B. Bayazeed, G. Hook, A. Johnson, J. Cronin, and A. D. Baron, "Endothelial dysfunction is associated with cholesterol levels in the high normal range in humans," *Circulation*, vol. 96, no. 10, pp. 3287–3293, 1997.

[9] D. Streja and E. Streja, *Management of Dyslipidemia in the Elderly*, 2014.

[10] F. Ahmed, C. Waslien, M. Al-Sumaie, and P. Prakash, "Trends and risk factors of hypercholesterolemia among Kuwaiti adults: National Nutrition Surveillance Data from 1998 to 2009," *Nutrition Journal* , vol. 28, no. 9, pp. 917–923, 2012.

[11] A. Al Riyami, M. A. Abd Elaty, M. Morsi, H. Al Kharusi, W. Al Shukaily, and S. Jaju, "Oman World Health Survey: part 1—methodology, sociodemographic profile and epidemiology of non-communicable diseases in Oman," *Oman Medical Journal*, vol. 27, no. 5, pp. 425–443, 2012.

[12] M. M. Al-Nozha, M. R. Arafah, M. A. Al-Maatouq et al., "Hyperlipidemia in Saudi Arabia," *Saudi Medical Journal*, vol. 29, no. 2, pp. 282–287, 2008.

[13] A. Mohammadbeigi, E. Moshiri, N. Mohammadsalehi, H. Ansari, and A. Ahmadi, "Dyslipidemia Prevalence in Iranian Adult Men: The Impact of Population-Based Screening on the Detection of Undiagnosed Patients," *The World Journal of Men's Health*, vol. 33, no. 3, p. 167, 2015.

[14] M. Basulaiman, C. El Bcheraoui, M. Tuffaha et al., "Hypercholesterolemia and its associated risk factors-Kingdom of Saudi Arabia, 2013," *Annals of Epidemiology*, vol. 24, no. 11, pp. 801–808, 2014.

[15] Y. F. Gebreyes, D. Y. Goshu, T. K. Geletew et al., "Prevalence of high bloodpressure, hyperglycemia, dyslipidemia, metabolic syndrome and their determinants in Ethiopia: Evidences from the National NCDs STEPS Survey, 2015," *PLoS ONE*, vol. 13, no. 5, p. e0194819, 2018.

[16] B. Jiang, A. J. Ma, H. Li et al., "Prevalence of hypercholesterolemia and influence factors in residents aged 18-65 years in Beijing," *Zhonghua liu xing bing xue za zhi= Zhonghua liuxingbingxue zazhi*, vol. 38, no. 7, pp. 938–943, 2017.

[17] Y. Karpov and Y. Khomitskaya, "PROMETHEUS: An observational, cross-sectional, retrospective study of hypertriglyceridemia in Russia," *Cardiovascular Diabetology*, vol. 14, no. 1, 2015.

[18] F. Bayram, D. Kocer, K. Gundogan et al., "Prevalence of dyslipidemia and associated risk factors in Turkish adults," *Journal of Clinical Lipidology*, vol. 8, no. 2, pp. 206–216, 2014.

3

Ceramide and Ischemia/Reperfusion Injury

Xingxuan He and Edward H. Schuchman ⓘ

Department of Genetics & Genomic Sciences, Icahn School of Medicine at Mount Sinai, 1425 Madison Avenue, New York, NY 10029, USA

Correspondence should be addressed to Edward H. Schuchman; edward.schuchman@mssm.edu

Academic Editor: Afaf El-Ansary

Ceramide, a bioactive membrane sphingolipid, functions as an important second messenger in apoptosis and cell signaling. In response to stresses, it may be generated by de novo synthesis, sphingomyelin hydrolysis, and/or recycling of complex sphingolipids. It is cleared from cells through the activity of ceramidases, phosphorylation to ceramide-1-phosphate, or resynthesis into more complex sphingolipids. Ischemia/reperfusion (IR) injury occurs when oxygen/nutrition is rapidly reintroduced into ischemic tissue, resulting in cell death and tissue damage, and is a major concern in diverse clinical settings, including organ resection and transplantation. Numerous reports show that ceramide levels are markedly elevated during IR. Mitochondria are major sites of reactive oxygen species (ROS) production and play a key role in IR-induced and ceramide-mediated cell death and tissue damage. During the development of IR injury, the initial response of ROS and TNF-alpha production activates two major ceramide generating pathways (sphingomyelin hydrolysis and de novo ceramide synthesis). The increased ceramide has broad effects depending on the IR phases, including both pro- and antiapoptotic effects. Therefore, strategies that reduce the levels of ceramide, for example, by modulation of ceramidase and/or sphingomyelinases activities, may represent novel and promising therapeutic approaches to prevent or treat IR injury in diverse clinical settings.

1. Introduction

Sphingolipids are essential structural components of all cell membranes and highly bioactive compounds that play important roles in signal transduction and numerous other cellular processes such as cell proliferation, differentiation, and apoptosis. Ceramide is a central component of sphingolipid structure and metabolism. There are several ways to generate ceramide in mammalian cells (Figure 1): hydrolysis of sphingomyelin, de novo synthesis from palmitoyl-CoA and serine, catabolism of glucosylceramide and galactosylceramide, synthesis from sphingosine and fatty acid, and dephosphorylation of ceramide-1-phosphate. However, these multiple pathways for ceramide generation do not contribute evenly, and there are many cell specific and other regulatory checkpoints that activate the specific pathways.

Over the past two decades, ceramide has been recognized as a key bioactive lipid and second messenger that mediates the proliferation, survival, and death of cells. Ceramide's role as a second messenger was first recognized in 1990 in the context of HL-60 cell proliferation [1]. In the late 1990s, further publications demonstrated the accumulation of ceramide in response to diverse cellular stresses, like infection, radiation, cytokines, death ligands, reactive oxygen species (ROS), and others [2–4]. Stress-induced ceramide accumulation leads to reorganization of the plasma membrane and formation of ceramide-rich platforms, often referred to as "rafts." These raft platforms recruit and cluster death receptors and signaling molecules at the cell membrane to facilitate amplification of signal transduction cascades and activation of cell death signaling pathways [5–7]. Increasing evidence also reveals that ceramide elevation is involved in diverse diseases, like diabetes [8], cardiovascular disease [9, 10], Alzheimer's disease [11, 12], and others.

In addition, the role of ceramide in the pathogenesis of ischemia/reperfusion (IR) injury has attracted considerable attention. IR injury occurs when the blood supply returns to tissues after a period of ischemia or lack of oxygen,

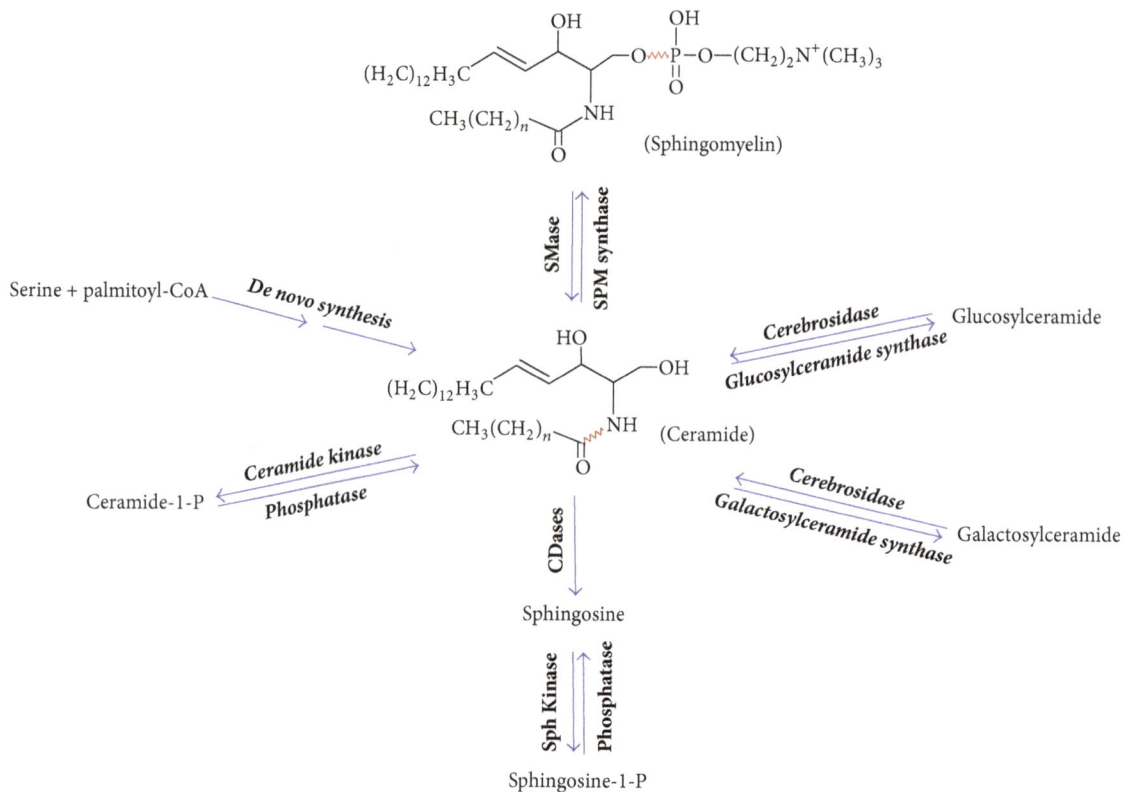

FIGURE 1: Scheme of ceramide metabolism.

resulting in cell death and tissue damage. Although there are no standard classifications of ischemia and IR, to better understand IR injury it can be classified into different phases or types according to the time and extent of the insults. For example, total or partial ischemia is defined as full obstruction of the blood vessel or blockage of a small area only, respectively. Brief/early phase or prolonged/later phase ischemia is defined by the length of time the tissues lack oxygen, from minutes to hours, respectively. Reperfusion results in a series of pathological changes associated with the time and extent of the ischemia. Mild or severe ischemia and IR are defined based on a combination of the area of blockage and time following reperfusion. In organ transplantation events, ischemia also can be classified as cold or warm ischemia. Generally, cold ischemia (4–7°C) can be protective due to reduced metabolic processes and cellular ATP demand at lower temperature, whereas warm ischemia (37°C) is usually harmful to cells and molecular pathways [13].

The incidence of IR injury is substantial. There are millions of individuals each year in the US suffering from cardiac infarction, stroke, thrombosis, blood vessel clamped surgery, and organ failure requiring transplantation. Restoration of blood supply should protect the tissues from damage, but reperfusion often leads to injury. Even though it is widely accepted that this IR injury results from the production of ROS, recruiting neutrophils, macrophages, and inflammatory mediators to the injured tissue, the mechanisms of IR injury

remain to be elucidated. In this review we will concentrate on the molecular mechanisms of ceramide elevation and tissue damage observed during IR injury.

The first report indicating the involvement of ceramide in IR injury was published by Bradham et al. in 1997 [14]. They demonstrated that there was a significant elevation of ceramide during liver transplantation (cold ischemia and warm reperfusion). In the same year, the accumulation of ceramide also was observed in both heart and renal IR injuries [15, 16]. After more than a decade of investigations, it has become clear that ceramide generation plays a key role during IR injury. A comprehensive understanding of the mechanism behind these changes has not yet been clarified, as several sphingolipid-metabolizing enzymes have been involved in ceramide generation during IR injury. A further understanding of these mechanisms could lead to more targeted therapies to prevent ceramide generation during IR.

2. The Sphingomyelin/Ceramide Signaling Pathway and IR

Two major pathways, sphingomyelin hydrolysis and de novo biosynthesis, have been implicated in the generation of ceramide. Both pathways may be activated separately or in parallel depending on stimuli or on the cell type [17]. Diverse oxidative stresses induce cell apoptosis or necrosis and tissue damage via activation of SMases, resulting in sphingomyelin hydrolysis with ceramide generation. The accumulation of ceramide has been reported in multiple models of ischemia, including rat cerebral cortex and gerbil hippocampus ischemia [18, 19], as well as in models of reperfusion injury, including artery occlusions in rat brain, liver, and heart [20–22]. In renal and cardiac IR injury models, activation of SMase and accumulation of ceramide were observed in the later phase of IR. Meanwhile, decreases in sphingomyelin corresponded to the increases in ceramide [23]. A few studies describe endogenous ceramide accumulation in brain via activation of a SMase leading to sphingomyelin hydrolysis during severe and lethal cerebral IR. In vitro, after hypoxia/reoxygenation of cardiac myocytes, the early responses (peaking at 10 min) included the activation of neutral SMase and low level ceramide accumulation [24]. Factor associated with neutral SMase activation (FAN), a protein that links neutral SMase to the tumor necrosis factor alpha (TNFα) receptor, mediates activation of neutral SMase and subsequent apoptosis. The expression of a dominant-negative FAN in rat cardiomyocytes almost completely abrogated hypoxia/reoxygenation-induced cell death, whereas overexpression of wild-type FAN led to an exacerbation of IR injury [25].

During ischemia, aerobic metabolism interrupts due to the lack of oxygen supply. Build-up of oxidative cell damage occurs during reperfusion to sites of ischemia, which is characterized by excess ROS generation and inflammatory cytokine recruitment [22, 26]. Several studies have shown that ceramide generation by SMase contributes to ROS and TNFα induced cell death and tissue damage [27–29]. For example, Wistar rats subjected to total liver ischemia followed by reperfusion had significant accumulation of TNFα and an increase of SMase activity that coincided with IR injury [30]. In an in vitro study, overexpression of acid ceramidase protected murine fibroblasts from TNFα-induced cell apoptosis by shifting elevated ceramide towards cell survival sphingosine-1-phosphate [31]. In TNFα gene knockout mice, IR-induced hepatic apoptosis was attenuated, and animal survival was prolonged compared to wild-type mice. These data have further identified TNFα as a critical mediator in hepatic IR injury [32]. Ceramide and TNFα are also known to induce ROS generation, which in turn amplifies ROS/TNFα-ceramide cycling and exacerbates IR injury [27, 33]. In contrast, in a monoamine oxidase-A deficient animal model the effects of ROS attenuated ceramide generation and IR injury were reduced [26, 34].

Finally, with administration of SMase inhibitors and SMase knockdown by siRNA, SMase knockout mice have reduced ceramide accumulation during IR and attenuated cell apoptosis and tissue damage through a mechanism that may involve the blockade of C-Jun N-terminal kinase (JNK) activation, the impairment of mitochondrial function, and activation of caspases [21, 22, 35–37]. Taking these data together, ceramide generated from SMases plays a key role in IR-induced later phase damage, and the modulation of ceramide may be an important therapeutic target.

3. The De Novo Ceramide Synthesis Pathway and IR

De novo ceramide biosynthesis occurs at the cytosolic side of the endoplasmic reticulum (ER) and mitochondrion and serves as a precursor for the synthesis of more complex sphingolipids, including sphingomyelin and glycosphingolipids, in the Golgi [38, 39]. Ceramide synthases are a family of key enzymes in de novo ceramide synthesis. There are six ceramide synthase isoforms that have been cloned and characterized [40]. Each of the six mammalian ceramide synthases appears to regulate the synthesis of a specific subset of ceramides and displays a unique substrate specificity profile for chain-length and/or saturation of the fatty acid acyl-CoA [41]. Increased ceramide synthesis occurred upon reperfusion in the ischemic area after coronary occlusion in mice, which correlated with the enhanced expression of serine palmitoyltransferase (SPT), the first key enzyme in de novo ceramide synthesis. Myriocin, an inhibitor of SPT, significantly protected the ischemic area from IR injury [37]. Dihydroceramide desaturase, an oxygen sensitive double bond generating enzyme, is the last key enzyme in de novo synthesis. The activity of dihydroceramide desaturase was significantly inhibited and dihydroceramide levels were increased during hypoxia. The elevated dihydrosphingolipids may be involved in exacerbating the IR injury [42]. In a mouse cerebral IR model, after 30 min of middle artery occlusion, followed by 24 hr reperfusion, the content of all ceramide species was elevated without any change in the content of sphingomyelin. Thus, the accumulation of ceramide was consistent with activation of ceramide synthase, rather than the activation of SMases. Moreover, IR-induced stimulation of ceramide synthase activity was very sensitive to the inhibitor fumonisin B1 (FB1) [43]. Studies in cell lines have shown that ceramide generation is involved in the activation of JNK and promotion of Bax translocation to the mitochondria, which also suggested that ceramide may signal through the mitochondrial cell death pathway in response to IR injury [44, 45]. Inhibition of ceramide synthase with both FB1 and JNK3 knockout reduced the accumulation of ceramide and decreased the size of brain infarct regions in a cerebral IR model [43].

4. Ceramide Clearance and IR

Interestingly, ceramide concentrations in the myocardium of rats had no apparent change during 30 min of ischemia, but following 3 hours of reperfusion there was a significant elevation. These increases in ceramide were not associated with SMase activity, but rather with reduced ceramidase activity [46]. Furthermore, short periods of anoxia (3 h) followed by reoxygenation (0–5 h) led to a time-dependent increase of caspase activity in human umbilical vein endothelial cells (HUVECs), which was associated with a significant decrease in glucosylceramide synthase mRNA levels and protein expression, but no changes in SMase. After 24 h middle cerebral artery occlusion in rats, increase of ceramide levels also coincided with the decease of glucosylceramide synthase activity in rat brain [47]. These in vivo and in vitro data suggest that the inhibition of ceramide clearance may also contribute to the IR-induced accumulation of ceramide and tissue damage and indicates that strategies to treat IR-induced tissue injury via ceramidase treatment or inhibition of glucosylceramide synthase may also be viable strategies.

5. Mitochondrial Damage and IR

Increasing evidence suggests that mitochondria are important intracellular compartments for sphingolipid metabolism, including sphingomyelin and ceramide [48]. Moreover, several enzymes engaged in ceramide metabolism have been identified in mitochondria. With their own set of ceramide synthesizing and hydrolyzing enzymes, mitochondria serve as a specialized compartment of ceramide metabolism in cells. For example, SMases in mitochondria have been identified from zebrafish, mouse, and rat [20, 49, 50]. Purified ceramide synthase from bovine liver mitochondria showed higher activity than that from the ER [51]. Further studies of submitochondrial localization revealed that both outer and inner mitochondrial membranes have enzymatic machinery that can synthesize ceramide [52]. Recent studies demonstrated that ceramide synthase is associated with adenine nucleotide translocase, the inner membrane component of the mitochondrial permeability transition pore (MPTP), and suggested that ceramide generation by ceramide synthase could mediate MPTP activity and mitochondrial $Ca2+$ homeostasis [53]. An additional source of ceramide in mitochondria is via the reverse activity of neutral ceramidase, and recent reports also describe ceramide formation from acyl-CoA and sphingosine mediated by the coupled activities of mitochondrial neutral ceramidase and thioesterase. Furthermore, mitochondria from neutral ceramidase deficient mice liver exhibited significantly reduced ceramide formation from sphingosine and palmitate, further implicating this "reverse reaction" [52, 54]. Ceramide also can be transported from the ER to mitochondria [55].

Many investigations have also shown a close connection between ceramide signaling and mitochondrial function and that mitochondria are the primary site of ROS production under normal physiologic conditions as well as during ischemia and IR insults [28, 56]. Regardless of the diverse pathways of IR-induced ceramide generation in mitochondria, ceramide-induced apoptosis has common consequences: suppression of the respiratory chain, elevation of ROS formation, discharge of membrane potential, opening of MPTP, and release of proapoptotic proteins [28, 57].

For over a decade multiple studies have shown that mitochondrial dysfunction appears to be one essential step in IR tissue damage, although the impact of ceramide on mitochondrial function during IR is not fully understood and may depend on cell type and stimuli. Indirectly, ceramide activates protein phosphatase 2A (PP2A), resulting in increases of the proapoptotic Bcl-2 family proteins by dephosphorylation of Bax (activation) and Bcl-2 (inactivation) [58]. In addition, ceramide-induced activation of protein phosphatases leads to inactivation of serine/threonine kinase Akt/PKB and activation of proapoptotic Bad [59]. Interestingly, ceramide can trigger Bax into an active conformation and lead to translocation from the cytosol to the mitochondrial membrane with release of cytochrome c during hypoxia/reoxygenation in neuronal cells. Knockdown of SMase or ceramide synthase attenuates Bax translocation [45].

Other indirect mechanisms linking ceramide and mitochondria include ceramide's interaction with protein kinase PKC and mitogen-activated protein kinase (MAPK). Several studies indicate that the increased ceramide levels in heart IR can target PKC δ, resulting in activation and mitochondria translocation of PKC δ accompanied by cytochrome c release and activation of caspase [60]. Members of the MAPK superfamily, p38 MAPK and JNK, can also be activated by both endogenous ceramide generation in liver IR and addition of exogenous ceramide, followed by translocation to mitochondria, activation/translocation of the Bcl-2 family proteins, initiation of cytochrome c release, and apoptosis [35, 61]. Ceramide can also trigger $Ca2+$ release from ER to mitochondria. Excessive accumulation of $Ca2+$ in mitochondria could trigger opening of the MPTP at a high conductance state and lead to cell death [62].

Recent studies have also shown that Sirtuin 3 induces mitochondrial dysfunction by enhancing ceramide biosynthesis via deacylation of ceramide synthase [63]. Ceramide can also suppress the respiratory chain in isolated mitochondria, resulting in increased production of ROS in endothelial cells after hypoxia/reoxygenation. Extensive studies using isolated mitochondria demonstrate that ceramide generation in the outer mitochondrial membrane leads to formation of large pore ceramide-rich rafts, opening of MPTP, and initiation of cytochrome c release [62, 64]. These extensive findings suggest that ceramide generated in both the cytosol and mitochondria may play a critical role in IR-induced mitochondrial injury, dysfunction, and tissue damage. Protection of mitochondrial function via modulation of ceramide could therefore be another essential strategy to prevent IR-induced injury.

6. The Protective Effect of Ceramide during Preconditioning

Taken together, the information above indicates that ceramide plays a central key in hypoxia/reoxygenation-induced cell death and IR-induced tissue damage. Paradoxically, in recent years several lines of evidence have also demonstrated that ceramide has a protective effect in IR injury when ischemia is used for preconditioning. Ischemic preconditioning (IPC) is a phenomenon whereby brief ischemia provides significant protection against subsequent severe ischemia and reperfusion injury in heart [65, 66], brain [67], and liver [68] (Figure 2).

Little is known about the molecular mechanisms involved in ceramide-induced protection on IR injury. In general, evidence has shown that low concentrations of ceramide promote cell survival while higher concentrations induce cell death [69, 70]. Earlier investigations indicated that IPC promoted a transient accumulation of specific ceramide species, while SM remained unchanged, and treatment with low doses of exogenous ceramide had a similar protective

Ischemia
(early stage or short time)

↓

ROS and TNFα
(low concentrations)

│ *De novo Synthesis*

↓

Ceramide
(low concentration)

│ *Kinase Activation*

↓

S1P

↓

Antiapoptosis and IR injury

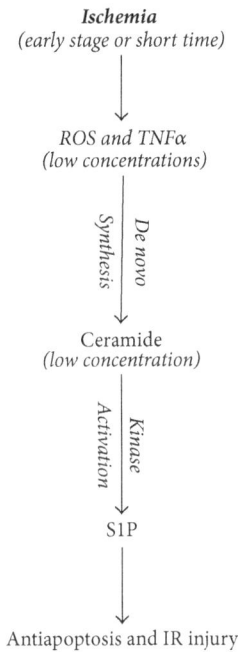

FIGURE 2: Scheme of ischemic preconditioning and sphingolipids.

effect on IR insult [65, 71]. Pretreatment with FB1, an inhibitor of ceramide synthase, in rat cortical neurons abolished the neuroprotective effect [71]. These data suggest that de novo ceramide synthesis contributes to IPC-induced ischemia tolerance. Although ROS generation and cytokine activation have been recognized to contribute to IR injury, it has also recently been suggested that ROS and TNFα at low doses may also enhance cytoprotective mechanisms [72, 73]. In vivo, IPC promotes ROS and TNFα production, and low doses of ROS and TNFα can mimic the IPC-induced protective effect in heart and liver [32]. Moreover, the protective effect of TNFα is associated with the release of ROS [74]. In addition, an antioxidant, N-2-mercaptopropionyl glycine (MPG), can block both the ROS and TNFα-induced cytoprotective effects [75]. The exact roles of ROS and TNFα in the cytoprotective effect of ceramide are still elusive. It is suggested that cell survival S1P, derived from increased ceramide hydrolysis, induces cardioprotection via activation of protein kinase C and that ceramide itself may lead to transmembrane receptor clustering and activation of a variety of kinases and phosphatases that regulate the cell antiapoptotic process [76, 77].

S1P is known to be a survival factor for a variety of cell types. A ceramide-S1P rheostat model demonstrated that increases in the concentration of proapoptotic ceramide can be countered by increases in the levels of antiapoptotic S1P [78]. There are two key limiting enzymes in this rheostat, ceramidase and sphingosine kinase (SPHK), regulating the levels of intracellular S1P derived from ceramide. It was reported that exogenous S1P can mimic the preconditioning protective phenotype in an ex vivo rat heart model in which infarct size was significantly reduced [79]. In mouse hearts, SPHK activity was increased following IPC. Further

evidence demonstrated that increased intracellular S1P levels induced by IPC resulted from PKC activation and SPHK phosphorylation [80, 81]. Moreover, the cardioprotective effect of S1P could be partially abolished by the sphingosine kinase (N,N-dimethylsphingosine, DMS) or ceramidase (N-oleoylethanolamine, NOE) inhibitors [79, 80], and sphingosine kinase 1 mutation sensitized mouse myocardium to IR injury [82]. Taken together, these studies strongly suggested that the ceramide-S1P rheostat plays a key role in this IPC effect. Strategies to enhance intracellular S1P levels by modulation of ceramidase and/or SPHK activities could be novel and promising therapeutic approaches to prevent tissues from IR injury during severe ischemia and following reperfusion.

The majority of S1P is stored in platelets and erythrocytes. It is secreted by an essential transporter [83] and plays numerous biological roles, such as antiapoptotic process and proliferation. However, a growing body of evidence has shown S1P can be pathogenic in various diseases [84]. Increased S1P levels can enhance inflammation and trigger the S1P-ceramide recycling pathway, causing apoptosis and tissue damage as well [85]. The pathogenic effect of S1P should be also considered in IR injury.

In summary, ceramide accumulation has been demonstrated in various models of IR and has been implicated as an important mediator of apoptosis in the injured tissue. Ceramide accumulation during IR could occur through a combination of mechanisms, and the effects may range from protective to damaging based on timelines, the extent of ceramide production, tissue type, and/or extent of ischemia and reperfusion. The IR-induced accumulation of ceramide appears to be a general phenomenon for many organs, including heart, kidney, liver, brain, and intestine, but with differences. Each of these organ systems will be discussed below.

7. Heart

Heart damage caused by IR is well recognized as a significant cause of morbidity and mortality (thrombolysis, myocardial infarction, cardiac surgery, primary percutaneous coronary intervention, etc.). The first study demonstrating that hypoxia/reoxgenation induced a progressive accumulation of ceramide was in cardiomyocytes [15] where in the rat heart left coronary occlusion model the ceramide content in ischemic myocardium was increased to 155% in early ischemia and further elevated to 250% after 3 h reperfusion. Further in vitro and in vivo findings suggested that ceramide was involved in cardiomyocyte apoptosis and IR-induced heart injury, although the mechanisms of IR-induced accumulation of ceramide are incompletely understood. Several lines of evidence indicate that sphingomyelin hydrolysis by SMases is responsible for the ceramide elevation in cardiac IR injury [25, 65, 66]. Pharmacological inhibitors of SMases, such as D609, amitriptyline, and desipramine, as well as the expression of domain-negative FAN, prevented IR-induced accumulation of ceramide and attenuated cardiomyocyte apoptosis and IR heart injury [20, 77, 86]. An interesting study in IR-induced mouse heart injury revealed ceramide elevation and inhibition of SPHK, which led to upregulation of the ceramide/S1P ratio resulting in cardiac tissue damage [34]. In addition of SMase activation, enhanced de novo ceramide synthesis also may be involved in IR heart injury in mice [37]. Another report showed that accumulation of ceramide in reperfused rat heart was associated with reduced ceramidase activity, rather than enhanced SMase activity [46]. Thus, multiple mechanisms appear to be involved.

There is also strong evidence demonstrating that IPC enables cardiomyocytes and heart to become more resistant to subsequent severe ischemia and reperfusion [65, 66], and it is known that ROS and antioxidant play key roles in the effect of IPC on ceramide-induced cardioprotection. Pre-treatment with antioxidants prevented ceramide generation and abrogated this effect [33, 74]. In addition, the IPC-induced cardioprotective effect may also result from PKC activation and SPHK activation to enhance intracellular S1P levels in rat heart [80]. This effect of S1P could be blocked by a sphingosine kinase inhibitor and by using a sphingosine kinase 1 mutant mouse model [82].

8. Brain

Stroke is a major cause of long-term disability. Ischemic stroke occurs when cerebral arteries are occluded or stenosed by emboli or local atherosclerotic disease. Restoration of blood flow following ischemic stroke can be achieved by means of thrombolysis or recanalization. However, in some clinical cases, reperfusion may exacerbate the injury initially caused by ischemia, producing a cerebral reperfusion injury. The signaling cascades activated by cerebral ischemia and IR that may promote neuronal death are not well understood.

Kubota et al. provided the first evidence that ceramide was increased in an ischemic human brain resulting from an acute case of cerebral occlusion [87]. Ceramide accumulation has been subsequently reported in multiple models of neural ischemia, including cerebral cortex and hippocampus ischemia, and in models of IR, including rat cerebral and coronary artery occlusions [19, 36, 88, 89]. There are diverse molecular mechanisms involved in ceramide-induced cerebral ischemia and IR injury, and it appears that the pathway of ceramide activation depends on the severity of the insult to the brain.

In severe cerebral ischemia and IR, ceramide accumulation resulted from activation of acid SMase and inhibition of glucosylceramide synthase [19, 47]. It is also known that integrin-associated Lyn kinase suppressed acid SMase activity, promoting cell survival, and after 1 h of middle cerebral artery occlusion and 48 h of reperfusion in the mouse brain the disruption of the intergrin-Lyn kinase complex led to activation of acid SMase and accumulation of ceramide [90]. Consistent with these results, severe ischemia-induced brain injury is decreased in acid SMase knockdown mice and following treatment of healthy mice with an acid SMase inhibitor [36].

In mild ischemia and IR, the de novo ceramide synthesis pathway also can contribute to ceramide accumulation [71, 89]. In a mouse cerebral IR model, de novo ceramide generation appeared to promote cell death by abrogating the mitochondrial respiratory chain [63], as inhibition of ceramide synthesis with FB1 reduced the size of cerebral infarct regions [43]. However, this model of ceramide action is complicated by the neuroprotective effects observed in IPC. Brief ischemia protects primary cultured neurons from hypoxia-induced cell death; ceramide levels were elevated while sphingomyelin remained unchanged following brief hypoxia. Treatment with exogenous, low dose ceramide had a similar neuroprotective effect [71]. This protective effect was abolished by pretreatment with FB1 [89]. Together, these later data suggest that the de novo ceramide synthesis pathway, rather than sphingomyelin/ceramide pathway, is involved in two stages: neuroprotection during brief ischemia and cerebral injury in mild ischemia and IR.

9. Liver

Hepatic IR damage, which can occur in diverse settings including liver resection and transplantation, trauma, hemorrhagic shock, or liver surgery, is a serious clinical complication that may compromise liver function because of extensive hepatocellular loss. Initial evidence regarding the role of ceramide showed that it was elevated in rat liver after cold ischemia and warm reperfusion during transplantation [14]. In the total rat liver ischemia/reperfusion model, both neutral and acid SMase activities were initially decreased during the early phase of ischemia, and acid SMase activity was increased during the later phase (over 1 h). The initial inhibition of SMase activities may contribute to the enhanced S1P levels due to a negative crosstalk between S1P and acid SMase [91, 92]. Accumulation of TNFα and activation of SMase also were observed during reperfusion of the ischemic lobe of rat livers [30]. TNFα gene knockout mice also exhibited hepatoprotection against IR-induced liver injury [32]. In a murine model of warm hepatic IR injury, an early phase of ceramide elevation was observed at 30 min after IR due to activation of acid SMase and inhibition of sphingomyelin synthase, followed by acid ceramidase stimulation. The later phase of ceramide elevation occurred at 6 h after IR due to only activation of acid SMase and unchanged acid ceramidase activity. Administration of SMase inhibitors decreased ceramide accumulation during hepatic IR and attenuated cell necrosis, cytochrome c release, and caspase activation [21]. In contrast, administration of a ceramidase inhibitor enhanced ceramide generation and exacerbated hepatic IR injury [35]. These results suggest that ceramide generated from TNFα mediated activation of acid SMase, irrespective of acid ceramidase, plays an important role in IR-induced liver damage and that the modulation of ceramide levels by inhibition of SMase and/or activation of ceramidase may be of therapeutic relevance.

10. Kidney

Ischemia/reperfusion injury is an unavoidable complication after kidney transplantation and is associated with delayed graft function and graft rejection. The first report implicating ceramide in renal IR injury was published by Zager et al. in 1997 [16]. In this whole mouse renal IR model, ceramide levels in kidney showed transient reduction after 45 min of ischemia, followed by a 2- to 3-fold increase during the reperfusion phase [93]. Of interest, the decrease of ceramide paralleled a decline of both acid and neutral SMase activities during the ischemia phase, but the increase during reperfusion was not associated with increased SMase activities, which in fact decreased further [23]. This SMase suppression may be accounted for by ceramide driven S1P production, resulting in inhibition of SMase activity [91, 92]. Using an in vivo model of unilateral renal occlusion, the renal injury was attenuated by inhibiting de novo ceramide synthesis with FB1, but not by suppressing SMase activity with the D609 inhibitor [94]. In the normal mouse renal cortex, C24, C22, and C16 ceramide are the main species, constituting 70%, 10%, and 20% of the total ceramide content, respectively. C16 ceramide was significantly increased, and all others mildly increased, after IR. Interestingly, IR induced an apparent shift towards unsaturated (versus saturated) fatty acids within the C22 and C24, but not the C16 ceramide pool [93]. These findings suggested that de novo ceramide biosynthesis plays a key role in IR-induced C16 ceramide accumulation and renal injury.

11. Intestine

Intestinal IR injury can occur in diverse conditions, including small bowel transplantation, acute mesenteric ischemia, hemorrhagic, and traumatic or septic shock [95]. IR injury is a major difficulty in small bowel transplantation. It is known that intestinal IR leads to severe destruction of distant tissues because of damage to the intestinal mucosal barrier, which causes a systemic inflammatory reaction and multiple organ dysfunctions [96]. A few reports have demonstrated that apoptosis was a major mode of rat and mouse small intestinal epithelial cell death induced by intestinal IR [97, 98]. The first study suggesting a role for ceramide in intestinal IR injury was published by Liu et al. in 2007 [99]. In the rat IR model of clamping superior mesenteric artery, intestinal IR caused intestinal mucosal epithelial apoptosis and accumulation of ceramide, followed by upregulation of SMase mRNA expression [100]. In the small rat bowel graft model following cold ischemia and subsequent warm reperfusion, an elevation of ceramide and extensive apoptosis were observed in the intestinal tissue [101]. To determine whether de novo ceramide synthesis was involved in intestinal IR injury, a rat splanchnic artery occlusion and reperfusion model was implemented. By clamping both the superior mesenteric artery and the celiac artery for 45 min followed by reperfusion, all rats died during a 4 h reperfusion period due to significant accumulation of ceramide, elevated production of TNFα, and extensive apoptosis in the intestine. Pretreatment with FB1 dramatically reduced these proapoptotic reactions [102]. These findings reveal that ceramide generated by SMase and do novo biosynthesis both contribute to the development of intestinal IR injury.

12. Summary

During ischemia and reperfusion, the generation of excess ROS and inflammatory cytokines in IR tissue is the initial

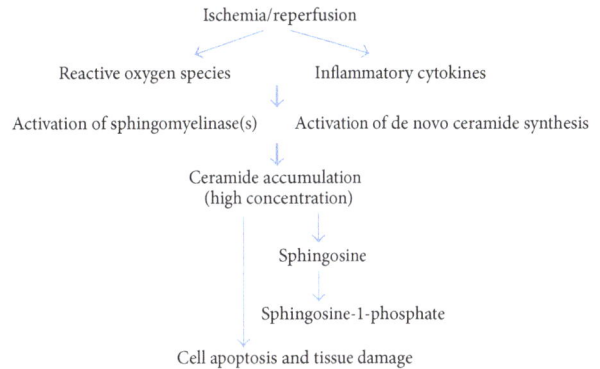

FIGURE 3: Flowchart of IR injury and sphingolipids.

key response. These oxidative and inflammatory stress stimuli activate two major ceramide generation pathways (sphingomyelin/ceramide and de novo synthesis), resulting in significant ceramide accumulation, cell apoptosis, and tissue damage (Figure 3). Regardless of the diverse pathways and complicated mechanisms underlying ceramide generation, it is widely accepted that the generation of ceramide is central to the pathogenesis in ischemia/reperfusion injury. Generally, the severity of IR-induced tissue injury depends on the phase and extent of ischemia and reperfusion, which in turn is associated with the amount and selective pathway of ceramide generation. Tissues also generate protective S1P, likely by activation of ceramidase and/or SPHK, during very early phases of ischemia, which may cause transient reduction of ceramide and S1P-induced inhibition of SMase. Following a brief ischemia, increased ROS generation induces small amounts of ceramide by activation of the de novo synthesis pathway. During this early phase of ischemia, ceramide generation is generally protective in nature. After prolonged ischemia following reperfusion, the massive production of ROS and TNFα results in significant accumulation of ceramide via activation of the sphingomyelin/ceramide signaling pathway, combined with alterations of downstream ceramide metabolism (e.g., glycosylceramide synthase, ceramidase, and SPHK), which in turn engages downstream pathways of inflammation and apoptosis contributing to mild to severe IR injuries. Administration of pharmacological inhibitors of SMase and ceramide synthase, genetic knockdown by siRNA, and use of SMase knockout mice reduced ceramide accumulation during IR and attenuated cell apoptosis and tissue damage. It is well known that mitochondria are a major site of ROS generation and have their own set of ceramide generating enzymes. Ceramide-induced mitochondrial dysfunction appears to be an essential step in IR-induced tissue damage. Although ceramide contributes to the development of IR-induced tissue injury, it presents differently between organs. The sphingomyelin/ceramide signaling pathway is dominant for ceramide-induced IR injury in heart, liver, brain, and intestine, whereas ceramide accumulation primarily results from the de novo biosynthesis pathway in IR-injured kidney. Further studies are required to further understand the role of ceramide in ischemic organs and IR injury, but based on extensive data accumulated over the past two decades it is clear that strategies that reduce ceramide by modulations of ceramidase and/or SMase activities may represent novel and promising therapeutic approaches to prevent or treat IR injury in diverse clinical settings.

References

[1] T. Okazaki, A. Bielawska, R. M. Bell et al., "Role of ceramide as a lipid mediator of 1 alpha, 25-dihydroxyvitamin D3-induced HL-60 cell differentiation," *The Journal of Biological Chemistry*, vol. 265, no. 26, pp. 15823–15831, 1990.

[2] E. Gulbins and R. Kolesnick, "Acid sphingomyelinase-derived ceramide signaling in apoptosis," *Subcellular Biochemistry*, vol. 36, pp. 229–244, 2002.

[3] Y. H. Zeidan and Y. A. Hannun, "The acid sphingomyelinase/ceramide pathway: Biomedical significance and mechanisms of regulation," *Current Molecular Medicine*, vol. 10, no. 5, pp. 454–466, 2010.

[4] X. He and E. H. Schuchman, "Potential role of acid sphingomyelinase in environmental health," *Zhong Nan Da Xue Xue Bao Yi Xue Ban*, vol. 37, no. 2, pp. 109–125, 2012.

[5] E. Gulbins and R. Kolesnick, "Raft ceramide in molecular medicine," *Oncogene*, vol. 22, no. 45, pp. 7070–7077, 2003.

[6] E. L. Smith and E. H. Schuchman, "The unexpected role of acid sphingomyelinase in cell death and the pathophysiology of common diseases," *The FASEB Journal*, vol. 22, no. 10, pp. 3419–3431, 2008.

[7] E. Gulbins, "Regulation of death receptor signaling and apoptosis by ceramide," *Pharmacological Research*, vol. 47, no. 5, pp. 393–399, 2003.

[8] J. A. Chavez and S. A. Summers, "A ceramide-centric view of insulin resistance," *Cell Metabolism*, vol. 15, no. 5, pp. 585–594, 2012.

[9] W. Doehner, A. C. Bunck, M. Rauchhaus et al., "Secretory sphingomyelinase is upregulated in chronic heart failure: A second messenger system of immune activation relates to body composition, muscular functional capacity, and peripheral blood flow," *European Heart Journal*, vol. 28, no. 7, pp. 821–828, 2007.

[10] W. Pan, J. Yu, R. Shi et al., "Elevation of ceramide and activation of secretory acid sphingomyelinase in patients with acute coronary syndromes," *Coronary Artery Disease*, vol. 25, no. 3, pp. 230–235, 2014.

[11] X. He, Y. Huang, B. Li, C.-X. Gong, and E. H. Schuchman, "Deregulation of sphingolipid metabolism in Alzheimer's disease," *Neurobiology of Aging*, vol. 31, no. 3, pp. 398–408, 2010.

[12] S. Patil, J. Melrose, and C. Chan, "Involvement of astroglial ceramide in palmitic acid-induced Alzheimer-like changes in primary neurons," *European Journal of Neuroscience*, vol. 26, no. 8, pp. 2131–2141, 2007.

[13] P. Biberthaler, B. Luchting, S. Massberg et al., "Ischemia at 4°C: A novel mouse model to investigate the effect of hypothermia on postiscnemic hepatic microcirculatory injury," *Research in Experimental Medicine*, vol. 200, no. 2, pp. 93–105, 2001.

[14] C. A. Bradham, R. F. Stachlewitz, W. Gao et al., "Reperfusion after liver transplantation in rats differentially activates the mitogen-activated protein kinases," *Hepatology*, vol. 25, no. 5, pp. 1128–1135, 1997.

[15] A. E. Bielawska, J. P. Shapiro, and L. Jiang, "Ceramide is involved in triggering of cardiomyocyte apoptosis induced by ischemia and reperfusion," *American Journal of Pathology*, vol. 151, no. 5, pp. 1257–1263, 1997.

[16] R. A. Zager, M. Iwata, D. S. Conrad, K. M. Burkhart, and Y. Igarashi, "Altered ceramide and sphingosine expression during the induction phase of ischemic acute renal failure," *Kidney International*, vol. 52, no. 1, pp. 60–70, 1997.

[17] W.-C. Liao, A. Haimovitz-Friedman, R. S. Persaud et al., "Ataxia telangiectasia-mutated gene product inhibits DNA damage-induced apoptosis via ceramide synthase," *The Journal of Biological Chemistry*, vol. 274, no. 25, pp. 17908–17917, 1999.

[18] M. Kubota, K. Narita, T. Nakagomi et al., "Sphingomyelin changes in rat cerebral cortex during focal ischemia," *Neurological Research*, vol. 18, no. 4, pp. 337–341, 1996.

[19] M. Nakane, M. Kubota, T. Nakagomi et al., "Lethal forebrain ischemia stimulates sphingomyelin hydrolysis and ceramide generation in the gerbil hippocampus," *Neuroscience Letters*, vol. 296, no. 2-3, pp. 89–92, 2000.

[20] I. Ramirez-Camacho, R. Bautista-Perez, F. Correa et al., "Role of sphingomyelinase in mitochondrial ceramide accumulation during reperfusion," *Biochimica et Biophysica Acta*, vol. 1862, no. 10, pp. 1955–1963, 2016.

[21] H. Tuzcu, B. Unal, E. Kırac et al., "Neutral sphingomyelinase inhibition alleviates apoptosis, but not ER stress, in liver ischemia–reperfusion injury," *Free Radical Research*, vol. 51, no. 3, pp. 253–268, 2017.

[22] B. Unal, F. Ozcan, H. Tuzcu, E. Kırac, G. O. Elpek, and M. Aslan, "Inhibition of neutral sphingomyelinase decreases elevated levels of nitrative and oxidative stress markers in liver ischemia–reperfusion injury," *Redox Report*, vol. 22, no. 4, pp. 147–159, 2017.

[23] R. A. Zager, S. Conrad, K. Lochhead, E. A. Sweeney, Y. Igarashi, and K. M. Burkhart, "Altered sphingomyelinase and ceramide expression in the setting of ischemic and nephrotoxic acute renal failure," *Kidney International*, vol. 53, no. 3, pp. 573–582, 1998.

[24] O. M. Hernandez, D. J. Discher, N. H. Bishopric, and K. A. Webster, "Rapid activation of neutral sphingomyelinase by hypoxia-reoxygenation of cardiac myocytes," *Circulation Research*, vol. 86, no. 2, pp. 198–204, 2000.

[25] N. W. O'Brien, N. M. Gellings, M. Guo, S. B. Barlow, C. C. Glembotski, and R. A. Sabbadini, "Factor associated with neutral sphingomyelinase activation and its role in cardiac cell death," *Circulation Research*, vol. 92, no. 6, pp. 589–591, 2003.

[26] P. H. Chan, "Reactive Oxygen Radicals in Signaling and Damage in the Ischemic Brain," *Journal of Cerebral Blood Flow & Metabolism*, vol. 21, no. 1, pp. 2–14, 2001.

[27] N. Suematsu, H. Tsutsui, J. Wen et al., "Oxidative stress mediates tumor necrosis factor-α-induced mitochondrial DNA damage and dysfunction in cardiac myocytes," *Circulation*, vol. 107, no. 10, pp. 1418–1423, 2003.

[28] J.-S. Won and I. Singh, "Sphingolipid signaling and redox regulation," *Free Radical Biology & Medicine*, vol. 40, no. 11, pp. 1875–1888, 2006.

[29] J. Fan, B. X. Wu, and C. E. Crosson, "Suppression of acid sphingomyelinase protects the retina from ischemic injury," *Investigative Ophthalmology & Visual Science*, vol. 57, no. 10, pp. 4476–4484, 2016.

[30] A. V. Alessenko, E. I. Galperin, L. B. Dudnik et al., "Role of

tumor necrosis factor alpha and sphingomyelin cycle activation in the induction of apoptosis by ischemia/reperfusion of the liver," *Biochemistry*, vol. 67, no. 12, pp. 1347–1355, 2002.

[31] A. Strelow, K. Bernardo, S. Adam-Klages et al., "Overexpression of acid ceramidase protects from tumor necrosis factor-induced cell death," *The Journal of Experimental Medicine*, vol. 192, no. 5, pp. 601–611, 2000.

[32] N. Teoh, J. Field, J. Sutton, and G. Farrell, "Dual Role of Tumor Necrosis Factor-α in Hepatic Ischemia-Reperfusion Injury: Studies in Tumor Necrosis Factor-α Gene Knockout Mice," *Hepatology*, vol. 39, no. 2, pp. 412–421, 2004.

[33] S. Lecour, P. Owira, and L. H. Opie, "Ceramide-induced preconditioning involves reactive oxygen species," *Life Sciences*, vol. 78, no. 15, pp. 1702–1706, 2006.

[34] D. Pchejetski, O. Kunduzova, A. Dayon et al., "Oxidative stress-dependent sphingosine kinase-1 inhibition mediates monoamine oxidase A-associated cardiac cell apoptosis," *Circulation Research*, vol. 100, no. 1, pp. 41–49, 2007.

[35] L. Llacuna, M. Mari, C. Garcia-Ruiz, J. C. Fernandez-Checa, and A. Morales, "Critical role of acidic sphingomyelinase in murine hepatic ischemia-reperfusion injury," *Hepatology*, vol. 44, no. 3, pp. 561–572, 2006.

[36] Z. F. Yu, M. Nikolova-Karakashian, D. Zhou, G. Cheng, E. H. Schuchman, and M. P. Mattson, "Pivotal role for acidic sphingomyelinase in cerebral ischemia-induced ceramide and cytokine production, and neuronal apoptosis," *Journal of Molecular Neuroscience*, vol. 15, no. 2, pp. 85–97, 2000.

[37] M. R. Reforgiato, G. Milano, G. Fabriàs et al., "Inhibition of ceramide de novo synthesis as a postischemic strategy to reduce myocardial reperfusion injury," *Basic Research in Cardiology*, vol. 111, no. 2, article no. 12, 2016.

[38] E. C. Mandon, I. Ehses, J. Rother, G. Van Echten, and K. Sandhoff, "Subcellular localization and membrane topology of serine palmitoyltransferase, 3-dehydrosphinganine reductase, and sphinganine N- acyltransferase in mouse liver," *The Journal of Biological Chemistry*, vol. 267, no. 16, pp. 11144–11148, 1992.

[39] A. H. Futerman, B. Stieger, A. L. Hubbard, and R. E. Pagano, "Sphingomyelin synthesis in rat liver occurs predominantly at the cis and medial cisternae of the Golgi apparatus," *The Journal of Biological Chemistry*, vol. 265, no. 15, pp. 8650–8657, 1990.

[40] A. H. Futerman and H. Riezman, "The ins and outs of sphingolipid synthesis," *Trends in Cell Biology*, vol. 15, no. 6, pp. 312–318, 2005.

[41] Y. Mizutani, A. Kihara, and Y. Igarashi, "Mammalian Lass6 and its related family members regulate synthesis of specific ceramides," *Biochemical Journal*, vol. 390, no. 1, pp. 263–271, 2005.

[42] F. D. Testai, J. P. Kilkus, E. Berdyshev, I. Gorshkova, V. Natarajan, and G. Dawson, "Multiple sphingolipid abnormalities following cerebral microendothelial hypoxia," *Journal of Neurochemistry*, vol. 131, no. 4, pp. 530–540, 2014.

[43] J. Yu, S. A. Novgorodov, D. Chudakova et al., "JNK3 signaling pathway activates ceramide synthase leading to mitochondrial dysfunction," *The Journal of Biological Chemistry*, vol. 282, no. 35, pp. 25940–25949, 2007.

[44] H.-P. Tian, T.-Z. Qiu, J. Zhao, L.-X. Li, and J. Guo, "Sphingomyelinase-induced ceramide production stimulate calcium-independent JNK and PP2A activation following cerebral ischemia," *Brain Injury*, vol. 23, no. 13-14, pp. 1073–1080, 2009.

[45] J. Jin, Q. Hou, T. D. Mullen et al., "Ceramide generated by

sphingomyelin hydrolysis and the salvage pathway is involved in hypoxia/reoxygenation-induced bax redistribution to mitochondria in NT-2 cells," *The Journal of Biological Chemistry*, vol. 283, no. 39, pp. 26509–26517, 2008.

[46] D. X. Zhang, R. M. Fryer, A. K. Hsu et al., "Production and metabolism of ceramide in normal and ischemic-reperfused myocardium of rats," *Basic Research in Cardiology*, vol. 96, no. 3, pp. 267–274, 2001.

[47] K. Takahashi, I. Ginis, R. Nishioka et al., "Glucosylceramide synthase activity and ceramide levels are modulated during cerebral ischemia after ischemic preconditioning," *Journal of Cerebral Blood Flow & Metabolism*, vol. 24, no. 6, pp. 623–627, 2004.

[48] D. Ardail, I. Popa, K. Alcantara et al., "Occurrence of ceramides and neutral glycolipids with unusual long-chain base composition in purified rat liver mitochondria," *FEBS Letters*, vol. 488, no. 3, pp. 160–164, 2001.

[49] T. Yabu, A. Shimuzu, and M. Yamashita, "A novel mitochondrial sphingomyelinase in zebrafish cells," *The Journal of Biological Chemistry*, vol. 284, no. 30, pp. 20349–20363, 2009.

[50] B. X. Wu, V. Rajagopalan, P. L. Roddy, C. J. Clarke, and Y. A. Hannun, "Identification and characterization of murine mitochondria-associated neutral sphingomyelinase (MA-nSMase), the mammalian sphingomyelin phosphodiesterase 5," *The Journal of Biological Chemistry*, vol. 285, no. 23, pp. 17993–18002, 2010.

[51] H. Shimeno, S. Soeda, M. Yasukouchi, N. Okamura, and A. Nagamatsu, "Fatty Acyl-Co A: Sphingosine acyltransferase in bovine brain mitochondria: Its solubilization and reconstitution onto the membrane lipid liposomes," *Biological & Pharmaceutical Bulletin*, vol. 18, no. 10, pp. 1335–1339, 1995.

[52] C. Bionda, J. Portoukalian, D. Schmitt, C. Rodriguez-Lafrasse, and D. Ardail, "Subcellular compartmentalization of ceramide metabolism: MAM (mitochondria-associated membrane) and/or mitochondria?" *Biochemical Journal*, vol. 382, no. 2, pp. 527–533, 2004.

[53] S. A. Novgorodov, D. A. Chudakova, B. W. Wheeler et al., "Developmentally regulated ceramide synthase 6 increases mitochondrial Ca^{2+} loading capacity and promotes apoptosis," *The Journal of Biological Chemistry*, vol. 286, no. 6, pp. 4644–4658, 2011.

[54] S. A. Novgorodov, B. X. Wu, T. I. Gudz et al., "Novel pathway of ceramide production in mitochondria: Thioesterase and neutral ceramidase produce ceramide from sphingosine and acyl-CoA," *The Journal of Biological Chemistry*, vol. 286, no. 28, pp. 25352–25362, 2011.

[55] J. Stiban, L. Caputo, and M. Colombini, "Ceramide synthesis in the endoplasmic reticulum can permeabilize mitochondria to proapoptotic proteins," *Journal of Lipid Research*, vol. 49, no. 3, pp. 625–634, 2008.

[56] M. P. Murphy, "How mitochondria produce reactive oxygen species," *Biochemical Journal*, vol. 417, no. 1, pp. 1–13, 2009.

[57] X. Zhang, B. Li, Y. Zhang et al., "Mitochondrial changes in apoptosis of HT-29 cells induced by C2-ceramide," *Wei Sheng Yan Jiu*, vol. 37, no. 5, pp. 558–559, 2008.

[58] M. Xin and X. Deng, "Protein phosphatase 2A enhances the proapoptotic function of Bax through dephosphorylation," *The Journal of Biological Chemistry*, vol. 281, no. 27, pp. 18859–18867, 2006.

[59] A. Garcia, X. Cayla, J. Guergnon et al., "Serine/threonine pro-

tein phosphatases PP1 and PP2A are key players in apoptosis," *Biochimie*, vol. 85, no. 8, pp. 721–726, 2003.

[60] C. L. Murriel, E. Churchill, K. Inagaki, L. I. Szweda, and D. Mochly-Rosen, "Protein kinase Cδ activation induces apoptosis in response to cardiac ischemia and reperfusion damage: a mechanism involving bad and the mitochondria," *The Journal of Biological Chemistry*, vol. 279, no. 46, pp. 47985–47991, 2004.

[61] C. R. Weston and R. J. Davis, "The JNK signal transduction pathway," *Current Opinion in Cell Biology*, vol. 19, no. 2, pp. 142–149, 2007.

[62] M. Di Paola, P. Zaccagnino, G. Montedoro, T. Cocco, and M. Lorusso, "Ceramide induces release of pro-apoptotic proteins from mitochondria by either a Ca2+-dependent or a Ca2+-independent mechanism," *Journal of Bioenergetics and Biomembranes*, vol. 36, no. 2, pp. 165–170, 2004.

[63] S. A. Novgorodov, C. L. Riley, J. A. Keffler et al., "SIRT3 deacetylates ceramide synthases: Implications for mitochondrial dysfunction and brain injury," *The Journal of Biological Chemistry*, vol. 291, no. 4, pp. 1957–1973, 2016.

[64] L. J. Siskind, R. N. Kolesnick, and M. Colombini, "Ceramide forms channels in mitochondrial outer membranes at physiologically relevant concentrations," *Mitochondrion*, vol. 6, no. 3, pp. 118–125, 2006.

[65] A. Beresewicz, A. Dobrzyn, and J. Gorski, "Accumulation of specific ceramides in ischemic/reperfused rat heart; effect of ischemic preconditioning," *Journal of Physiology and Pharmacology*, vol. 53, no. 3, pp. 371–382, 2002.

[66] L. Argaud, A. Prigent, L. Chalabreysse, J. Loufouat, M. Lagarde, and M. Ovize, "Ceramide in the antiapoptotic effect of ischemic preconditioning," *American Journal of Physiology-Heart and Circulatory Physiology*, vol. 286, no. 1, pp. H246–H251, 2004.

[67] F. C. Barone, R. F. White, P. A. Spera et al., "Ischemic preconditioning and brain tolerance: temporal histological and functional outcomes, protein synthesis requirement, and interleukin-1 receptor antagonist and early gene expression," *Stroke*, vol. 29, no. 9, pp. 1937–1951, 1998.

[68] Z. Cao, Y. Yuan, G. Jeyabalan et al., "Preactivation of NKT cells with α-GalCer protects against hepatic ischemia-reperfusion injury in mouse by a mechanism involving IL-13 and adenosine A2A receptor," *American Journal of Physiology-Gastrointestinal and Liver Physiology*, vol. 297, no. 2, pp. G249–G258, 2009.

[69] J. Mitoma, M. Ito, S. Furuya, and Y. Hirabayashi, "Bipotential roles of ceramide in the growth of hippocampal neurons: Promotion of cell survival and dendritic outgrowth in dose- and developmental stage-dependent manners," *Journal of Neuroscience Research*, vol. 51, no. 6, pp. 712–722, 1998.

[70] S. Willaime, P. Vanhoutte, J. Caboche, Y. Lemaigre-Dubreuil, J. Mariani, and B. Brugg, "Ceramide-induced apoptosis in cortical neurons is mediated by an increase in p38 phosphorylation and not by the decrease in ERK phosphorylation," *European Journal of Neuroscience*, vol. 13, no. 11, pp. 2037–2046, 2001.

[71] M. I. H. Bhuiyan, M. N. Islam, S. Y. Jung, H. H. Yoo, Y. S. Lee, and C. Jin, "Involvement of ceramide in ischemic tolerance induced by preconditioning with sublethal oxygen-glucose deprivation in primary cultured cortical neurons of rats," *Biological & Pharmaceutical Bulletin*, vol. 33, no. 1, pp. 11–17, 2010.

[72] D. M. Yellon and J. M. Downey, "Preconditioning the myocardium: from cellular physiology to clinical cardiology," *Physiological Reviews*, vol. 83, no. 4, pp. 1113–1151, 2003.

[73] S. Lecour, M. N. Sack, and L. H. Opie, "Sphingolipid signaling: a potential pathway for TNF-alpha induced preconditioning," *Annales De Cardiologie Et D'Angeiologie*, vol. 52, no. 6, pp. 363–369, 2003.

[74] S. Lecour, L. Rochette, and L. Opie, "Free radicals trigger TNFα-induced cardioprotection," *Cardiovascular Research*, vol. 65, no. 1, pp. 239–243, 2005.

[75] C. P. Baines, M. Goto, and J. M. Downey, "Oxygen radicals released during ischemic preconditioning contribute to cardioprotection in the rabbit myocardium," *Journal of Molecular and Cellular Cardiology*, vol. 29, no. 1, pp. 207–216, 1997.

[76] M. Das, J. Cui, and D. K. Das, "Generation of survival signal by differential interaction of p38MAPKα and p38MAPKβ with caveolin-1 and caveolin-3 in the adapted heart," *Journal of Molecular and Cellular Cardiology*, vol. 42, no. 1, pp. 206–213, 2007.

[77] P. Der, J. Cui, and D. K. Das, "Role of lipid rafts in ceramide and nitric oxide signaling in the ischemic and preconditioned hearts," *Journal of Molecular and Cellular Cardiology*, vol. 40, no. 2, pp. 313–320, 2006.

[78] S. Spiegel and S. Milstien, "Sphingosine-1-phosphate: an enigmatic signalling lipid," *Nature Reviews Molecular Cell Biology*, vol. 4, no. 5, pp. 397–407, 2003.

[79] S. Lecour, R. M. Smith, B. Woodward, L. H. Opie, L. Rochette, and M. N. Sack, "Identification of a novel role for sphingolipid signaling in TNFα and ischemic preconditioning mediated cardioprotection," *Journal of Molecular and Cellular Cardiology*, vol. 34, no. 5, pp. 509–518, 2002.

[80] Z.-Q. Jin, E. J. Goetzl, and J. S. Karliner, "Sphingosine kinase activation mediates ischemic preconditioning in murine heart," *Circulation*, vol. 110, no. 14, pp. 1980–1989, 2004.

[81] Y. Nishino, I. Webb, and M. S. Marber, "Sphingosine kinase isoforms and cardiac protection," *Cardiovascular Research*, vol. 76, no. 1, pp. 3-4, 2007.

[82] Z.-Q. Jin, J. Zhang, Y. Huang, H. E. Hoover, D. A. Vessey, and J. S. Karliner, "A sphingosine kinase 1 mutation sensitizes the myocardium to ischemia/reperfusion injury," *Cardiovascular Research*, vol. 76, no. 1, pp. 41–50, 2007.

[83] T. M. Vu, A. Ishizu, J. C. Foo et al., "Mfsd2b is essential for the sphingosine-1-phosphate export in erythrocytes and platelets," *Nature*, vol. 550, no. 7677, pp. 524–528, 2017.

[84] I. Karunakaran and G. van Echten-Deckert, "Sphingosine 1-phosphate – A double edged sword in the brain," *Biochimica et Biophysica Acta (BBA) - Biomembranes*, vol. 1859, no. 9, pp. 1573–1582, 2017.

[85] J. Qin, E. Berdyshev, J. Goya, V. Natarajan, and G. Dawson, "Neurons and oligodendrocytes recycle sphingosine 1-phosphate to ceramide: significance for apoptosis and multiple sclerosis," *The Journal of Biological Chemistry*, vol. 285, no. 19, pp. 14134–14143, 2010.

[86] E. Usta, M. Mustafi, F. Artunc et al., "The challenge to verify ceramide's role of apoptosis induction in human cardiomyocytes - a pilot study," *Journal of Cardiothoracic Surgery*, vol. 6, no. 1, article no. 38, 2011.

[87] M. Kubota, S. Kitahara, H. Shimasaki, and N. Ueta, "Accumulation of ceramide in ischemic human brain of an acute case of cerebral occlusion," *The Japanese Journal of Experimental Medicine*, vol. 59, no. 2, pp. 59–64, 1989.

[88] I. Herr, A. Martin-Villalba, E. Kurz et al., "FK506 prevents stroke-induced generation of ceramide and apoptosis signaling," *Brain Research*, vol. 826, no. 2, pp. 210–219, 1999.

[89] J. Liu, I. Ginis, M. Spatz, and J. M. Hallenbeck, "Hypoxic preconditioning protects cultured neurons against hypoxic stress via TNF-α and ceramide," *American Journal of Physiology-*

Cell Physiology, vol. 278, no. 1, pp. C144–C153, 2000.

[90] D. A. Chudakova, Y. H. Zeidan, B. W. Wheeler et al., "Integrin-associated Lyn kinase promotes cell survival by suppressing acid sphingomyelinase activity," *The Journal of Biological Chemistry*, vol. 283, no. 43, pp. 28806–28816, 2008.

[91] J. K. Yun and M. Kester, "Regulatory role of sphingomyelin metabolites in hypoxia-induced vascular smooth muscle cell proliferation," *Archives of Biochemistry and Biophysics*, vol. 408, no. 1, pp. 78–86, 2002.

[92] A. Gomez-Munoz, J. Kong, B. Salh, and U. P. Steinbrecher, "Sphingosine-1-phosphate inhibits acid sphingomyelinase and blocks apoptosis in macrophages," *FEBS Letters*, vol. 539, no. 1-3, pp. 56–60, 2003.

[93] T. Kalhorn and R. A. Zager, "Renal cortical ceramide patterns during ischemic and toxic injury: Assessments by HPLC-mass spectrometry," *American Journal of Physiology-Renal Physiology*, vol. 277, no. 5, pp. F723–F733, 1999.

[94] Y. Itoh, T. Yano, T. Sendo et al., "Involvement of de novo ceramide synthesis in radiocontrast-induced renal tubular cell injury," *Kidney International*, vol. 69, no. 2, pp. 288–297, 2006.

[95] S. Homer-Vanniasinkam, J. N. Crinnion, and M. J. Gough, "Post-ischaemic organ dysfunction: A review," *European Journal of Vascular and Endovascular Surgery*, vol. 14, no. 3, pp. 195–203, 1997.

[96] H. Mitsuoka, E. B. Kistler, and G. W. Schmid-Schönbein, "Protease inhibition in the intestinal lumen: Attenuation of systemic inflammation and early indicators of multiple organ failure in shock," *Shock*, vol. 17, no. 3, pp. 205–209, 2002.

[97] Z. Sun, X. Wang, X. Deng et al., "The influence of intestinal ischemia and reperfusion on bidirectional intestinal barrier permeability, cellular membrane integrity, proteinase inhibitors, and cell death in rats," *Shock*, vol. 10, no. 3, pp. 203–212, 1998.

[98] T. Noda, R. Iwakiri, K. Fujimoto, S. Matsuo, and T. Y. Aw, "Programmed cell death induced by ischemia-reperfusion in rat intestinal mucosa," *American Journal of Physiology-Gastrointestinal and Liver Physiology*, vol. 274, no. 2, pp. G270–G276, 1998.

[99] K.-X. Liu, W. He, T. Rinne, Y. Liu, M.-Q. Zhao, and W.-K. Wu, "The effect of *Ginkgo biloba* extract (EGb 761) pretreatment on intestinal epithelial apoptosis induced by intestinal ischemia/reperfusion in rats: role of ceramide," *American Journal of Chinese Medicine*, vol. 35, no. 5, pp. 805–819, 2007.

[100] K.-X. Liu, S.-Q. Chen, W.-Q. Huang, Y.-S. Li, M. G. Irwin, and Z. Xia, "Propofol pretreatment reduces ceramide production and attenuates intestinal mucosal apoptosis induced by intestinal ischemia/reperfusion in rats," *Anesthesia & Analgesia*, vol. 107, no. 6, pp. 1884–1891, 2008.

[101] L. Wei, A. Wedeking, R. Büttner, J. C. Kalff, R. H. Tolba, and G. Van Echten-Deckert, "A natural tetrahydropyrimidine protects small bowel from cold ischemia and subsequent warm in vitro reperfusion injury," *Pathobiology*, vol. 76, no. 4, pp. 212–220, 2009.

[102] S. Cuzzocrea, R. Di Paola, T. Genovese et al., "Anti-inflammatory and anti-apoptotic effects of fumonisin B1, an inhibitor of ceramide synthase, in a rodent model of splanchnic ischemia and reperfusion injury," *The Journal of Pharmacology and Experimental Therapeutics*, vol. 327, no. 1, pp. 45–57, 2008.

4

Serum Fatty Acids, Traditional Risk Factors, and Comorbidity as Related to Myocardial Injury in an Elderly Population with Acute Myocardial Infarction

Kristian Laake,[1,2,3] Ingebjørg Seljeflot,[1,2,3] Erik B. Schmidt,[4] Peder Myhre,[1,3,5] Arnljot Tveit,[6] Harald Arnesen,[1,2,3] and Svein Solheim[1,3]

[1]Center for Clinical Heart Research, Department of Cardiology, Oslo University Hospital, Ullevål, 0450 Oslo, Norway
[2]Faculty of Medicine, University of Oslo, 0316 Oslo, Norway
[3]Center for Heart Failure Research, University of Oslo, 0316 Oslo, Norway
[4]Department of Cardiology, Aalborg University Hospital, 9000 Aalborg, Denmark
[5]Department of Cardiology, Akershus University Hospital HF, 1478 Lørenskog, Norway
[6]Department of Medical Research, Vestre Viken Hospital Trust, Bærum Hospital, 1346 Rud, Norway

Correspondence should be addressed to Kristian Laake; kristian.laake@ous-hf.no

Academic Editor: Gerhard M. Kostner

Background. Epidemiological and randomized clinical trials indicate that marine polyunsaturated n-3 fatty acids (n-3 PUFAs) may have cardioprotective effects. *Aim.* Evaluate the associations between serum fatty acid profile, traditional risk factors, the presence of cardiovascular diseases (CVD), and peak Troponin T (TnT) levels in elderly patients with an acute myocardial infarction (AMI). *Materials and Methods.* Patients ($n = 299$) consecutively included in the ongoing Omega-3 fatty acids in elderly patients with myocardial infarction (OMEMI) trial were investigated. Peak TnT was registered during the hospital stay. Serum fatty acid analysis was performed 2–8 weeks later. *Results.* No significant correlations between peak TnT levels and any of the n-3 PUFAs were observed. However, patients with a history of atrial fibrillation had significantly lower docosahexaenoic acid levels than patients without. Significantly lower peak TnT levels were observed in patients with a history of hyperlipidemia, angina, MI, atrial fibrillation, intermittent claudication, and previous revascularization (all $p < 0.02$). *Conclusions.* In an elderly population with AMI, no association between individual serum fatty acids and estimated myocardial infarct size could be demonstrated. However, a history of hyperlipidemia and the presence of CVD were associated with lower peak TnT levels, possibly because of treatment with cardioprotective medications.

1. Background

Coronary heart disease (CHD) including myocardial infarction (MI) is one of the leading causes of mortality in the Western world and the incidence increases with advancing age [1]. The cardioprotective effects of marine polyunsaturated n-3 fatty acids (n-3 PUFA) have been extensively studied, and earlier epidemiological [2] and large scale clinical trials [3, 4] have shown beneficial effects, although these are not without controversy [5]. In the landmark GISSI-Prevenzione-trial [3], secondary prevention with 1 g/day of n-3 PUFA supplementation resulted in a 45% decrease in sudden death, but with

no effect regarding nonfatal MI. Later trials have contradicted these results [6, 7] showing no effect of n-3 PUFA on clinical endpoints; however they have been criticized for lacking statistical power [8] and for using inadequate treatment dosages [9]. The combined treatment with both statins and n-3 PUFA after MI has shown improved cardiovascular (CV) outcomes compared to treatment with statins alone [10], and treatment with n-3 PUFA within 14 days of a first MI has been associated with a 32% risk reduction of all-cause mortality in patients on concurrent lipid-lowering, antihypertensive, and antiplatelet treatment [11]. N3-PUFA supplementation has also shown improved outcome in heart failure [12, 13].

Mozaffarian et al. examined a large cohort of elderly subjects without prevalent CHD and found that higher levels of very long chain marine n-3 PUFAs in plasma were associated with a lower total mortality and especially fewer cardiac deaths [14]. There are, however, few published studies on circulating fatty acids concentrations and their association to CV events, and some studies are based on diet-questionnaires for estimation of intakes of marine n-3 PUFA [15, 16]. Infarct size is important for mortality and morbidity, and while there are no data in humans regarding the effect of n-3 PUFA on infarct size, one study has shown that rats fed with dietary fish oil for 8 weeks had a significantly reduced infarct size [17].

Studies on mechanisms have revealed n-3 PUFAs triglyceride lowering and anti-inflammatory and antiarrhythmic and platelet inhibiting and blood pressure lowering effects, including improved endothelial function and plaque stability [18]. The Troponin T cardiac biomarker (TnT) has been shown to correlate with infarct size [19] and peak TnT levels have been shown to be strongly correlated with infarct size assessed by single-photon emission computed tomography [20].

The aim of our study was to evaluate the serum fatty acid profile, traditional risk factors, and relevant cardiovascular disease states as related to myocardial injury estimated by peak TnT in an elderly population of patients that have experienced an acute MI. Our hypothesis was that the pattern of serum fatty acids as well as traditional risk factors and previous cardiovascular disease states would influence peak TnT levels. To our knowledge this has not been previously examined.

2. Materials and Methods

2.1. Study Design. The present study is a substudy of the OMEMI trial with a design that has previously been described in detail [21]. In short, the study is a prospective randomized placebo controlled multicenter trial evaluating the effect of a 2-year intervention with n-3 PUFA supplementation (1.8 g/d) on cardiovascular endpoints in elderly patients, age 70–82 years, that have suffered an acute MI (type 1–4), and without comorbidity thought to be incompatible with study drugs or a 2 year follow-up. Compliance is evaluated by serum levels of fatty acids recorded at baseline and at 24 months.

Hypertension (HT) was defined as previously known systolic blood pressure >140 and/or diastolic blood pressure >90 mmHg or treatment for HT. Diabetes was defined as being insulin dependent or noninsulin dependent. Hyperlipidemia was defined as being previously treated for/or diagnosed with dyslipidemia. Atrial fibrillation (AF) was defined as a history of ECG-documented paroxystic, persistent, or permanent AF and smokers were defined as current smokers. The study was carried out in compliance with the Helsinki Declaration and approved by the Regional Ethics Committee. All subjects gave their written informed consent to participate (ClinicalTrials.gov, NCT01841944).

The present results are based on participants included from November 2012 to October 2014.

2.2. Laboratory Methods. Peak TnT levels were registered during the index MI. Further data collection and serum fatty acid analyses were performed in blood samples obtained 2–8 weeks later at inclusion in the study. Blood samples were obtained in fasting condition (>10 h) by standard venipuncture between 08:00 and 11:00 am, before daily intake of medication. Serum was prepared by centrifugation within one hour at 2500 g for 10 min and kept frozen (−80°C) until analyzed. Routine blood samples were determined with conventional methods. Electrochemiluminescence technology for quantitative measurement was used for repeated measures of TnT (3rd-generation cTroponinT, Elecsys 2010, Roche, Mannheim, Germany) with interassay coefficient of variation (CV) of 7%. NT-proBNP was measured in serum using Elecsys proBNP sandwich immunoassay on Elecsys 2010 (Roche Diagnostics, Indianapolis, USA) with CV of 7%.

Serum phospholipids fatty acid composition was analyzed by gas chromatography at the Lipid Research Laboratory, Aalborg University Hospital, Denmark. Briefly, serum lipids were extracted by the Folch procedure [22]. Serum 500 μL was mixed with 5 mL chloroform-methanol 2 : 1 containing 50 μL/mL butylated hydroxytoluene as antioxidants and shaken. After addition of 750 μL 0.9% sodium chloride, the tubes were mixed and centrifuged at 3220 g for 10 minutes. The lower organic phase containing total lipids was collected and a second extraction was performed on the remaining protein disc. The organic phase was then dried under nitrogen for 45 min at 40°C and dissolved in 1 mL chloroform. Separation of phospholipids fatty acid fraction from total lipids was performed by the method of Burdge et al. [23], with the modification that all types of phospholipids were sampled (see also below). The tube containing the organic phase was transferred to Agilent Bond Elut NH2 column 200 mg (Agilent Technologies, US) preconditioned with 4 mL hexane. The phospholipids fatty acid fraction was eluted with 2 mL chloroform-methanol 3 : 2 followed by 2 mL of methanol and then dried under nitrogen for 1 hour at 40°C. Methylation of the fatty acids was performed before being analyzed by gas chromatography using Varian 3900 gas chromatograph equipped with a CP-8400 autosampler, a flame ionization detector, and a high-polarity polyester CP-Sil 88 60 m ×0.25 mm capillary column (Varian, Middleburg, Netherlands).

The serum content of linoleic acid (LA) (18:2, n-6), arachidonic acid (AA) (20:4, n-6), eicosapentaenoic acid (EPA) (20:5, n-3), and docosahexaenoic acid (DHA) (22:6 n-3) was expressed as a percent of total fatty acids (wt%) and the CVs were 0.4%, 0.6%, 1.1%, and 1.8%, respectively.

2.3. Statistics. As most data had a skewed distribution, the results are presented as median values (25 and 75 percentiles) unless otherwise stated. Nonparametric statistics were used throughout. For group comparison, Mann-Whitney U test was used for continuous variables. Analyses of correlations were performed with Spearman rho. A two-tailed value of $p \leq 0.05$ was considered statistically significant. The statistical analyses were performed with IBM SPSS Statistics, version 21.0.0.2 (IBM, New York, USA).

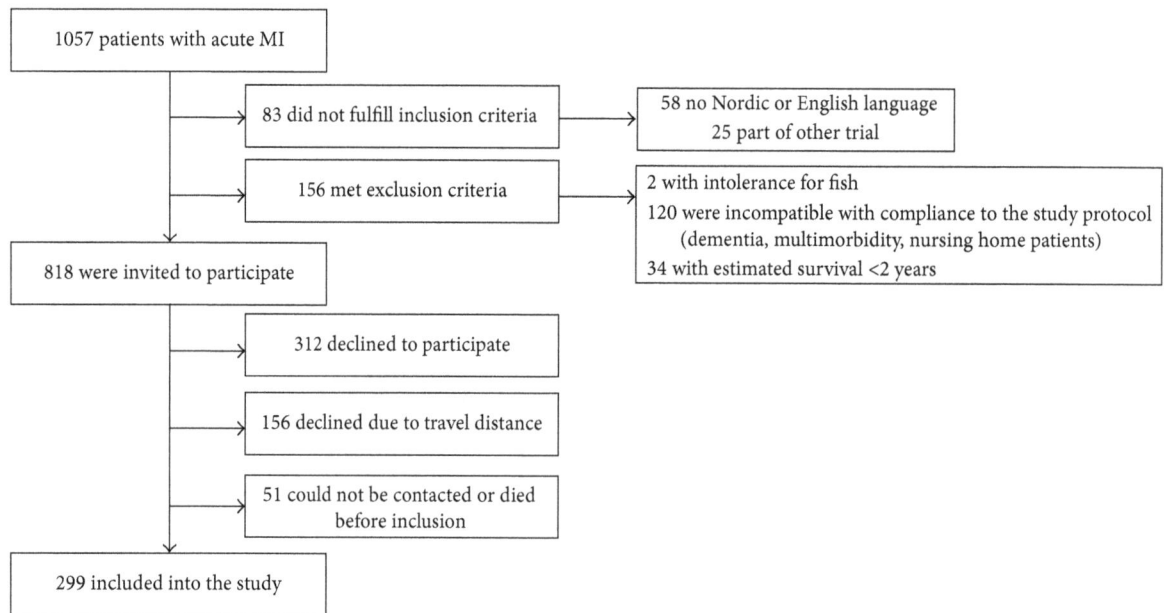

FIGURE 1: Flowchart of patients.

3. Results

Flowchart for the inclusion of patients in the present cohort is shown in Figure 1. Characteristics of the population at inclusion are shown in Table 1. Median age was 75 years, 74% were male, 61% had hypertension, 48% had hyperlipidemia, 30% had a previous myocardial infarction, 23% had diabetes, 14% were current smokers, and 13% had atrial fibrillation.

3.1. Serum Fatty Acids in relation to Myocardial Injury. We focused primarily on the levels of AA and LA n-6 and marine EPA and DHA n-3 fatty acids, as well as the AA + LA/EPA + DHA (n-6/n-3) ratio, as these are the most relevant for CHD. Median values are shown in Table 2.

We found no significant correlations between peak TnT levels and any of the fatty acids or the n-6/n-3 ratio (Table 3). We also evaluated any difference in peak TnT levels between the lowest (10th) and highest (90th percentile) of serums EPA + DHA and AA + LA and n-6/n-3 ratio, but the observed results were insignificant (all $p > 0.7$). There was also no correlation between any of the fatty acids or the n-6/n-3 ratio and NT-proBNP, an indicator of heart failure.

Significantly higher serum levels of EPA and DHA in participants who reported previous intake of n-3 PUFA supplements ($n = 135$) compared to those who did not ($n = 161$) were observed, resulting in a significant difference in the n-6/n-3 ratio (3.0 versus 4.1, $p < 0.001$). However, no difference in peak TnT levels between these two groups (644 versus 711 ng/L, $p = 0.908$) was revealed. Furthermore, there were no significant differences in levels of different fatty acids between patients with ST-segment elevation myocardial infarction and non-ST-segment elevation myocardial infarction.

In patients with known atrial fibrillation ($n = 38$), significantly lower DHA levels (5.1 versus 5.7 wt%, $p = 0.005$)

TABLE 1: Baseline characteristics of the study cohort. Data presented as percentages or median values (25 and 75 percentiles).

Age (y)	75 (72, 78)
Gender (male/female) (%)	73.9/26.1
Smoker (current/previous) (%)	13.7/46.8
Body mass index (kg/m^2)	25.6 (23.8, 28.3)
STEMI (%)	31.4
3-vessel disease (%)	21.3
Peak Troponin T level (ng/L) @ index MI	700 (153, 2500)
NT-proBNP ($n = 173$)	75.0 (33.0, 162.5)
History of hypertension (%)	182 (60.9)
History of hyperlipidemia (%)	143 (47.8)
History of atrial fibrillation (%)	38 (12.7)
Previous myocardial infarction (%)	90 (30.1)
Previous heart failure (%)	16 (5.4)
Previous diabetes (%)	69 (23.1)
Medication @ index MI ($n = 134$):	
Aspirin (%)	64 (47.8)
Other platelet inhibitors (%)	7 (5.2)
Anticoagulation (%)	19 (14.1)
Beta blocker (%)	61 (45.5)
ACE-I/AT II blocker (%)	61 (45.5)
Calcium channel blocker (%)	30 (22.4)
Statin (%)	68 (50.7)
Diuretic (%)	38 (28.4)
Nitrates (%)	8 (6.0)
n-3 PUFA supplements (%) ($n = 296$)	135 (45.6)

ACE-I/AT II: angiotensin-converting enzyme inhibitors/angiotensin II receptor blockers; STEMI: ST-segment elevation myocardial infarction.

and a higher n-6/n-3 ratio (4.2 versus 3.5, $p = 0.028$) were observed.

TABLE 2: Median values of selected n-3 fatty acids (% of total fatty acids in serum phospholipids) at baseline.

Eicosapentaenoic acid (EPA) n-3 (20:5) (wt%)	2.4
Docosahexaenoic acid (DHA) n-3 (22:6) (wt%)	5.6
Linoleic acid (LA) n-6 (18:2) (wt%)	19.0
Arachidonic acid (AA) n-6 (20:4) (wt%)	9.6
n6/n3 ratio	3.6

TABLE 3: Coefficients of correlations[1] between the selected fatty acids (% of total fatty acids in serum phospholipids) and peak Troponin T (ng/L) during index myocardial infarction.

Eicosapentaenoic acid (EPA)	$r = -.052$
n-3 (20:5)	$p = 0.369$
Docosahexaenoic acid (DHA)	$r = .060$
n-3 (22:6)	$p = 0.303$
Linoleic acid (LA)	$r = .021$
n-6 (18:2)	$p = 0.717$
Arachidonic acid (AA)	$r = .014$
n-6 (20:4)	$p = 0.808$
n6/n3 ratio	$r = .002$
	$p = 0.966$

[1]Spearman's rho are given.

There were no significant differences in any of the n-6 or n-3 fatty acid levels between genders or in patients with or without a previous MI or smokers and nonsmokers. Lower levels of AA were found in diabetics versus nondiabetics (10.7 versus 9.2, $p < 0.001$).

3.2. Peak TnT Levels in relation to Traditional Risk Factors and Cardiovascular Disease States. Comparisons between peak TnT (ng/L) levels during index hospitalization and CVD risk factors and relevant disease entities are shown in Table 4.

Significantly lower TnT levels were observed in patients with a history of hyperlipidemia ($p = 0.018$), angina pectoris ($p < 0.001$), previous MI ($p = 0.003$), atrial fibrillation ($p = 0.007$), and claudication ($p = 0.015$) and also for participants that have previously undergone revascularization procedures ($p = 0.003$). However, no significant differences were observed between genders or in patients with or without hypertension or diabetes or smokers versus nonsmokers. We found no correlations between infarct size estimated by peak TnT and age or body mass index (BMI). However, a strong positive correlation between infarct size and NT-proBNP ($r = 0.57$, $p < 0.001$) measured at inclusion was found.

With regard to previous CVD states and concurrent medical treatment (data on medication use available in 134), we found that, for previous angina, 77% were on aspirin, 67% were on statins, and 72% on were beta blockers; for previous MI, 76% were on aspirin, 79% were on statins, and 79% were on beta blockers; and, for previous AF, 59% were treated with oral anticoagulants.

4. Discussion

The main finding in the present study was that no significant association could be demonstrated between estimated infarct

TABLE 4: Comparisons of peak Troponin T (ng/L) levels and the presence of cardiovascular risk factors and relevant disease entities.

	−	+	p
Previous diabetes ($n = 69$)	732.5	664.0	0.892
History of hyperlipidemia ($n = 143$)	916.5	496.0	**0.018**
History of hypertension ($n = 182$)	1009.0	584.0	0.057
Previous angina ($n = 88$)	923.0	251.0	**0.001**
Previous claudication ($n = 27$)	795.0	237.0	**0.015**
Previous myocardial infarction ($n = 90$)	910.0	344.0	**0.003**
Previous revascularization ($n = 83$)	874.5	297.0	**0.003**
History of atrial fibrillation ($n = 38$)	778.0	153.0	**0.007**
Current smoker ($n = 41$)	767.5	519.0	0.292
Previous n-3 PUFA supplementation ($n = 135$)	711.0	644.0	0.908

+ denotes presence of risk factors or disease entities.
− denotes absence of risk factors or disease entities.

size by peak TnT and the relative content of long chained marine n-3 PUFAs in serum phospholipids in elderly subjects with an acute MI. There was, however, a significant inverse correlation between serum levels of DHA and the history of atrial fibrillation, and patients with a history of hyperlipidemia and relevant CVD entities presented with significantly lower peak TnT levels. Finally, a strong correlation was observed between peak TnT during the index infarction and NT-proBNP at inclusion.

Long chain marine n-3 PUFAs have several different effects that together are considered cardioprotective [18, 24]. There is limited data on their ability to modulate infarct size and protect ischemic myocardium during an acute MI. Even though previous animal studies may suggest that n-3 PUFAs have the ability to reduce infarct size, it could not be demonstrated in our study. It has been shown that cardiomyocytes generate significant amounts of reactive oxygen species (ROS) during ischemia, which may contribute significantly to cellular injury [25]. Therefore, the possibility for n3-PUFAs to provide acute cardiovascular protection and reduce the extent of myocardial injury is suggested to be through their anti-inflammatory and antioxidant functions [26–28].

An earlier study revealed that EPA significantly reduced infarct size in rabbit hearts, mainly dependent on its ability to increase nitric oxide (NO) production and opening of calcium-activated potassium channels [29]. It has also been shown that ischemia-reperfusion injury was reduced in rats after 8 weeks of n-3 supplementation [30]. A study administering DHA acutely before induction of coronary ischemia in Sprague-Dawley rats has revealed similar results, with lowering of lipid peroxides and a reduction in myocardial infarct size and creatine kinase release [31]. In a recent study it has also been demonstrated that Resolvin D1, a DHA metabolite, confers myocardial protection and reduces infarct size via mechanisms involving the PI3K/Akt signaling pathway and attenuation of caspase-3 and caspase-8 activation [32]. In two human trials, cardiac patients pretreated with a high-dose of n-3 PUFAs before planned invasive treatment did display lower serum levels of cardiac biomarkers following percutaneous coronary intervention (PCI) [33] and coronary artery bypass surgery [34] compared to controls. Our population had an intake of dietary marine n-3 PUFAs prior to enrolment of a median n-6/n-3 ratio of 3.6, which is considered to be very low. It has been noted that background n-3 fatty acid intake could change the cardiovascular outcomes in trials with different populations receiving the same treatment doses [35] and that a different n-6/n-3 ratio would influence infarct size [36]. Although we found no significant difference in peak TnT levels between those with lowest (10th) versus highest (90th) percentiles of n-6 and/or n-3 fatty acids, it might be suggested that the results would be different in a population with a higher n-6/n-3 ratio. Almost half of our study population was on aspirin and statin treatment; thus any effect of n3-PUFAs on infarct size may have been masked.

Interestingly, significantly lower levels of DHA in patients with a history of atrial fibrillation were found. In the landmark GISSI-Prevenzione-trial, it was shown that n-3 fatty acid supplementation led to a reduction in sudden cardiac death, pointing to n-3 PUFAs antiarrhythmic potential [3, 37]. Our findings are in accordance with Wu et al. who found that higher circulating levels of DHA in older adults were associated with lower risk of incident AF [38]. Furthermore, in a large, albeit younger, Danish cohort, data showed an inverse, however, not statistically significant, association between the development of AF and n-3 PUFAs in adipose tissue [39]. In this cohort a U-shaped association was established between incident AF and marine n-3 fatty acid consumption, especially for EPA and DHA, with lowest risk close to the median intake [40]. McLennan did reveal that DHA is increasingly accumulated in the myocardial cell membrane over EPA, which might point to DHA as the most potent fatty acid regarding the antiarrhythmic effects of n-3 PUFAs [41]. Still, its role in AF seems unclear. A meta-analysis evaluating 16 studies did not find any benefit of n-3 PUFA supplementation on secondary prevention or incidence of new AF after cardiovascular surgery [42], and a recent trial randomizing patients with previous known atrial fibrillation to 4 g/day n-3 PUFA or control did not show any reduction in AF recurrence [43].

We also found significantly lower TnT levels in those with a history of hyperlipidemia, especially for angina and other relevant CVD states. Kloner et al. did show that patients experiencing an AMI with a history of angina were more likely to have smaller creatine kinase-determined infarcts compared to those without [44]. The authors consider this to be related to the effect of ischemic preconditioning, which could explain our findings. It should also be noted that patients with previous diseases like hyperlipidemia, MI, and intermittent claudication in our study are likely to use both platelet inhibitors and statins, which could affect the size and extent of myocardial infarction. A retrospective study has shown that recent aspirin use was associated with smaller myocardial infarct size [45] and similar effects have been

reported after clinical trials with beta blockers [46]. Furthermore, pretreatment with statins (rosuvastatin) prior to PCI is shown to reduce periprocedural myocardial injury in patients with ACS and ST-segment elevation myocardial infarction [47, 48]. Data on medication use prior to index infarction was available in almost half of our population, and we found that patients with previous CVD were well treated with the majority on aspirin, statin, and beta blocker. We could therefore speculate that previous drug treatment may have limited the infarct size in our participants with risk factors for CVD and relevant disease states.

Finally, the rather strong correlation between peak TnT during the index infarction and NT-proBNP at inclusion 2–8 weeks later is to be expected as TnT as a marker of infarct size should be expected to associate with NT-proBNP as a predictor of myocardial injury and long-term prognosis [49].

5. Strengths and Limitations

The strengths of the study are attributed to the rather large population and the serum phospholipids analysis of n-3 fatty acids which is considered an important addition to dietary questionnaires. It could be argued that our measure of serum phospholipids, 2–8 weeks after index hospitalization, would not provide a correct reflection of the situation during the AMI. Although we did find significantly higher levels in patients previously reporting intake of n-3 PUFA supplements versus those who did not, confirming our belief that elderly patients would not largely change their dietary habits in a relatively short time, therefore our results should be considered representative. Our population also has a previously high intake of marine n-3 fatty acids. Also, we determined the relative fatty acid composition of phospholipids in serum and not absolute concentrations of individual fatty acids. Lastly, although peak TnT has shown strong correlation with infarct size assessed by radionuclide imaging, it is not considered the gold standard for this estimation. Also, peak TnT levels may not be the absolute peak as the blood samples have been taken according to clinical guidelines during the MI, and not through a standardized method in the protocol.

6. Conclusions

In an elderly population with AMI, no significant association between serum fatty acids and estimated infarct size could be demonstrated. However, an inverse correlation was noted between serum levels of DHA and a history of atrial fibrillation. Hyperlipidemia and the presence of CVD were associated with lower peak TnT levels, probably because of treatment with cardioprotective medication at the time of the AMI.

Acknowledgments

The authors thank the staff at Center for Clinical Heart Research for excellent assistance. The work was supported by Stein Erik Hagen Foundation for Clinical Heart Research, Oslo, Norway, and Olav Thons Foundation, Oslo, Norway.

References

[1] C. E. D. Saunderson, R. A. Brogan, A. D. Simms, G. Sutton, P. D. Batin, and C. P. Gale, "Acute coronary syndrome management in older adults: guidelines, temporal changes and challenges," *Age and Ageing*, vol. 43, no. 4, Article ID afu034, pp. 450–455, 2014.

[2] D. Kromhout, E. B. Bosschieter, and C. D. L. Coulander, "The inverse relation between fish consumption and 20-year mortality from coronary heart disease," *The New England Journal of Medicine*, vol. 312, no. 19, pp. 1205–1209, 1985.

[3] Gruppo Italiano per lo Studio della Sopravvivenza nell'Infarto Miocardico, "Dietary supplementation with n-3 polyunsaturated fatty acids and vitamin E after myocardial infarction: results of the GISSI-Prevenzione trial," *The Lancet*, vol. 354, no. 9177, pp. 447–455, 1999.

[4] M. L. Burr, A. M. Fehily, J. F. Gilbert et al., "Effects of changes in fat, fish, and fibre intakes on death and myocardial reinfarction: diet and reinfarction," *The Lancet*, vol. 2, no. 8666, pp. 757–761, 1989.

[5] L. Hooper, R. L. Thompson, R. A. Harrison et al., "Risks and benefits of omega 3 fats for mortality, cardiovascular disease, and cancer: systematic review," *The British Medical Journal*, vol. 332, no. 7544, pp. 752–755, 2006.

[6] B. Rauch, R. Schiele, S. Schneider et al., "OMEGA, a randomized, placebo-controlled trial to test the effect of highly purified omega-3 fatty acids on top of modern guideline-adjusted therapy after myocardial infarction," *Circulation*, vol. 122, no. 21, pp. 2152–2159, 2010.

[7] D. Kromhout, E. J. Giltay, and J. M. Geleijnse, "n-3 fatty acids and cardiovascular events after myocardial infarction," *The New England Journal of Medicine*, vol. 363, no. 21, pp. 2015–2026, 2010.

[8] D. Kromhout, S. Yasuda, J. M. Geleijnse, and H. Shimokawa, "Fish oil and omega-3 fatty acids in cardiovascular disease: do they really work?" *European Heart Journal*, vol. 33, no. 4, pp. 436–443, 2012.

[9] J. J. DiNicolantonio, J. H. O'Keefe, and C. J. Lavie, "The big ones that got away: omega-3 meta-analysis flawed by excluding the biggest fish oil trials," *Archives of Internal Medicine*, vol. 172, no. 18, article 1427, 2012.

[10] A. Macchia, M. Romero, A. D'Ettorre, G. Tognoni, and J. Mariani, "Exploratory analysis on the use of statins with or without n-3 PUFA and major events in patients discharged for acute myocardial infarction: an observational retrospective study," *PLoS ONE*, vol. 8, no. 5, Article ID e62772, 2013.

[11] C. D. Poole, J. P. Halcox, S. Jenkins-Jones et al., "Omega-3 fatty acids and mortality outcome in patients with and without type 2 diabetes after myocardial infarction: a retrospective, matched-cohort study," *Clinical Therapeutics*, vol. 35, no. 1, pp. 40–51, 2013.

[12] L. Tavazzi, A. P. Maggioni, R. Marchioli et al., "Effect of n-3 polyunsaturated fatty acids in patients with chronic heart failure (the GISSI-HF trial): a randomised, double-blind, placebo-controlled trial," *The Lancet*, vol. 372, no. 9645, pp. 1223–1230, 2008.

[13] D. Mozaffarian, R. N. Lemaitre, I. B. King et al., "Circulating long-chain ω-3 fatty acids and incidence of congestive heart failure in older adults: the cardiovascular health study: a cohort study," *Annals of Internal Medicine*, vol. 155, no. 3, pp. 160–170, 2011.

[14] D. Mozaffarian, R. N. Lemaitre, I. B. King et al., "Plasma phospholipid long-chain ω-3 fatty acids and total and cause-specific mortality in older adults, a cohort study," *Annals of Internal Medicine*, vol. 158, no. 7, pp. 515–525, 2013.

[15] E. Strand, E. R. Pedersen, G. F. T. Svingen et al., "Dietary intake of n-3 long-chain polyunsaturated fatty acids and risk of myocardial infarction in coronary artery disease patients with or without diabetes mellitus: a prospective cohort study," *BMC Medicine*, vol. 11, no. 1, article 216, 2013.

[16] F. Marangoni, G. Novo, G. Perna et al., "Omega-6 and omega-3 polyunsaturated fatty acid levels are reduced in whole blood of Italian patients with a recent myocardial infarction: the AGE-IM study," *Atherosclerosis*, vol. 232, no. 2, pp. 334–338, 2014.

[17] B.-Q. Zhu, R. E. Sievers, Y.-P. Sun, N. Morse-Fisher, W. W. Parmley, and C. L. Wolfe, "Is the reduction of myocardial infarct size by dietary fish oil the result of altered platelet function?" *American Heart Journal*, vol. 127, no. 4, pp. 744–755, 1994.

[18] E. B. Schmidt, H. Arnesen, R. De Caterina, L. H. Rasmussen, and S. D. Kristensen, "Marine n-3 polyunsaturated fatty acids and coronary heart disease. Part I. Background, epidemiology, animal data, effects on risk factors and safety," *Thrombosis Research*, vol. 115, no. 3, pp. 163–170, 2005.

[19] M. Panteghini, C. Cuccia, G. Bonetti, R. Giubbini, F. Pagani, and E. Bonini, "Single-point cardiac troponin T at coronary care unit discharge after myocardial infarction correlates with infarct size and ejection fraction," *Clinical Chemistry*, vol. 48, no. 9, pp. 1432–1436, 2002.

[20] D. Tzivoni, D. Koukoui, V. Guetta, L. Novack, and G. Cowing, "Comparison of Troponin T to creatine kinase and to radionuclide cardiac imaging infarct size in patients with ST-elevation myocardial infarction undergoing primary angioplasty," *The American Journal of Cardiology*, vol. 101, no. 6, pp. 753–757, 2008.

[21] K. Laake, P. Myhre, L. M. Nordby et al., "Effects of omega 3 supplementation in elderly patients with acute myocardial infarction: design of a prospective randomized placebo controlled study," *BMC Geriatrics*, vol. 14, no. 1, article 74, 2014.

[22] J. Folch, M. Lees, and G. H. Sloane Stanley, "A simple method for the isolation and purification of total lipides from animal tissues," *The Journal of Biological Chemistry*, vol. 226, no. 1, pp. 497–509, 1957.

[23] G. C. Burdge, P. Wright, A. E. Jones, and S. A. Wootton, "A method for separation of phosphatidylcholine, triacylglycerol, non-esterified fatty acids and cholesterol esters from plasma by solid-phase extraction," *British Journal of Nutrition*, vol. 84, no. 5, pp. 781–787, 2000.

[24] Y. Adkins and D. S. Kelley, "Mechanisms underlying the cardioprotective effects of omega-3 polyunsaturated fatty acids," *Journal of Nutritional Biochemistry*, vol. 21, no. 9, pp. 781–792, 2010.

[25] T. L. Vanden Hoek, C. Li, Z. Shao, P. T. Schumacker, and L. B. Becker, "Significant levels of oxidants are generated by isolated cardiomyocytes during ischemia prior to reperfusion," *Journal of Molecular and Cellular Cardiology*, vol. 29, no. 9, pp. 2571–2583, 1997.

[26] E. Giordano and F. Visioli, "Long-chain omega 3 fatty acids: molecular bases of potential antioxidant actions," *Prostaglandins Leukotrienes and Essential Fatty Acids*, vol. 90, no. 1, pp. 1–4, 2014.

[27] D. Richard, K. Kefi, U. Barbe, P. Bausero, and F. Visioli, "Polyunsaturated fatty acids as antioxidants," *Pharmacological Research*, vol. 57, no. 6, pp. 451–455, 2008.

[28] A. Leaf and P. C. Weber, "Cardiovascular effects of n-3 fatty acids," *The New England Journal of Medicine*, vol. 318, no. 9, pp. 549–557, 1988.

[29] H. Ogita, K. Node, H. Asanuma et al., "Eicosapentaenoic acid reduces myocardial injury induced by ischemia and reperfusion in rabbit hearts," *Journal of Cardiovascular Pharmacology*, vol. 41, no. 6, pp. 964–969, 2003.

[30] R. L. Castillo, C. Arias, and J. G. Farías, "Omega 3 chronic supplementation attenuates myocardial ischaemia-reperfusion injury through reinforcement of antioxidant defense system in rats," *Cell Biochemistry and Function*, vol. 32, no. 3, pp. 274–281, 2014.

[31] D. Richard, F. Oszust, C. Guillaume et al., "Infusion of docosa-hexaenoic acid protects against myocardial infarction," *Prostaglandins Leukotrienes and Essential Fatty Acids*, vol. 90, no. 4, pp. 139–143, 2014.

[32] K. Gilbert, J. Bernier, V. Bourque-Riel, M. Malick, and G. Rousseau, "Resolvin D1 reduces infarct size through a phosphoinositide 3-kinase/protein kinase B mechanism," *Journal of Cardiovascular Pharmacology*, vol. 66, no. 1, pp. 72–79, 2015.

[33] F. Foroughinia, J. Salamzadeh, and M. H. Namazi, "Protection from procedural myocardial injury by omega-3 polyunsaturated fatty acids (PUFAs): is related with lower levels of creatine kinase-MB (CK-MB) and troponin I?" *Cardiovascular Therapeutics*, vol. 31, no. 5, pp. 268–273, 2013.

[34] M. Veljovic, A. Popadic, Z. Vukic et al., "Myocardial protection during elective coronary artery bypasses grafting by pretreatment with omega-3 polyunsaturated fatty acids," *Vojnosanitetski Pregled*, vol. 70, no. 5, pp. 484–492, 2013.

[35] R. De Caterina, "n-3 fatty acids in cardiovascular disease," *The New England Journal of Medicine*, vol. 364, no. 25, pp. 2439–2450, 2011.

[36] I. Rondeau, S. Picard, T. M. Bah, L. Roy, R. Godbout, and G. Rousseau, "Effects of different dietary omega-6/3 polyunsaturated fatty acids ratios on infarct size and the limbic system after myocardial infarction," *Canadian Journal of Physiology and Pharmacology*, vol. 89, no. 3, pp. 169–176, 2011.

[37] T. A. Rix, J. H. Christensen, and E. B. Schmidt, "Omega-3 fatty acids and cardiac arrhythmias," *Current Opinion in Clinical Nutrition and Metabolic Care*, vol. 16, no. 2, pp. 168–173, 2013.

[38] J. H. Y. Wu, R. N. Lemaitre, I. B. King et al., "Association of plasma phospholipid long-chain omega-3 fatty acids with incident atrial fibrillation in older adults: the cardiovascular health study," *Circulation*, vol. 125, no. 9, pp. 1084–1093, 2012.

[39] T. A. Rix, A. M. Joensen, S. Riahi, S. Lundbye-Christensen, K. Overvad, and E. B. Schmidt, "Marine n-3 fatty acids in adipose tissue and development of atrial fibrillation: a Danish cohort study," *Heart*, vol. 99, no. 20, pp. 1519–1524, 2013.

[40] T. A. Rix, A. M. Joensen, S. Riahi et al., "A U-shaped association between consumption of marine n-3 fatty acids and development of atrial fibrillation/atrial flutter—a Danish cohort study," *Europace*, vol. 16, no. 11, pp. 1554–1561, 2014.

[41] P. L. McLennan, "Myocardial membrane fatty acids and the antiarrhythmic actions of dietary fish oil in animal models," *Lipids*, vol. 36, supplement 1, pp. S111–S114, 2001.

[42] J. Mariani, H. C. Doval, D. Nul et al., "N-3 polyunsaturated fatty acids to prevent atrial fibrillation: updated systematic review and meta-analysis of randomized controlled trials," *Journal of the American Heart Association*, vol. 2, no. 1, Article ID e005033, 2013.

[43] A. Nigam, M. Talajic, D. Roy et al., "Fish oil for the reduction of atrial fibrillation recurrence, inflammation, and oxidative stress," *Journal of the American College of Cardiology*, vol. 64, no. 14, pp. 1441–1448, 2014.

[44] R. A. Kloner, T. Shook, K. Przyklenk et al., "Previous angina alters in-hospital outcome in TIMI 4. A clinical correlate to preconditioning?" *Circulation*, vol. 91, no. 1, pp. 37–45, 1995.

[45] K. J. Mukamal, M. A. Mittleman, M. Maclure, J. B. Sherwood, R. J. Goldberg, and J. E. Muller, "Recent aspirin use is associated with smaller myocardial infarct size and lower likelihood of Q-wave infarction," *American Heart Journal*, vol. 137, no. 6, pp. 1120–1128, 1999.

[46] C. A. Campbell, K. Przyklenk, and R. A. Kloner, "Infarct size reduction: a review of the clinical trials," *Journal of Clinical Pharmacology*, vol. 26, no. 5, pp. 317–329, 1986.

[47] K. H. Yun, M. H. Jeong, S. K. Oh et al., "The beneficial effect of high loading dose of rosuvastatin before percutaneous coronary intervention in patients with acute coronary syndrome," *International Journal of Cardiology*, vol. 137, no. 3, pp. 246–251, 2009.

[48] J. W. Kim, K. H. Yun, E. K. Kim et al., "Effect of high dose rosuvastatin loading before primary percutaneous coronary intervention on infarct size in patients with ST-segment elevation myocardial infarction," *Korean Circulation Journal*, vol. 44, no. 2, pp. 76–81, 2014.

[49] S. K. James, B. Lindahl, A. Siegbahn et al., "N-terminal pro-brain natriuretic peptide and other risk markers for the separate prediction of mortality and subsequent myocardial infarction in patients with unstable coronary artery disease: a global utilization of strategies to open occluded arteries (GUSTO)-IV substudy," *Circulation*, vol. 108, no. 3, pp. 275–281, 2003.

5

Mitochondrial-Targeted Antioxidant MitoQ Prevents *E. coli* Lipopolysaccharide-Induced Accumulation of Triacylglycerol and Lipid Droplets Biogenesis in Epithelial Cells

Ekaterina Fock, Vera Bachteeva, Elena Lavrova, and Rimma Parnova ⓘ

I. M. Sechenov Institute of Evolutionary Physiology and Biochemistry of the Russian Academy of Sciences, Saint-Petersburg, Russia

Correspondence should be addressed to Rimma Parnova; rimma_parnova@mail.ru

Academic Editor: Clifford A. Lingwood

The effect of bacterial lipopolysaccharide (LPS) on eukaryotic cell could be accompanied by a significant metabolic shift that includes accumulation of triacylglycerol (TAG) in lipid droplets (LD), ubiquitous organelles associated with fatty acid storage, energy regulation and demonstrated tight spatial and functional connections with mitochondria. The impairment of mitochondrial activity under pathological stimuli has been shown to provoke TAG storage and LD biogenesis. However the potential mechanisms that link mitochondrial disturbances and TAG accumulation are not completely understood. We hypothesize that mitochondrial ROS (mROS) may play a role of a trigger leading to subsequent accumulation of intracellular TAG and LD in response to a bacterial stimulus. Using isolated epithelial cells from the frog urinary bladder, we showed that LPS decreased fatty acids oxidation, enhanced TAG deposition, and promoted LD formation. LPS treatment did not affect the mitochondrial membrane potential but increased cellular ROS production and led to impairment of mitochondrial function as revealed by decreased ATP production and a reduced maximal oxygen consumption rate (OCR) and OCR directed at ATP turnover. The mitochondrial-targeted antioxidant MitoQ at a dose of 25 nM did not prevent LPS-induced alterations in cellular respiration, but, in contrast to nonmitochondrial antioxidant α-tocopherol, reduced the effect of LPS on the generation of ROS, restored the LPS-induced decline of fatty acids oxidation, and prevented accumulation of TAG and LD biogenesis. The data obtained indicate the key signaling role of mROS in the lipid metabolic shift that occurs under the impact of a bacterial pathogen in epithelial cells.

1. Introduction

Bacterial lipopolysaccharide (LPS), the main membrane component of Gram-negative bacteria, is one of the most important pathogen-associated molecular patterns, which elicits the host innate immune response as well as inflammation. The effect of LPS on eukaryotic cells could be accompanied by a significant metabolic shift including accumulation of triacylglycerol (TAG) deposited in lipid droplets (LD), ubiquitous organelles that are associated with fatty acid storage, energy regulation, and control of bioactive lipid mediator production [1, 2]). Both *in vitro* and *in vivo*, LD-associated accumulation of TAG in response to LPS has been shown mainly in immune cells such as macrophages [3, 4], leukocytes [5], and microglia [6]. Systemic administration of LPS has been shown to cause an increase in the TAG content in the kidney, liver, and heart [7–10].

The effect of LPS on intracellular TAG accumulation has been evidenced to be based on multifaceted and highly cell-type specific pathways. Among them are the increase of CD36-mediated uptake of fatty acids and their incorporation into TAG [3, 4], the decrease of adipose triglyceride lipase- (ATGL-) mediated TAG lipolysis [4], the impairment of fatty acids oxidation (FAO), and downregulation of expression of the transcriptional factor PPARα and its downstream genes that are required for FAO [7, 8, 11, 12]. However, these effects can be triggered by earlier step(s) in LPS signaling, initiating alterations in the expression and activity of proteins involved in cellular lipid metabolism. These steps are still poorly understood.

LD has tight spatial and functional connections with mitochondria, and impairment of mitochondrial activity provokes TAG storage and LD biogenesis [13–16]. In different cell types, challenge with LPS causes an increase in reactive oxygen species (ROS) generation, a decline in mitochondrial membrane potential (MPP) and respiratory complexes activity, and a decrease in the oxygen consumption rate (OCR) and ATP production [17–21]. Mitochondria, especially complexes I-III of the electron transport chain (ETC), are the predominant cellular source of ROS which are important for cellular signaling and are tightly regulated by the endogenous antioxidant scavenging system [22, 23]. The link between ETC disturbances, mitochondrial ROS (mROS) generation, and TAG accumulation was revealed from data that showed that antimycin, an inhibitor of respiratory complex III, whose effect may be coupled to mROS generation, causes a decrease of FAO and stimulates TAG accumulation [10, 13, 24]. The amount of LD has been shown to be increased in the glia of mitochondrial mutants with elevated level of ROS, and reduction of ROS prevents LD accumulation [16]. These data prompted us to suggest that mROS trigger subsequent accumulation of intracellular TAG and LD in response to a bacterial stimulus.

To clarify the involvement of mROS in the LPS-induced shift of lipid metabolism, in this study we used MitoQ, a ubiquinone derivative that is covalently attached to a lipophilic triphenylphosphonium cation. Such a structure and a high potential across the inner membrane of mitochondria allow MitoQ and other structurally similar compounds to be extremely highly concentrated in mitochondria matrix scavenging active radicals [25]. The protective effect of mitochondrial-targeted antioxidants against LPS-induced inflammation or even acute sepsis has been shown in different in vivo models [26, 27]. In in vitro experiments with LPS, mitochondrial-targeted antioxidants have been shown to prevent the increase of proinflammatory cytokine production in macrophages [28] and suppressed NF-kB and MAPKs activation in microglial cells [29]. However, the potential link between mROS generation and LD biogenesis in response to LPS has not been studied yet.

As a cellular model, we used epithelial cells from the frog urinary bladder mucosa (FUBEC, *rog rinary ladder pithelial ells*). Epithelium of the urinary bladder forms a barrier to pathogen entry and is the first line of defense against penetrating microorganisms. For this reason, uroepithelia of different animal species possess an arsenal of tools for the innate immune defense, including the recognition of pathogen factors and TLR-triggered generation of a variety of inflammatory mediators [30–32]. As we reported previously, FUBEC express TLR4 and respond to LPS via a cascade of inflammatory signaling events leading to the increase of iNOS expression and PGE_2 synthesis [30]. Importantly, as in other cells, FUBEC accumulate LD in response to exogenously added fatty acids demonstrating the existence of LD-biogenesis machinery [33].

In this study, we characterized the effects of LPS on the rate of fatty acid oxidation (FAO), TAG storage, and LD accumulation and analyzed LPS-induced alterations in mitochondrial function. With the use of the mitochondrial-targeted antioxidant MitoQ, we tried to demonstrate whether mROS play the trigger role in LPS-induced lipid metabolic shift.

2. Materials and Methods

2.1. Reagents. E. coli LPS (serotype 0127:B8), Leibovitz L15 medium, oligomycin, FCCP (carbonyl cyanide-p- trifluoromethoxyphenylhydrazone), rotenone, antimycin, myxothiazol, α-tocopherol, 3-(4,5-dimethylthiazol-2-yl)-2,5-diphenyltetrazolium bromide (MTT), 2,7-dichlorofluorescein-diacetate (DCF-DA), and lipid standards were from Sigma-Aldrich (St. Louis, MO, USA). Nile Red was from Invitrogen (Carlsbad, CA, USA). [9, 10- ^{3}H(N)]-oleic acid was from Perkin Elmer (Boston, MA, USA). 5,5',6,6'-Tetrachloro-1,1',3,3'-tetraethylbenzimi-dazolylcarbocyanine iodide (JC-1) was purchased from Molecular Probes (Eugene, OR, USA). MitoQ was a kind gift from Dr. Michael Murphy (Cambridge, UK).

2.2. Animals. Male frogs *Rana temporaria L.*, which originated from the wild population in the Northern European region of Russia, were kept for 2-4 weeks in a hemiaquatic bath at $+5°C$. All procedures using animals were performed in accordance with the European Communities Council Directive (24th November 1986; 86/609/EEC) and were approved by the local Institutional Animal Care and Use Committee.

2.3. Culturing of FUBEC. The experiments were carried out on frog urinary bladder epithelial cells isolated as described previously [34]. Cells were washed with sterile amphibian Ringer solution (ARS), containing 85 mM NaCl, 4 mM KCl, 17.5 mM $NaHCO_3$, 0.8 mM KH_2PO4, 2 mM glucose, 1.5 mM $CaCl_2$, 0.8 mM $MgCl_2$, and 40 μg/ml gentamycin at pH 7.6, and they were then resuspended in Leibovitz L-15 medium diluted with ultrapure water for adaptation to frog osmolality (230 mOsmol/kg H_2O), and finally they were supplemented with 40 μg/ml gentamycin. Cells were incubated in 24-well plates at 23°C in a humid chamber (1.5–2 × 10^6 cells/250 μl per well).

2.4. Cell Viability Assay. Cells were seeded onto 96-well plates at a density of 1.5 x 10^5 cells/well and were incubated for 21 h with different concentration of MitoQ. MTT-reagent (at a final concentration of 0.5 mg/ml) was added 3 h before the end of the incubation, and then cells were lysed by a mixture of isopropanol: HCl (100: 1). The colored product of the MTT reduction was scanned with a microplate reader at 570 nm using 620 nm as the reference wavelength. The results were expressed as a percentage of the optical density, with the control taken as 100%. Since MitoQ was dissolved in 96% ethanol, corresponding volumes of the solvent were added. The ethanol concentration was less than 0.5%.

2.5. Lipid Droplets Staining. The stock Nile Red solution in DMSO (1 mg/ml) was diluted *ex tempore* 1000-fold with L-15 medium. At the end of incubation with or without LPS, 100 μl of the diluted Nile Red solution was added to 250 μl of the cell suspension for 10 min. Then, the suspension was transported to a confocal camera and microphotographs were obtained with a Leica TCS SP5 MP microscope (λ_{ex} = 488 nm, λ_{ex} = 510 – 560 nm, dry objective x 40).

2.6. Oxygen Consumption Rate (OCR) Analysis. OCR (nmol per minute per 10^6 cells) was measured using a polarographic oxygen Clark-type electrode (Econix-Expert Ltd, Russia) at $23°C$ under constant stirring. At the end of the appropriate incubation, 1.3 ml of the cell suspension containing 5-7 x 10^6 cells was placed in a polarographic chamber. Respiration was allowed to stabilize before any additions. Then, oligomycin (the ATP-synthase inhibitor, 1.5 μM) was added to estimate respiration independent of ATP synthesis. To evaluate the maximal capacity of ETC, the protonophore FCCP was titrated at different concentrations (0.75-1.25 μM) until a maximal respiration level was reached. The respiration was then inhibited by adding 2 μM rotenone, a complex I inhibitor, and 2 μM antimycin and 1 μM myxothiazol, complex III inhibitors. Finally, the addition of 2 mM KCN enabled the measurement of nonmitochondrial OCR. The basal OCR was calculated as the OCR before any additions minus the nonmitochondrial OCR. OCR_{ATP} was determined as the basal OCR minus OCR after oligomycin addition.

2.7. ROS Measurement. Cells were seeded onto 96-well plates at a density of $1.5-1.8 \times 10^5$ cells per well and incubated for 2 h with or without MitoQ followed by exposure to LPS for 1 h. The fluorescent dye DCF-DA was added to the incubation medium, at a final concentration of 10 μM, 20 min before the end of the incubation with LPS. The fluorescence of the reaction product of ROS with DCF-DA was estimated by a Fluoroscan FL (Thermo Fisher Scientific, Waltham, MA) at $\lambda_{em} = 538$ nm and $\lambda_{ex} = 485$ nm. The ROS content was expressed in arbitrary units.

2.8. Mitochondrial Membrane Potential Evaluation and ATP Assay. At the end of the appropriate incubation, cells were rinsed with ARS, resuspended in 50 μl of the same solution, incubated with JC-1 at a concentration of 2.5 μg/ml for 20 min in the dark, and analyzed on a Navios flow cytometer (Beckman Coulter) at FL1 (525 ± 40 nm) and FL2 (575 ± 30 nm), with $\lambda_{ex} = 488$ nm. The change in color from red to green was quantified and analyzed.

Cellular ATP production was evaluated by the luciferin-luciferase method using a commercial kit (Lumteck, Moscow) according to the manufacturer's protocol.

2.9. Lipid Extraction and Separation of Lipid Classes by TLC. FUBEC were incubated with or without LPS for 21 h in 24-well plates at a density of 1.5 x 10^6 cells/well. At the end of the incubation, cells were harvested, washed, and subjected to lipid extraction by a chloroform/methanol (2:1) mixture. Lipid extracts were washed with 0.2 vol of a 0.75% KCl solution and centrifuged for 5 min at 250 g. The lower phase was evaporated until dry, and the sediment dissolved in a chloroform-methanol mixture (2:1, v/v) was applied to a chromatographic plate DC-Alufolien (Merck, Germany). Lipid classes were separated in a solvent system containing hexane-diethyl ether-acetic acid (33:11.3:1, v/v). The plates were sprayed with 20% H_2SO_4 in methanol and heated at $150°C$. Lipid spots were identified with the use of corresponding standards. For quantitative analysis, the

plates were scanned and the densities of the lipid spots were measured by ImageJ and Microsoft Office Excel. The absolute values of TAG were calculated based on the optical density of a known concentration of triolein solutions applied on the same plate.

2.10. Incorporation of [^{3}H]-Oleic Acid into Lipids and Evaluation of FAO. Freshly isolated and washed cells (1.2 x 10^6 in each sample) were incubated with 26 pmol of [9,10-^{3}H(N)]-oleic acid (specific activity of 45.5 Ci/mmol) for 1 h at $23°C$. After that cells were washed with ARS, resuspended in culture medium, and incubated for 21 h with or without LPS. Where appropriate, before LPS, cells were incubated for 2 h with 25 nM MitoQ. At the end of the incubation, cells were centrifuged for 10 min at 100 g and the supernatant was gathered. The pellet was subjected to the lipid extraction procedure followed by separation of lipid classes by TLC as described above. The plate was developed in iodine vapor. After iodine evaporation, zones corresponding to lipid classes were cut out and their radioactivity was measured by an LKB 1209/1215 Rack-Beta counter.

For FAO evaluation, at the end of the incubation, aliquots of extracellular medium were mixed with 4 vol of chloroform-methanol (2:1). The oleic acid oxidation rate was evaluated by measurement of the radioactivity of the water phase. The scintillation count was normalized to 10^6 cells.

2.11. Statistics. The results are presented as the means ± SE. Statistical analysis was performed with the help of Microsoft Office Excel and the statistical software package AtteStat, version 13.1. The Shapiro-Wilk test was used to check samples for normality. The statistical significance of differences was determined by Student's t-test for paired samples or one-way ANOVA where appropriate. Differences between tests and controls were considered statistically significant at P value <0.05.

3. Results

3.1. LPS-Stimulated Lipid Droplet Biogenesis and Intracellular Accumulation of TAG. The neutral lipid fluorescent dye Nile Red revealed the presence of LD at different quantities in virtually all control FUBEC, nonuniformly distributed in the cytoplasm, sometimes forming aggregates (Figure 1(a)). Cells treated for 21 h with LPS displayed a significantly elevated number and size of LD (Figures 1(a) and 1(b)). The increase of Nile Red fluorescence in the selected range of wavelengths may have been caused by augmentation of both of TAG and the cholesterol ester (CE) content. However, densitometric analysis of the thin-layer chromatograms of the total lipid extracts displayed a dose-dependent increase of the TAG content in the presence of LPS (Figure 1(c)), whereas CE deposition was not promoted (data are not shown).

3.2. LPS Inhibited FAO and TAG Breakdown. To understand the metabolic origin of TAG accumulation in response to LPS, we preincubated FUBEC with [^{3}H]-oleic acid before

(a)

(b)

(c)

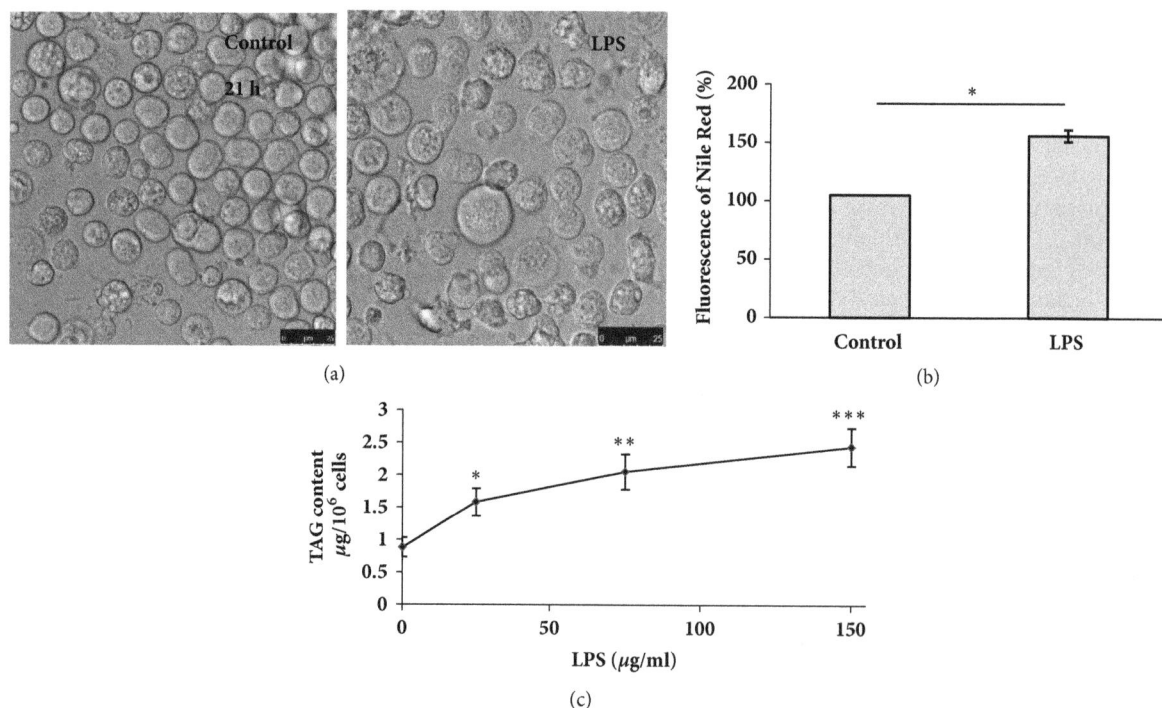

FIGURE 1: LPS stimulates lipid droplet biogenesis (a, b) and intracellular accumulation of TAG (c). (a, b) FUBEC incubated with LPS (25 μg/ml) or without it for 21 h were stained with Nile Red for 10 min. The suspension was then transferred to the confocal camera, and microphotographs were obtained with a Leica TCS SP5 MP microscope (λ_{ex} = 488 nm, λ_{ex} = 510 – 560 nm, dry objective x40). ∗ p<0.05 (n=4). (c) Dose dependency of LPS effect on the intracellular TAG content. Cells were incubated with or without LPS for 21 h, harvested, and subjected to lipid extraction and TLC. n=6 independent experiments. ∗ p<0.05; ∗∗ p< 0.01; ∗∗∗ p<0.001 versus control (the zero point on the x-axis).

challenge with LPS and then examined the incorporation of the radioactive label into the main lipid classes.

LPS caused an increase of radioactivity in TAG, diacylglycerol, and free oleic acid, as well as a reduction of radioactivity in CE (Figures 2(a) and 2(b)). Evaluation of FAO showed that LPS caused a decrease in [^{3}H]-oleic acid oxidation (Figure 2(a)). These results are in a good agreement with data obtained by other authors. Impaired FAO due to LPS was observed in macrophages [3, 4], dendritic cells [19], and AC16 cells [11], as well as in the liver, kidney, and heart following systemic administration of LPS [7–9].

Because LPS was added after the removal of nonincorporated [^{3}H]-oleate from the extracellular medium, the increase of radioactivity in TAG could be caused by the redistribution of the label between lipid classes. The labelling of PLs was unchanged whereas FAO was reduced and the content of [3H]-oleic acid was significantly elevated (Figures 2(a) and 2(b)). These data allow us to suggest that the reduction of FAO is the main reason for the increased amount of TAG in LPS-stimulated cells.

This suggestion was confirmed by a time-course analysis of the effect of LPS. During the first 5 hours of incubation, both control and stimulated cells consumed TAG and oxidized fatty acids at a similar rate. Then, control cells continued TAG consumption and oxidation of fatty acids at a nearly constant rate (Figures 2(c) and 2(d)), whereas in the presence of LPS these processes were significantly slowed. After 21 h of

incubation, the differences became twofold (Figures 2(c) and 2(d)).

3.3. Effect of LPS on OCR, Mitochondrial Membrane Potential, and Production of ATP. Based on the results that LPS inhibits FAO, next we tested whether LPS affected mitochondrial function in FUBEC. OCR measurements revealed that application of LPS at a dose that declined FAO (25 μg/ml) already after 2.5 h of incubation led to the reduction of OCR$_{max}$ in the presence of the uncoupler FCCP (Figures 3(a) and 3(b)) indicating decreased mitochondrial effectiveness in the presence of LPS. After 21 h of incubation with LPS, besides the decline of the uncoupler effect, a significant decrease in basal OCR (LPS versus control) was observed (Figures 3(c) and 3(d)). OCR in the presence of oligomycin (ATP-synthase inhibitor) did not differ in control and LPS-stimulated cells. Thus, the oxygen consumption required for ATP synthesis (OCR$_{ATP}$) in LPS-stimulated cells was lower than that in control cells whereas H$^+$-leaking remained unchanged. It should be noted that the addition of oligomycin in a FCCP background did not lower OCR (data are not shown) indicating the absence of other effects of oligomycin on respiration besides ATP-synthase inhibition. Nonmitochondrial respiration was the same for control and LPS-stimulated cells, independent of the duration of incubation (Figures 3(a)–3(d)).

As shown in Figure 3(e), a 21 h incubation with LPS resulted in a small in magnitude but statistically significant

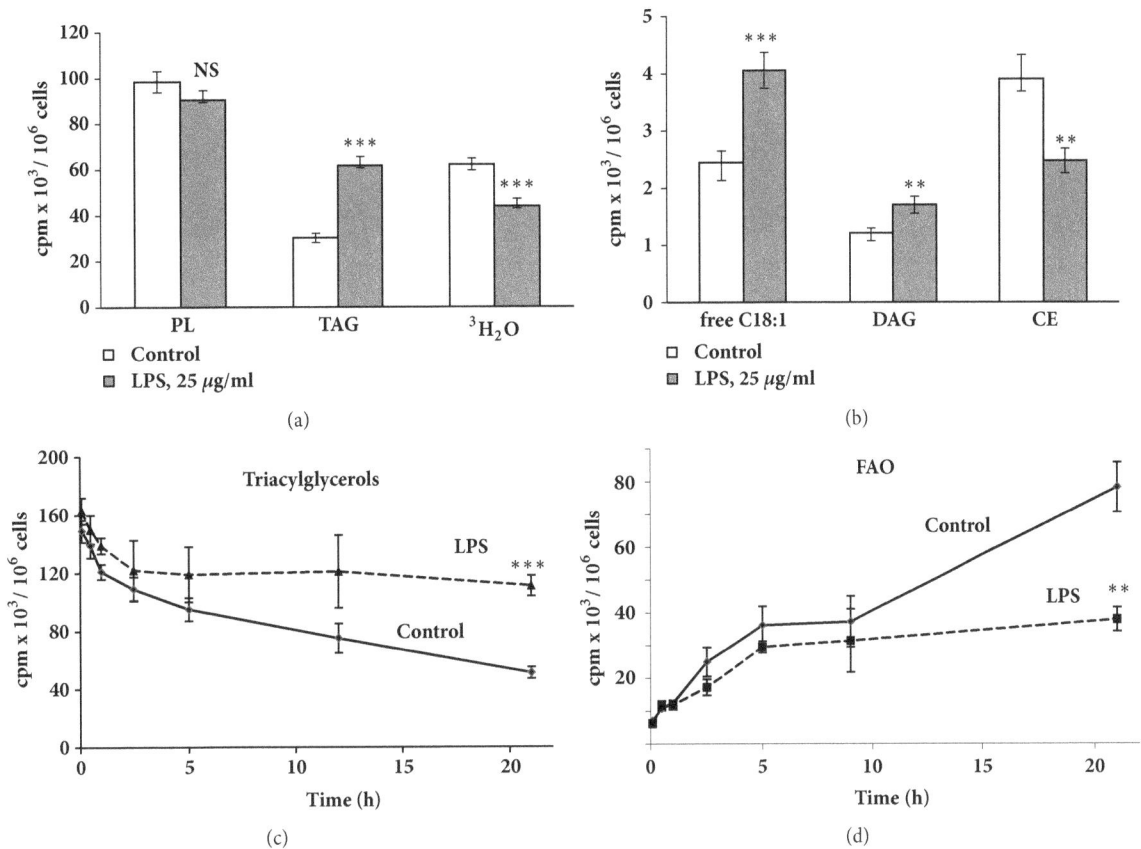

FIGURE 2: Incubation of FUBEC with LPS changes the distribution of [^{3}H]-C18:1 between lipid classes. (a, b) Effect of LPS on the metabolism of different lipid classes. PL: phospholipids, TAG: triacylglycerols, DAG: diacylglycerols, CE: cholesterol esters. ((c), (d)) Time-course of the effect of LPS on [^{3}H]-TAG level and on [^{3}H]-C18:1 oxidation. Cells were preloaded with [^{3}H]-C18:1 for 1 h and then rinsed and incubated with LPS (25 μg/ml, 21 h). At the end of the incubation, cells were harvested and total lipid extraction and TLC were performed. n=6 independent experiments. ∗∗ p< 0.01; ∗∗∗ p<0.001 versus control.

decrease of the ATP level, indicating an energy deficiency in FUBEC after prolonged LPS challenge.

MPP is another indicator of mitochondrial bioenergetics function. FUBEC staining with JC-1 revealed that LPS treatment did not induce any detectable changes of MPP (Figure 4) indicating preservation of the mitochondrial integrity in the presence of LPS.

3.4. Selection of the MitoQ Concentration with the Use of a Cytotoxicity Test and OCR Measurement.

Taking into account the fact that mitochondria are one of the main sources of ROS and that LPS in FUBEC targets mitochondrial function, the mitochondrial-targeted antioxidant MitoQ was chosen for use in the following experiments. Even if its protective effect against different pathological stimuli was demonstrated both in vivo [26, 27] and in vitro [24, 28], MitoQ can possess prooxidant properties stimulating superoxide and H_2O_2 production [35, 36] and decrease OCR_{ATP}, OCP_{max}, and MMP [35, 37]. Additionally, it should be mentioned that the concentration of MitoQ used in in vitro experiments varies greatly in the literature—from 1 nM [36] to 300-500 nM [35, 38] or even 1 μM [24, 38, 39]. In this context, it was necessary to choose

concentration of MitoQ that was suitable for our cellular model.

The toxicity of MitoQ in a concentration range from 25 nM to 1 μM was evaluated by the MTT-test. The results showed that MitoQ had a dose-dependent toxic effect in FUBEC—1 μM significantly decreased the MTT-test indexes whereas 25 nm had practically no effect (Figure 5(a)). Analysis of OCR by FUBEC that were incubated for 3 h with MitoQ revealed that 1 μM increased H^+ leak and decreased OCR_{max} in the presence of FCCP, indicating a detrimental effect of this dose on mitochondrial function (Figures 5(b) and 5(c)). MitoQ at a dose of 100 nM caused decrease of OCR_{max}, whereas 25 nM had only a tendency to decrease it, indicating that MitoQ at doses higher than 25 nM inhibited the ETC. Based on these data, the following experiments were performed with 25 nM MitoQ.

3.5. MitoQ Did Not Prevent the LPS-Induced Decline of OCR_{max} but Decreased Basal and LPS-Stimulated ROS Production.

To test whether MitoQ was able to influence the LPS-induced decline of OCR, we preincubated cells with 25 nM MitoQ for 2 h followed by a 21 h incubation with LPS. The measurement of OCR revealed that MitoQ did not prevent

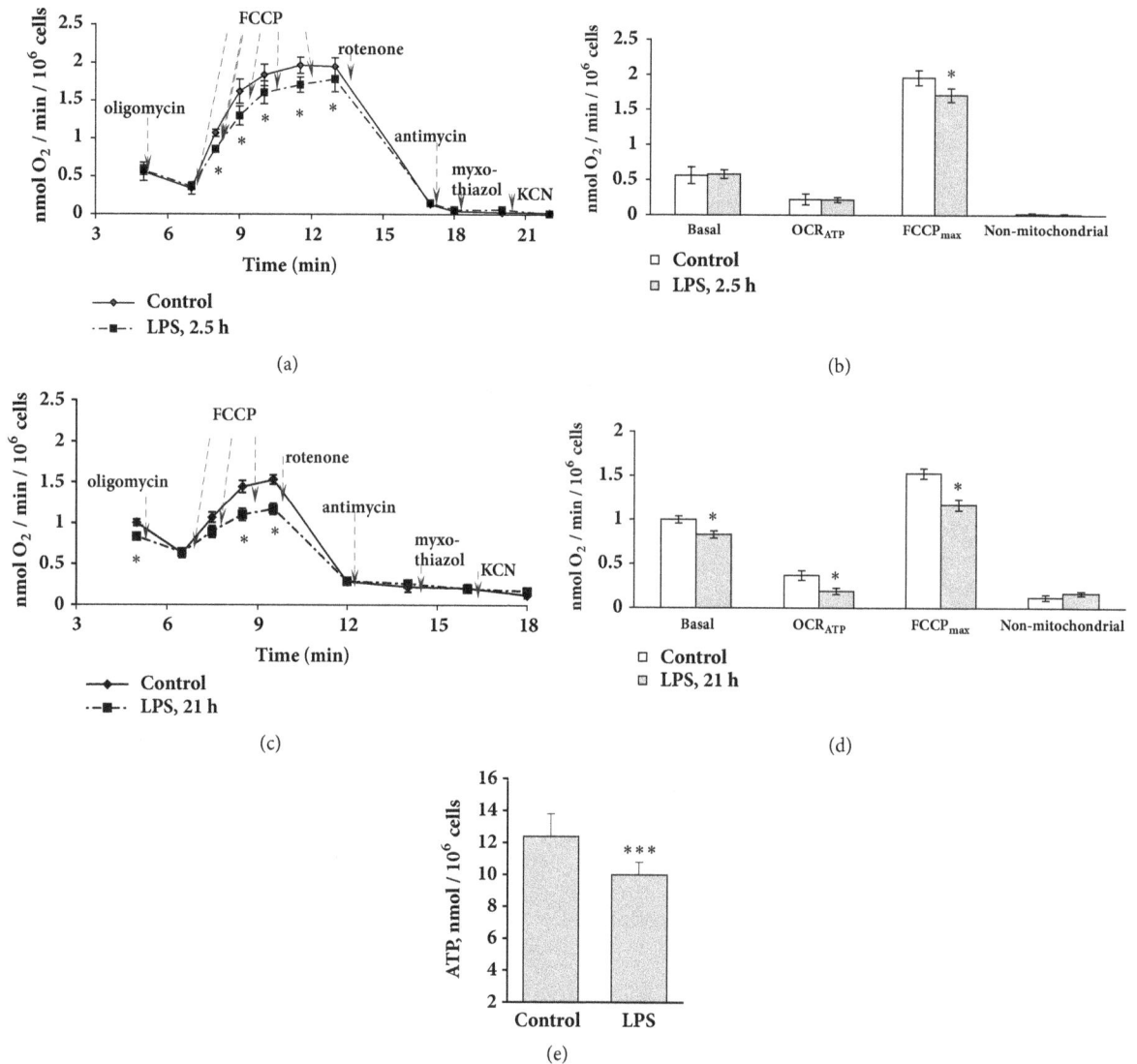

FIGURE 3: Effect of LPS on the oxygen consumption rate under different respiratory conditions and on ATP synthesis. OCR by FUBEC incubated with or without LPS (25 μg/ml) for 2.5 h (**a, b**) and 21 h (**c, d**) was analyzed. ((a), (c)) Respirometry experiments; ((b), (d)) quantification of data. Arrows indicate the addition of drugs. n=6 independent experiments. (**e**) ATP production by FUBEC, incubated with or without LPS (25 μg/ml, 21 h), n=22. *p <0.05; ***p <0.001 versus control.

the LPS-induced decrease of OCR_{max}, and even there was a trend toward an additive effect for the two drugs (data not shown).

To analyze the antioxidant capacity of MitoQ, FUBEC were preincubated for 2 h with 25 nM MitoQ prior to a 1 h incubation with LPS. Antioxidant treatment resulted in the suppression of both basal and LPS-stimulated ROS production (Figure 6(a)). α-Tocopherol, a nonmitochondrial antioxidant, at concentrations 10 and 50 μM, did not reduce LPS-stimulated ROS production, and, by itself, demonstrated rather weak prooxidant properties (data not shown).

3.6. MitoQ Prevented LPS-Induced FAO Decline, TAG Accumulation, and LD Formation. To examine the involvement of mROS in LPS-induced lipid metabolic shift, we analyzed the effect of MitoQ on the LPS-stimulated changes of lipid

metabolism in cells preincubated with [3H]-oleate. While LPS led to a significant increase in TAG radioactivity and a decrease in the level of FAO, MitoQ significantly suppressed both effects (Figures 6(b) and 6(c)). α-Tocopherol (10 and 50 μM), which did not possess antioxidant properties in FUBEC, also had no effect on LPS-induced lipid metabolism alterations (data not shown). Nile Red staining of FUBEC preincubated with 25 nM MitoQ for 2 h revealed that MitoQ *per se* did not change the number or size of LD but almost completely eliminated the effect of LPS on LD biogenesis (Figures 6(d)–6(h)).

4. Discussion

This study was designed to clarify whether mROS contribute to the LPS-induced shift of lipid metabolism in

(a)

(b)

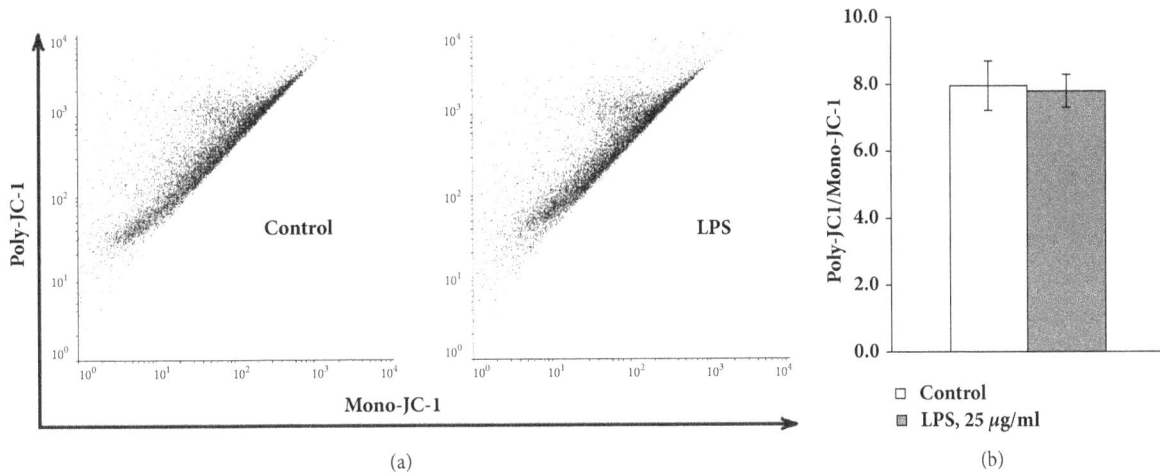

FIGURE 4: LPS does not change MPP in FUBEC. Cells were incubated with 25 μg/ml LPS for 21 h followed by staining with JC-1 and red/green image density measurements. The green fluorescence density indicated JC-1 monomers while the red fluorescence density JC-1-polymers. n=4 independent experiments.

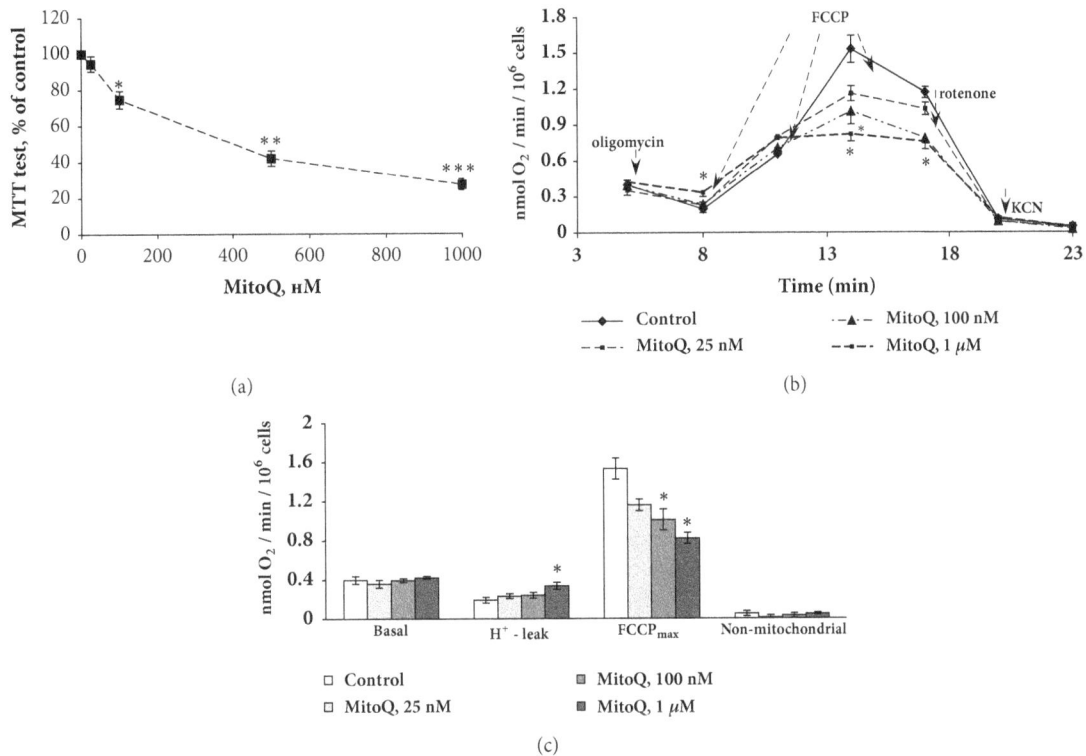

(a)

(b)

(c)

FIGURE 5: Evaluation of MitoQ toxicity on FUBEC. (a) MTT-test, 21 h incubation with MitoQ. (b, c) Oxygen consumption rate (3 h incubation with MitoQ). (b) Respirometry experiments; (c) quantification of data. n=4 independent experiments; $p < 0.05$; $**$ $p < 0.01$; $***$ $p < 0.001$ versus control.

epithelial cells. First, we obtained data that accumulation of TAG in FUBEC could be attributed to an LPS-induced decline of mitochondrial FAO leading to subsequent TAG retention. This suggestion is based on the LPS-induced intracellular accumulation of nonoxidized fatty acids and inhibition of FAO and on a time-course of the effect of LPS on the rate of FAO and TAG breakdown.

We further demonstrated that LPS treatment actually increased cellular ROS production and led to the impairment of mitochondrial function, which was revealed by reduced OCR_{max} and OCR_{ATP} as well as by the decrease of ATP production. The data suggest that LPS affects the respiratory chain, decreasing oxidative phosphorylation. However, LPS treatment did not appear to dramatically change the respiratory function of FUBEC mitochondria, since the effect of

(a)

(b)

(c)

(d)

(e)

(f)

(g)

(h)

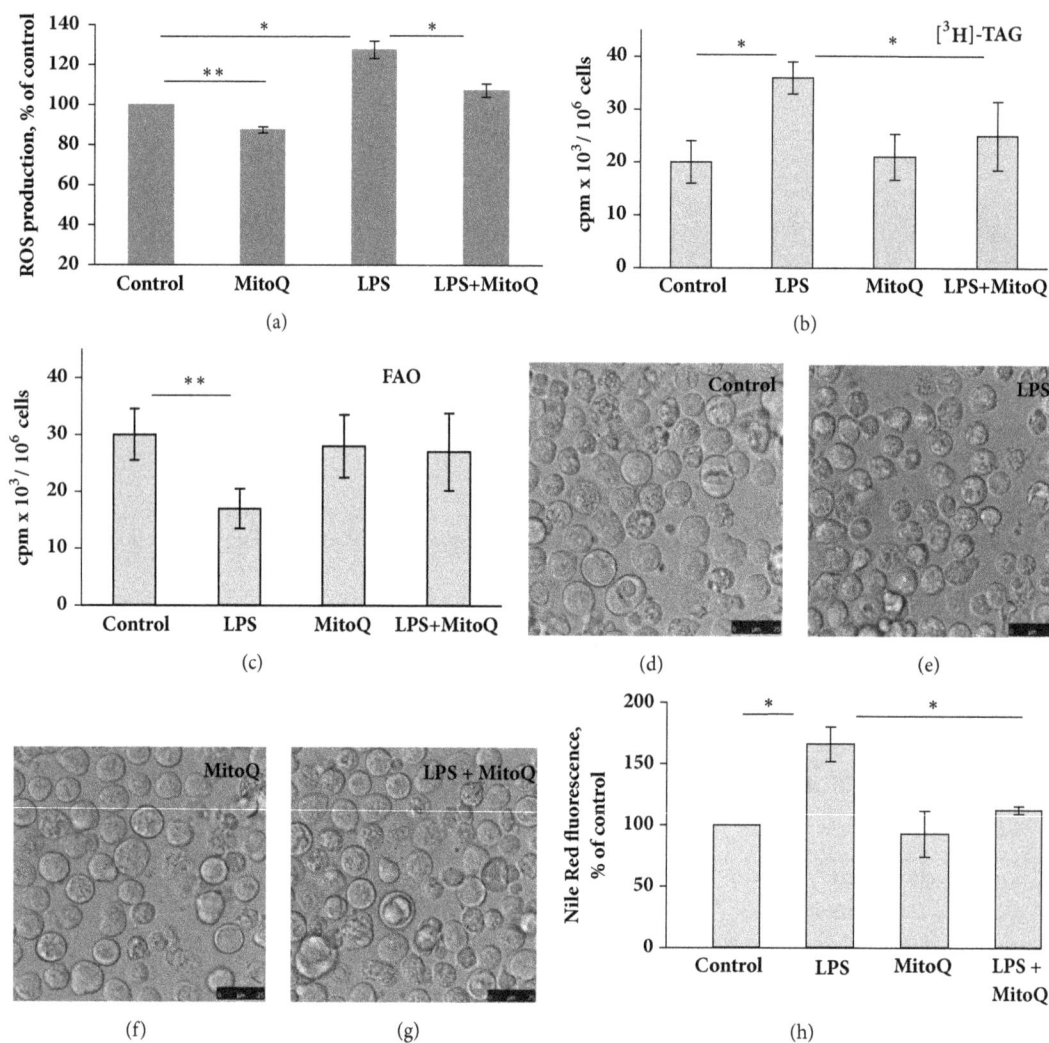

FIGURE 6: MitoQ decreases basal and LPS-stimulated ROS production (a) and prevented LPS-induced TAG accumulation (b), decline of FAO (c), and LD biogenesis (d-h). (a) Cells were pretreated with MitoQ for 2 h and then exposed to 25 μg/ml LPS for 1 h. The fluorescent dye, DCF-DA, was added to the incubation medium at a final concentration of 10 μM, 20 min before the end of incubation with LPS, n = 6. ((b)-(h)) Cells were pretreated for 2 h with MitoQ (25 nM) and then incubated for 21 h with 25 μg/ml LPS. ((b), (c)) Cells were preloaded with [³H]-C18:1 during 1 h, n=6 independent experiments; ((d)-(h)) Cells were stained with Nile Red, n=4 independent experiments. * p< 0.05; ** p< 0.01.

LPS on OCR was not accompanied by an increase of H⁺ leak or a decrease of MMP, indicating the preservation of mitochondrial integrity. Next, we determined the appropriate concentration of MitoQ by testing its toxicity with a MTT-test and OCR evaluation. The data obtained indicated that FUBEC were highly sensitive to the toxic effect of MitoQ and that the dose applied in our study (25 nM) was significantly less than was generally used in most in vitro works [28, 35, 38, 39]. MitoQ did not prevent the LPS-induced decrease of OCR$_{max}$ but reduced the effect of LPS on ROS, indicating the existence of different targets and mechanisms of its action within mitochondria. The study of LPS-induced lipid metabolic changes in the presence of MitoQ revealed that a mitochondrial-targeted antioxidant restored the LPS-induced decline of FAO and prevented accumulation of TAG and LD biogenesis.

Since the ROS measurement was performed with the use of DCF-DA, the data obtained do not allow for differentiation between ROS of mitochondrial and nonmitochondrial origin. However it seems to be more important that the LPS-stimulated production of ROS and lipid metabolic shift were prevented by a mitochondrial-targeted antioxidant, but not a nonmitochondrial one (α-tocopherol). Given the multiplicity of ROS-producing sources in mitochondria, various mechanisms of LPS-induced ROS generation inside mitochondria could be proposed. For example, LPS was shown to reduce complex I activity, leading to increased production of superoxide and H$_2$O$_2$ [40]. Complex I is one of the main sources of mitochondrial ROS [22, 41], and its dysfunction can be a trigger for inflammatory responses. The intrinsic mechanism of LPS-induced generation of mitochondrial ROS coupled with TRAF6-mediated mitochondrial complex

I impairment has been revealed in macrophages [42]. However, whether such a mechanism exists in other cell types is unclear.

The potential mechanisms of mROS-mediated inhibition of FAO and TAG retention in response to LPS could not be completely understood in the frame of the present study. The enzymatic activity of carnitine palmitoyltransferase I (CPT-1), the rate-limiting enzyme for the β-oxidation of fatty acids, can be significant and rapidly (within 30 min) downregulated by ROS [43]. CPT-1 may be subjected to posttranslational modifications by ROS-mediated lipid peroxidation products resulting in a sharp decrease of its activity [44]. However, our data on a time-course analysis of the LPS effect indicate that statistically significant inhibition of FAO was observed much later (at least, after 9 h of incubation) than the LPS effects on ROS production (1 h) or influences on OCR (2.5 h). These data suggest that the effect of LPS is rather related to mROS-triggered control of expression of proteins which contributed to FAO regulation or TAG breakdown. Unfortunately, we failed to demonstrate convincingly the influence of LPS and MitoQ on CPT-1 expression by immunoblotting due to insufficient specificity of commercial antibodies for FUBEC. However, both *in vivo* and *in vitro*, it was shown that LPS reduced the expression of CPT-1 as well as transcription factors PPARα and PGC-1α [3, 7, 8, 11, 12]. Reduced PPARγ expression was shown to mediate TAG accumulation caused by antimycin-induced mitochondrial dysfunction [13]. The high level of mROS has been reported to promote activation of the redox-sensitive transcriptional factor NF-kB, which plays a central role in LPS signaling, in cells challenged with LPS [45, 46]. Mitochondrial-targeted antioxidants were shown to reduce the LPS-triggered inflammatory response by regulating the NF-κB pathway in HUVEC [47], microglia cells [31], and C6 glioma cells [48]. Taking into account the extremely broad spectrum of ROS signaling, all of the above-mentioned mechanisms of mitochondrial FAO control could be targeted by mROS.

Regardless of the mechanism, our data demonstrate, for the first time, the key role of mROS in inhibition of mitochondrial FAO, TAG retention, and LD biogenesis under the influence of a bacterial pathogen. Our results are in a good agreement with the data of Boren and Brindle [24] who demonstrated a key role of mROS in the decline of FAO in etoposide-induced apoptosis of murine lymphoma cells. It is important to note that the mucosal surface of the amphibian urinary bladder is commonly exposed to Gram-negative bacteria and that isolated uroepithelial cells demonstrate a relatively high tolerance to LPS. In contrast to many other cell types, LPS, at least at the dose applied in this study, does not induce apoptosis in isolated FUBEC [34]. In this context, generation of mROS in response to the LPS action on mitochondria and activation of downstream signaling pathways that mediate FAO decline and TAG retention could be

rather considered as a mechanism of metabolic downregulation, representing an adaptive cellular response to bacterial pathogens.

Abbreviations

ARS: Amphibian Ringer solution
CE: Cholesterol esters
ETC: Electron transport chain
FAO: Fatty acid oxidation
FUBEC: Frog urinary bladder epithelial cells
LD: Lipid droplets
LPS: Lipopolysaccharide
MPP: Mitochondrial membrane potential
mROS: Mitochondrial reactive oxygen species
OCR: Oxygen consumption rate
ROS: Reactive oxygen species
TAG: Triacylglycerol
TLC: Thin-layer chromatography.

Acknowledgments

The authors thank Dr. M. Murphy (MRC Mitochondrial Biology Unit, Cambridge, United Kingdom) for MitoQ. This study was supported by the Russian Federal Agency of Scientific Organizations (Grant no. AAAA-A18-118012290371-3). The results were reported at the 42nd FEBS Congress (Jerusalem, 2017).

References

[1]   R. V. Farese Jr. and T. C. Walther, "Lipid droplets finally get a little R-E-S-P-E-C-T," *Cell*, vol. 139, no. 5, pp. 855–860, 2009.

[2]   M. A. Welte, "Expanding roles for lipid droplets," *Current Biology*, vol. 25, no. 11, pp. R470–R481, 2015.

[3]   K. R. Feingold, J. K. Shigenaga, M. R. Kazemi et al., "Mechanisms of triglyceride accumulation in activated macrophages," *Journal of Leukocyte Biology*, vol. 92, no. 4, pp. 829–839, 2012.

[4]   Y.-L. Huang, J. Morales-Rosado, J. Ray et al., "Toll-like receptor agonists promote prolonged triglyceride storage in macrophages," *The Journal of Biological Chemistry*, vol. 289, no. 5, pp. 3001–3012, 2014.

[5]   P. Pacheco, F. A. Bozza, R. N. Gomes et al., "Lipopolysaccharide-induced leukocyte lipid body formation in vivo: Innate immunity elicited intracellular loci involved in eicosanoid metabolism," *The Journal of Immunology*, vol. 169, no. 11, pp. 6498–6506, 2002.

[6]   A. Khatchadourian, S. D. Bourque, V. R. Richard, V. I. Titorenko, and D. Maysinger, "Dynamics and regulation of lipid droplet formation in lipopolysaccharide (LPS)-stimulated microglia," *Biochimica et Biophysica Acta (BBA) - Molecular and Cell Biology of Lipids*, vol. 1821, no. 4, pp. 607–617, 2012.

[7]   K. R. Feingold, Y. Wang, A. Moser, J. K. Shigenaga, and C. Grunfeld, "LPS decreases fatty acid oxidation and nuclear

hormone receptors in the kidney," *Journal of Lipid Research*, vol. 49, no. 10, pp. 2179–2187, 2008.

[8] U. Maitra, S. Chang, N. Singh, and L. Li, "Molecular mechanism underlying the suppression of lipid oxidation during endotoxemia," *Molecular Immunology*, vol. 47, no. 2-3, pp. 420–425, 2009.

[9] X. Wang and R. D. Evans, "Effect of endotoxin and platelet-activating factor on lipid oxidation in the rat heart," *Journal of Molecular and Cellular Cardiology*, vol. 29, no. 7, pp. 1915–1926, 1997.

[10] R. A. Zager, A. C. Johnson, and S. Y. Hanson, "Renal tubular triglyercide accumulation following endotoxic, toxic, and ischemic injury," *Kidney International*, vol. 67, no. 1, pp. 111–121, 2005.

[11] K. Drosatos, Z. Drosatos-Tampakaki, R. Khan et al., "Inhibition of c-Jun-N-terminal kinase increases cardiac peroxisome proliferator-activated receptor α expression and fatty acid oxidation and prevents lipopolysaccharide-induced heart dysfunction," *The Journal of Biological Chemistry*, vol. 286, no. 42, pp. 36331–36339, 2011.

[12] Y. Wang, A. H. Moser, J. K. Shigenaga, C. Grunfeld, and K. R. Feingold, "Downregulation of liver X receptor-α in mouse kidney and HK-2 proximal tubular cells by LPS and cytokines," *Journal of Lipid Research*, vol. 46, no. 11, pp. 2377–2387, 2005.

[13] S. Vankoningsloo, M. Piens, C. Lecocq et al., "Mitochondrial dysfunction induces triglyceride accumulation in 3T3-L1 cells: role of fatty acid β-oxidation and glucose," *Journal of Lipid Research*, vol. 46, no. 6, pp. 1133–1149, 2005.

[14] H. Zirath, A. Frenzel, G. Oliynyk et al., "MYC inhibition induces metabolic changes leading to accumulation of lipid droplets in tumor cells," *Proceedings of the National Academy of Sciences of the United States of America*, vol. 110, no. 25, pp. 10258–10263, 2013.

[15] S.-J. Lee, J. Zhang, A. M. K. Choi, and H. P. Kim, "Mitochondrial Dysfunction Induces Formation of Lipid Droplets as a Generalized Response to Stress," *Oxidative Medicine and Cellular Longevity*, vol. 2013, Article ID 327167, 10 pages, 2013.

[16] L. Liu, K. Zhang, H. Sandoval et al., "Glial Lipid Droplets and ROS Induced by Mitochondrial Defects Promote Neurodegeneration," *Cell*, vol. 160, no. 1-2, pp. 177–190, 2015.

[17] P. E. James, S. K. Jackson, O. Y. Grinberg, and H. M. Swartz, "The effects of endotoxin on oxygen consumption of various cell types in vitro: An EPR oximetry study," *Free Radical Biology & Medicine*, vol. 18, no. 4, pp. 641–647, 1995.

[18] Y.-C. Chuang, J.-L. Tsai, A. Y. W. Chang, J. Y. H. Chan, C.-W. Liou, and S. H. H. Chan, "Dysfunction of the mitochondrial respiratory chain in the rostral ventrolateral medulla during experimental endotoxemia in the rat," *Journal of Biomedical Science*, vol. 9, no. 5-6, pp. 542–548, 2002.

[19] C. M. Krawczyk, T. Holowka, J. Sun et al., "Toll-like receptor-induced changes in glycolytic metabolism regulate dendritic cell activation," *Blood*, vol. 115, no. 23, pp. 4742–4749, 2010.

[20] C. Quoilin, A. Mouithys-Mickalad, J. Duranteau, B. Gallez, and M. Hoebeke, "Endotoxin-induced basal respiration alterations of renal HK-2 cells: A sign of pathologic metabolism downregulation," *Biochemical and Biophysical Research Communications*, vol. 423, no. 2, pp. 350–354, 2012.

[21] L. A. Voloboueva, J. F. Emery, X. Sun, and R. G. Giffard, "Inflammatory response of microglial BV-2 cells includes a glycolytic shift and is modulated by mitochondrial glucose-regulated protein 75/mortalin," *FEBS Letters*, vol. 587, no. 6, pp. 756–762, 2013.

[22] M. P. Murphy, "How mitochondria produce reactive oxygen species," *Biochemical Journal*, vol. 417, no. 1, pp. 1–13, 2009.

[23] V. P. Skulachev, "Cationic antioxidants as a powerful tool against mitochondrial oxidative stress," *Biochemical and Biophysical Research Communications*, vol. 441, no. 2, pp. 275–279, 2013.

[24] J. Boren and K. M. Brindle, "Apoptosis-induced mitochondrial dysfunction causes cytoplasmic lipid droplet formation," *Cell Death & Differentiation*, vol. 19, no. 9, pp. 1561–1570, 2012.

[25] G. F. Kelso, C. M. Porteous, C. V. Coulter et al., "Selective targeting of a redox-active ubiquinone to mitochondria within cells: Antioxidant and antiapoptotic properties," *The Journal of Biological Chemistry*, vol. 276, no. 7, pp. 4588–4596, 2001.

[26] D. A. Lowes, N. R. Webster, M. P. Murphy, and H. F. Galley, "Antioxidants that protect mitochondria reduce interleukin-6 and oxidative stress, improve mitochondrial function, and reduce biochemical markers of organ dysfunction in a rat model of acute sepsis," *British Journal of Anaesthesia*, vol. 110, no. 3, pp. 472–480, 2013.

[27] G. S. Supinski, M. P. Murphy, and L. A. Callahan, "MitoQ administration prevents endotoxin-induced cardiac dysfunction," *American Journal of Physiology-Regulatory, Integrative and Comparative Physiology*, vol. 297, no. 4, pp. R1095–R1102, 2009.

[28] B. Kelly, G. M. Tannahill, M. P. Murphy, and L. A. J. O'Neill, "Metformin inhibits the production of reactive oxygen species from NADH:ubiquinone oxidoreductase to limit induction of interleukin-1β (IL-1β) and boosts interleukin-10 (IL-10) in lipopolysaccharide (LPS)-activated macrophages," *The Journal of Biological Chemistry*, vol. 290, no. 33, pp. 20348–20359, 2015.

[29] J. Park, J.-S. Min, B. Kim et al., "Mitochondrial ROS govern the LPS-induced pro-inflammatory response in microglia cells by regulating MAPK and NF-κB pathways," *Neuroscience Letters*, vol. 584, pp. 191–196, 2015.

[30] J. D. Schilling, S. M. Martin, D. A. Hunstad et al., "CD14- and toll-like receptor-dependent activation of bladder epithelial cells by lipopolysaccharide and type 1 piliated *Escherichia coli*," *Infection and Immunity*, vol. 71, no. 3, pp. 1470–1480, 2003.

[31] J. Song and S. N. Abraham, "TLR-mediated immune responses in the urinary tract," *Current Opinion in Microbiology*, vol. 11, no. 1, pp. 66–73, 2008.

[32] S. Nikolaeva, V. Bachteeva, E. Fock et al., "Frog urinary bladder epithelial cells express TLR4 and respond to bacterial LPS by increase of iNOS expression and L-arginine uptake," *American Journal of Physiology-Regulatory, Integrative and Comparative Physiology*, vol. 303, no. 10, pp. R1042–R1052, 2012.

[33] E. V. Fedorova, E. M. Fok, V. T. Bakhteeva, E. A. Lavrova, and R. G. Parnova, "Deposition of exogenous and endogenously generated unsaturated fatty acids in lipid droplets triacylglycerol as a mechanism of its sequestration in epithelial cells," *Ross Fiziol Zh Im I M Sechenova*, vol. 100, pp. 964–978, 2014.

[34] S. Nikolaeva, L. Bayunova, T. Sokolova et al., "GM1 and GD1a gangliosides modulate toxic and inflammatory effects of E. coli lipopolysaccharide by preventing TLR4 translocation into lipid rafts," *Biochimica et Biophysica Acta (BBA) - Molecular and Cell Biology of Lipids*, vol. 1851, no. 3, pp. 239–247, 2015.

[35] B. D. Fink, J. A. Herlein, M. A. Yorek, A. M. Fenner, R. J. Kerns, and W. I. Sivitz, "Bioenergetic effects of mitochondrial-targeted coenzyme Q analogs in endothelial cells," *The Journal of Pharmacology and Experimental Therapeutics*, vol. 342, no. 3, pp. 709–719, 2012.

Mitochondrial-Targeted Antioxidant MitoQ Prevents E. coli Lipopolysaccharide-Induced Accumulation...

45

[36] L. Plecitá-Hlavatá, J. Ježek, and P. Ježek, "Pro-oxidant mito-chondrial matrix-targeted ubiquinone MitoQ10 acts as anti-oxidant at retarded electron transport or proton pumping within Complex I," *The International Journal of Biochemistry & Cell Biology*, vol. 41, no. 8-9, pp. 1697–1707, 2009.

[37] J. Trnka, M. Elkalaf, M. Anděl, and F. Gallyas Jr., "Lipophilic Triphenylphosphonium Cations Inhibit Mitochondrial Elec-tron Transport Chain and Induce Mitochondrial Proton Leak," *PLoS ONE*, vol. 10, no. 4, p. e0121837, 2015.

[38] M. J. McManus, M. P. Murphy, and J. L. Franklin, "Mito-chondria-derived reactive oxygen species mediate caspase-dependent and -independent neuronal deaths," *Molecular and Cellular Neuroscience*, vol. 63, pp. 13–23, 2014.

[39] D. A. Lowes, C. Wallace, M. P. Murphy, N. R. Webster, and H. F. Galley, "The mitochondria targeted antioxidant MitoQ protects against fluoroquinolone-induced oxidative stress and mitochondrial membrane damage in human Achilles tendon cells," *Free Radical Research*, vol. 43, no. 4, pp. 323–328, 2009.

[40] V. Vanasco, N. D. Magnani, M. C. Cimolai et al., "Endotoxemia impairs heart mitochondrial function by decreasing electron transfer, ATP synthesis and ATP content without affecting membrane potential," *Journal of Bioenergetics and Biomem-branes*, vol. 44, no. 2, pp. 243–252, 2012.

[41] W. J. H. Koopman, L. G. J. Nijtmans, C. E. J. Dieteren et al., "Mammalian mitochondrial complex I: biogenesis, regulation, and reactive oxygen species generation," *Antioxidants & Redox Signaling*, vol. 12, no. 12, pp. 1431–1470, 2010.

[42] A. P. West, I. E. Brodsky, C. Rahner et al., "TLR signalling aug-ments macrophage bactericidal activity through mitochondrial ROS," *Nature*, vol. 472, no. 7344, pp. 476–480, 2011.

[43] D. Setoyama, Y. Fujimura, and D. Miura, "Metabolomics reveals that carnitine palmitoyltransferase-1 is a novel target for oxida-tive inactivation in human cells," *Genes to Cells*, vol. 18, no. 12, pp. 1107–1119, 2013.

[44] G. Serviddio, A. M. Giudetti, F. Bellanti et al., "Oxidation of hepatic carnitine palmitoyl transferase-I (CPT-I) impairs fatty acid beta-oxidation in rats fed a methionine-choline deficient diet," *PLoS ONE*, vol. 6, no. 9, 2011.

[45] G. Gloire, S. Legrand-Poels, and J. Piette, "NF-κB activation by reactive oxygen species: fifteen years later," *Biochemical Pharmacology*, vol. 72, no. 11, pp. 1493–1505, 2006.

[46] I. Russo, A. Luciani, P. de Cicco, E. Troncone, and C. Ciacci, "Butyrate attenuates lipopolysaccharide-induced inflammation in intestinal cells and Crohn's mucosa through modulation of antioxidant defense machinery," *PLoS ONE*, vol. 7, no. 3, Article ID e32841, 2012.

[47] D. A. Lowes, A. M. Almawash, N. R. Webster, V. L. Reid, and H. F. Galley, "Melatonin and structurally similar compounds have differing effects on inflammation and mitochondrial function in endothelial cells under conditions mimicking sepsis," *British Journal of Anaesthesia*, vol. 107, no. 2, pp. 193–201, 2011.

[48] RB. Barhoumi Faske Liu Tjalkens, "Manganese potentiates lipopolysaccharide-induced expression of NOS2 in C6 glioma cells through mitochondrial-dependent activation of nuclear factor kappaB," *Brain Res Mol Brain Res*, vol. 122, pp. 167–179, 2004.

6

Erythrocyte Omega-3 Fatty Acid Content in Elite Athletes in Response to Omega-3 Supplementation

Franchek Drobnic,[1,2] Félix Rueda,[3] Victoria Pons,[1] Montserrat Banquells,[1] Begoña Cordobilla,[3] and Joan Carles Domingo[3]

[1]*Departamento de Investigación del CAR, Av. Alcalde Barnils 3, 08173 Sant Cugat del Vallés, Spain*
[2]*Servicios Médicos del FC Barcelona, Av. del Sol, s/n, Sant Joan Despí, 08970 Barcelona, Spain*
[3]*Departamento de Bioquímica y Biología Molecular, Facultad de Biología, Av. Diagonal 643, 08028 Barcelona, Spain*

Correspondence should be addressed to Franchek Drobnic; drobnic@car.edu

Academic Editor: Kamal A. Amin

Introduction. Supplementation of Omega-3 fatty acids (n-3FA) in athletes is related to the anti-inflammatory and/or antioxidant effect and consequently its action on all the processes of tissue restoration and adaptation to physical stress. *Objective.* Evaluate the Omega-3 Index (O3Ix) response, in red blood cells, to supplemental EPA + DHA intake in the form of high purity and stable composition gums (G), in elite summer athletes. *Method.* Twenty-four summer sport athletes of both sexes, pertaining to the Olympic Training Center in Spain, were randomized to two groups (2G = 760 or 3G = 1140 mg of n-3 FA in Omegafort OKids, Ferrer Intl.) for 4 months. Five athletes and four training staff volunteers were control group. *Results.* The O3Ix was lower than 8% in 93.1% of all the athletes. The supplementation worked in a dose-dependent manner: 144% for the 3G dose and 135% for the 2G, both $p < 0.001$, with a 3% significant decrease of Omega-6 FAs. No changes were observed for the control group. *Conclusions.* Supplementation with n-3FA increases the content of EPA DHA in the red blood cells at 4 months in a dose-dependent manner. Athletes with lower basal O3Ix were more prone to increment their levels. The study is registered with Protocol Registration and Results System (ClinicalTrials.gov) number NCT02610270.

1. Introduction

Docosahexaenoic acid (DHA C22: 6) and eicosapentaenoic acid (EPA C20: 5) are the most important polyunsaturated fatty acids (FAs) known as long-chain Omega-3 (n-3). Both are considered essential FA and are important components of the lipid bilayer of cell membranes. For its incorporation, they should be synthesized from essential fatty acid, alpha-linolenic acid (ALA), or taken directly preformed in the diet. At present, it is suggested that the administration of purified EPA and/or DHA can offer a wide range of beneficial effects [1], ranging from the plasticity, neuronal development, and functionality of the central nervous system [2] to the treatment and prevention of chronic diseases with an inflammatory component [3]. The indication of diet supplementation in sport activities is due to the anti-inflammatory and/or antioxidant effect and consequently its action on all the processes of tissue restoration and adaptation to the physical stress and training, from the connective tissue to the neural development [4]. However, it has been shown that DHA is more effective than EPA in modulating specific markers of inflammation as well as blood lipids [5]. The epidemiological studies related to the impact of the supplementation on the physical activities are focused on supposed actions of the n-3 FA on muscle metabolism and tissue recovery [6], functional performance, and inflammation [7] and, with a very specified indications on sport, as exercise induced asthma [8], traumatic brain injury [9] or injury recovery, training adaptation, and sarcopenia [10]. Those studies report the use of different doses, concentration, and duration, and they do not always reference the previous state of the amount of these FA in cell membranes or usual diet.

The use of biomarker-based approaches has made it possible to study and evaluate with criteria the Omega-3 Index (O3Ix), which is defined as the sum of EPA+DHA content in red blood cell (RBC) membranes and has been considered a risk factor for death from coronary heart disease and as a biomarker of n-3 FA status [11]. An O3Ix of ≥8% has been recommended for its cardiovascular protective effect [12] and has been postulated to be adequate also in the elite athlete [13]. Until now, Von Schacky et al.'s study [13] is the only one that targets elite athlete as reference values for O3Ix. Surprisingly, those subjects, with a diet geared toward better performance, demonstrate not only a low consumption of fish but also a low level O3Ix, far from the desired range over 8%. Well conducted studies have confirmed that dietary or supplemental intake of EPA + DHA is associated with higher levels of the O3Ix [14–16].

In the present study, the objective was to model the O3Ix response to supplemental EPA + DHA intake within attainable dietary ranges in athletes. We had the primary hypothesis that if adherence to treatment is adequate and diet is maintained, a change in the O3Ix would be observed depending on the dosage of the supplementation. This information will be important for making better EPA + DHA recommendations to achieve a target O3Ix for future studies on the evaluation of the effect of n-3 FAs in the different physical exercise activities.

2. Methods

This project has tried to follow the guidelines in the design, conduct, and reporting of studies of human health benefits of foods, summarized by Welch et al. [17].

2.1. Participants. Eligibility criteria were based on selecting healthy athletes of both sexes, from a specific age range and committed to the project. Athletes belonged to different summer sport federations and lived at the Olympic Training Center (OTC) of Sant Cugat del Vallés (Spain). Exclusion criteria included any type of inflammatory process or the use of anti-inflammatory medications, consumption of n-3 FA supplements and n-3 FA-supplemented foods in the past 3 months. planning to change dietary habits, or training schedule. In order to establish a control group (C), three athletes and four members of the technical staff met the same requirements for maintaining diet, weight, and daily activity level. All of them had their official sport preparticipation screening evaluation, signed informed consent, and went through a medical examination that included medical history, physical examination, anthropometric measurements, complete blood count, and standard chemistry panel to rule out the presence of any newly developed illness or inflammatory process. Diet analysis was conducted to evaluate, through a week registration, all the nutritional components of their diet. The study protocol was approved by the Ethical Committee of the Sports Council of the Generalitat de Catalunya 08/2014/CEIEGC. All procedures followed were in accordance with the ethical standards of the Helsinki Declaration of 1975, revised in 2000.

2.2. Intervention. This was a randomized, parallel-group with control subjects study. Participants ($n = 24$), 13 women (55%), were randomized to take two different dosages of n-3 FAs daily, as fish oil supplements in the form of gums (Omegafort OKids, Ferrer Intl.) for 4 months, the approximate time of lifespan of the RBC [17] and time that it takes the composition in FA of cell membranes to reach a new steady state [18]. To ensure even distribution among treatment groups, a computer randomization scheme was used, which is stratified by sex and age and balanced the size of the two blocks. Eligible participants were assigned to treatment group 2G, two gums daily, or 3G, three gums daily (i.e., 2G = 760 mg or 3G = 1140 mg of n-3 FA). All researchers and clinicians, except the Head Dept., and participants, were blinded to treatment assignment. Analysis of the fish oil gums verified that they contained 35,7% DHA, 27,7% EPA, 3.32% docosapentaenoic acid, 18,5% oleic acid, 3,1% vaccenic acid, 1,4% stearic acid, 1,2% palmitic acid, 1,7% arachidonic acid, and small amounts of other fatty acids (Table 1).

All participants were instructed to maintain their training schedule, diet and activity level, and their usual consumption of fatty fish as well as their no consumption of any supplementation during the study course. The participants were contacted monthly to ensure gummies intake compliance and to discuss any difficulties in following the treatment. Also, participants reported back to the Research Department after 8 weeks to return the gum boxes and to receive new supplies.

2.3. Blood Sample Collection. Blood samples collection was performed in the fasting state by venipuncture and harvested in K_2-EDTA-containing tubes before and after the intervention (12 hours without any intake with the exception of water, 48 hours without alcohol, and 12 hours without perform vigorous exercise). After each blood sample collection, a complete blood count and a general biochemistry profile were obtained. Whole blood was centrifuged at 1500 ×g for 15 minutes at 4°C. Except for assays that required unfrozen specimens, samples were stored at −80°C until they were analyzed.

2.4. RBC Fatty Acid Analysis. Erythrocytes were separated from the plasma by centrifugation (3000 rpm, 1500 ×g, for 10 min) and washed with an equal volume of saline. These erythrocytes resuspended with saline were stored in a freshly 0.01% butylated hydroxyl toluene- (BHT-) treated Eppendorf vials at −80°C. The fatty acids composition was determined using the method by Lepage and Roy [19]; erythrocyte's membranes were extracted from aliquots of 200 μL of erythrocyte suspensions and the fatty acids converted to methyl esters by reaction with acetyl chloride for 60 min at 100°C. Methyl ester fatty acids (FAME) were separated and analyzed by gas chromatography performed on a Shimadzu GCMS-QP2010 Plus gas chromatograph/mass spectrometer (Shimadzu, Kyoto, Japan) and peaks were identified through mass spectra and by comparing with respect to a reference FAME mixture (GLC-744 Nu-Chek Prep. Inc., Elysian MN, USA) the elution

TABLE 1: Fatty acid composition of the diet supplement (gums).

Fatty acid	Composition* (%)	1 gum (mg)	2 gums (mg)	3 gums (mg)
MUFA				
Oleic (18:1, n-9)	18,5	111	222	334
Vaccenic acid (18:1, n-7)	3,1	19	37	56
Other MUFA	1,4	8	17	25
SFA				
Stearic acid (18:0)	1,4	8	17	25
Palmitic acid (16:0)	1,2	7	14	22
Other SFA	0,5	3	6	9
PUFA n-6				
Arachidonic acid (20:4, n-6)	1,7	10	20	30
Other n-6	1,9	12	23	35
PUFA n-3				
DHA (22:6, n-3)	35,7	214	428	643
EPA (20:5, n-3)	27,7	166	332	499
DPA (22:5, n-3)	3,3	20	40	60
Other n-3	3,5	21	42	63
Total DHA + EPA		380	761	1141

SFA, saturated fatty acid; MUFA, monounsaturated fatty acid; PUFA, polyunsaturated fatty acid; n-6, Omega-6; n-3, Omega-3; DHA, docosahexaenoic acid, EPA, eicosapentaenoic acid; DPA, docosapentaenoic acid. *To determine the quantitative fatty acid (FA) composition, FAs were analyzed by gas chromatography-mass spectrometry. The results express in molar % of total fatty acids.

pattern and relative retention times of FAME. The O3Ix was calculated as erythrocyte (EPA + DHA)/(total fatty acids) × 100% (percentage molar of total fatty acids) [11].

2.5. Statistical Analysis. Mean changes from baseline to 4 months were calculated and compared between groups using paired t-test. Differences among groups were tested by analysis of variance using a general linear model. All statistical tests were performed at a significance level of 0.05. Adjusted $p < 0.05$ was considered significant. Continuous data are reported as the mean ± SD. For descriptive purposes, categorical data are presented as percentages. The statistical software program SPSS for Windows, version 13.0, was used for all data analysis.

3. Results

A total of 41 individuals were screened between May and June 2014; twenty-four of them met the inclusion criteria and were randomly assigned to any treatment group. Besides, seven subjects were selected as control subjects. Since the study is to determine the level of impregnation in the tissue not a placebo but a control that maintained the same diet was considered, all baseline measurements were completed during the selection process. One subject withdrew from the study between baseline and the final point due to an injury during training, and he voluntarily dropped out from the study. In addition, two subjects from the 2G and one from the 3G groups comment the lack of compliance at the first control and desired to abandon the study. Among study completers, compliance was presumably total in all groups. Anyway, real compliance was assessed by interrogation, by counting

returned capsules, and by analysis of red cell phospholipid fatty acid composition, which reflects dietary fatty acid composition. Those athletes, whose percentage of EPA in red cell membrane fatty acids differed, ≥2 Standard Deviations from the mean of the respective treatment group, were also considered noncompliant [20]. Noncompliant patients were excluded from the valid case analysis, but RBC analysis was performed confirming the absence of increase of EPA and consequently in O3Ix (Figure 1). Finally, from the 24 volunteers included in groups 2G and 3G, and once applied compliance criteria, the sample was reduced to nine and eleven subjects in both groups, respectively. Adherence to the therapy was 82% for 2G and 92% for 3G groups.

No significant differences were found between groups of participants with respect to baseline and diet characteristics. Nevertheless, the control group presented some differences related to age, height, and weight and slightly to the caloric intake, basically referred to as the technical staff. RBC FA content was similar between groups (Table 2). The mean O3Ix at study entry (mean ± SD) was 5.1 ± 1.0% with a range of 3.3% to 7.8%. On average, there were no gender differences in relation to O3Ix. Body weight, BMI, blood pressure, and heart rate did not change significantly during the study.

Distribution of O3Ix shows a Gauss distribution where a 93.1% of the athletes had values lower than 8% after EPA + DHA supplementation (Figure 2). The EPA + DHA supplementation increased the O3Ix in a dose-dependent manner (Table 2), affecting both EPA and DHA which resulted in a significant increase in O3Ix of 144% (116–157%) for the 3G dose and 135% (120–149%) for the 2G dose ($p < 0.001$). Participants who had lower basal O3Ix were more prone to increment their levels (Figure 1). Omega-3 FAs increase was

TABLE 2: Erythrocyte fatty acids profile of the participants as a function of the dose through the study.

Mean (SD)	2G (N:9)			3G (N:11)			Control (N:6)			Control versus 2G/3G	
	Basal	Post	$p\leq$	Basal	Post	$p\leq$	Basal	Post	$p\leq$	Basal	Post
SFA	44.7 (1.1)	45.6 (1.5)	NS	45.2 (0.7)	45.8 (0.8)	0.05	45.4 (0.9)	45.8 (0.8)	NS	NS	NS
MUFA	20.9 (1.6)	21.3 (1.2)	NS	20.0 (0.9)	21.0 (1,0)	0.05	21.0 (1.8)	21.9 (1.8)	NS	NS	NS
PUFA	34.4 (1.5)	33.6 (1.5)	NS	34.8 (0.9)	34.0 (1.2)	0.05	33.5 (1.4)	32.3 (1.4)	0.05	NS	0,05
n-6	27.9 (1.8)	25.2 (1.4)	0.01	28.6 (1.4)	25.4 (1.6)	0.001	27.5 (1.6)	26.7 (1.2)	NS	NS	0,05
n-3	6.4 (1.1)	8.4 (0.9)	0.001	6.3 (1.2)	8.6 (1.0)	0.001	6.1 (0.9)	5.6 (1.0)	NS	NS	0,001
n-6/n-3	4.5 (1.1)	3.0 (0.4)	0.01	4.7 (1.0)	3.0 (0.5)	0.001	4.6 (1.0)	4.9 (1.1)	NS	NS	0,01
O3Ix	5.0 (0.8)	6.8 (0.7)	0.001	4.9 (1,0)	7.0 (0,8)	0.001	4.8 (0,9)	4.4 (0,9)	NS	NS	0,001

Group 2G, two oil gums daily; 3G, three oil gums daily. SFA, saturated fatty acid; MUFA, monounsaturated fatty acid; PUFA, polyunsaturated fatty acid; n-6, Omega-6; n-3, Omega-3; and O3Ix, Omega-3 Index.

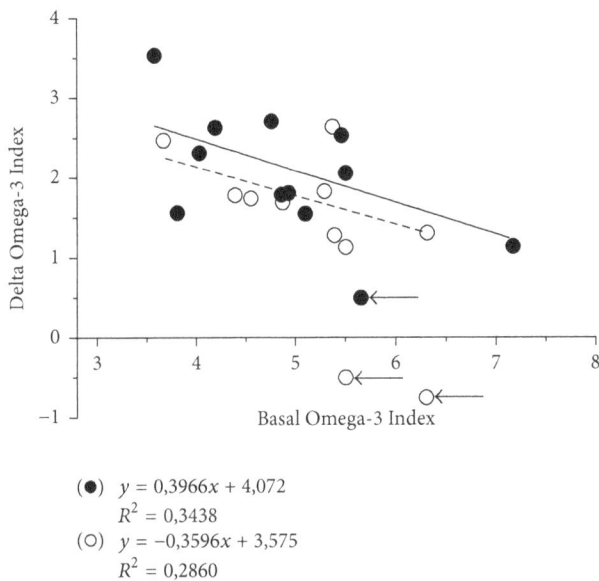

(●) $y = 0,3966x + 4,072$
$R^2 = 0,3438$
(○) $y = -0,3596x + 3,575$
$R^2 = 0,2860$

FIGURE 1: Changes on Omega-3 Index in athletes after 4 months of supplementation in function of the dose: (○) group 2G, two oil gums daily; (●) group 3G, three oil gums daily. The arrows show those subjects that were unable to comply with the treatment. These data were not added to the statistical evaluation; it is only to show the place in this figure. The line of tendency reflects the delta change of all subjects depending on the basal level and different doses.

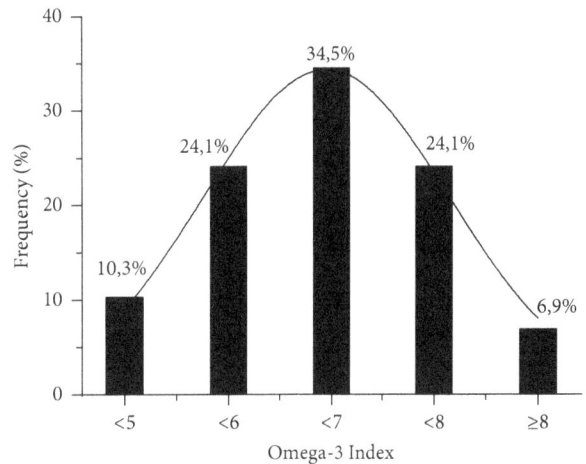

FIGURE 2: Distribution of the basal Omega-3 Index in the population of athletes of this study.

accompanied by a significant decrease in total Omega-6 (n-6) FAs in both intervention groups from 28.0 to 25.2% (2G, $p < 0.01$) and 28.6 to 25.4% (3G, $p < 0,001$), respectively. No change in O3Ix was observed for the control group from baseline (93%). No significance changes were observed on the other FA except for MUFA, in the 3G group (20.0 ± 0.9 to 21.0 ± 1.0%).

4. Discussion

The present study evaluated the effect of EPA + DHA supplementation in athletes on the O3Ix response. The required amount of Omega-3 intake is not clearly defined, although there are certain recommendations based on individualized dietary patterns by state and age [21]. The World Health Organization establishes a need for consumption of 250–500 mg/day of EPA + DHA [22], while the International Society for the Study of Fatty Acids and Lipids [23] adjusts it to 500 mg/day. Certain effects as cardioprotective [24] or triglyceride decrease [25] at doses of 1 to 3–5 g/day of EPA + DHA are advised. In sport activities, similar dose also has been recommended [26]. However, the actual consumption of fatty fish does not become desirable for the different population groups and, as a result, the Omega-3 daily intake from the diet is insufficient. In Spain, the consumption of Omega-3 is 1.5 g/day with an average consumption of 0.2–2 g/day and it is estimated that Americans consume <100 mg/day of EPA + DHA [27]. While it may be considered satisfactory, the type of n-3 FAs consumed is basically ALA. As they have not the same biological effects and conversion of ALA to EPA and DHA is not carried out effectively, the EPA and DHA ingested are well below the recommendations of 0.25 to 0.5% of the daily energy, only reaching 0.05%. Moreover, the Omega-6/Omega-3 ratio in the same population is 15-16/1, well above the 4-5/1 that is considered suitable [28]. Therefore, the need for supplementation with n-3 FAs is real in all population age ranging from children to the elderly, and athletes are not out of this population.

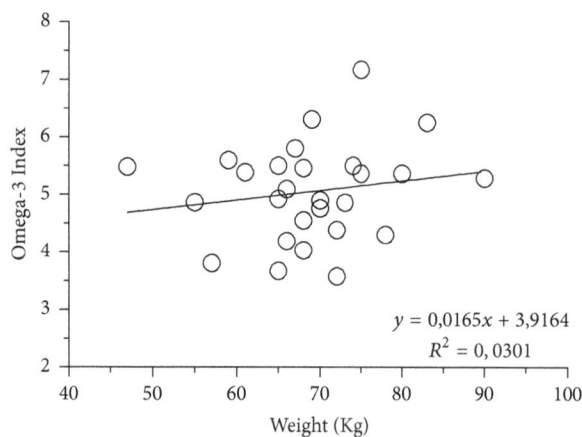

FIGURE 3: Relation of Omega-3 Index with body weight. There is a low correlation of the O3Ix distribution related to body weight in the population of this study.

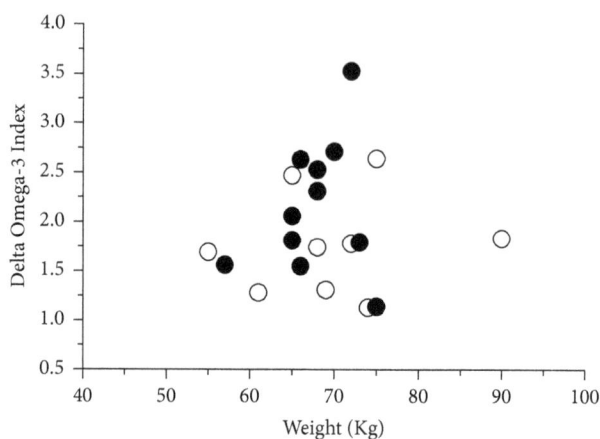

FIGURE 4: Delta change of the Omega-3 Index related to body weight and dosage: (O) group 2G, two oil gums daily; (●) group 3G, three oil gums daily. The changes on values of the O3Ix are more related to dosage than the previous body weight.

We have found an average baseline O3Ix of 5.0% in the participants of the present study; this level is consistent with previous studies of adults' subjects reporting to be low habitual fish consumers [29] and in the same range of the athletes evaluated by Von Schacky et al. [13]. Our results suggest that athletes, with low fish intake who increased their dietary intake by 760 to 1140 mg/day of EPA + DHA, would experience an increase in O3Ix values of about 1.8% to 2.1% by mean, lower results than those observed when period of treatment is longer [30–32] over 3.5% and near to 5% increase (see Supplementary Material available online at https://doi.org/10.1155/2017/1472719). From these, it can be estimated that an average healthy adult with a low O3Ix (i.e., 5,0%) would require 1.5–2 g/day of EPA + DHA for 4 months or more to increase 2 index points and bring it to the desired 8%. Our different response could be related to the different ratio of EPA/DHA (44% EPA/56% DHA) as it has been argued by the different velocity of incorporation depending on that percentage [33]. That asseveration is not in agreement with Browning et al. study [30], with similar relation between DHA/EPA and fish oil content. Under this perspective, more concern has to be considered with the amount of Omega-3 offered and treatment duration, considering the different quantity and quality of products, which present different bioavailability [34].

Body weight does not explain the variability of O3Ix in response to EPA + DHA supplementation in our data (Figures 3 and 4). Flock et al. [15] demonstrated a greater tendency to respond to a given EPA + DHA intake in individuals with lower body weight, suggesting that, to achieve a target O3Ix, consumption recommendations of EPA + DHA should be made on the basis of body weight, in a similar way as it happens with current dietary protein requirements. This discrepancy with our data could be due to the fact that the weight of the athletes is usually adequate to their physical activity needs and their caloric and nutrient intake, in all cases under their daily requirements (except for PUFA) [35]. In Folk et al.'s study [15], it was estimated that the requirements to increase its O3Ix from 4.3% to 8% in individuals weighing

75 kg were about 1.2 g/day of EPA + DHA; 0.9 g/day if they weighed 55 kg; or 1.5 g/day for individuals weighing 95 kg. As it can be observed in Figures 3 and 4, that observation does not correlate similarly in the subjects of our intervention. It seems that this adjustment of doses is not needed in the athletes if their weight is the expected for performance and nourishment is adequate. Possibly in athletes a higher dose needs to be administrated to achieve the desired 8% in O3Ix.

We also found that the EPA + DHA incorporation into the membranes of RBC follows a dose-dependent increase in both groups assayed, which is potentially saturable. This suggests that EPA + DHA concentrations in the membrane of RBC could be regulated to some degree and it could reach a saturation point. This finding is consistent with previous observation that individuals, with higher content of EPA + DHA in their RBC membranes, incorporate, at a slower rate, additional EPA + DHA than those presenting a lower baseline level [36].

We did not find that women on average had a higher O3Ix than men at study entry. This relationship between sex and O3Ix seems not to be related only to the difference in body weight [37] as it appears when this factor was accounted for in the model by adjusting the dose per unit of body weight [38]. In the same study, the participants, that are more physically active, tended to experience greater elevation in O3Ix as dose increased, suggesting that something related to exercise may enhance incorporation of EPA + DHA in RBC membranes in individuals taking fish oil supplements.

With respect to the MUFA change in the 3G group, with lower levels from the beginning, it can be attributed to the 334 mg of MUFA ingested daily, but it was not statistically different. MUFA consumption can be beneficial when replacing carbohydrate and saturated fat in the diet but not when replacing PUFAs. Although MUFA showed to have positive impact on surrogate markers, the potential impact of MUFA intake alone on disease outcomes, such as CVD or diabetes, remains unclear. Therefore, the role of MUFAs on health and disease when consumed as an eating pattern (i.e.,

Mediterranean diet) should be more studied [39]. Maybe the explanation of the favorable properties lies in the oleic acid, the MUFA most abundant fatty acid found in food, or, more particularly, in its minor though highly bioactive molecules, the phenolic compounds, which have been associated with the prevention of the main chronic diseases [40].

The results presented here show that elite athletes, despite following a diet presumably healthy and suitable for their sporting activity, have an O3Ix well below that recommended and that, despite a strict follow-up of the DHA + EPA supplement in the diet during the study period, they do not reach the desired levels. Given the potential role of DHA and EPA in cardiovascular protection, preservation of the central nervous system, repair of the musculoskeletal system, and significant influence on cellular behavior and responsiveness to signals [41], it would be advisable to increase the intake of these Omega-3 FAs in the diet of these athletes. A study of its long-term benefits is guaranteed.

5. Strengths and Limitations

The strengths of this study were as follows: sample of subjects with elevated physical activity who were under a controlled and healthy diet, single blind study design that compared two doses of EPA + DHA with respect to a control group with a high adherence to the treatment, adequate duration of supplementation, 4 months, and the use of validated analytical methods to determine biomarker response to treatment.

Limitations include the scarce sample, the homogeneity of white, young, and healthy population, and the lack of background genetic data.

6. Conclusions and Further Research

It is confirmed that athletes even with a presumably healthy diet have low O3Ix. The marine-derived n-3 FA supplementation increases the RBC EPA + DHA content in a dose-time related manner. Future studies need to assess how EPA or DHA individually or different ratios of both affect O3Ix response and to clarify the potential correlation between changes in the O3Ix and its effect on prevalence and severity of the injury recovery.

Authors' Contributions

Drobnic Franchek, Pons Victoria, and Banquells Montserrat carried out the design, selection of individuals, follow-up of the diets, statistic evaluation, and discussion of the results. Cordobilla Begoña, Rueda Félix, and Domingo Joan Carles carried out the methodology design of blood extraction, Omega-3 blood analysis, and discussion of the results. Banquells Montserrat carried out the blood analysis samples and manipulation; Pons Victoria evaluated the daily diet of the athletes. Drobnic Franchek and Domingo Joan Carles wrote the article. All authors read and approved the final manuscript.

Acknowledgments

The authors are thankful to all the participants who contributed to their research especially the athletes participating in the study for their commitment and the coaches and technical staff for their support and involvement. They also are very grateful to the nursing and clinician staff of the Physiology Department of the Olympic Training Center and of the Departamento de Bioquímica i Biologia Molecular de la Facultat de Biologia de la Universitat de Barcelona. This study was supported by Ferrer International Laboratories sources.

References

[1] C. I. F. Janssen and A. J. Kiliaan, "Long-chain polyunsaturated fatty acids (LCPUFA) from genesis to senescence: the influence of LCPUFA on neural development, aging, and neurodegeneration," *Progress in Lipid Research*, vol. 53, no. 1, pp. 1–17, 2014.

[2] D. W. Luchtman and C. Song, "Cognitive enhancement by omega-3 fatty acids from child-hood to old age: findings from animal and clinical studies," *Neuropharmacology*, vol. 64, pp. 550–565, 2013.

[3] C. M. Yates, P. C. Calder, and G. Ed Rainger, "Pharmacology and therapeutics of omega-3 polyunsaturated fatty acids in chronic inflammatory disease," *Pharmacology and Therapeutics*, vol. 141, no. 3, pp. 272–282, 2014.

[4] R. J. Shei, M. R. Lindley, and T. D. Mickleborough, "Omega-3 polyunsaturated fatty acids in the optimization of physical performance," *Military Medicine*, vol. 179, no. 11S, pp. 144–156, 2014.

[5] J. Allaire, P. Couture, M. Leclerc et al., "A randomized, crossover, head-to-head comparison of eicosapentaenoic acid and docosahexaenoic acid supplementation to reduce inflammation markers in men and women: The Comparing EPA to DHA (ComparED) Study," *American Journal of Clinical Nutrition*, vol. 104, no. 2, pp. 280–287, 2016.

[6] C. J. Gerling, J. Whitfield, K. Mukai, and L. L. Spriet, "Variable effects of 12 weeks of omega-3 supplementation on resting skeletal muscle metabolism," *Applied Physiology, Nutrition and Metabolism*, vol. 39, no. 9, pp. 1083–1091, 2014.

[7] E. J. H. Lewis, P. W. Radonic, T. M. S. Wolever, and G. D. Wells, "21 days of mammalian omega-3 fatty acid supplementation improves aspects of neuromuscular function and performance in male athletes compared to olive oil placebo," *Journal of the International Society of Sports Nutrition*, 2015.

[8] T. D. Mickleborough and K. W. Rundell, "Dietary polyunsaturated fatty acids in asthma- and exercise-induced bronchoconstriction," *European Journal of Clinical Nutrition*, vol. 59, no. 12, pp. 1335–1346, 2005.

[9] E. C. Barrett, M. I. McBurney, and E. D. Ciappio, "ω-3 fatty acid supplementation as a potential therapeutic aid for the recovery from mild traumatic brain injury/concussion," *Advances in Nutrition*, vol. 5, no. 3, pp. 268–277, 2014.

[10] F. G. Di Girolamo, R. Situlin, S. Mazzucco, R. Valentini, G. Toigo, and G. Biolo, "Omega-3 fatty acids and protein metabolism: Enhancement of anabolic interventions for sarcopenia," *Current Opinion in Clinical Nutrition and Metabolic Care*, vol. 17, no. 2, pp. 145–150, 2014.

[11] W. S. Harris and C. Von Schacky, "The omega-3 index: a new risk factor for death from coronary heart disease?" *Preventive Medicine*, vol. 39, no. 1, pp. 212–220, 2004.

[12] C. von Schacky, "Omega-3 index and cardiovascular health," *Nutrients*, vol. 6, no. 2, pp. 799–814, 2014.

[13] C. Von Schacky, M. Kemper, R. Haslbauer, and M. Halle, "Low omega-3 index in 106 german elite winter endurance athletes: A pilot study," *International Journal of Sport Nutrition and Exercise Metabolism*, vol. 24, no. 5, pp. 559–564, 2014.

[14] F. Delodder, L. Tappy, L. Liaudet, P. Schneiter, C. Perrudet, and M. M. Berger, "Incorporation and washout of n-3 PUFA after high dose intravenous and oral supplementation in healthy volunteers," *Clinical Nutrition*, vol. 34, no. 3, pp. 400–408, 2015.

[15] M. R. Flock, A. C. Skulas-Ray, W. S. Harris, T. D. Etherton, J. A. Fleming, and P. M. Kris-Etherton, "Determinants of erythrocyte omega-3 fatty acid content in response to fish oil supplementation: a dose-response randomized controlled trial," *The Journal of the American Heart Association*, vol. 2, no. 6, Article ID e000513, 2013.

[16] C. G. Walker, L. M. Browning, A. P. Mander et al., "Age and sex differences in the incorporation of EPA and DHA into plasma fractions, cells and adipose tissue in humans," *British Journal of Nutrition*, vol. 111, no. 4, pp. 679–689, 2014.

[17] R. W. Welch, J.-M. Antoine, J.-L. Berta et al., "Guidelines for the design, conduct and reporting of human intervention studies to evaluate the health benefits of foods," *British Journal of Nutrition*, vol. 106, no. 2, pp. S3–S15, 2011.

[18] D. Shemin and D. Rittenberg, "The life span of the human red blood cell," *The Journal of biological chemistry*, vol. 166, no. 2, pp. 627–636, 1946.

[19] M. B. Katan, J. P. Deslypere, A. P. J. M. van Birgelen, M. Penders, and M. Zegwaard, "Kinetics of the incorporation of dietary fatty acids into serum cholesteryl esters, erythrocyte membranes, and adipose tissue: an 18-month controlled study," *Journal of Lipid Research*, vol. 38, no. 10, pp. 2012–2022, 1997.

[20] C. Von Schacky, P. Angerer, W. Kothny, K. Theisen, and H. Mudra, "The effect of dietary ω-3 fatty acids on coronary atherosclerosis. A randomized, double-blind, placebo-controlled trial," *Annals of Internal Medicine*, vol. 130, no. 7, pp. 554–562, 1999.

[21] J. Aranceta and C. Pérez-Rodrigo, "Recommended dietary reference intakes, nutritional goals and dietary guidelines for fat and fatty acids: A systematic review," *British Journal of Nutrition*, vol. 107, no. 2, pp. S8–S22, 2012.

[22] B. Burlingame, C. Nishida, R. Uauy, and R. Weisell, *Fats and Fatty Acids in Human Nutrition*, vol. 55, 1–3, S. Karger AG, 2009.

[23] J. A. Nettleton, P. Legrand, and R. P. Mensink, "ISSFAL 2014 debate: it is time to update saturated fat recommendations," *Annals of Nutrition and Metabolism*, vol. 66, no. 2-3, pp. 104–108, 2015.

[24] T. L. Psota, S. K. Gebauer, and P. Kris-Etherton, "Dietary Omega-3 Fatty Acid Intake and Cardiovascular Risk," *American Journal of Cardiology*, vol. 98, no. 4, pp. 3–18, 2006.

[25] A. Saremi and R. Arora, "The utility of omega-3 fatty acids in cardiovascular disease," *American Journal of Therapeutics*, vol. 16, no. 5, pp. 421–436, 2009.

[26] J. A. Villegas, M. T. Martínez, F. J. López et al., "Ácidos grasos omega 3 en las lesiones deportivas. ¿Una posible ayuda terapéutica?" *Archivos de Medicina del Deportea*, vol. 104, pp. 529–532, 2004.

[27] *Dietary Reference Intakes Research Synthesis*, National Academies Press, Washington, D.C., 2006.

[28] C. Gómez Candela, L. M. Bermejo López, and V. Loria Kohen, "Importance of a balanced omega 6/omega 3 ratio for the maintenance of health: nutritional recommendations," *Nutricion Hospitalaria*, vol. 26, no. 2, pp. 323–329, 2011.

[29] A. C. Skulas-Ray, P. M. Kris-Etherton, W. S. Harris, J. P. Vanden Heuvel, P. R. Wagner, and S. G. West, "Dose-response effects of omega-3 fatty acids on triglycerides, inflammation, and endothelial function in healthy persons with moderate hypertriglyceridemia," *American Journal of Clinical Nutrition*, vol. 93, no. 2, pp. 243–252, 2011.

[30] L. M. Browning, C. G. Walker, A. P. Mander et al., "Incorporation of eicosapentaenoic and docosahexaenoic acids into lipid pools when given as supplements providing doses equivalent to typical intakes of oily fish1-4," *American Journal of Clinical Nutrition*, vol. 96, no. 4, pp. 748–758, 2012.

[31] A. M. Hill, J. D. Buckley, K. J. Murphy, and P. R. C. Howe, "Combining fish-oil supplements with regular aerobic exercise improves body composition and cardiovascular disease risk factors," *The American Journal of Clinical Nutrition*, vol. 85, no. 5, pp. 1267–1274, 2007.

[32] B. Tartibian, B. H. Maleki, and A. Abbasi, "The effects of omega-3 supplementation on pulmonary function of young wrestlers during intensive training," *Journal of Science and Medicine in Sport*, vol. 13, no. 2, pp. 281–286, 2010.

[33] L. M. Arterburn, E. B. Hall, and H. Oken, "Distribution, interconversion, and dose response of n-3 fatty acids in humans," *The American Journal of Clinical Nutrition*, vol. 83, no. 6S, pp. 1467s–1476s, 2006.

[34] F. Drobnic, B. Cordobilla, F. Rueda, and J. C. Domingo, "Characterization of different supplements of omega-3 acids in its application in the pediatric age," *Acta Pediatrica Espanola*, vol. 71, no. 11, pp. e353–e357, 2013.

[35] R. J. Maughan and L. M. Burke, "Practical nutritional recommendations for the athlete," *Nestle Nutrition Institute Workshop Series*, vol. 69, pp. 131–149, 2011.

[36] J. Cao, K. A. Schwichtenberg, N. Q. Hanson, and M. Y. Tsai, "Incorporation and clearance of omega-3 fatty acids in erythrocyte membranes and plasma phospholipids," *Clinical Chemistry*, vol. 52, no. 12, pp. 2265–2272, 2006.

[37] T. Ogura, H. Takada, M. Okuno et al., "Fatty acid composition of plasma, erythrocytes and adipose: Their correlations and effects of age and sex," *Lipids*, vol. 45, no. 2, pp. 137–144, 2010.

[38] M. R. Flock, W. S. Harris, and P. M. Kris-Etherton, "Long-chain omega-3 fatty acids: Time to establish a dietary reference intake," *Nutrition Reviews*, vol. 71, no. 10, pp. 692–707, 2013.

[39] R. Estruch, E. Ros, J. Salas-Salvadó et al., "Primary prevention of cardiovascular disease with a Mediterranean diet," *The New England Journal of Medicine*, vol. 368, no. 14, pp. 1279–1290, 2013.

[40] A. Cárdeno, M. Sánchez-Hidalgo, and C. Alarcón-De-La-Lastra, "An up-date of olive oil phenols in inflammation and cancer: Molecular mechanisms and clinical implications," *Current Medicinal Chemistry*, vol. 20, no. 37, pp. 4758–4776, 2013.

[41] P. C. Calder, "Docosahexaenoic Acid," *Annals of Nutrition & Metabolism*, vol. 69, no. 1, pp. 7–21, 2016.

7

Effect of *Helicobacter pylori* Infection on Serum Lipid Profile

Mohamadreza Haeri,[1] Mahmoud Parham,[2] Neda Habibi,[3] and Jamshid Vafaeimanesh ⓘ[3,4]

[1]*Department of Biochemistry, Qom University of Medical Sciences, Qom, Iran*
[2]*Clinical Research Development Center, Qom University of Medical Sciences, Qom, Iran*
[3]*Gastroenterology and Hepatology Disease Research Center, Qom University of Medical Sciences, Qom, Iran*
[4]*Gastrointestinal and Liver Diseases Research Center, Iran University of Medical Sciences, Tehran, Iran*

Correspondence should be addressed to Jamshid Vafaeimanesh; jvafaeemanesh@yahoo.com

Academic Editor: Javier S. Perona

Background. Some studies suggest a significant relationship between *Helicobacter pylori* infection and atherogenesis; but the mechanism of the relationship is almost unknown. The current study aimed at evaluating the relationship between *H. pylori* infection and serum lipid profile. *Patients and Methods.* The current study was conducted on 2573 patients, from 2008 to 2015. The serum anti-*Helicobacter pylori* antibody titer and serum lipid profile were assessed in the study population; data were statistically analyzed by SPSS version 16. P values < 0.05 were considered significant. *Results.* In the current study, 66.5% of the cases were serologically positive for *H. pylori*. Among male cases, the level of low density lipoprotein (LDL) was higher in patients with *H. pylori* infection, compared with that of the ones without the infection (P = 0.03); although level of triglyceride (TG) was higher and the level of high density lipoprotein (HDL) was lower in the cases with *H. pylori* infection; there was no statistically significant difference between the cases with and without *H. pylori* infection regarding the level of HDL and TG. Among female cases, the level of TG was significantly lower in patients with *H. pylori* infection, compared with that of the ones without the infection (P = 0.001); but there was no significant difference between the cases with and without *H. pylori* infection regarding the level of LDL and HDL. The mean fasting blood sugar (FBS) in the cases with *H. pylori* infection was significantly higher than that of the ones without the infection (P = 0.04). *Conclusion.* According to the results of the current study, the levels of LDL and FBS were high among the male cases with *H. pylori* infection. However, in females with *H. pylori* infection the level of TG was low; hence, it seems that the atherogenicity of *H. pylori* affected the level of blood sugar more.

1. Introduction

Coronary artery disease is one of the main causes of mortality in the developed countries. Diabetes, lipid profile disorders, hypertension, and smoking are the common risk factors of atherogenesis, which result in the coronary artery disease [1].

Although knowledge on the relationship between the classical risk factors of coronary artery disease has increased, the mechanisms involved in the mortality of the disease were not explained completely. Therefore, studying the relationship between the disease and other risk factors is of great importance.

Different studies investigated the relationship between ischemic heart disease and chronic infection. Some studies reported *H. pylori* infection as a risk factor for coronary artery disease [2, 3].

Helicobacter pylori infection is still one of the most common infections worldwide. A significant relationship was reported among *H. pylori* infection, metabolic risk factors, atherogenesis, and cardiovascular diseases in some of the studies [3–5].

It is important to find the mechanism of *H. pylori* pathogenicity in the incidence of cardiovascular diseases. Some of the animal model studies reported some findings in this regard; for example, Elizalde et al. indicated that *H. pylori* infection caused platelet aggregation in rats [6].

Also, Byrne et al. showed that *H. pylori* binds to von Willebrand factor (vWF), which causes platelet aggregation following the interaction with glycoprotein Ib [7].

TABLE 1: Demographic and Clinical Data of Female and Male Cases.

Variable	Male	Female	P value
Age (year)	42.5 ± 3	45 ± 4	0.002
FBS (mg/dL)	105 ± 3	106 ± 2	0.6
TG (mg/dL)	161 ± 3.5	135 ± 4	0.001
Cholesterol (mg/dL)	191.5 ± 10	191.8 ± 8	0.9
LDL (mg/dL)	111 ± 3	112 ± 4	0.89
HDL (mg/dL)	38.3 ± 2	43.7 ± 3	0.001
H. pylori seropositive (%)	68	65	0.8

FBS, fasting blood sugar; TG, triglyceride; LDL, low density lipoprotein; HDL, high density lipoprotein.

There are some other reports on the relationship between *H. pylori* infection and chronic progression of atherogenesis; but the mechanism is not identified perfectly yet [2].

Among metabolic risk factors, changes in serum lipid profiles such as low density lipoprotein (LDL) cholesterol and high density lipoprotein (HDL) cholesterol are known as the risk factors for cardiovascular diseases [8].

Different studies showed a significant relationship between the chronic infections and changes in the metabolism of lipids [9, 10]; some other studies also stated that *H. pylori* infection may change the level of serum lipid profile and increase the risk factors of atherogenesis [11].

Can *H. pylori* infection increase the risk of coronary artery diseases by changing serum lipid profile?

Some researchers reported findings contrary to the developed hypothesis [12, 13]. Considering the high prevalence of *H. pylori* infection in Iran [14], investigations on this issue are of great importance; also, because the relationship between lipid profile and *H. pylori* infection is not investigated precisely in the developing countries, the current study was designed, to determine whether Helicobacter pylori infection can be a risk factor for coronary heart disease by affecting the blood lipid profile.

2. Materials and Methods

The current study was conducted on 2573 Iranian cases without the history of *H. pylori* eradication therapy, with coronary artery disease, from 2008 to 2015. The age range of cases was 21 to 97 years (44.13 ± 15.48). Totally, 238 cases (98 male and 140 female) were excluded from the study as their serum triglyceride (TG) was > 400 mg/dL; 157 cases (69 male and 88 female) were also excluded from the study due to receiving cholesterol-lowering medication. Finally, results of 2178 cases were analyzed. The current study was approved in the Ethic Committee of Qom University of Medical Sciences, Qom, Iran, and the written consent was obtained from all cases. The case group was infected with Helicobacter pylori infection and the control group was non-infected.

2.1. Data Collection. To collect data in the current study, a questionnaire was used; the questionnaire included medical history, smoking status, exercising program, menstrual status, and alternative hormone replacement therapy. Smoking status included the continuous, sometimes, and non-smoker

options. Antecubital vein blood samples were drained during the morning fasting; serum samples were separated after centrifugation. HDL-cholesterol was measured by a commercial kit including magnesium chloride and phosphotungstate, after sedimentation of lipoproteins including apo B. TG was measured by an enzymatic method. Fasting blood sugar (FBS) was also measured employing the enzymatic method of hexokinase. LDL-cholesterol profile was calculated based on Friedewald formula [15].

C-reactive protein (CPR) was measured by nephelometry technique; amounts were measured to the nearest 0.004 mg/dL. Blood pressure was measured by a trained nurse using a mercury manometer, after 5-minute resting in sitting position. Serum *H. pylori* IgG titer was measured by an enzyme-linked immunosorbent assay (ELISA) kit (Padtan Elm Co., Iran); Ig G titers ≥ 22 U/mL were reported as positive. Statistical analysis was performed by SPSS version 16; P values < 0.05 were considered significant.

3. Results

The current study was conducted on 2178 cases from 2008 to 2015. Among the cases, 1205 were male and 1132 female. Demographic and clinical data of the cases are shown in Table 1.

According to Table 2, the mean FBS in male cases was significantly higher among the ones with *H. pylori* infection, compared with that of the ones without the infection [105 $\pm$ 8 mg/dL versus 101 $\pm$ 4 mg/dL] (P value = 0.04). The level of LDL was significantly higher among the case with *H. pylori* infection, compared to that of the ones without the infection (P value = 0.03); although higher levels of TG and lower levels of HDL were reported among *H. pylori* seropositive cases, there was no significant difference between the infected and non-infected cases regarding the level of TG and HDL (Table 2).

According to Table 3, in female cases, the level of FBS was significantly higher among *H. pylori* seropositive cases, compared with that of seronegative ones (P value = 0.04). Also, the level of TG was significantly lower among cases with *H. pylori* infection, compared with the ones without such infection (P value = 0.001); but the difference between *H. pylori* seropositive and seronegative cases, regarding the level of LDL and HDL, was insignificant (Table 3).

TABLE 2: Results of *Helicobacter pylori* Seropositive and Seronegative Male Cases.

Variable	*H. pylori* Seropositive Cases	H. pylori Seronegative Cases	P value
Age (year)	45 ± 3	40 ± 2	0.001
FBS (mg/dL)	105 ± 8	101 ± 4	0.04
TG (mg/dL)	171 ± 9	168 ± 6	0.7
Cholesterol (mg/dL)	192 ± 11	187 ± 9	0.07
LDL (mg/dL)	113 ± 4	109 ± 6	0.03
HDL (mg/dL)	38.5 ± 3	38.7 ± 4	0.7

FBS, fasting blood sugar; TG, triglyceride; LDL, low density lipoprotein; HDL, high density lipoprotein.

TABLE 3: Results of *Helicobacter pylori* in Seropositive and Seronegative Female Cases.

Variable	*H. pylori* Seropositive Cases	H. pylori Seronegative Cases	P value
Age (year)	48 ± 6	41 ± 4	0.001
FBS (mg/dL)	106 ± 9	102 ± 5	0.04
TG (mg/dL)	127 ± 6	155 ± 10	0.001
Cholesterol (mg/dL)	187 ± 12	186 ± 9	0.001
LDL (mg/dL)	110 ± 6	115 ± 8	0.06
HDL (mg/dL)	44 ± 4	43 ± 5	0.15

FBS, fasting blood sugar; TG, triglyceride; LDL, low density lipoprotein; HDL, high density lipoprotein.

4. Discussion

Helicobacter pylori infection is the most common bacterial infection worldwide, especially in the developing countries; its prevalence varies in different countries, 30% in the developed countries versus 80% in the developing countries [16]. The prevalence of *H. pylori* among the current study population was 66.5%.

Helicobacter pylori infection may develop many extra-intestinal complications. During the recent years, many studies were conducted on the relationship between *H. pylori* infection and atherosclerotic diseases such as ischemic heart disease [3].

Helicobacter pylori may play a role in the development of ischemic heart disease through different methods such as colonization of endothelial cells, changes in lipid profile, hypercoagulation, platelet aggregation, induction of molecular mimicry mechanisms, and progression of low-grade systemic inflammation [17].

The current study aimed at evaluating the relationship between *H. pylori* serological status and serum lipid profile. According to the results of the present study, 66.5% of total population was *H. pylori* seropositive; the prevalence of infection among male and female cases was 68% and 65%, respectively; it was 46.8% and 39.6% in a similar study [18].

According to the results of the current study, in male cases, *H. pylori* seropositivity resulted in significant increase of serum LDL (P value = 0.03); although higher levels of TG and lower levels of HDL were observed among *H. pylori* seropositive cases, there was no significant difference between seropositive and seronegative cases regarding the level of HDL and TG. Results of a study conducted in Japan indicated that *H. pylori* infection in Japanese male patients indirectly caused changes in serum lipid profile

including an increase in LDL-cholesterol and a decrease in HDL-cholesterol [18]. Also, previously performed studies showed that *H. pylori* infection is associated with lower levels of HDL-cholesterol in Europeans living in the USA [19, 20].

Results of a large epidemiological survey by Laurilaet et al. showed that the level of TG and total cholesterol was significantly higher and the level of HDL-cholesterol was significantly lower in male patients with *H. pylori* infection, compared with the ones without such infection [10]. In a similar study conducted in South Korea, the relationship between *H. pylori* infection and higher levels of LDL-cholesterol was reported [21].

According to the results of the current study, the level of TG was significantly lower in female cases with *H. pylori* infection than the female cases without the infection (P value = 0.001); however, the difference between *H. pylori* seropositive and seronegative female cases regarding the level of LDL and HDL was insignificant; the results were consistent with those of a study in Japan [18].

Results of a study showed that in patients with frequent infection with *H. pylori*, the level of LDL and HDL increases and decreases from the base levels, respectively, compared to those of the healthy cases [22]. Although result of a recently conducted study showed that successful eradication of *H. pylori* increases the risk of increased LDL and decreased HDL, it does not reduce the risk of cardiovascular diseases [22].

However, results of the current study and those of some other similar studies to some extent indicated the effect of *H. pylori* on lipid profile in patients with *H. pylori* infection, although results of the current study were not to the extent of justifying the atherogenic effects of *H. pylori*, and it seems that the bacteria induce their possible atherogenic effects through other mechanisms. Results of the current study indicated that the mean FBS in male cases with *H. pylori* infection was significantly higher than that of the ones without the infection [105 ± 8 mg/dL versus 101 ± 4 mg/dL] (*P* value = 0.04). Also, the FBS level was significantly higher in *H. pylori* seropositive female cases, compared with the seronegative ones (*P* value = 0.04); the current study results were consistent with those of other studies [23]. It seems that the bacteria induce insulin resistance through stimulation of systemic inflammation in patients with *H. pylori* infection; it can be considered as a mechanism in which *H. pylori* induces its atherogenic effects [24].

However, the mechanism in which accordingly *H. pylori* infection causes the increase in lipid profile is not identified completely. Results of an in vitro study (1992) showed that *H. pylori* can increase the absorption of cholesterol from serum and egg yolk; hence, it can be concluded that cholesterol binding to *H. pylori* can reduce absorption of dietary cholesterol [25].

Results of an experimental study showed that interleukin-(IL-) 8 is produced in the *H. pylori* infected mucosa more than normal range [26]. Production of IL-8 in *H. pylori* infection results in the stimulation of mucosa by oxidized LDL and monocytes, and then increases the immigration of lymphocytes-T to smooth muscle cells, and consequently leads to the production of plaque thrombosis [27].

Interleukin-10 (IL-10) is produced by mononuclear cells after the incidence of inflammation. HDL-cholesterol can regulate the production of cytokines by itself. Inflammation, by changing the level of HDL-cholesterol, stimulates the production of IL-10 by circulating mononuclear cells [28].

Some studies showed the positive effect of *H. pylori* eradication therapy on lipid profile; for example, successful eradication of *H. pylori* can reduce the risk of high LDL and low HDL [22].

The *H. pylori* eradication therapy increases apo A and HDL-cholesterol, while total cholesterol and LDL are not changed [28, 29]. A randomized clinical trial evaluated the long-term effect of garlic and nutritional supplements, with a 2-week antibiotic therapy regimen on the treatment of *h. pylori* infection, and lipoproteins and cholesterol profiles [30].

Briefly, at least some patients with *H. pylori* infection showed permanent or long-term complications of atherogenic lipid profile, which can accelerate the incidence of atherogenesis and many other complicated clinical diseases such as coronary heart disease, brain stroke, and peripheral artery occlusive disease [31].

5. Conclusion

According to the results of the current study, *H. pylori* can increase the level of LDL and FBS in seropositive male patients; the bacteria also play a role in the incidence of coronary artery disease through affecting atherogenic lipid profile and blood sugar level; but, in seropositive female patients, considering the lower levels of TG, it seems that the atherogenic effect of *H. pylori* mostly affects the level of blood sugar.

References

[1] A. Onat, I. Sari, G. Hergenç et al., "Predictors of abdominal obesity and high susceptibility of cardiometabolic risk to its increments among Turkish women: a prospective population-based study," *Metabolism - Clinical and Experimental*, vol. 56, no. 3, pp. 348–356, 2007.

[2] M. Berrutti, R. Pellicano, S. Fagoonee et al., "Potential relationship between Helicobacter pylori and ischemic heart disease: Any pathogenic model?" *Panminerva Medica*, vol. 50, no. 2, pp. 161–163, 2008.

[3] J. Vafaeimanesh, S. F. Hejazi, V. Damanpak, M. Vahedian, M. Sattari, and M. Seyyedmajidi, "Association of *Helicobacter pylori* infection with coronary artery disease: is *Helicobacter pylori* a risk factor?" *The Scientific World Journal*, vol. 2014, Article ID 516354, 6 pages, 2014.

[4] P. Libby, D. Egan, and S. Skarlatos, "Roles of infectious agents in atherosclerosis and restenosis: an assessment of the evidence and need for future research," *Circulation*, vol. 96, no. 11, pp. 4095–4103, 1997.

[5] G. H. Dahlén, J. Boman, L. S. Birgander, and B. Lindblom, "Lp(a) lipoprotein, IgG, IgA and IgM antibodies to Chlamydia pneumoniae and HLA class II genotype in early coronary artery disease," *Atherosclerosis*, vol. 114, no. 2, pp. 165–174, 1995.

[6] J. I. Elizalde, J. Gómez, J. Panés et al., "Platelet activation in mice and human Helicobacter pylori infection," *The Journal of Clinical Investigation*, vol. 100, no. 5, pp. 996–1005, 1997.

[7] M. F. Byrne, S. W. Kerrigan, P. A. Corcoran et al., "Helicobacter pylori binds von Willebrand factor and interacts with GPIb to induce platelet aggregation," *Gastroenterology*, vol. 124, no. 7, pp. 1846–1854, 2003.

[8] J. Sasaki, T. Kita, H. Mabuchi et al., "Gender difference in coronary events in relation to risk factors in Japanese hypercholesterolemic patients treated with low-dose simvastatin," *Circulation Journal*, vol. 70, no. 7, pp. 810–814, 2006.

[9] F. J. Nieto, P. Sorlie, G. W. Comstock et al., "Cytomegalovirus infection, lipoprotein(a), and hypercoagulability: An atherogenic link?" *Arteriosclerosis, Thrombosis, and Vascular Biology*, vol. 17, no. 9, pp. 1780–1785, 1997.

[10] A. Laurila, A. Bloigu, S. Näyhä, J. Hassi, M. Leinonen, and P. Saikku, "Association of Helicobacter pylori infection with elevated serum lipids," *Atherosclerosis*, vol. 142, no. 1, pp. 207–210, 1999.

[11] P. Cullen and G. Assmann, "High risk strategies for atherosclerosis," *Clinica Chimica Acta*, vol. 286, pp. 31–45, 1999.

[12] J. Danesh and R. Peto, "Risk factors for coronary heart disease and infection with Helicobacter pylori: Meta-analysis of 18 studies," *BMJ*, vol. 316, no. 7138, pp. 1130–1132, 1998.

[13] J. Zhu, A. A. Quyyumi, J. B. Muhlestein et al., "Lack of association of Helicobacter pylori infection with coronary artery disease and frequency of acute myocardial infarction or death," *American Journal of Cardiology*, vol. 89, no. 2, pp. 155–158, 2002.

[14] M. Moosazadeh, K. Lankarani, and M. Afshari, "Meta-analysis of the prevalence of helicobacter pylori infection among children and adults of Iran," *International Journal of Preventive Medicine*, vol. 7, no. 1, p. 48, 2016.

[15] J. Knopfholz, C. C. D. Disserol, A. J. Pierin et al., "Validation of the friedewald formula in patients with metabolic syndrome," *Cholesterol*, vol. 2014, Article ID 261878, 5 pages, 2014.

[16] J. C. Atherton, "The pathogenesis of *Helicobacter pylori*-induced gastro-duodenal diseases," *Annual Review of Pathology: Mechanisms of Disease*, vol. 1, pp. 63–96, 2006.

[17] E. Vizzardi, I. Bonadei, B. Piovanelli et al., "Helicobacter Pylori and ischemic heart disease," *Panminerva Medica*, vol. 53, no. 3, pp. 193–202, 2011.

[18] H. Satoh, Y. Saijo, E. Yoshioka, and H. Tsutsui, "Helicobacter Pylori Infection is a Significant Risk for Modified Lipid Profile in Japanese Male Subjects," *Journal of Atherosclerosis and Thrombosis*, vol. 17, no. 10, pp. 1041–1048, 2010.

[19] A. Hoffmeister, D. Rothenbacher, G. Bode et al., "Current infection with helicobacter pylori, but not seropositivity to Chlamydia pneumoniae or Cytomegalovirus, is associated with an atherogenic, modified lipid profile," *Arteriosclerosis, Thrombosis, and Vascular Biology*, vol. 21, no. 3, pp. 427–432, 2001.

[20] G. Chimienti, F. Russo, B. L. Lamanuzzi et al., "Helicobacter pylori is associated with modified lipid profile: Impact on Lipoprotein(a)," *Clinical Biochemistry*, vol. 36, no. 5, pp. 359–365, 2003.

[21] H.-L. Kim, H. H. Jeon, I. Y. Park, J. M. Choi, J. S. Kang, and K.-W. Min, "Helicobacter pylori infection is associated with elevated low density lipoprotein cholesterol levels in elderly Koreans," *Journal of Korean Medical Science*, vol. 26, no. 5, pp. 654–658, 2011.

[22] S. Y. Nam, K. H. Ryu, B. J. Park, and S. Park, "Effects of *Helicobacter pylori* infection and its eradication on lipid profiles and cardiovascular diseases," *Helicobacter*, vol. 20, no. 2, pp. 125–132, 2015.

[23] J. Vafaeimanesh, M. Parham, and M. Bagherzadeh, "Helicobacter pylori infection prevalence: Is it different in diabetics and nondiabetics?" *Indian Journal of Endocrinology and Metabolism*, vol. 19, no. 3, pp. 364–368, 2015.

[24] J. Vafaeimanesh, M. Parham, M. Seyyedmajidi, and M. Bagherzadeh, "*Helicobacter pylori* infection and insulin resistance in diabetic and nondiabetic population," *The Scientific World Journal*, vol. 2014, Article ID 391250, 5 pages, 2014.

[25] R. Ansorg, K.-D. Müller, G. Von Recklinghausen, and H. P. Nalik, "Cholesterol binding of Helicobacter pylori," *Zentralblatt für Bakteriologie*, vol. 276, no. 3, pp. 323–329, 1992.

[26] J. E. Crabtree, J. I. Wyatt, L. K. Trejdosiewicz et al., "Interleukin-8 expression in Helicobacter pylori infected, normal, and neoplastic gastroduodenal mucosa," *Journal of Clinical Pathology*, vol. 47, no. 1, pp. 61–66, 1994.

[27] Y. Liu, L. M. Hultén, and O. Wiklund, "Macrophages isolated from human atherosclerotic plaques produce IL-8, and oxysterols may have a regulatory function for IL-8 production," *Arteriosclerosis, Thrombosis, and Vascular Biology*, vol. 17, no. 2, pp. 317–323, 1997.

[28] M. T. Coronado, A. O. Pozzi, M. A. Punchard, P. González, and P. Fantidis, "Inflammation as a modulator of the HDL cholesterol-induced inteleukin-10 production by human circulating mononuclear cells," *Atherosclerosis*, vol. 202, no. 1, pp. 183-184, 2009.

[29] H. Scharnagl, M. Kist, A. B. Grawitz, W. Koenig, H. Wieland, and W. März, "Effect of Helicobacter pylori eradication on high-density lipoprotein cholesterol," *American Journal of Cardiology*, vol. 93, no. 2, pp. 219-220, 2004.

[30] L. Zhang, M. H. Gail, Y.-Q. Wang et al., "A randomized factorial study of the effects of long-term garlic and micronutrient supplementation and of 2-wk antibiotic treatment for Helicobacter pylori infection on serum cholesterol and lipoproteins," *American Journal of Clinical Nutrition*, vol. 84, no. 4, pp. 912–919, 2006.

[31] G. M. Buzás, "Metabolic consequences of Helicobacter pylori infection and eradication," *World Journal of Gastroenterology*, vol. 20, no. 18, pp. 5226–5234, 2014.

8

The Correlation of Dyslipidemia with the Extent of Coronary Artery Disease in the Multiethnic Study of Atherosclerosis

Moshrik Abd alamir ⓘ,[1] Michael Goyfman,[2] Adib Chaus,[1] Firas Dabbous,[3] Leslie Tamura,[1] Veit Sandfort,[4] Alan Brown,[1] and Mathew Budoff ⓘ[5]

[1]Advocate Lutheran General Hospital, Parkside Ste B-01, 1775 Dempster St, Park Ridge, IL 60068, USA
[2]Stony Brook University Hospital, 101 Nicolls Rd, Stony Brook, NY 11794, USA
[3]James R. & Helen D. Russell Institute for Research & Innovation, Advocate Lutheran General Hospital,
  Center for Advanced Care, 1700 Luther Lane, Suite 1410, Park Ridge, IL 60068, USA
[4]National Institute of Health, 30 Convent Dr, Bethesda, MD 20892, USA
[5]Harbor UCLA Cardiology, 1000 W. Carson St, Torrance, CA 90509, USA

Correspondence should be addressed to Moshrik Abd alamir; moshrik.abdalamir@advocatehealth.com

Academic Editor: Gerhard M. Kostner

Background. The extent of coronary artery calcium (CAC) improves cardiovascular disease (CVD) risk prediction. The association between common dyslipidemias (combined hyperlipidemia, simple hypercholesterolemia, metabolic Syndrome (MetS), isolated low high-density lipoprotein cholesterol, and isolated hypertriglyceridemia) compared with normolipidemia and the risk of multivessel CAC is underinvestigated. *Objectives.* To determine whether there is an association between common dyslipidemias compared with normolipidemia, and the extent of coronary artery involvement among MESA participants who were free of clinical cardiovascular disease at baseline. *Methods.* In a cross-sectional analysis, 4,917 MESA participants were classified into six groups defined by specific LDL-c, HDL-c, or triglyceride cutoff points. Multivessel CAC was defined as involvement of at least 2 coronary arteries. Multivariate Poisson regression analysis evaluated the association of each group with multivessel CAC after adjusting for CVD risk factors. *Results.* Unadjusted analysis showed that all groups except hypertriglyceridemia had statistically significant prevalence ratios of having multivessel CAC as compared to the normolipidemia group. The same groups maintained statistical significance prevalence ratios with multivariate analysis adjusting for other risk factors including Agatston CAC score [combined hyperlipidemia 1.41 (1.06–1.87), hypercholesterolemia 1.55 (1.26–1.92), MetS 1.28 (1.09–1.51), and low HDL-c 1.20 (1.02–1.40)]. *Conclusion.* Combined hyperlipidemia, simple hypercholesterolemia, MetS, and low HDL-c were associated with multivessel coronary artery disease independent of CVD risk factors and CAC score. These findings may lay the groundwork for further analysis of the underlying mechanisms in the observed relationship, as well as for the development of clinical strategies for primary prevention.

1. Introduction

The total CAC score (Agatston score) using noncontrast computed tomography is a recognized estimation of atherosclerosis in asymptomatic adults with at least moderate risk of cardiovascular disease [1–3]. When compared with a risk-stratification tool such as the Framingham Risk Score (FRS), the CAC score has been shown to have a superior role in predicting future cardiac events and all-cause mortality [2, 4].

Previous studies demonstrated that measurement of coronary artery calcium stratified patient risk for cardiovascular disease regardless of dyslipidemia burden or definition [5]. However, the relationship between subclinical atherosclerosis and dyslipidemia type has been investigated using either the prevalent CAC or carotid thickness as surrogate markers [5, 6]. Although the traditional Agatston CAC score is a powerful predictor of mortality, there is an emerging evidence that extent of subclinical atherosclerosis, as indicated by the number of vessels with CAC, further improves cardiovascular risk prediction [2, 7]. With the current ability to identify those with multivessel disease at an asymptomatic stage using the CAC, the healthcare providers could potentially tailor the diagnostic and therapeutic approach for individual patients based on the extent of CAC. Therefore, the purpose of this study was to determine whether there is an association between common dyslipidemias (combined hyperlipidemia, simple hypercholesterolemia, MetS, isolated low high-density lipoprotein cholesterol and isolated hypertriglyceridemia) compared with normolipidemia, and the extent of coronary artery involvement (multivessel CAC), among MESA participants who were free of clinical cardiovascular disease at baseline. It remains to be seen whether the association between dyslipidemia types and the CAD extent still exists after controlling for the absolute calcium score.

2. Methods

2.1. Study Cohort. The Multiethnic Study of Atherosclerosis (MESA) is a prospective observational evaluation of 6,814 men and women, aged 45 to 84 years, from four racial/ethnic groups (White, Asian, Hispanic, and Black) in the United States. At the time of enrollment, participants had no known cardiovascular disease.

Participants were enrolled between July 2000 and September 2002 at field centers located in Forsyth County, North Carolina; St. Paul, Minnesota; Chicago; New York City; Baltimore, Maryland; and Los Angeles. Institutional review boards approved the study protocol at each study center. Further details regarding the MESA study have been detailed elsewhere [8].

Study participants provided information about cardiovascular risk factors. A central laboratory (University of Vermont, Burlington, Vermont) measured levels of total and high-density lipoprotein cholesterol, triglycerides, plasma glucose, and high-sensitivity C-reactive protein (CRP) after a 12-hour fast. The coronary calcium scan was done twice in each participant to increase accuracy. For each scan, the participant was asked to remain still and momentarily hold his/her breath twice, each time for 20 to 30 seconds, in order to get good quality pictures. Participants' CAC scores were reported as average CAC Agatston scores. Vessel-specific CAC scores were calculated in 6,540 MESA participants (96%). Of those, 6,479 MESA participants with coronary calcium CT scans at baseline and vessel-specific CAC distribution (left main, left anterior descending, left circumflex, and right) were screened. Included in the final study cohort were 4,917 participants, after exclusion of 1562 participants who were taking lipid-lowering medication as well as diabetics and those who lacked measurements of dyslipidemia and/or CAC. Individuals with triglycerides > 500 were not specifically excluded from the analysis. However, in those cases it is usually not possible to calculate LDL-c if using the Friedewald formula, and patients without LDL-c values would have been excluded from our analysis.

Cardiovascular risk factors were measured or collected, and included height, weight, and waist circumference, medical history including presence of diabetes (using the 2003 American Diabetes Association criteria), hypertension (defined as systolic blood pressure > 139 mm Hg at baseline visit, or diastolic blood pressure > 89 mm Hg, or by a history of physician diagnosed hypertension and taking a medication for hypertension), and assessment of personal habits such as alcohol and tobacco use [9–11].

2.2. Exposure Variables. The following cardiovascular risk factors were collected at MESA field centers: height, weight, waist circumference, alcohol and tobacco use, family history of heart attack, CAC score, diabetes, and hypertension (systolic blood pressure ≥ 140 mmHg at baseline visit, diastolic blood pressure ≥ 90 mmHg, or a history of taking an antihypertensive medication). Age and race/ethnicity were self-reported. Lipids, including total and high-density lipoprotein cholesterol, triglycerides, inflammatory markers, and glucose levels, were measured from fasting plasma samples in a central laboratory (University of Vermont, Burlington, VT). Venous blood samples were collected after a 12 h fast by certified technicians using standardized venipuncture procedures. Samples were then centrifuged at 2000*g* for 15 min at 4°C within 30 min of collection. EDTA plasma samples were aliquoted on ice, stored at −70°C, and then shipped on dry ice to the MESA central laboratory at University of Vermont [12].

2.3. Dyslipidemia. Table 1 identifies the various HDL-c, LDL-c, and triglyceride categories defining the six mutually exclusive dyslipidemias, including normolipidemia as a reference group.

We created these dyslipidemia groups using criteria based on current National Cholesterol Education Program (NCEP)/Adult Treatment Panel- (ATP-) III guidelines that define LDL-C, HDL-C, and triglyceride thresholds as abnormal [13].

Participants were classified based on the most severe dyslipidemia. For example, someone would be classified in the only MetS if the person had low HDL-c and elevated triglycerides; the person would not be in the low HDL-c group. To define the subtype of dyslipidemia appropriately, we had to exclude diabetes and lipid-lowering therapy from this analysis. Diabetes is considered a coronary heart disease risk equivalent and there is a strong independent relationship of diabetes with low HDL-c, elevated triglycerides, and CAC extent. Participants receiving lipid-lowering therapy were excluded because lipid lowering has a substantial impact on all lipid parameters as well as on CAC. Fasting triglycerides were measured in plasma using a glycerol blanked enzymatic method developed by Trig/GB (Roche Diagnostics, Indianapolis, Indiana). Cholesterol and HDL-c were measured in plasma on the Hitachi 911 using a cholesterol esterase, cholesterol oxidase reaction (Chol R1, Roche Diagnostics). For triglyceride levels < 400 mg/dL, the LDL-c was calculated using the Friedewald formula; otherwise nuclear magnetic resonance spectroscopy was used for triglycerides > 400 mg/dl [9, 14]. Serum glucose was measured by rate reflectance spectrophotometry using thin film adaptation of the glucose oxidase method on the Vitros analyzer (Johnson & Johnson Clinical Diagnostics Inc., Rochester, NY). The laboratory analytical

TABLE 1: Lipid categories defined by HDL, LDL, and triglyceride levels.

Dyslipidemia category	HDL (mg/dL)	LDL-c (mg/dL)	Triglycerides (mg/dL)
Normolipidemia	>40 men >50 women	<160	<150
Combined hyperlipidemia	No cutoff	≥160	≥150
Hypercholesterolemia	No cutoff	≥160	<150
Dyslipidemia compatible with metabolic syndrome (MetS)	≤40 men ≤50 women	<160	≥150
Low HDL-c	≤40 men ≤50 women	<160	<150
Hypertriglyceridemia	>40 men >50 women	<160	≥150

CV was 1.1%. Insulin was determined by a radioimmunoassay method using the Linco Human Insulin Specific RIA Kit (Linco Research Inc., St. Charles, MO). The laboratory analytical CV was 4.9%. CRP was measured in the Laboratory for Clinical Biochemistry Research at University of Vermont (Burlington, VT). CRP was determined using the BNII nephelometer (N High-Sensitivity CRP and N Antiserum to Human Fibrinogen; Dade Behring Inc., Deerfield, IL) [12].

2.4. Coronary Artery Calcium (CAC) Extent. Electron-beam computed tomography (EBT) or multidetector row helical computed tomography (MDCT) was used to measure CAC, defined by a minimum of 130 Hounsfield units [10]. The effective dose of radiation for CAC scoring was approximately 1 mSv [11]. The protocol and interpretation of CAC scans in the MESA study have been reported previously [10]. Interobserver and intraobserver agreement were high [15]. The Agatston scoring method quantified baseline CAC [16], and scores were adjusted with a standard biweekly phantom scan to ensure equivalence among sites [15].

Extent of CAC was analyzed according to the number of main coronary arteries (left main, left anterior descending, left circumflex, and right) with calcification ranging from 0 to 4. Multivessel CAC was defined as involvement of at least 2 coronary arteries. This included three-vessel CAC, which was defined as involvement of the left main or left anterior descending coronary artery in addition to CAC in the left circumflex and the right coronary arteries. Single vessel CAC was classified as a distinct entity apart from no CAC and from multivessel disease. Our statistical analysis focused on the relationship between dyslipidemias and specifically multivessel disease, as relationships between dyslipidemias and CAC score in general have previously been described [6].

2.5. Statistical Analysis. A cross-sectional sample of participants from the MESA cohort was classified into 6 mutually exclusive dyslipidemia categories (including "normal" as reference group) based on their levels of LDL-c, HDL-c, and triglycerides (Table 1). Differences in baseline demographic and cardiac risk factor data (age, race/ethnicity, gender, clinical site, education, history of hypertension, current smoking status, alcohol use, waist circumference, fasting glucose, fasting insulin, CRP, and creatinine) were evaluated across the 6 lipid categories, using the Chi-Square test for categorical variables and ANOVA for continuous numerical variables. The latter were reported as mean/standard deviation, with the exception of skewed variables including CAC score, serum insulin, CRP, and triglyceride levels, which were reported as median/interquartile range. Multivariate Poisson regression analysis was performed to assess the relationship between multivessel CAC and the type of dyslipidemia (including normal). This was performed both unadjusting and adjusting for the aforementioned demographic and cardiac risk factor data, including the total phantom-adjusted coronary artery calcium (CAC) score. We chose widely accepted parameters associated with cardiovascular risk that were also available in the MESA database. Final model was adjusted for age, gender, race, high-school education, smoking, hypertension, waist circumference, serum glucose level, serum insulin, serum CRP level, and Agatston's calcium score. All variables were adjusted simultaneously. The model based on the chosen variables reflects the well-recognized risk factors associated with CAC score. All the current variables in our final model are literature-derived risk factors. We performed the forward and backward selection to determine the composition of variables for a good model fit and there was no difference with either method. In addition, the *p* value for interaction between different variables such as gender, race/ethnicity, and CRP and lipid variables was not significant for common or internal CIMT or prevalent CAC in the previously published MESA analysis by Paramsothy et al. [6]. A two-sided *p* value < 0.05 was considered to be significant. Data were analyzed using SAS version 9.3 (SAS Institute, Cary, NC).

3. Results

3.1. Baseline Characteristics. The baseline characteristics of the study cohort are shown in Table 2.

The majority of the cohort consisted of white (39%) and female (53%) participants, with an average age of 61.6 years. Most participants completed at least a high-school level of education (83%) and currently used alcohol (58%). Only 14% of participants were current smokers. Approximately 39% of the cohort had a history of hypertension, and 53% of the cohort had one of the five types of dyslipidemia.

TABLE 2: Baseline characteristics and cardiac risk factors of MESA cohort excluding those on lipid-lowering medications and diabetics (collected 2000–2002).

Variable	All	Normal	Combined	HC	MetS	Low HDL	HTG	p value
N (%)	4917	2329 (47.4)	172 (3.5)	345 (7.0)	793 (16.1)	907 (18.5)	371 (7.6)	
Age (SD)*	61.6 (10.3)	62.3 (10.5)	61.9 (9.9)	61.4 (9.5)	60.8 (10.3)	60.8 (10.6)	61.5 (9.4)	0.015
Male gender (%)*	2311 (47.0)	1054 (45.3)	74 (43.0)	155 (44.9)	438 (55.2)	406 (44.8)	184 (49.6)	<0.001
Race/ethnicity*								
White (%)	1930 (39.3)	960 (41.2)	70 (40.7)	135 (39.1)	311 (39.2)	282 (31.1)	172 (46.4)	<0.001
Asian (%)	616 (12.5)	279 (12.0)	20 (11.6)	31 (9.0)	126 (15.9)	117 (12.9)	43 (11.6)	
Black (%)	1285 (26.1)	700 (30.1)	24 (14.0)	113 (32.8)	96 (12.1)	303 (33.4)	49 (13.2)	
Hispanic (%)	1086 (22.1)	390 (16.8)	58 (33.7)	66 (19.1)	260 (32.8)	205 (22.6)	107 (28.8)	
Completed high school (%)*	4079 (83.0)	2029 (87.1)	136 (79.1)	285 (82.6)	609 (76.8)	730 (80.5)	290 (78.2)	<0.001
Current smoker (%)*	690 (14.0)	296 (12.7)	21 (12.2)	46 (13.3)	143 (18.0)	129 (14.2)	55 (14.8)	0.011
Hypertension (%)*	1903 (38.7)	874 (37.5)	62 (36.1)	127 (36.8)	339 (42.8)	336 (37.1)	165 (44.5)	0.014
Waist circumference, cm (SD)*	96.8 (14.0)	93.6 (14.4)	100.2 (12.4)	97.0 (12.7)	101.4 (12.6)	99.3 (13.5)	99.0 (13.1)	<0.001
Fasting glucose, mg/dL (SD)*	91.1 (19.2)	88.5 (13.6)	95.8 (26.0)	92.0 (21.1)	96.4 (27.9)	91.8 (18.9)	92.0 (19.3)	<0.001
Insulin, mU/L*	7.9 (11.4–5.8)	6.7 (9.1–5.0)	9.1 (13.6–6.9)	7.7 (11.0–5.8)	10.9 (14.9–8.0)	9.1 (12.6–6.5)	8.6 (12.5–6.5)	<0.001
CRP, mg/L*	1.9 (4.2–0.8)	1.5 (3.7–0.7)	2.4 (5.7–1.2)	1.9 (4.2–1.0)	2.3 (4.4–1.1)	2.2 (4.8–0.9)	2.4 (5.8–1.1)	0.0001
Creatinine, mg/dL (SD)	0.95 (0.22)	0.94 (0.20)	0.94 (0.19)	0.96 (0.19)	0.96 (0.24)	0.94 (0.26)	0.94 (0.23)	0.352
Total cholesterol, mg/dL (SD)*	197.1 (35.3)	190.4 (27.1)	262.9 (23.2)	250.0 (22.2)	197.1 (33.4)	175.7 (26.5)	212.0 (30.6)	<0.0001
LDL, mg/dL (SD)*	120.4 (31.0)	112.7 (24.4)	177.2 (18.4)	177.2 (18.8)	113.8 (25.7)	114.6 (24.0)	116.3 (25.1)	<0.0001
HDL, mg/dL (SD)*	51.5 (15.0)	60.7 (14.2)	44.7 (8.9)	51.9 (11.6)	37.4 (6.0)	40.3 (6.1)	54.2 (10.4)	<0.0001
Triglycerides, mg/dL*	108 (156–76)	82 (105–63)	189 (216–171)	106 (126–85)	203 (255–172)	105 (127–83)	185 (221–162)	<0.0001
Median CAC score with range	0 (0–6252)	0 (0–6252)	4.08 (0–2791)	3.74 (0–2348)	0 (0–2946)	0 (0–3358)	0 (0–2867)	<0.001
CAC score > 0(%)*	2778 (56.50)	1397 (59.9)	82 (47.6)	161 (46.6)	412 (51.9)	516 (56.8)	210 (56.6)	<0.001
Number of calcified vessels								
0 (%)*	2778 (56.4)	1397 (59.9)	82 (47.6)	161 (46.6)	412 (51.9)	516 (56.8)	210 (56.6)	<0.0001
1 (%)*	889 (18.0)	401 (17.2)	36 (20.9)	79 (22.9)	139 (17.5)	158 (17.4)	76 (20.4)	
2 (%)*	622 (12.6)	264 (11.3)	18 (10.4)	54 (15.6)	118 (14.8)	131 (14.4)	37 (9.9)	<0.0001
3 (%)*	480 (9.7)	204 (8.7)	27 (15.7)	42 (12.1)	89 (32.8)	82 (9.0)	36 (9.7)	
4 (%)*	148 (3.0)	63 (2.7)	9 (5.2)	9 (2.6)	35 (4.4)	20 (2.2)	12 (3.2)	
Multivessel CAC (%)*	1250 (25.4)	531 (22.8)	54 (31.4)	105 (30.4)	242 (30.5)	233 (25.6)	85 (22.9)	<0.0001

* $p < 0.05$ for categorical variables using Chi-Squared test or for continuous variables using ANOVA. Percentages may not add up to 100% due to rounding. Continuous variables reported using mean (standard deviation) or median (interquartile range). Normal = normolipidemia, Combined = combined hyperlipidemia, HC = hypercholesterolemia, MetS = metabolic syndrome dyslipidemia, and HTG = hypertriglyceridemia.

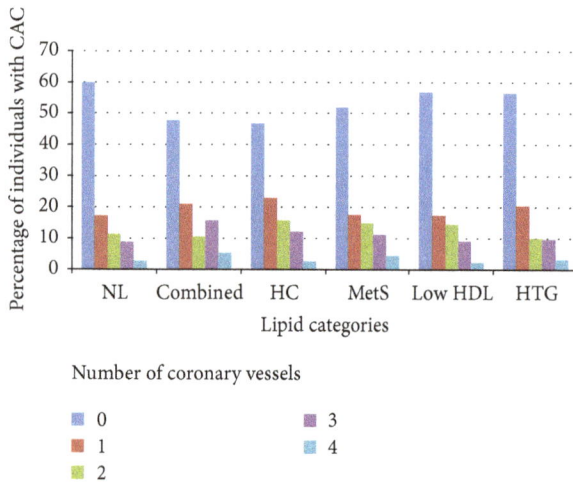

FIGURE 1: Number of coronary vessels with calcification as a function of lipid profiles. The between-group differences in the number of affected vessels with CAC, showing significant between-group differences ($p < 0.001$). NL = normolipidemia, Combined = combined hyperlipidemia, HC = hypercholesterolemia, MetS = dyslipidemia compatible with metabolic syndrome, HTG = hypertriglyceridemia, and CAC = coronary artery calcification.

TABLE 3: Unadjusted prevalence ratio of multivessel CAC as a function of lipid groups.

Lipid group	PR (95% CI)
Normolipidemia	Ref group
Combined hyperlipidemia*	1.37 (1.04–1.82)
Hypercholesterolemia*	1.33 (1.08–1.64)
Metabolic syndrome dyslipidemia*	1.33 (1.14–1.55)
Low HDL-c	1.12 (0.96–1.31)
Hypertriglyceridemia	1.00 (0.79–1.26)

*Statistically significant at $p < 0.05$.

3.2. Dyslipidemia. The low HDL-c dyslipidemia group was the most common type of dyslipidemia, followed by the MetS group. The latter had the largest waist circumference (Table 2).

3.3. CAC Findings. A comparison was performed using Chi-Square test between the lipid groups and the number of affected vessels with CAC, showing significant between-group differences ($p < 0.001$, Figure 1).

The normolipidemia group had the highest percentage of individuals with zero affected vessels (60%), whereas the combined hyperlipidemia group had the highest percentage of those with three- and four-vessel calcification (16% and 5%, resp.).

Unadjusted univariate Poisson regression analysis showed that combined hyperlipidemia, simple hypercholesterolemia, and dyslipidemia of metabolic syndrome had a statistically significant likelihood of having a multivessel CAC as compared to the normolipidemia reference group (Table 3).

By contrast, there was no statistically significant difference between the low HDL-c, hypertriglyceridemia, and

normolipidemia groups in the prevalence of multivessel CAC in the unadjusted model. Subsequent multivariate Poisson regression analysis adjusting for the demographic and cardiac risk factors, including Agatston calcium score, showed that the same lipid groups (combined hyperlipidemia, simple hypercholesterolemia, and dyslipidemia of metabolic syndrome) and the multivessel CAC maintained statistical significance in the model, compared to the normolipidemia reference group (Table 4). Interestingly, the previously nonsignificant HDL have become significant in the adjusted model. Furthermore, higher Agatston calcium score, age, male gender, Asian race, currently smoking, hypertension, and waist circumference were all significantly associated with multivessel disease. The other demographic and cardiac risk factors previously mentioned were not found to have a significant relationship with multivessel disease.

4. Discussion

Previous literature has shown that different types of dyslipidemia have varying association with CAC scores [2, 3, 6, 17]; however, our study elucidates a relationship between the different dyslipidemia types and the extent of coronary artery disease, defined as the rate of multivessel CAC. We focus on whether or not the heterogeneities of baseline subclinical atherosclerotic extent have any relationship with dyslipidemia type and, if so, which dyslipidemia is associated with increased rates of multivessel CAC. As expected, the majority of the normolipidemia group had no CAC-affected vessels. In comparison, all dyslipidemia groups except for the hypertriglyceridemia were associated with higher rates of multivessel CAC. With further adjustment for demographic and cardiac risk factors, the same dyslipidemia groups were still associated with multivessel CAC, even after controlling for the total CAC score. An earlier study by Paramsothy et al. showed that, of 4,795 MESA participants, without known clinical cardiovascular disease, those with combined hyperlipidemia, hypercholesterolemia, and MetS had increased relative risk for prevalent CAC [6] compared with normolipidemia participants when adjusting for demographic and CVD risk factors.

Similar to Paramsothy et al., we found that isolated hypertriglyceridemia was not associated with CAC extent [6]. These findings suggest that an isolated elevation in triglyceride levels may not be a pathologic risk factor for subclinical atherosclerosis, although hypertriglyceridemia may still be an important factor in cardiovascular disease. The hypertriglyceridemia group in this study also had a relatively high HDL-c of 53.6 mg/DL, which may have inversely affected overall CAC prevalence [6].

In contrast to the Paramsothy study where no significant association was found between low HDL-c and prevalent CAC, we found an association between low HDL-c and increased extent of CAC. The prevalence of low HDL-c group and adjusted risk factors were identical in both studies. In their study of 6093 participants, Allison and Wright showed that the individuals with an HDL-c level < 40 mg/dl had significantly higher calcium scores while increases in HDL-c were associated with a significant reduction in risk for

TABLE 4: Adjusted prevalence ratio of multivessel CAC as a function of lipid groups.

	Adjusted PR (95% CI)
Lipid group	
Normolipidemia	Ref group
Combined hyperlipidemia*	1.41 (1.06–1.87)
Hypercholesterolemia*	1.55 (1.26–1.92)
Metabolic syndrome dyslipidemia*	1.28 (1.09–1.51)
Low HD-c*	1.20 (1.02–1.40)
Hypertriglyceridemia	1.05 (0.83–1.33)
*Age (years)**	1.05 (1.05–1.06)
*Male sex**	1.71 (1.49–1.96)
Race	
White	Ref group
Asian*	1.23 (1.04–1.45)
Black	1.11 (0.88–1.39)
Hispanic	0.89 (0.74–1.07)
High school education	1.05 (0.89–1.23)
*Current smoker**	1.32 (1.12–1.55)
*Hypertension**	1.19 (1.05–1.34)
*Waist circumference (cm)**	1.01 (1.0004–1.010757)
Serum glucose level (mg/dL)	0.99 (0.99–1.00)
Serum insulin level (mu/L)	1.00 (0.99–1.01)
Serum CRP level (mg/L)	1.00 (0.99–1.01)
Serum creatine level (mg/dL)	0.96 (0.74–1.24)
*Agatston's calcium score**	1.00 (1.0004–1.0005)

Final model adjusting for age, gender, race, high school education, smoking, hypertension, waist circumference, serum glucose level, serum insulin, serum CRP level, and Agatston's calcium score. All variables are adjusted simultaneously. *Significant at $p < 0.01$.

the presence of any calcified plaque. Multivariate logistic regression revealed that HDL-c is predictive of calcified plaque development independent of LDL-c. However, sensitivity and positive predictive values for HDL-c were low [18]. Furthermore, Noda et al. studied two hundred and eighty-nine consecutive patients who underwent 64-slice multidetector CT for suspected coronary artery disease. They found that HDL-c cholesterol levels were more accurate for diagnosing the presence of high-risk coronary plaque with areas under receiver operating curve (AUC) of 0.840 in patients with CCS 0, than with AUC of 0.633 in those with CCS 1 to 10, 0.605 in those with CCS 11 to 100, 0.591 in those with CCS 101 to 400, and 0.571 in those with CCS > 400 [19]. Therefore, in participants with lower CAC score, low HDL-c may have more influence on the extent of CAC than those with higher calcium score. This may explain the significance of HDL-c levels found in our study, since, unlike Paramsothy et al., we controlled for CAC score and focused on the extent of CAC.

Combined hyperlipidemia, simple hypercholesterolemia, and participants with MetS had significantly increased risk of multivessel involvement of the coronary arteries in patients with subclinical CAD [20]. This is in line with Paramsothy et al.'s findings where the same dyslipidemia groups were associated with prevalent CAC. Given these findings, elevated LDL-c seems to be the principal determinant of CAC prevalence and its extent, with high TG levels having less influence [6, 17, 20, 21]. Of notice, in the multivariate analysis, the confidence interval of the combined dyslipidemia group is rather wide because the numbers of participants with this disorder are relatively small, 172 (3.5%), influencing the precision of the population estimate.

Atherogenic dyslipidemia is largely underdiagnosed and undertreated in clinical practice per the findings of Fruchart et al. in the Residual Risk Reduction Initiative (R3i). These are prevalent in patients with type 2 diabetes, metabolic syndrome, and/or established CVD. Especially in MetS, these patients commonly had elevated ApoB levels, smaller LDL particle size, and elevated ApoCIII levels and as such metabolic syndrome is associated with residual CVD risk [22].

Gender may also contribute to CAC extent, as we found CAC extent was significantly increased among men. Male gender is a recognized independent risk factor for coronary heart disease by the Framingham Risk Score. In addition, in their study of 6814 participants, McClelland et al. found that men had greater calcium levels than women [15]. The same study showed that calcium amount and prevalence were steadily higher with increasing age and in Whites, whereas Asians had the highest ratio of multivessel CAC in our study.

Coronary artery calcium, a known marker of coronary atherosclerotic plaque, has been consistently associated with cardiovascular morbidity and mortality [2, 15, 17, 23, 24]. The clinical significance of the demonstrated association between dyslipidemia types and the extent of CAC, controlling for CAC score, remains unclear.

Although the absence of CAC is not reassuring in symptomatic patients, the CAC score may be associated with myocardial perfusion defects in asymptomatic patients. Studies have shown that CAC burden and extent predict future coronary revascularization procedures [24]. As Budoff et al. first demonstrated, both the number of calcified vessels and general CAC prevalence are independently associated with increased likelihood of significant angiographic disease [25]. Moreover, CAC prevalence, adjusting for CAC score, was predictive of the mode of revascularization. Independent predictors of coronary artery bypass graft versus percutaneous coronary intervention included three- or four-vessel CAC, higher CAC burden, and involvement of the left main coronary artery.

The significant association between dyslipidemia types, except hypertriglyceridemia, and the extent of CAC may prompt more aggressive treatment and prevention of elevated LDL-c, low HDL-c, combined hyperlipidemia, and MetS. Simple hypercholesterolemia may have the greatest impact in determining the severity of atherosclerotic disease, especially in those with diabetes and taking lipid-lowering medications. As others have shown, LDL-c is the dominant lipid determinant of atherosclerotic disease [6, 17].

As the MESA study gathered data from a large, multiethnic population, the results may be widely generalizable. The imaging and laboratory procedures were standardized at a common institution. Our study also has several limitations. Although we attempted to adjust for all possible confounding factors in our model, the residual confounding by unevaluated factors cannot be completely ruled out. As we examined cross-sectional associations, the possibility of temporal and selection biases may exist. Participants on statin therapy and with diabetes were excluded because these factors could potentially misclassify the lipid categories and confound the relationship between dyslipidemia and CAC. We used the term dyslipidemia compatible with metabolic syndrome instead of dyslipidemia of metabolic syndrome. To define this category, we did not factor in obesity or blood pressure to isolate the impact of dyslipidemia on the extent of subclinical atherosclerosis.

The results of this study further elucidate the role of CAC in cardiovascular disease. Clinicians already use the CAC score to predict future cardiovascular morbidity and mortality in asymptomatic patients with moderate risk factors. Callister et al. [26] showed that CAC scoring might help assess the effects of statin therapy. Budoff et al. illustrated a relationship between coronary calcium extent and CABG [25]. Recognizing the extent of coronary artery calcium may provide additional information beyond whether or not atherosclerotic disease is present. The association of dyslipidemias with CAC extent is further evidence in support of the importance of dyslipidemia, especially simple hypercholesterolemia, as a target for therapy.

The association between multivessel coronary artery disease and the cardiovascular risk factors, including dyslipidemia, had been widely studied in patients with documented clinical CVD, using invasive coronary angiography [27, 28]. Our study is novel to address this relationship, assessed by cardiac computed tomography, in a large multiethnic cohort free of clinical CVDs at baseline. We believe the underlying mechanisms of these associations should be relevant to disease prevention and require further investigation. Furthermore, the current findings may explain the difference when comparing the results of different studies. For instance, in patients with low HDL-c, although the CVD risk is high independently of other cardiovascular risk factors, the clinical trials have shown lack of improvement in the cardiovascular outcome when using drug therapies to boost the level of HDL-c. Our observation highlights the fact that the relative impact of low HDL-c, and potentially the targeting pharmacotherapy, on CAD extent may vary depending on the population selected, low versus high calcium score. Similarly, CVD outcome studies with triglyceride-lowering agents have produced inconsistent results, meaning that no convincing evidence is available that lowering triglycerides by any approach can reduce mortality. The association between plasma TG levels (both fasting and nonfasting) and CV risk is often attenuated once adjusting for other lipid parameters. However, it is worthy to notice the lack of association between high TG level and CAC extent existing in our study even prior to adjusting for other lipid parameters. A number of studies have found that the association between plasma TG levels (both fasting and nonfasting) and CV risk is often attenuated once adjusting for other lipid parameters, including HDL-c and non-HDL-c. An analysis conducted by the Emerging Risk Factors Collaboration demonstrated that there was a significant and stepwise association between fasting and nonfasting TG levels and CVD risk. However, this association was no longer significant after adjustment for HDL-c and non-HDL-c [29]. Elevated TG levels are closely associated with higher levels of non-HDL-c and apoB and low levels of HDL-c, and this may explain why this association is weakened after adjustment for these parameters. This is inherent in the study design since we classified the study cohort into 6 mutually exclusive dyslipidemia categories.

Further research is needed in examining coronary artery calcium extent, not simply the score, as a potential tool in prognosticating, treating, and perhaps preventing subclinical atherosclerosis in otherwise low-risk populations. Other lipid parameters such as non-HDL-c and lipoprotein(a) (Lp(a)) have been proposed as independent risk factors for coronary heart disease and their impact on the extent of multivessel CAC needs to be explored further. A previous study using the MESA population showed that non-HDL-c is independently associated with increased CAC in patient populations without CAC at baseline and especially non-HDL-c > 190 is associated with a significant progression of CAC [30]. Nonetheless, we have identified the 5 major dyslipidemia groups in our analysis based on current National Cholesterol Education Program (NCEP)/Adult Treatment Panel- (ATP-) III guidelines that define LDL-C, HDL-C, and triglyceride thresholds as abnormal. These groups account for the complexity of having more than 1 abnormal lipid parameter in the individual patients. It is challenging to create another mutually exclusive group by using non-HDL-c cutoff because this parameter encompasses all of the circulating atherogenic lipoproteins, including LDL-c. Furthermore, the aforementioned study included participants with diabetes and lipid-lowering medications, whom we excluded in the current analysis, to avoid the effect of these 2 factors on lipid parameters and CAC.

Genetic studies and multiple epidemiologic studies have identified Lp(a) as a risk factor for atherosclerotic diseases such as coronary heart disease and stroke [31]. In a previous study of 410 outpatients, Jug et al. showed that LP(a) is an independent predictor of coronary calcium (odds ratio 7.81, 95% confidence interval 1.41 to 43.5) [32]. In another study by Cho et al., participants with Lp(a) level ≥ 50 mg/dL had an odds ratio of 1.333 (95% CI 1.027–1.730) for CAC progression compared to those with Lp(a) < 50 mg/dL after adjusting for confounding factors [33]. Future investigations testing the association between the Lp(a) and CAC extent in population free of CVD at baseline are warranted. In addition, a published study of MESA showed that the measures of CAC burden and distribution each are independently predicting need for percutaneous coronary intervention versus CABG over an 8.5-year follow-up [2]. The new ACC/AHA 2013 guidelines support intensifying statin therapy when the CAC score is ≥75th percentile for age, gender, and ethnicity/race or when the calcium score is ≥300 Agatston units [34]. Whether adding the extent of CAC, single versus multivessel CAC, to the absolute calcium score poses any incremental therapeutic value needs to be tested in a large population-based cohort.

5. Conclusion

In a population-based cohort, the extent of multivessel CAC was associated with different dyslipidemia types except for hypertriglyceridemia. The results were still significant even after controlling for CAC score. Future research should focus on the mechanistic understanding of this relationship for disease prevention and investigate the association between other promising lipid parameters and CAC extent in asymptomatic adults with subclinical atherosclerosis.

Abbreviations

AUC: Areas under receiver operating curve
CAC: Coronary artery calcium
CAD: Coronary artery disease
CCS: Coronary calcium score
CIMT: Carotid intima media thickness
CRP: C-reactive protein
CVD: Cardiovascular disease
(EBT):   Electron-beam computed tomography
FRS: Framingham Risk Score
HDL-c: High-density lipoprotein cholesterol
Lp(a):   Lipoprotein(a)
LDL-c: Low density lipoprotein cholesterol
MDCT: Multidetector row helical computed tomography
MetS: Dyslipidemia compatible with metabolic syndrome
MESA: Multiethnic Study of Atherosclerosis
TG: Triglycerides.

Authors' Contributions

Moshrik Abd alamir is the principal investigator. Michael Goyfman contributed to study design and is an editor. Adib Chaus contributed to literature research and manuscript format. Firas Dabbous contributed to statistics. Leslie Tamura contributed to drafting initial manuscript. Veit Sandfort contributed to study design and is an editor. Alan Brown is the lipid expert of the study. Mathew Budoff is the senior author.

Acknowledgments

This research was supported by the National Heart, Lung, and Blood Institute (R01-HL-071739 and Contracts N01-HC-95159 through N01-HC-95165 and N01-HC-95169). The authors thank the other investigators, the staff, and the participants of the MESA study for their valuable contributions. A full list of participating MESA investigators and institutions can be found at https://www.mesa-nhlbi.org.

References

[1] A. J. Taylor et al., "ACCF/SCCT/ACR/AHA/ASE/ASNC/NASCI/SCAI/SCMR 2010 Appropriate Use Criteria for Cardiac Computed Tomography. A Report of the American College of Cardiology Foundation Appropriate Use Criteria Task Force, the Society of Cardiovascular Computed Tomography, the American College of Radiology, the American Heart Association, the American Society of Echocardiography, the American Society of Nuclear Cardiology, the North American Society for Cardiovascular Imaging, the Society for Cardiovascular Angiography and Interventions, and the Society for Cardiovascular Magnetic Resonance," *Journal of Cardiovascular Computed Tomography*, vol. 4, no. 6, p. 407, 2010.

[2] M. G. Silverman, J. R. Harkness, R. Blankstein et al., "Baseline subclinical atherosclerosis burden and distribution are associated with frequency and mode of future coronary revascularization: Multi-ethnic study of atherosclerosis," *JACC: Cardiovascular Imaging*, vol. 7, no. 5, pp. 476–486, 2014.

[3] P. K. Shah, "Screening Asymptomatic Subjects for Subclinical Atherosclerosis. Can We, Does It Matter, and Should We?" *Journal of the American College of Cardiology*, vol. 56, no. 2, pp. 98–105, 2010.

[4] M. S. Lauer, "Primary prevention of atherosclerotic cardiovascular disease: The high public burden of low individual risk," *Journal of the American Medical Association*, vol. 297, no. 12, pp. 1376–1378, 2007.

[5] S. S. Martin, M. J. Blaha, R. Blankstein et al., "Dyslipidemia, coronary artery calcium, and incident atherosclerotic cardiovascular disease: implications for statin therapy from the multi-ethnic study of atherosclerosis," *Circulation*, vol. 129, no. 1, pp. 77–86, 2014.

[6] P. Paramsothy, R. H. Knopp, A. G. Bertoni et al., "Association of combinations of lipid parameters with carotid intima-media thickness and coronary artery calcium in the MESA (Multi-Ethnic Study of Atherosclerosis)," *Journal of the American College of Cardiology*, vol. 56, no. 13, pp. 1034–1041, 2010.

[7] R. Tota-Maharaj, P. H. Joshi, M. J. Budoff et al., "Usefulness of regional distribution of coronary artery calcium to improve the prediction of all-cause mortality," *American Journal of Cardiology*, vol. 115, no. 9, pp. 1229–1234, 2015.

[8] D. E. Bild, D. A. Bluemke, G. L. Burke et al., "Multi-ethnic study of atherosclerosis: objectives and design," *American Journal of Epidemiology*, vol. 156, no. 9, pp. 871–881, 2002.

[9] G. R. Warnick, R. H. Knopp, V. Fitzpatrick, and L. Branson, "Estimating low-density lipoprotein cholesterol by the Friedewald equation is adequate for classifying patients on the basis of nationally recommended cutpoints," *Clinical Chemistry*, vol. 36, no. 1, pp. 15–19, 1990.

[10] J. J. Carr, J. C. Nelson, N. D. Wong et al., "Calcified coronary artery plaque measurement with cardiac CT in population-based studies: Standardized protocol of Multi-Ethnic Study of Atherosclerosis (MESA) and Coronary Artery Risk Development in Young Adults (CARDIA) study," *Radiology*, vol. 234, no. 1, pp. 35–43, 2005.

[11] B. Messenger, D. Li, K. Nasir, J. J. Carr, R. Blankstein, and M. J. Budoff, "Coronary calcium scans and radiation exposure in the multi-ethnic study of atherosclerosis," *The International Journal of Cardiovascular Imaging*, vol. 32, no. 3, pp. 525–529, 2016.

[12] P. Holvoet, N. S. Jenny, P. J. Schreiner, R. P. Tracy, and D. R. Jacobs, "The relationship between oxidized LDL and other cardiovascular risk factors and subclinical CVD in different ethnic groups: The Multi-Ethnic Study of Atherosclerosis (MESA)," *Atherosclerosis*, vol. 194, no. 1, pp. 245–252, 2007.

[13] Expert Panel on Detection, Evaluation, and Treatment of High Blood Cholesterol in Adults, "Executive Summary of the Third Report of the National Cholesterol Education Program (NCEP) Expert Panel on Detection, Evaluation, and Treatment of High Blood Cholesterol in Adults (Adult Treatment Panel III)" *JAMA*, vol. 285, no. 19, pp. 2486–2497, 2001.

[14] J. D. Otvos, "Measurement of lipoprotein subclass profiles by nuclear magnetic resonance spectroscopy," *Clinical Laboratory*, vol. 48, no. 3-4, pp. 171–180, 2002.

[15] R. L. McClelland, H. Chung, R. Detrano, W. Post, and R. A. Kronmal, "Distribution of coronary artery calcium by race, gender, and age: results from the Multi-Ethnic Study of Atherosclerosis (MESA)," *Circulation*, vol. 113, no. 1, pp. 30–37, 2006.

[16] A. S. Agatston, W. R. Janowitz, F. J. Hildner, N. R. Zusmer, M. Viamonte Jr., and R. Detrano, "Quantification of coronary artery calcium using ultrafast computed tomography," *Journal of the American College of Cardiology*, vol. 15, no. 4, pp. 827–832, 1990.

[17] C. W. Tsao, S. R. Preis, G. M. Peloso et al., "Relations of long-term and contemporary lipid levels and lipid genetic risk scores with coronary artery calcium in the Framingham Heart Study," *Journal of the American College of Cardiology*, vol. 60, no. 23, pp. 2364–2371, 2012.

[18] M. A. Allison and C. M. Wright, "A comparison of HDL and LDL cholesterol for prevalent coronary calcification," *International Journal of Cardiology*, vol. 95, no. 1, pp. 55–60, 2004.

[19] Y. Noda, R. Matsutera, Y. Kohama et al., "Impact of coronary calcium score on the relation between high-density lipoprotein cholesterol levels and the presence of high-risk coronary plaque detected by coronary computed tomography angiography," *Journal of the American College of Cardiology*, vol. 63, no. 12, p. A1064, 2014.

[20] H. C. J. McGill and C. A. McMahan, "Determinants of atherosclerosis in the young. Pathobiological Determinants of Atherosclerosis in Youth (PDAY) Research Group," *American Journal of Cardiology*, vol. 82, no. 10B, pp. 30T–36T, 1998.

[21] E. Van Craeyveld, F. Jacobs, Y. Feng et al., "The relative atherogenicity of VLDL and LDL is dependent on the topographic site," *Journal of Lipid Research*, vol. 51, no. 6, pp. 1478–1485, 2010.

[22] J.-C. Fruchart, F. M. Sacks, M. P. Hermans et al., "The residual risk reduction initiative: a call to action to reduce residual vascular risk in dyslipidaemic patients," *Diabetes & Vascular Disease Research*, vol. 5, no. 4, pp. 319–335, 2008.

[23] A. Bellasi, C. Lacey, A. J. Taylor et al., "Comparison of prognostic usefulness of coronary artery calcium in men versus women (results from a meta- and pooled analysis estimating all-cause mortality and coronary heart disease death or myocardial infarction)," *American Journal of Cardiology*, vol. 100, no. 3, pp. 409–414, 2007.

[24] J. D. Schuijf et al., "A comparative regional analysis of coronary atherosclerosis and calcium score on multislice CT versus myocardial perfusion on SPECT," *The Journal of Nuclear Medicine*, vol. 47, no. 11, pp. 1749–1755, 2006.

[25] M. J. Budoff, D. Georgiou, A. Brody et al., "Ultrafast computed tomography as a diagnostic modality in the detection of coronary artery disease: A multicenter study," *Circulation*, vol. 93, no. 5, pp. 898–904, 1996.

[26] T. Q. Callister, P. Raggi, B. Cooil, N. J. Lippolis, and D. J. Russo, "Effect of HMG-CoA reductase inhibitors on coronary artery disease as assessed by electron-beam computed tomography," *The New England Journal of Medicine*, vol. 339, no. 27, pp. 1972–1978, 1998.

[27] J. E. Digby, N. Ruparelia, and R. P. Choudhury, "Niacin in cardiovascular disease: Recent preclinical and clinical developments," *Arteriosclerosis, Thrombosis, and Vascular Biology*, vol. 32, no. 3, pp. 582–588, 2012.

[28] M. J. Landray, R. Haynes, J. C. Hopewell et al., "Effects of extended-release niacin with laropiprant in high-risk patients," *The New England Journal of Medicine*, vol. 371, no. 3, pp. 203–212, 2014.

[29] E. Di Angelantonio, N. Sarwar, P. Perry et al., "Major lipids, apolipoproteins, and risk of vascular disease," *Journal of the American Medical Association*, vol. 302, no. 18, pp. 1993–2000, 2009.

[30] S. K. Zalawadiya, V. Veeranna, S. Panaich, A. Kottam, and L. Afonso, "Non-high-density lipoprotein cholesterol and coronary artery calcium progression in a multiethnic US population," *American Journal of Cardiology*, vol. 113, no. 3, pp. 471–474, 2014.

[31] M. Bucci, C. Tana, M. A. Giamberardino, and F. Cipollone, "Lp(a) and cardiovascular risk: Investigating the hidden side of the moon," *Nutrition, Metabolism & Cardiovascular Diseases*, vol. 26, no. 11, pp. 980–986, 2016.

[32] B. Jug, J. Papazian, R. Lee, and M. J. Budoff, "Association of lipoprotein subfractions and coronary artery calcium in patient at intermediate cardiovascular risk," *American Journal of Cardiology*, vol. 111, no. 2, pp. 213–218, 2013.

[33] J. H. Cho, D. Y. Lee, E. S. Lee et al., "Increased risk of coronary artery calcification progression in subjects with high baseline Lp(a) levels: The Kangbuk Samsung Health Study," *International Journal of Cardiology*, vol. 222, pp. 233–237, 2016.

[34] N. J. Stone, "ACC/AHA guideline on the treatment of blood cholesterol to reduce atherosclerotic cardiovascular risk in adults: a report of the American College of Cardiology/American Heart Association Task Force on Practice Guidelines," *Journal of the American College of Cardiology*, vol. 63, no. 25, pp. 2889–2934, 2013.

9

Developing a Modified Low-Density Lipoprotein (M-LDL-C) Friedewald's Equation as a Substitute for Direct LDL-C Measure in a Ghanaian Population

Richard K. D. Ephraim ⓘ,[1] Emmanuel Acheampong ⓘ,[2,3] Swithin M. Swaray,[1] Enoch Odame Anto ⓘ,[2,3] Hope Agbodzakey,[2] Prince Adoba ⓘ,[2] Bright Oppong Afranie,[2] Emmanuella Nsenbah Batu,[2,4] Patrick Adu ⓘ,[1] Linda Ahenkorah Fondjo ⓘ,[2] Samuel Asamoah Sakyi ⓘ,[2] and Beatrice Amoah[2]

[1]Department of Medical Laboratory Science, School of Allied Health Sciences, University of Cape Coast, Ghana
[2]Department of Molecular Medicine, School of Medical Sciences, Kwame Nkrumah University of Science and Technology, Ghana
[3]School of Medical and Health Sciences, Edith Cowan University, Western Australia, Australia
[4]Department of Biochemistry, Dalian Medical University, China

Correspondence should be addressed to Emmanuel Acheampong; emmanuelachea1990@yahoo.com

Academic Editor: Gerhard M. Kostner

Despite the availability of several homogenous LDL-C assays, calculated Friedewald's LDL-C equation remains the widely used formula in clinical practice. Several novel formulas developed in different populations have been reported to outperform the Friedewald formula. This study validated the existing LDL-C formulas and derived a modified LDL-C formula specific to a Ghanaian population. In this comparative study, we recruited 1518 participants, derived a new modified Friedewald's LDL-C (M-LDL-C) equation, evaluated LDL-C by Friedewald's formula (F-LDL-C), Martin's formula (N-LDL-C), Anandaraja's formula (A-LDL-C), and compared them to direct measurement of LDL-C (D-LDL-C). The mean D-LDL-C (2.47 ± 0.71 mmol/L) was significantly lower compared to F-LDL-C (2.76 ± 1.05 mmol/L), N-LDL-C (2.74 ± 1.04 mmol/L), A-LDL-C (2.99 ± 1.02 mmol/L), and M-LDL-C (2.97 ± 1.08 mmol/L) $p < 0.001$. There was a significantly positive correlation between D-LDL-C and A-LDL-C ($r=0.658$, $p<0.0001$), N-LDL-C ($r=0.693$, $p<0.0001$), and M-LDL-C ($r=0.693$, $p<0.0001$). M-LDL-c yielded a better diagnostic performance [(area under the curve (AUC)=0.81; sensitivity (SE) (60%) and specificity (SP) (88%)] followed by N-LDL-C [(AUC=0.81; SE (63%) and SP (85%)], F-LDL-C [(AUC=0.80; SE (63%) and SP (84%)], and A-LDL-C (AUC=0.77; SE (68%) and SP (78%)] using D-LDL-C as gold standard. Bland–Altman plots showed a definite agreement between means and differences of D-LDL-C and the calculated formulas with 95% of values lying within ±0.50 SD limits. The modified LDL-C (M-LDL-C) formula derived by this study yielded a better diagnostic accuracy compared to A-LDL-C and F-LDL-C equations and thus could serve as a substitute for D-LDL-C and F-LDL-C equations in the Ghanaian population.

1. Background

Cardiovascular diseases (CVDs) are the leading cause of morbidity and mortality globally [1, 2]. High LDL-C is of longstanding clinical and research interest as it is an independent risk factor for cardiovascular events and coronary heart diseases (CHDs) [3, 4]. The advent of lipoprotein cholesterol measurement led to epidemiologic analyses that considered the potential effects of the various particles on cardiovascular (CVD) risk [5]. Previous studies have clearly incriminated high levels of LDL-C as atherogenic and have established a link between elevated LDL-C and cardiovascular events [6, 7].

Since the inception of Friedewald's LDL-C equation in 1972, it has been used most widely to estimate LDL-C in clinical practice as well as in health screenings [8]. The American National Cholesterol Education Program (NCEP)

working group on lipoprotein measurements recommended that LDL-C concentration be determined with a total analytical error not exceeding ±12% (≤4% imprecision and ≤4% inaccuracy) to guarantee correct patient classification into NCEP risk categories [8]. However there have been numerous attempts to improve its accuracy and reliability in population-based studies [9–12]. Regardless of the importance of accurate evaluation for LDL-C, the Friedewald's formula with its inherent remarkable deviation and limitation continues to be used in clinical and research settings as a cost-effective method to estimate LDL-C when triglyceride levels are less than 4.52 mmol/l [13].

Ultracentrifugation and beta-quantitation are the gold standards for LDL-C measurement [14]. Other methods include direct measurement of LDL-C using a homogenous assay. These methods are expensive, inconvenient, and not readily available in most routine laboratories. Not only is ultracentrifugation as a separation method tedious and time-consuming, but high salt concentrations and centrifugal forces can substantially alter high labile lipoproteins [15, 16]. Furthermore, the diagnostic performance of direct measurement of LDL-C is limited by high triglycerides (TG) levels [17]. In addition, the concentrations of TGs in the various lipoprotein fractions are known to be heterogeneous and therefore change with lipid disorders and other conditions [17, 18].

Previous studies have shown that the formula underrates LDL-C and CV risk stratification even when triglyceride levels are below 4.52 mmol/l [13, 19]. Following the aforementioned drawbacks, Anandaraja et al. and Vojovic et al. have attempted to derive formulas that are specific for Indian and Serbian population, respectively [11, 20]. Martin et al. also provided an alternative to improve LDL-C estimation in a South African population but proposed that external validation and further modifications were needed to improve its utilization [9].

The heterogeneity of population, as well as differences in dietary habits, calls for a more population-specific LDL-C formula that will be generic, accurate, and precise. The scarcity of literature on modified LDL-C formula in a West African population makes it necessary that we begin to document and validate existing formulas in our setting. Research in this direction will provide the breakthrough in combating the burden of atherosclerosis and serve as a useful guide for stakeholders in the management and control of cardiovascular diseases in Ghana. Using fasting lipid profile data from patients who visited the laboratory department of the National Cardiothoracic Centre, this study validated the existing LDL-C formulas and derived a modified LDL-C formula specific to a Ghanaian population.

2. Materials and Methods

2.1. Study Design/Site. This was a comparative study for the estimation of LDL-C using three different formulas and direct estimation by a homogenous assay. Data was collected for the lipid profile samples received in the laboratory unit of the National Cardiothoracic Centre (NCC) in Accra from December 2016 to April 2017. The NCC in Korle Bu, Accra,

is one of the few functioning referral centres in West Africa where complete evaluation of cardiothoracic diseases is not only possible but very safe and of a standard comparable internationally.

2.2. Ethical Consideration. This study was approved by the Committee on Human Research, Publication and Ethics (CHRPE) of School of Medical Sciences, KNUST. The subjects were adequately informed of the purpose, procedures, nature, risks, and minimal discomfort of the study. Participants were coded and assured of strict anonymity, confidentiality, and the freedom to exit or decline participation at any time without penalty.

2.3. Sample Size and Selection of Participants. Samples for lipid profile analysis were collected from patients visiting the laboratory unit of the NCC over the period stated. This was after participants had given their consent. Of a total of 1540 samples analysed, 22 were excluded because they had triglyceride levels greater than 4.52mmol/l (400mg/dl). The sample size of 1518 (N=1518) was comprised of 782 males and 736 females.

2.4. Inclusion and Exclusion Criteria. Participants with no evidence of metabolic conditions (diabetes, chronic renal failure) as per clinical history and had observed at least ten (10) hours of overnight fasting were included. Samples with triglyceride levels greater than 4.52mmol/l (400mg/dl) were excluded.

2.5. Sample Collection and Biochemical Assays. A volume of at least 3 mL of venous blood was taken into plain tubes after a 12-hour or minimum of 8-hour overnight fast via phlebotomy. The blood could clot, and serum was separated by centrifugation (2000g for 10mins) and analysed on URIT 8210 automatic chemistry analyser by TC and TG were measured enzymatically by CHOD-PAP and Glycerol phosphate peroxidase-PAP methods, respectively, using reagent kit obtained from Human Diagnostic Worldwide, Germany. TC and TG were calibrated using general chemistry calibrator provided by Human Diagnostic. The reagent kit for direct LDL-C assay, the Chema LDL direct FL test, was manufactured by Hospitex Diagnostics, Italy. HDL-c measurement was performed using a direct homogenous method without precipitation with the use of enzymatic colorimetric assay provided by Human Diagnostic, Germany. LDL-C concentration was directly determined by enzymatic assays.

LDL-C concentrations were also calculated by Friedewald's, Anandaraja's and Martin's formula as follows:

$$F\text{-}LDL\text{-}C = TC - HDL - (TG/5)$$

$$A\text{-}LDL\text{-}C = (0.9TC) - (0.9XTG/5) - 28$$

$$N\text{-}LDL\text{-}C = TC - HDL\text{-}C - TG/Novel\ Factor \text{ (all}$$
calculated in mg/dL).

2.6. Derivation of M-LDL-c in a Ghanaian Population. Re-examination of Friedewald's formula for LDL-C determination in our setting is based on the current results, following the procedure which led to Friedewald's formula derivation. Factor for VLDL-C concentration estimation was recalculated. Total cholesterol, TG, LD-C, and HDL-C concentration measurements were used in the initial group to calculate the VLDL-C/TG ratio for a Ghanaian population. The sum of HDL-C and LDL-C was subtracted from TC for each person. This accounted for the assessment of VLDL-C concentration for each person. Afterwards, to determine the mean of the ratio, the TG concentration was divided by the corresponding calculated VLDL-C. The mean ratio, TG/VLDL, was 4 compared to 2.2 according to Friedewald, M-LDL-C (mmol/L) =TC-HDL-C-TG/4.0 [11]

2.7. Data Analysis. Data collected were stored in MS Excel spread sheet. Using the Statistical Package for Social Sciences program (SPSS, version 21.0 for Windows) and GraphPad prism (Version 5 for windows, Inc. 2007), statistical analyses were carried out. Results are expressed as means ± SD and percentages in parenthesis. Unpaired t-test and one-way ANOVA were used to compare mean values of continuous variables for two and more than two categories. Person's correlation analysis was used to determine the association between directly measured LDL-C and calculated LDL-C. The Bland-Altman plots for comparison were used to determine level of bias and agreement of the calculated LDL-C to direct LDL-C. Linear regression analysis was used to generate linear models for the estimation of LDL-C. For all statistical comparisons, a $P < 0.05$ was considered as statistically significant.

3. Results

Lipoprotein concentrations and their distributions in the validation group are given in Table 1. The D-LDL-C values were significantly lower than F-LDL-C, ALDL-C, N-ALDL-C, and M-ALDL-C values (p < 0.001). The mean absolute bias amongst calculated LDL-Cs compared to the direct method was 0.29 ± 0.34 mmol/L for Friedewald's formula, 0.27 ± 0.33 mmol/L for Martin's formula, 0.57 ± 0.31 mmol/L for Anandaraja's formula, and 0.50 ± 0.34 mmol/L for modified formula.

The mean %ΔLDL between calculated LDL-Cs compared to the direct method was 12.39 ± 27.34% for Friedewald's formula, 11.74 ± 29.25% for Martin's formula, 23.54 ± 30.45% for Anandaraja's formula, and 21.33± 28.43% for modified formula (Table 2).

The ability of the formulas to correctly classify subjects at the clinical decision cut-off points in specific subgroups is shown in Table 3. Separate analysis was done for subgroups defined by cut-off values (ranges) for TC, TG, and D-LDL-C values provided by NCEP ATPIII guidelines. There were significantly lower mean values of F LDLC in the first quartile of TC compared to D-LDLC and higher mean values in the rest of the quartiles for TC, and few ranges of TG and of LDL-C. A-LDL-C were significantly higher (p < 0.001) than most ranges of TC in the D-LDL-C. A-LDL-C showed no significant difference compared to D-LDL-C in TG and LDL-C. N-LDL-C was significantly lower in the first range of TC in the D-LDL-C but significantly higher in the rest of the ranges of TC in D-LDL-C (p < 0.001). N-LDL-C showed no

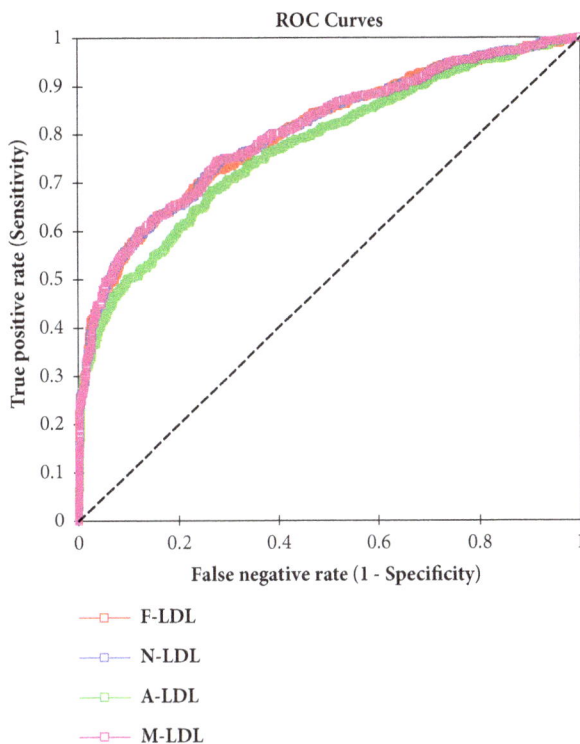

FIGURE 1: ROC curves depicting the accuracy of the different forms of LDL-C measurements.

significant difference compared to D-LDL-C in TG and LDL-C except 3.10±0.93 and 2.31±0.65, respectively (p < 0.001). M-LDL-C was significantly higher (p < 0.001) compared to D-LDLC in all ranges of TC and in most ranges of TG and of LDL-C (Table 3).

From Table 4, a cut-off value of 2.92 mmol/L F-LDL-C had a sensitivity of 0.63 and specificity of 0.84 with negative predictive value of 0.87 and positive predictive 0.74. With N-LDL-C, cut-off value 2.92 was used; this yielded a sensitivity of 0.63 and specificity of 0.85 for detection of LDL-C with positive predictive value of 0.78 and negative predictive value of 0.87. A-LDL-C had sensitivity of 0.68 and specificity of 0.73. M-LDL-C had a sensitivity of 0.60 and specificity of 0.88 with a cut-off value of 3.23mmol/L.

Figure 1 shows receiver operating characteristics (ROC) curve analyses for depicting the accuracy of Friedewald's formula (F-LDL-C), Martin's formula (N-LDL-C), and Anandaraja's formula (A-LDL-C) and M-LDL-C, LDL-C calculated by modified formula. Area under curve (AUC) was 0.80 for F-LDL-C, 0.81 for N-LDL-C, 0.77 for A-LDL-C, and 0.81 for M-LDL-C.

Correlation analysis between various formulas is used to estimate the LDL-C concentrations. With respect to the Pearson correlation coefficient (r) and coefficient of determination (R^2), various formulas were strongly positive correlated with each other (p<0.0001) (Table 5).

Figure 2 shows Bland-Altman plots for direct LDL-C against Friedewald's formula (F-LDL-C), Martin's formula (N-LDL-C), and Anandaraja's formula (A-LDL-C) and M-LDL-C, LDL-C calculated by modified formula. The mean

TABLE 1: Distribution of basic lipoprotein measurements.

	TC mmol/L	TG mmol/L	HDL mmol/L	Non-HDL mmol/L	D-LDL-C mmol/L	F-LDL-C mmol/L	N-LDL-C mmol/L	A-LDL-C mmol/L	M-LDL-C mmol/L
Mean	4.60	1.02	1.38	3.22	2.47	2.76**	2.74**	2.99**	2.97**
SD	1.20	0.57	0.36	1.13	0.71	1.05	1.04	1.02	1.08
1st Quartile	3.77	0.65	1.13	2.45	1.94	2.03	2.03	2.30	2.23
Median	4.45	0.88	1.33	3.09	2.43	2.64	2.63	2.87	2.64
3rd Quartile	5.21	1.23	1.59	3.78	2.90	3.28	3.26	3.54	3.28

**p<0.001: significant in comparison to D-LDL-C. TC: total cholesterol; TG: triglycerides; HDL-C: high-density lipoprotein cholesterol; LDL-C: low-density lipoprotein cholesterol; D-LDL-C: directly measured LDL-C; F-LDL-C LDL-C calculated by Friedewald's formula; N-LDL-C, LDL-C calculated by Martin's formula; A-LDL-C, LDL-C calculated by Anandaraja's formula; M-LDL-C, LDL-C calculated by modified formula.

TABLE 2: Mean percentage difference between D-LDL-C and calculated LDL-C.

	Δ F-LDL-C (%)	Δ N-LDL-C (%)	ΔA-LDL (%)	ΔM-LDL-C (%)
Mean	12.39	11.74	23.54	21.33
SD	27.34	29.25	30.45	28.43
1st Quartile	4.33	5.07	1.39	3.63
Median	13.79	13.51	23.84	23.8
3rd Quartile	29.63	28.55	44.81	39.11

SD: standard deviation; ΔA-LDL-C: mean percentage difference for Friedwald's formula; ΔA-LDL-C: mean percentage difference for Martin formula; ΔA-LDL-C: mean percentage difference for Anandaraja's formula; ΔA-LDL-C: mean percentage difference for modified formula. Mean percentage difference was calculated as [(calculated LDL-C)-(D-LDL-C)]/D-LDL-C∗100].

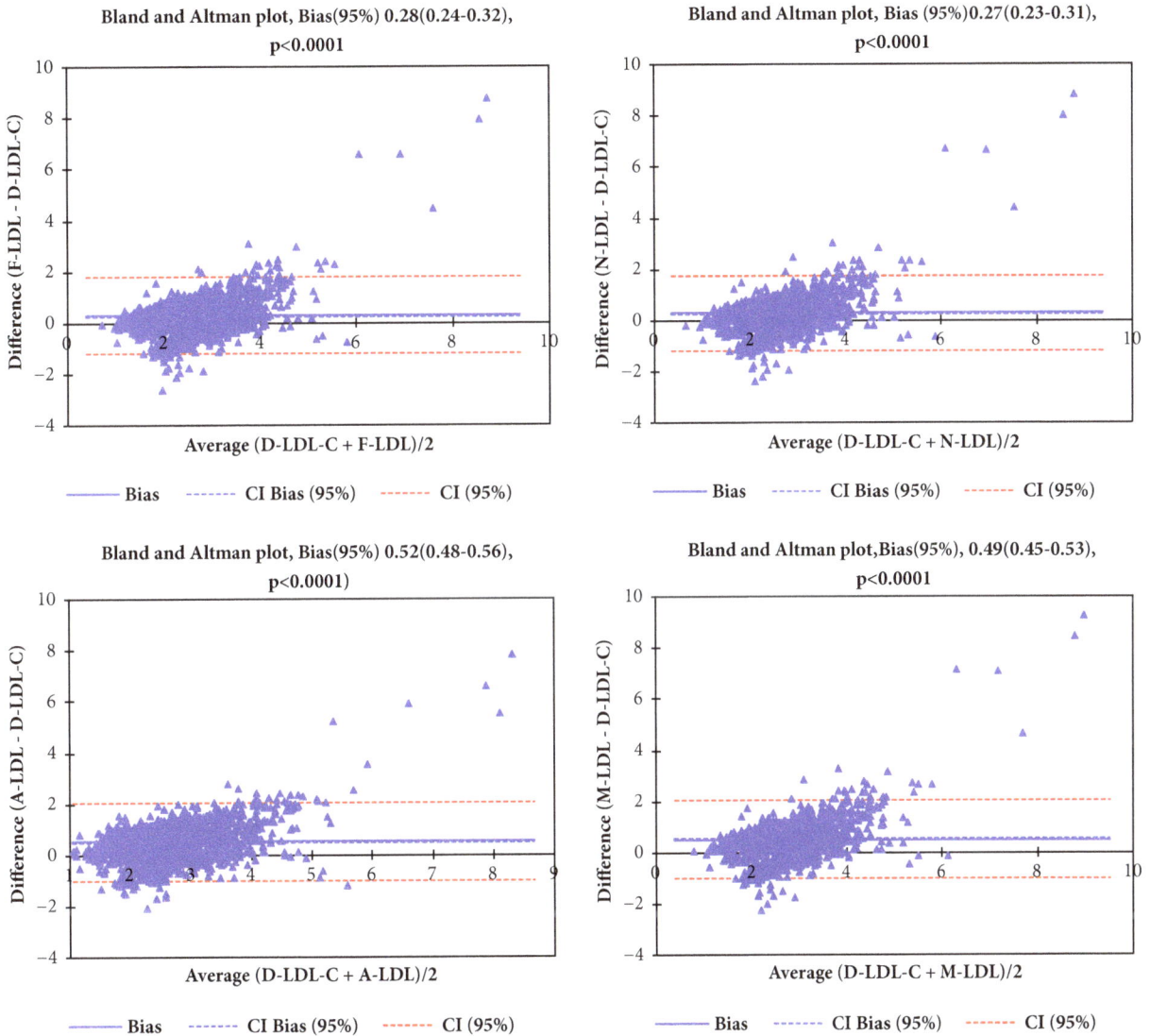

FIGURE 2: Bland and Altman plot of the different forms of LDL-C measurements (F-LDL-C, N-LDL-C, A-LDL-C, and M-LDL-C).

bias for F-LDL-C was 0.28(0.24-0.32), 0.27(0.23-90.31) for N-LDL-C, 0.52(0.48-0.56) for A-LDL-C, and 0.49(0.45-0.53) for M-LDL-C.

4. Discussion

Currently, the NCEP guidelines focus on diagnosis and treatment of TC and LDL-C. It is therefore relevant to accurately estimate LDL-C, as it has significant implications on cardiovascular risk stratification and can affect therapy and outcomes. The gold standard methods for quantifying LDL-borne cholesterol in serum are laborious and thus poorly suited to the modern laboratory [21]. Furthermore, many kinds of equipment and tubes are used, making conditions difficult to reproduce from one laboratory to another and

TABLE 3: Means ± SDs and percentages of correctly classified subjects in risk categories regarding TC, TG and D-LDL-C concentrations given by NCEP ATP III.

							% of subjects correctly classified by the following formulas			
	n	D-LDL-C	F-LDL-C	N-LDL-C	A-LDL-C	M-LDL-C	F	N	A	M
TC (mmol/L)										
≤4.13	586	2.01±0.53	1.91±0.40*	1.89±0.40**	2.11±0.38*	2.09±0.41*	77.6%	77.9%	92.6%	86.4%
4.14-5.16	536	2.49±0.51	2.79±0.41**	2.77±0.39**	3.04±0.31**	2.99±0.40**	78.1%	77.9%	84.1%	78.9%
5.17-6.20	272	2.92±0.50	3.52±0.45**	3.52±0.41**	3.80±0.36**	3.78±0.42**	72.4%	76.4%	65.6%	64.6%
6.21-7.24	89	3.44±0.65	4.45±0.49**	4.43±0.44**	4.66±0.38**	4.72±0.45**	65.8%	70.3%	56.9%	48.1%
≥7.25	35	3.90±0.66	6.16±2.18**	6.17±2.18**	6.37±1.84**	6.53±2.21**	63.8%	68.9%	59.6%	50.8%
TG (mmol/L)										
≤1.13	1055	2.40±0.69	2.63±0.90**	2.57±0.89*	2.92±0.91	2.78±0.91**	75.2%	79.1%	73.0%	80.6%
1.14-1.69	324	2.61±0.70	2.96±0.90**	3.10±0.93**	3.10±0.99	3.24±0.96**	30.2%	32.4%	29.3%	29.6%
1.70-2.25	76	2.75±0.69	3.18±1.20	3.30±0.89*	3.29±1.20	3.58±1.19*	19.7%	23.7%	25.0%	22.4%
2.26-2.82	40	2.68±0.75	3.49±2.82	3.74±2.74	3.51±2.49	4.01±2.82	11.4%	5.0%	2.5%	10.0%
2.83-4.52	23	2.74±1.01	3.00±1.05	3.43±0.96	3.03±1.02	3.75±1.07	4.3%	8.7%	8.7%	17.4%
LDL-C (mmol/L)										
≤2.59	903	2.01±0.38	2.33±0.65**	2.31±0.65**	2.61±0.68	2.53±0.67**	80.0%	80.4%	80.0%	81.4%
2.60-3.35	447	2.92±0.22	3.05±0.90	3.04±0.89	3.23±0.88	3.27±0.92**	41.4%	42.9%	31.7%	34.1%
3.36-4.12	142	3.62±0.18	3.98±0.91*	3.95±0.92*	4.18±0.93	4.22±0.97	31.9%	31.4%	19.2%	22.6%
4.13-4.89	21	4.36±0.21	5.68±2.54	5.64±2.57	5.72±2.26	5.94±2.26*	8.2%	9.5%	9.8%	42.5%
≥4.90	5	5.54±0.44	6.08±2.12	6.06±2.69	6.07±2.69	6.40±2.06	10.6%	8.8%	3.5%	7.7%

*$p<0.05$, **$p<0.0001$: significant compared to D-LDL-C.

TC: total cholesterol; TG: triglycerides; HDL-C: high-density lipoprotein cholesterol; LDL-C: low-density lipoprotein cholesterol; D-LDL-C: directly measured LDL-C; FLDL-C, LDL-C calculated by Friedewald's formula; A-LDL-C, LDL-C calculated by Anandaraja's formula; M-LDL-C, LDL-C calculated by modified formula. N-LDL-C, LDL-C calculated by Martin's formula; A-LDL-C, LDL-C calculated by Anandaraja's formula; M-LDL-C, LDL-C calculated by modified formula.

TABLE 4: Diagnosis performances of the various formulas.

Formulas	Cut-off point	AUC	95% CI	Sensitivity (95% CI)	Specificity (95%CI)	NPV	PPV
F-LDL-C	2.92	0.80	0.78-.83	0.63(0.60-0.67)	0.84(0.82-0.87)	0.87	0.74
N-LDL-C	2.92	0.81	0.78-.83	0.63(0.60-0.66)	0.85(0.82-0.87)	0.87	0.78
A-LDL-C	2.96	0.77	.75-.78	0.68(0.65-0.72)	0.73(0.70-0.77)	0.68	0.73
M-LDL-C	3.23	0.81	.78-.83	0.60(0.56-0.64)	0.88(0.85-0.90)	0.72	0.72

AUC: area under curve; CI: confidence interval; NPV: negative predictive value; PPV: positive predictive value; FLDL-C, LDL-C calculated by Friedewald's formula; N-LDL-C, LDL-C calculated by Martin's formula; A-LDL-C, LDL-C calculated by Anandaraja's formula; M-LDL-C, LDL-C calculated by modified formula.

TABLE 5: Pearson correlation coefficient (r) (bold) and coefficient of determination (R^2) (italics) between formulas.

Formulas	F-LDL	N-LDL	A-LDL	M-LDL	D-LDL-C
F-LDL-C (mmol/L)		**0.995**	**0.947**	**0.994**	**0.693**
		<0.0001	*<0.0001*	*<0.0001*	*<0.0001*
N-LDL-C (mmol/L)	*0.991*		**0.937**	**0.999**	**0.693**
	<0.0001		*<0.0001*	*<0.0001*	*<0.0001*
A-LDL-C (mmol/L)	*0.898*	*0.878*		**0.935**	**0.658**
	<0.0001	*<0.0001*		*<0.0001*	*<0.0001*
M-LDL-C (mmol/L)	*0.989*	*0.999*	*0.8743*		**0.693**
	<0.0001	*<0.0001*	*<0.0001*		*<0.0001*
D-LDL-C (mmol/L)	*0.481*	*0.480*	*0.434*	*0.481*	
	<0.0001	*<0.0001*	*<0.0001*	*<0.0001*	

FLDL-C, LDL-C calculated by Friedewald's formula; N-LDL-C, LDL-C calculated by Martin's formula; A-LDL-C, LDL-C calculated by Anandaraja's formula; M-LDL-C, LDL-C calculated by modified formula.

consistent separations highly dependent on the skills and care of the technician [16]. Nevertheless, ultracentrifugation remains the classic comparison method and is the basis for the accepted reference methods [16, 17, 21]. Numerous studies have been conducted to derive more precise formulas for LDL-C calculations in the past decades in different populations compared to the globally used Friedewald's formula [11, 20–22]. However, some of these modifications were not found to be suitable replacements of the Friedewald formula. This present study sought to compare the diagnostic performances of the Friedewald's (F-LDL-C), Martin's (N-LDL-C), and Anandaraja's (A-LDL-C) formulas to directly measured LDL-C in the Ghanaian population and to derive possible predictive equation for calculating LDL-C among our study population.

Lipoprotein concentrations and their distributions were analyzed in this study and we found significant differences between the mean values of F-LDL-C, N-LDL-C, A-LDL-C, and M-LDL-C with respect to D-LDL-C mean values. M-LDL-C values were significantly higher compared to the rest of the formulas except A-LDL-C (Table 1). This finding is not agreement with reports from a study conducted by Vujovic et al. in a Serbian population [11]. According to their work, mean LDL-C of participants directly measured was higher than those of calculated formulas; they reported a percentage difference of -6.9% for F-LDL-C and -3.9% of A-LDL-C [11]. We found a mean percentage differences of 12.39% and 23.54% for F-LDL-C and A-LDL-C, respectively (Table 2). In addition, Gupta and colleagues also found measured LDL-C to be higher than that obtained with the calculated formulas

[23]. On the other side, Boshtam et al. found that mean levels of D-LDL-C were lower than F-LDL-C in Iranian population [24] which is consistent with results in this study. Our results showed that mean difference between the two methods was statistically significant ($P<0.0001$). In parallel, some studies have demonstrated similar trend with higher results with calculated LDL-C for F-LDL-C as compared to directly measured LDL-C [25, 26].

Reports from the original study for the development of Friedewald's formula provided a simple division of blood plasma TG by 5 for mg/dL or 2.2 for mmol/L [22]; however this formula does not provide accurate estimation of VLDL-C. In our course to evaluate the reliability of Friedewald's, Martin's, and Anandaraja's formula, we developed a new modified formula, which closely resembles Friedewald's that exhibited a simple division of patient's plasma TG by 4. Several studies have suggested alterative calculation in different populations which include TG/2.2, TG/2.5. TG/2.8, TG/3.0, TG/3.3, and TG/3.9 (mmol/L) [27, 28].

The difference between calculated LDL-C and directly measured LDL-C results can be important regarding risk classification for coronary heart disease among patients [29]. In 2008, a study done by Jun et al., among Koreans, showed that F-LDL-C was significantly different from D-LDL-C over the concentration ranges of both TC and TG. In the same study, the mean %ΔLDL was -9.1% and it was anticipated that this difference was critical for the evaluation of patients with hyperlipidemia [30]. In this study, mean values of F-LDL-C were lower than that of D-LDL-C in the first quartile of TC with a mean %ΔLDL-C of -5.0%. However, F-LDL-C values were higher compared to the directly measured LDL-C in the rest of the quartiles for TC and few ranges of TG and LDL-C (Table 3). Of note, LDL-C levels calculated with our modified formula (M-LDL-C) were statistically significantly higher compared to D-LDL-C in all ranges of TC and in most ranges of TC and that of LDL-C to correctly classify subjects at the clinical decision cut-off points in specific subgroups.

F-LDL-C had a sensitivity of 63.0% and specificity of 84.0% with negative predictive value of 87.0% and positive predictive 74.0% with a cut-off of 2.92 mmol/L in this study. M-LDL-C had a sensitivity of 0.60 and specificity of 0.88 with a cut-off value of 3.23mmol/L. These findings contrast with reports from a study by Martin et al. in South Africa. In their work, they recorded a higher sensitivity and specificity with a cut-off of 2.5 mmol/L [9]. The possible reason for these inconsistencies might be that F-LDL-c is poor in assessing direct LDL-C when the LDL-C values are high and the different study populations in the studies [31].

In general, there were strong correlations among the various formulas for estimating LDL-C concentrations. However, moderate correlations were observed between the directly measured LDL-C and the various methods. Among the three formulas used in this study, the Anandaraja's formula showed the least correlation with the directly measured LDL-C (Table 5). The observed correlations are lower compared with reports from previous studies [25, 32–34]. These studies reported correlations of 0.88 [25] and 0.86 [33] and 0.786 [32], respectively. Another study conducted among Japanese also found a positive correlation between F-LDL-C and D-LDL-C [34]. Furthermore, Anandaraja and colleagues also reported a Pearson's correlation coefficient of 0.97 between LDL-C measured by their formula and D-LDL-C which was better as compared to that for F-LDL-C [20]. Conversely, Kapoor et al. found a lower correlation between A-LDL-C and D-LDL-C which is in relative agreement with the lower correlation coefficient of 0.658 observed in this present study [35].

Bland-Altman graphs showed a clear relationship between both the directly measured LDL-C and the calculation formulas. The observed low bias can be well appreciated in all plots though the bias between N-LDL-C and D-LDL-C was the lowest. This indicates that the calculation formulas and the directly measured LDL-C methods are systematically producing similar results (Figure 2). Some previous studies have reported that Friedewald calculation demonstrates better agreement with directly measured LDL-C [23, 36].

This study has strength being the first study to compare different methods of estimating LDL-c concentration in Ghana and the West African subregion. However, our study is limited by the fact that both derived models must be further scrutinized and validated bearing in mind the differences in race and the specific character of the applied method of measurement.

5. Conclusion

The modified LDL-c (M-LDL-c) formula derived by this study yielded a better diagnostic accuracy compared to A-LDL-c and F-LDL-c equation and thus could serve as a substitute for D-LDL-c and F-LDL-c equation in the Ghanaian population. Taking into consideration the racial variances as well as the specific character of the applied method of measurement, the study findings underscore the need for scrutiny, validation, and reliability evaluations of the generated models, to ascertain their clinical use. Further work should also examine the performance of rick calculations by the various formulae.

Acknowledgments

The authors would like to thank the patients, management, and staff of laboratory unit of the National Cardiothoracic Centre (NCC) Korle-bu Teaching Hospital Accra for making this a successful work.

References

[1] D. Mozaffarian, E. J. Benjamin, A. S. Go et al., "Heart disease and stroke statistics—2016 update: a report from the American Heart Association," *Circulation*, vol. 133, no. 4, pp. e38–e360, 2016.

[2] P. K. A. Abanilla, K.-Y. Huang, D. Shinners et al., "Cardiovascular disease prevention in Ghana: Feasibility of a faith-based organizational approach," *Bulletin of the World Health Organization*, vol. 89, no. 9, pp. 648–656, 2011.

[3] N. J. Stone, J. G. Robinson, A. H. Lichtenstein et al., "2013 ACC/AHA guideline on the treatment of blood cholesterol to reduce atherosclerotic cardiovascular risk in adults: a report of the American college of cardiology/American heart association task force on practice guidelines," *Journal of the American College of Cardiology*, vol. 63, no. 25, pp. 2889–2934, 2014.

[4] R. S. Rosenson, H. B. Brewer, B. J. Ansell et al., "Dysfunctional HDL and atherosclerotic cardiovascular disease," *Nature Reviews Cardiology*, vol. 13, no. 1, pp. 48–60, 2016.

[5] T. A. Jacobson, K. C. Maki, C. E. Orringer et al., "National lipid association recommendations for patient-centered management of dyslipidemia: Part 2," *Journal of Clinical Lipidology*, vol. 9, no. 6, pp. S1–S122, 2015.

[6] C. J. Girman, T. Rhodes, M. Mercuri et al., "The metabolic syndrome and risk of major coronary events in the Scandinavian Simvastatin Survival Study (4S) and the Air Force/Texas Coronary Atherosclerosis Prevention Study (AFCAPS/TexCAPS)," *American Journal of Cardiology*, vol. 93, no. 2, pp. 136–141, 2004.

[7] P. M. Ridker, M. J. Stampfer, and N. Rifai, "Novel risk factors for systemic atherosclerosis: a comparison of C-reactive protein, fibrinogen, homocysteine, lipoprotein(a), and standard cholesterol screening as predictors of peripheral arterial disease," *The Journal of the American Medical Association*, vol. 285, no. 19, pp. 2481–2485, 2001.

[8] P. S. Bachorik and J. W. Ross, "National cholesterol education program recommendations for measurement of low-density lipoprotein cholesterol: Executive summary," *Clinical Chemistry*, vol. 41, no. 10, pp. 1414–1420, 1995.

[9] J. Martins, S. A. S. Olorunju, L. M. Murray, and T. S. Pillay, "Comparison of equations for the calculation of LDL-cholesterol in hospitalized patients," *Clinica Chimica Acta*, vol. 444, pp. 137–142, 2015.

[10] P. Krishnaveni and V. M. Gowda, "Assessing the validity of friedewald's formula and anandraja's formula for serum LDL-cholesterol calculation," *Journal of Clinical and Diagnostic Research*, vol. 9, no. 12, pp. BC01–BC04, 2015.

[11] A. Vujovic, J. Kotur-Stevuljevic, S. Spasic et al., "Evaluation of different formulas for LDL-C calculation," *Lipids in Health and Disease*, vol. 9, article no. 27, 2010.

[12] R. Gasko, "Low-density lipoprotein cholesterol estimation by a new formula? Confirmation," *International Journal of Cardiology*, vol. 119, no. 2, pp. 242-243, 2007.

[13] H. Scharnagl, M. Nauck, H. Wieland, and W. März, "The friedewald formula underestimates LDL cholesterol at low concentrations," *Clinical Chemistry and Laboratory Medicine*, vol. 39, no. 5, pp. 426–431, 2001.

[14] A. Sawle, M. K. Higgins, M. P. Olivant, and J. A. Higgins, "A rapid single-step centrifugation method for determination of HDL, LDL, and VLDL cholesterol, and TG, and identification of predominant LDL subclass," *Journal of Lipid Research*, vol. 43, no. 2, pp. 335–343, 2002.

[15] T. G. Cole, C. A. Ferguson, D. W. Gibson, and W. L. Nowatzke, "Optimization of β-quantification methods for high-throughput applications," *Clinical Chemistry*, vol. 47, no. 4, pp. 712–721, 2001.

[16] M. Nauck, G. R. Warnick, and N. Rifai, "Methods for measurement of LDL-cholesterol: a critical assessment of direct measurement by homogeneous assays versus calculation," *Clinical Chemistry*, vol. 48, no. 2, pp. 236–254, 2002.

[17] W. G. Miller, G. L. Myers, I. Sakurabayashi et al., "Seven direct methods for measuring HDL and LDL cholesterol compared with ultracentrifugation reference measurement procedures," *Clinical Chemistry*, vol. 56, no. 6, pp. 977–986, 2010.

[18] H. Brewer, "Hypertriglyceridemia: changes in the plasma lipoproteins associated with an increased risk of cardiovascular disease," *American Journal of Cardiology*, vol. 83, no. 9, pp. 3–12, 1999.

[19] D. M. Waterworth, S. L. Ricketts, K. Song et al., "Genetic variants influencing circulating lipid levels and risk of coronary artery disease," *Arteriosclerosis, Thrombosis, and Vascular Biology*, vol. 30, no. 11, pp. 2264–2276, 2010.

[20] S. Anandaraja, R. Narang, R. Godeswar, R. Laksmy, and K. K. Talwar, "Low-density lipoprotein cholesterol estimation by a new formula in Indian population," *International Journal of Cardiology*, vol. 102, no. 1, pp. 117–120, 2005.

[21] C. M. M. de Cordova and M. M. de Cordova, "A new accurate, simple formula for LDL-cholesterol estimation based on directly measured blood lipids from a large cohort," *Annals of Clinical Biochemistry*, vol. 50, no. 1, pp. 13–19, 2013.

[22] W. T. Friedewald, R. I. Levy, and D. S. Fredrickson, "Estimation of the concentration of low-density lipoprotein cholesterol in plasma, without use of the preparative ultracentrifuge," *Clinical Chemistry*, vol. 18, no. 6, pp. 499–502, 1972.

[23] S. Gupta, M. Verma, and K. Singh, "Does LDL-C estimation using Anandaraja's formula give a better agreement with direct LDL-C estimation than the Friedewald's formula?" *Indian Journal of Clinical Biochemistry*, vol. 27, no. 2, pp. 127–133, 2012.

[24] M. Boshtam, M. A. Ramezani, G. Naderi, and N. Sarrafzadegan, "Is friedewald formula a good estimation for low density lipoprotein level in iranian population?" *Journal of Research in Medical Sciences*, vol. 17, no. 6, pp. 519–522, 2012.

[25] S. Sahu, R. Chawla, and B. Uppal, "Comparison of two methods of estimation of low density lipoprotein cholesterol, the direct versus Friedewald estimation," *Indian Journal of Clinical Biochemistry*, vol. 20, no. 2, pp. 54–61, 2005.

[26] R. Gasko, "Low-density lipoprotein cholesterol estimation by the Anandaraja's formula - Confirmation," *Lipids in Health and Disease*, vol. 5, no. 1, p. 18, 2006.

[27] D. M. Delong, E. R. Delong, P. D. Wood, K. Lippel, and B. M. Rifkind, "A Comparison of Methods for the Estimation of Plasma Low- and Very Low-Density Lipoprotein Cholesterol: The Lipid Research Clinics Prevalence Study," *Journal of the American Medical Association*, vol. 256, no. 17, pp. 2372–2377, 1986.

[28] N. Nakanishi, Y. Matsuo, H. Yoneda, K. Nakamura, K. Suzuki, and K. Tatara, "Validity of the conventional indirect methods including Friedewald method for determining serum low-density lipoprotein cholesterol level: Comparison with the direct homogeneous enzymatic analysis," *Journal of Occupational Health*, vol. 42, no. 3, pp. 130–137, 2000.

[29] S. M. Grundy, D. Becker, L. T. Clark et al., "Detection, evaluation, and treatment of high blood cholesterol in adults (Adult Treatment Panel III)," *Circulation*, vol. 106, no. 25, pp. 3143–3421, 2002.

[30] K. Ran Jun, H.-I. Park, S. Chun, H. Park, and W.-K. Min, "Effects of total cholesterol and triglyceride on the percentage difference between the low-density lipoprotein cholesterol concentration measured directly and calculated using the Friedewald formula," *Clinical Chemistry and Laboratory Medicine*, vol. 46, no. 3, pp. 371–375, 2008.

[31] S.-A. Ahmadi, M.-A. Boroumand, K. Gohari-Moghaddam, P. Tajik, and S.-M. Dibaj, "The impact of low serum triglyceride on LDL-cholesterol estimation," *Archives of Iranian Medicine*, vol. 11, no. 3, pp. 318–321, 2008.

[32] A. H. Kamal, M. Hossain, S. Chowdhury, and N. U. Mahmud, "A Comparison of Calculated with Direct Measurement of Low Density Lipoprotein Cholesterol Level," *Journal of Chittagong Medical College Teachers' Association*, vol. 20, no. 2, pp. 19–23, 2009.

[33] F. H. Fass, A. Earleywine, W. G. Smith, and D. L. Simmons, "How should low-density lipoprotein cholesterol concentration be determined?" *Journal of Family Practice*, vol. 51, no. 11, pp. 972–975, 2002.

[34] F. Kamezaki, S. Sonoda, S. Nakata, and Y. Otsuji, "A direct measurement for LDL-cholesterol increases hypercholesterolemia prevalence: Comparison with Friedewald calculation," *Journal of UOEH*, vol. 32, no. 3, pp. 211–220, 2010.

[35] R. Kapoor, M. Chakraborty, and N. Singh, "A leap above Friedewald formula for calculation of low-density lipoprotein-cholesterol," *Journal of Laboratory Physicians*, vol. 7, no. 1, p. 11, 2015.

[36] C. P. Onyenekwu, M. Hoffmann, F. Smit, T. E. Matsha, and R. T. Erasmus, "Comparison of LDL-cholesterol estimate using the Friedewald formula and the newly proposed de Cordova formula with a directly measured LDL-cholesterol in a healthy South African population," *Annals of Clinical Biochemistry*, vol. 51, no. 6, pp. 672–679, 2014.

10

High-Density Lipoprotein Binds to *Mycobacterium avium* and Affects the Infection of THP-1 Macrophages

Naoya Ichimura,[1,2] **Megumi Sato,**[1] **Akira Yoshimoto,**[1] **Kouji Yano,**[1] **Ryunosuke Ohkawa,**[1] **Takeshi Kasama,**[3] **and Minoru Tozuka**[1]

[1]*Analytical Laboratory Chemistry, Field of Applied Laboratory Science, Graduate School of Health Care Sciences, Tokyo Medical and Dental University, 1-5-45 Yushima, Bunkyo-Ku, Tokyo 113-8519, Japan*
[2]*Department of Clinical Laboratory, Medical Hospital of Tokyo Medical and Dental University, 1-5-45 Yushima, Bunkyo-Ku, Tokyo 113-8519, Japan*
[3]*Instrumental Analysis Research Division, Research Center for Medical and Dental Sciences, Tokyo Medical and Dental University, 1-5-45 Yushima, Bunkyo-Ku, Tokyo 113-8519, Japan*

Correspondence should be addressed to Minoru Tozuka; mtozuka.alc@tmd.ac.jp

Academic Editor: Shinichi Oikawa

High-density lipoprotein (HDL) is involved in innate immunity toward various infectious diseases. Concerning bacteria, HDL is known to bind to lipopolysaccharide (LPS) and to neutralize its physiological activity. On the other hand, cholesterol is known to play an important role in mycobacterial entry into host cells and in survival in the intracellular environment. However, the pathogenicity of *Mycobacterium avium* (*M. avium*) infection, which tends to increase worldwide, remains poorly studied. Here we report that HDL indicated a stronger interaction with *M. avium* than that with other Gram-negative bacteria containing abundant LPS. A binding of apolipoprotein (apo) A-I, the main protein component of HDL, with a specific lipid of *M. avium* might participate in this interaction. HDL did not have a direct bactericidal activity toward *M. avium* but attenuated the engulfment of *M. avium* by THP-1 macrophages. HDL also did not affect bacterial killing after ingestion of live *M. avium* by THP-1 macrophage. Furthermore, HDL strongly promoted the formation of lipid droplets in *M. avium*-infected THP-1 macrophages. These observations provide new insights into the relationship between *M. avium* infection and host lipoproteins, especially HDL. Thus, HDL may help *M. avium* to escape from host innate immunity.

1. Introduction

High-density lipoprotein (HDL), known as antiatherosclerotic lipoprotein, is also involved in innate immunity [1]. For instance, HDL binds to lipopolysaccharides (LPS) and lipoteichoic acid (LTA) derived from microorganisms and neutralizes the physiological activity of these molecules [2, 3]. Additionally, HDL is known to have a bactericidal activity toward *Yersinia enterocolitica* serotype O:3; this activity is mediated by the complement system [4].

Mycobacterium avium (*M. avium*) is a species of nontuberculous mycobacteria and causes opportunistic infections in immunocompromised hosts [5]. Although the incidence of *M. avium* infection is increasing worldwide [6], its pathogenicity remains poorly understood unlike that of *M. tuberculosis*. Mycobacterial infection is known to be affected by host lipid metabolism. In particular, cholesterol performs an important function in the invasion and survival of mycobacteria inside macrophages [7–11]. Cholesterol accumulation in the host cell membrane is observed at the entry point of *M. tuberculosis* and of other mycobacteria, and cholesterol depletion inhibits the invasion of cells by mycobacteria. After phagocytosis of mycobacteria, they are enveloped by the cell membrane containing cholesterol-rich domains; tryptophan aspartate-containing coat protein (TACO) is recruited to the phagosomes and prevents the fusion of these organelles with lysosomes [12]. Consequently, the engulfed mycobacteria evade degradation by lysosomes and survive inside the host cell. The invading mycobacteria modulate lipid metabolism in the host cell and promote formation of lipid droplets (LDs), which are mainly composed of neutral lipids such as triacylglycerol and cholesteryl ester [13]. LDs physiologically contribute to lipid storage and lipid metabolism and are available to mycobacteria as a carbon source [10, 14].

Scavenger receptor type B1 (SR-B1), a receptor of apolipoprotein A-I (apoA-I) present in HDL as a main component, is generally known to perform the function of transferring esterified cholesterol from matured HDL into the cytosol in accordance with the cholesterol gradient [15]. SR-B1 was also identified as a nonopsonic phagocytic receptor for mycobacteria because suppression of SR-B1 expression attenuates phagocytosis of mycobacteria [16, 17].

We screened various bacteria for binding with HDL and found that *M. avium* exhibits a stronger interaction with HDL than do other Gram-negative bacilli. The fact that HDL binds to *M. avium* has never been reported and the effect for innate immunity is unclear. The aim of the present study was to elucidate the molecular mechanism of HDL's binding to *M. avium* and the physiological meaning of this interaction.

2. Materials and Methods

2.1. Bacterial Strains and Growth Conditions. *M. avium* (ATCC 700737) was cultured for 3 weeks in Middlebrook 7H9 broth (Difco) supplemented with the oleic acid-albumin-dextrose complex (OADC; Becton Dickinson). Gram-negative bacteria (isolated at the Clinical Laboratory, Medical Hospital of Tokyo Medical and Dental University) were cultured on trypticase-soy-agar with 5% sheep blood (Becton Dickinson) before use.

2.2. Cell Culture, Differentiation, and Infection. The THP-1 cell line was obtained from ATCC (Manassas, VA) and maintained at $2–10 \times 10^5$ cells/mL in the RPMI 1640 medium (Sigma-Aldrich) supplemented with 10% heat-inactivated fetal bovine serum (Invitrogen), with a penicillin-streptomycin-L-glutamine solution (Wako), and 1x nonessential amino acids (GIBCO). The THP-1 cells were induced to differentiate into THP-1 macrophages by phorbol myristate acetate (PMA, Sigma-Aldrich) for 72 h. After washing with phosphate-buffered saline (PBS), the THP-1 macrophages were incubated in the serum-free RPMI 1640 supplemented with Nutridoma-SP (Roche) and then infected with *M. avium* (multiplicity of infection [MOI] 20 : 1) with or without HDL (50 μg protein/mL) for 24 h at 35°C in a humidified atmosphere containing 5% CO_2.

2.3. Isolation of HDL and Purification of apoA-I. HDL (1.063–1.210 g/mL) was isolated by means of ultracentrifugation from pooled serum samples obtained from healthy subjects. ApoA-I was purified from HDL according to the previously described method [18]. HDL and apoA-I were dialyzed against PBS and stored at 4°C and −20°C, respectively, until use.

2.4. Western Blot Analysis. Bacteria (3×10^8) were washed with PBS and then incubated with 40 μL of normal human serum (NHS), heat-inactivated serum (HIS), HDL, apoA-I, or PBS for 10 min at 37°C. After washing with PBS, the bacteria were mixed with lysis buffer (50 mM Tris-HCl pH 8.0, 150 mM NaCl, 0.1% SDS, and 0.5% sodium deoxycholate). The lysates were then analyzed by electrophoresis on 12.5% SDS-polyacrylamide gels, and the separated proteins were transferred onto polyvinylidene fluoride (PVDF) membranes (Millipore). ApoA-I was detected with a goat anti-apoA-I polyclonal antibody (Academy Bio-Medical Company) followed by a peroxidase- (POD-) conjugated rabbit anti-goat IgG antibody (Medical & Biological Laboratories). Finally, apoA-I was visualized using hydroperoxide and 3,3′-diaminobenzidine as the substrate.

2.5. Dot Blot Analysis. This analysis was carried out using a previously described method [19] with a slight modification. Lipids were extracted from the outer cell wall of *M. avium* by means of Folch's extraction procedure [20] and were then spotted onto a nitrocellulose membrane. The membrane was sequentially incubated with 5% (w/v) skim milk and 50 μg/mL HDL in 10 mM Tris-HCl (pH 8.0) containing 140 mM NaCl and 0.1% Tween 20 (TBS-T) for 1 h and 5 h, respectively, at room temperature. The membrane was washed with TBS-T and then incubated with a POD-conjugated anti-apoA-I polyclonal antibody (Cosmo Bio). ApoA-I was visualized as described above.

2.6. Thin Layer Chromatography (TLC) and TLC Blot Analysis. The lipids obtained from *M. avium* were spotted onto three Silica Gel 60 plates (Merck Millipore). The TLC plates were simultaneously developed with $CHCl_3/CH_3OH$ (95 : 5, v/v). After the run was completed, one of the plates was treated with 20% sulfuric acid solution to visualize the separated lipids, and the other 2 were subjected to TLC blot analysis according to a previously described method [21]. Briefly, the TLC plates were dipped into isopropyl alcohol containing 0.2% aqueous $CaCl_2$. The separated lipids were thermally transferred onto a PVDF membrane. One membrane was incubated with 1% bovine serum albumin (BSA) in TBS-T and subsequently incubated with 83 μg/mL of apoA-I in TBS-T at 4°C overnight. The membrane was washed with TBS-T and incubated with the POD-conjugated anti-apoA-I polyclonal antibody (Binding Site). ApoA-I was visualized as described above. To analyze by matrix-assisted laser desorption ionization time-of-flight mass spectrometry (MALDI-TOF MS), the lipid bound to apoA-I was extracted from the other membrane by means of $CHCl_3/CH_3OH$ (2 : 1, v/v) according to comparison with the position of the apoA-I spot on the membrane visualized by the TLC blotting described above.

2.7. Growth Curve Assay. Live *M. avium* cells (3×10^8/mL) were briefly sonicated to disperse clumps and mixed with the Middlebrook 7H9 broth supplemented with OADC with or without HDL (50 μg protein/mL). Absorbance of the medium at 530 nm was monitored for 144 h.

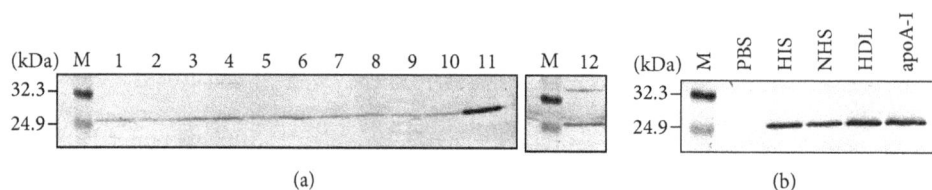

(a)

(b)

FIGURE 1: Interaction between bacteria and HDL. (a) The bacterial samples (lane 1: *Pseudomonas aeruginosa*, lane 2: *Aeromonas* species, lane 3: *Klebsiella oxytoca*, lane 4: *Klebsiella pneumoniae*, lane 5: *Serratia marcescens*, lane 6: *Citrobacter koseri*, lane 7: *Pseudomonas aeruginosa*, lane 8: *Enterobacter cloacae*, lane 9: *Acinetobacter baumannii*, lane 10: *Stenotrophomonas maltophilia*, lane 11: *Mycobacterium avium,* and lane 12: *Escherichia coli*) were mixed with normal human serum (NHS). The NHS conjugated bacteria were washed and solubilized by means of lysis buffer. The lysates were separated by 12.5% SDS-PAGE and immunoblotted with anti-apoA-I antibody. The molecular weight markers are also shown (M). (b) *M. avium* was incubated with NHS, heat-inactivated serum (HIS), HDL, and apoA-I for 10 min at 37°C. After washing with PBS, *M. avium* was solubilized and subjected to immunoblotting for apoA-I as described above.

2.8. Phagocytosis Assay Using Flow Cytometry. This assay was carried out according to the previously described method [22, 23]. Briefly, *M. avium* was autoclaved and stained with 0.5 mg/mL fluorescein isothiocyanate (FITC, DOJIN laboratories) in PBS for 30 min. After exhaustive washing, we resuspended the stained *M. avium* at 3×10^8/mL in PBS and then stored the suspension at −80°C until use. THP-1 macrophages (10^6/well) were cultured with the FITC-conjugated *M. avium* in the presence of HDL (50 μg protein/mL), apoA-I (50 μg/mL), or BSA (50 μg/mL). After quenching extracellular fluorescence using trypan blue (1.2 mg/dL), we fixed the cells with CellFIX (Becton Dickinson) and subjected them to flow cytometric analysis on a Navios flow cytometer (Beckman Coulter). The data were analyzed in the Kaluza software (Beckman Coulter). The FITC-conjugated *E. coli*, which was prepared by the similar method described above, was also analyzed by the phagocytosis assay as the reference.

2.9. Assay of Colony-Forming Units (CFUs). THP-1 macrophages (10^6/well) were infected with live *M. avium* (MOI 20 : 1) with or without HDL (50 μg protein/mL) for 24 h. The cells engulfing *M. avium* were washed three times with PBS and further cultured in the absence of *M. avium* and HDL for 0, 24, and 48 h. Then, THP-1 macrophages were lysed with PBS containing 0.1% SDS to recover the engulfed *M. avium*. The lysates diluted 100-fold with PBS were cultured on Middlebrook 7H11 agar plates (Difco) supplemented with OADC. CFUs were counted after 3-week incubation at 37°C.

2.10. Staining and Quantification of LDs. THP-1 macrophages (5×10^5/well) were infected with live *M. avium* (MOI 20 : 1) with or without HDL (50 μg protein/mL) for 24 h. After being fixed with a 10% formaldehyde solution for 15 min, they are stained with Oil red O (ORO) for 5 min. The cells were counterstained with Mayer's hematoxylin for 2 min in samples to be examined under a light microscope. Lipid bodies were counted for 50 cells. For quantification of LDs amount, infected THP-1 macrophages (5×10^5/well) were stained with only ORO as described above. ORO was then extracted in 500 μL of 100% isopropyl alcohol, and absorbance was measured at 540 nm as described previously [24].

2.11. Statistical Analysis. All of the values were shown by the mean ± standard error of the mean (SEM). Statistical analysis was performed by R software. CFUs, the number of LDs, and the others were statistically analyzed by paired Student's *t*-test, Kruskal-Wallis test, and multifactorial analysis of variance (ANOVA) followed by Tukey post hoc tests, respectively.

2.12. Ethics Statement. Blood samples were obtained from healthy donors who had provided written informed consent. The protocol of this study was approved by the Research Ethics Committee of Tokyo Medical and Dental University (decision number 1491).

3. Results

3.1. Interaction of M. avium with HDL through apoA-I. To screen the bacteria for interaction with HDL, some species of bacteria that were incubated with normal human serum (NHS) were analyzed by western blotting with an anti-apoA-I antibody (Figure 1(a)). The apoA-I signal was well visible in the lysate of *M. avium* incubated with NHS but only barely visible in samples from other Gram-negative bacteria. To confirm the direct interaction between apoA-I and *M. avium*, we carried out western blotting of the lysate of *M. avium* incubated with heat-inactivated serum (HIS), NHS, isolated HDL, or purified apoA-I (Figure 1(b)). The apoA-I signal was well visible after incubation not only with NHS but also with the HIS, HDL, and apoA-I, indicating that HDL bound directly to *M. avium* via apoA-I.

3.2. ApoA-I Binds to the Outer Cell Wall Lipid of M. avium. We hypothesized that apoA-I interacts with the abundant lipids on the *M. avium* cell wall because of amphipathic α-helices present in the apoA-I molecule. To test this hypothesis, we performed a western blot assay of the lipids extracted from the outer cell wall of *M. avium* (Figure 2(a)). ApoA-I was detected in the spots of the extracted lipids incubated with HDL and the intensity of signals increased with the amount of spotted lipids (at a constant HDL concentration). The TLC blot assay was carried out to determine whether apoA-I binds to a specific lipid molecule of *M. avium* (Figure 2(b)). Several lipid spots on the TLC plate were nonspecifically visualized by a charring reagent (Figure 2(b), left). One

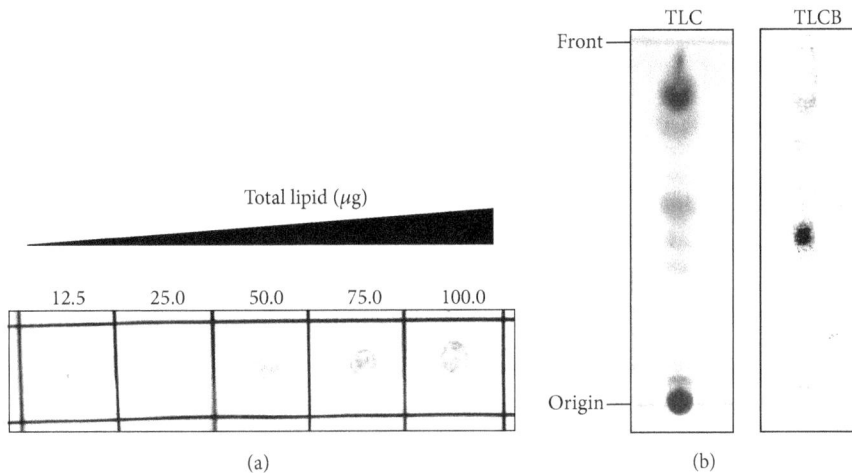

(a)

(b)

FIGURE 2: Interaction between lipids of *M. avium* and HDL. (a) The lipids extracted from *M. avium* were spotted in the indicated amount (12.5–100.0 μg) on a nitrocellulose membrane. The membrane was then incubated with HDL (50 μg protein/mL), and HDL bound to the lipids was visualized using an anti-apoA-I antibody. (b) TLC plates with the spotted lipids (150 μg each sample) extracted from *M. avium* were developed using $CHCl_3/CH_3OH$ (95 : 5, v/v). One plate was treated with a 20% H_2SO_4 solution to visualize the fractionated lipids (TLC). The fractionated lipids on the other plate were thermally transferred onto a PVDF membrane, which was then incubated with HDL (83 μg protein/mL). The lipid that bound to HDL was visualized using an anti-apoA-I antibody (TLCB).

of those spots visibly interacted with apoA-I by TLC blot analysis (Figure 2(b), right), indicating that apoA-I formed a complex with this specific lipid of *M. avium*.

The lipids extracted from the apoA-I-positive spot of the TLC blot were analyzed by MALDI-TOF MS (Supplemental Figure 1 in Supplementary Material available online at http://dx.doi.org/10.1155/2016/4353620). Two prominent peaks at m/z 1243.8 and m/z 1271.8 were observed along with the several peaks with an interval of 28 Da, suggesting that the lipid that bound to apoA-I included a variety of fatty acid moieties.

3.3. Analysis of Bactericidal Activity of HDL.
The growth curve analysis of *M. avium* in the presence or absence of HDL was performed to determine whether HDL directly kills *M. avium* (Figure 3). No significant difference was observed in the absorbance of the medium with or without HDL for 6 days. This finding indicated that HDL itself had no bactericidal activity toward *M. avium*.

3.4. Effects of HDL on Engulfment of M. avium by THP-1 Macrophages.
To find out whether HDL participates in the nonopsonic phagocytosis of *M. avium* by THP-1 macrophages, we used flow cytometry to analyze the effect of HDL on the number of THP-1 macrophages engulfing FITC-conjugated *M. avium* (Figure 4(a)). The percentage of FITC-positive cells decreased with the addition of HDL and apoA-I but did not change significantly after addition of BSA. The percentage of FITC-positive cells which is obtained using FITC-conjugated *E. coli* indicated a similar result to that of FITC-conjugated *M. avium*; however the addition of HDL did not decrease the percentage of FITC-positive cells. When the number of FITC-positive cells of control samples was set

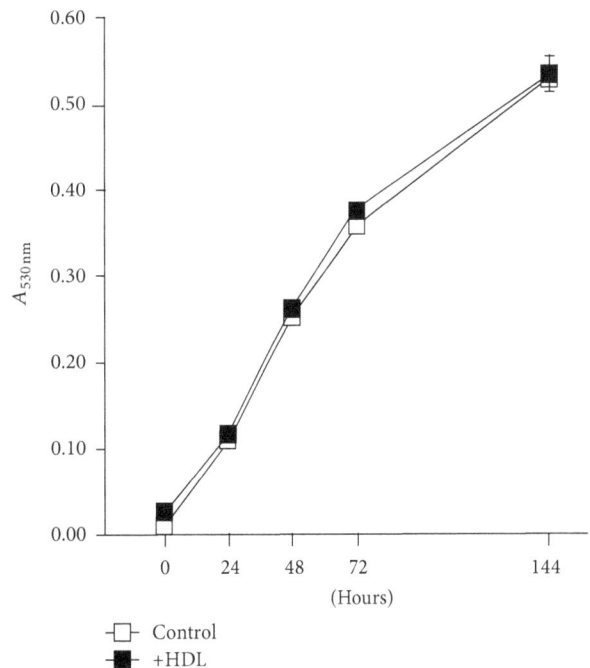

FIGURE 3: Bactericidal activity of HDL. *M. avium* was cultured with (+HDL; 50 μg protein/mL) or without (control) HDL and absorbance (at 530 nm) of the medium was measured at 0, 24, 48, 72, and 144 h. The values were indicated by mean ± SEM ($n = 3$).

to 100%, HDL and apoA-I decreased the number of FITC-positive cells to 68.7% ± 1.5% and 83.3% ± 0.4%, respectively (Figure 4(b)). BSA did not affect the engulfment of *M. avium* by THP-1 macrophages (101.8% ± 2.0%). In contrast, FITC-positive cells did not decrease, if anything increased, in the case of FITC-conjugated *E. coli*.

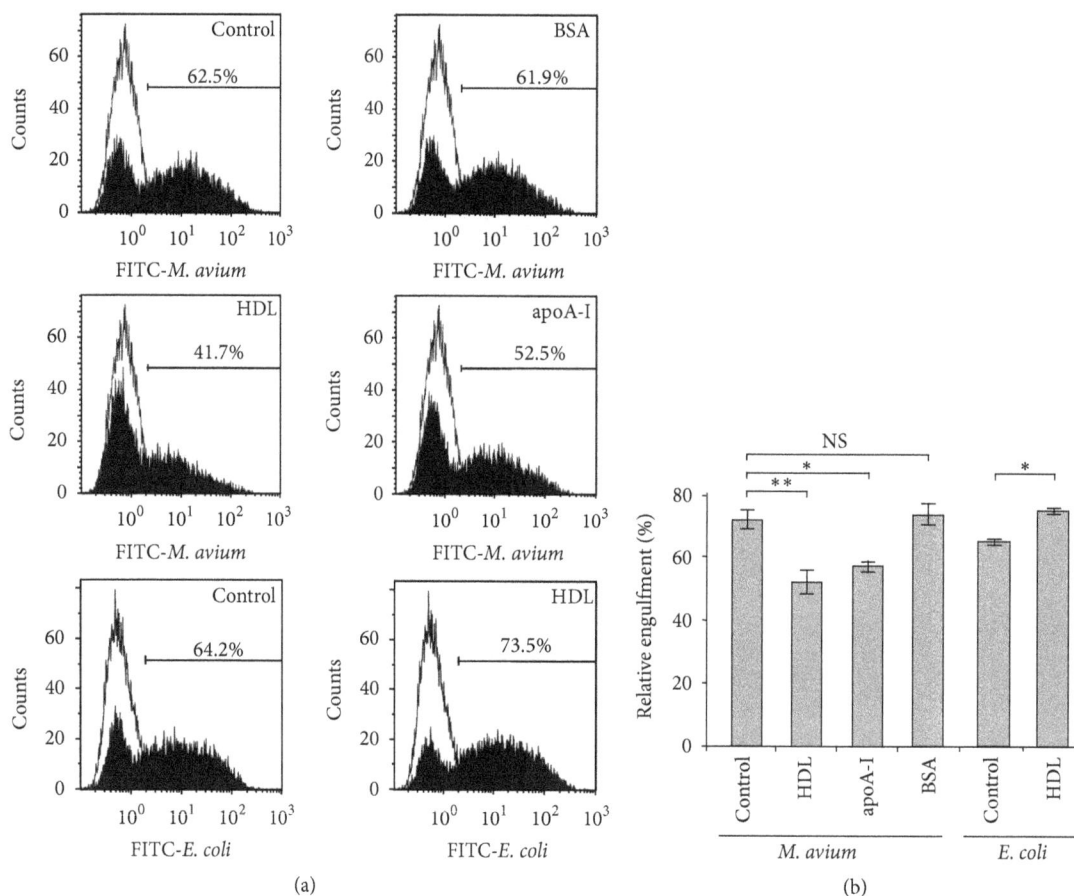

FIGURE 4: Effects of HDL on the engulfment of *M. avium* by THP-1 macrophages. (a) THP-1 macrophages (10^6/well) were cultured with FITC-conjugated *M. avium* or *E. coli* in the presence or absence (control) of HDL (50 μg protein/mL), apoA-I (50 μg/mL), or BSA (50 μg/mL) for 24 h. After extensive washing and quenching of extracellular bacteria, the cells were subjected to flow cytometric analysis. Representative scatter profiles are presented. (b) The percentages of FITC-positive cells compared to the control using FITC-*M. avium* or FITC-*E. coli* were quantitatively compared. The values were indicated by mean ± SEM ($n = 3$, $^*P < 0.05$, $^{**}P < 0.01$). NS: not significant.

3.5. Effects of HDL on Bactericidal Activity of THP-1 Macrophages. Assay of CFUs was performed to assess the number of live *M. avium* inside THP-1 macrophages by means of *M. avium* recovered from the lysed THP-1 macrophages (Figure 5(a)). The results that were expressed in CFU/well were $(8.5 ± 1.2) × 10^4$ and $(11.7 ± 0.6) × 10^4$ (mean ± SE) in the presence and the absence of HDL, respectively. This observation is consistent with the result obtained by the flow cytometric analysis (Figure 4), namely, that HDL attenuated the engulfment of *M. avium* by THP-1 macrophages. To assess the survival rate of *M. avium* inside THP-1 macrophages, the infected THP-1 macrophages were further cultured in the absence of *M. avium* and HDL for 24 and 48 h. The number of CFUs per well indicated a tendency to decrease in a time-dependent manner (Figure 5(b)). However, no significant difference was observed in the decreasing profiles between the presence and the absence of HDL, suggesting that the coexistence of HDL could not affect bacterial killing after ingestion by THP-1 macrophage.

3.6. Effects of HDL on LDs Formation. After the infection of THP-1 macrophages with live *M. avium*, LDs inside the cells were stained with Oil red O (ORO) (Figure 6(a)). The numbers of LDs per cell were significantly increased by the coexistence of live *M. avium* and HDL (Figure 6(b)). LDs amounts were also estimated by the absorbance at 540 nm of the extracted ORO (Figure 6(c)). The existence of HDL or *M. avium* alone indicated no significant effect in the quantity of ORO compared with the control. However, the coexistence of HDL and *M. avium* caused an increase in the LDs amount.

4. Discussion

In this study, we found that HDL binds to *M. avium* via the interaction between apoA-I (the main structural component of HDL) and a specific lipid molecule of *M. avium*. The specific lipid could be extremely nonpolar considering its behavior during TLC analysis. Although a structure of the specific lipid molecule was analyzed by MALDI-TOF-MS/MS (data not shown), we failed to identify it among the known lipid molecules of *M. avium*.

It should be noted that *M. avium* causes inflammation and increases permeability of blood vessels, increasing the chance of interaction with leaked HDL. If this specific lipid of *M.*

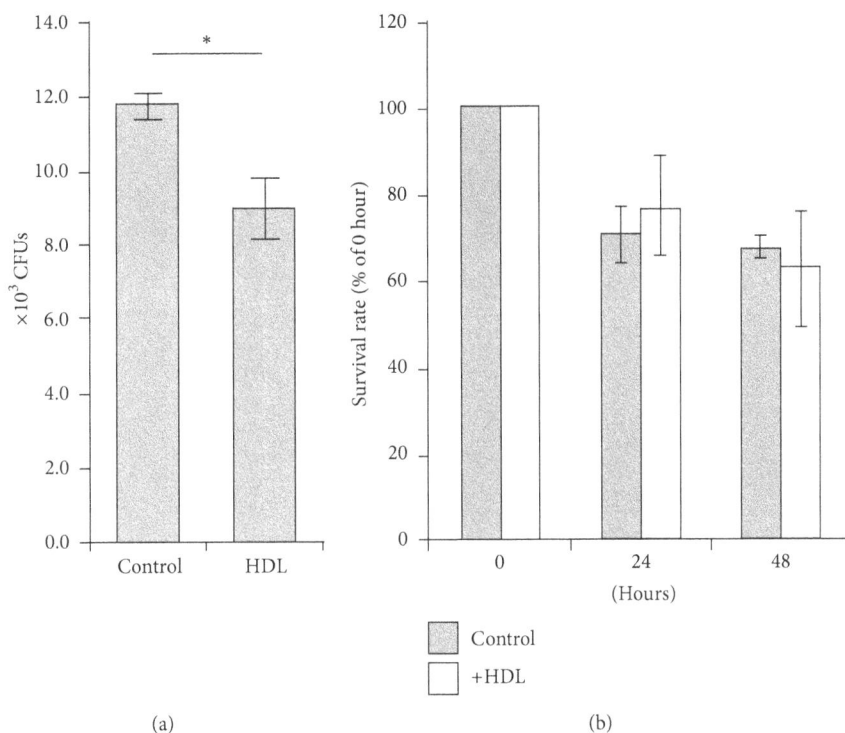

FIGURE 5: Effects of HDL on the viability of *M. avium* engulfed by THP-1 macrophages. (a) THP-1 macrophages (10^6/well) were infected with live *M. avium* in the presence or absence (control) of HDL (50 μg protein/mL) for 24 h. *M. avium* was recovered from lysed THP-1 macrophages for an assay of colony-forming units (CFUs; 3 weeks culture). (b) A part of infected THP-1 macrophages was further cultured for 24 and 48 h in a fresh medium. CFUs were also determined as described above. The data were indicated as the percentages against CFUs obtained without further cultivation. The values were indicated by mean ± SEM ($n = 3$, $^*P < 0.05$).

avium plays a role in host pathogenesis, HDL may neutralize the activity of this lipid, just as in the case of LPS and LTA. It could be important to identify the lipid molecule in order to elucidate a participation of HDL in *M. avium* infection.

We explored the influence of HDL's binding to *M. avium* on an innate immunity in experiments with THP-1 macrophages. HDL does not directly kill *M. avium*, but the presence of HDL significantly decreases the number of THP-1 macrophages engulfing *M. avium*, and HDL indicated no effect on bacterial killing after ingestion. It suggests that HDL allows *M. avium* to escape engulfment by THP-1 macrophages. SR-B1, known as an apoA-I receptor, also plays a role in nonopsonic phagocytosis of mycobacteria [16, 17]. According to these observations, we can hypothesize that HDL competitively inhibits the binding of *M. avium* to SR-B1. In this study, we did not analyze SR-B1, and further research is needed to assess the interaction between *M. avium* and HDL, including the participation of SR-B1.

When mycobacteria are recognized by Toll-like receptors 2 or 6 (TLR2/6), LDs formation is enhanced by TLR signaling [25, 26]. Moreover, some researchers proposed a mechanism behind the enhancement of LDs formation during mycobacterial infection: upregulation of a series of scavenger receptors including SR-B1 [27]. In our study, *M. avium* infection did not apparently upregulate LDs formation under serum-free condition; however, the coexistence of

HDL increased the number and the amount of LDs in THP-1 macrophages, suggesting that HDL-cholesterol is utilized as a raw material for LDs formation. Two mechanisms may explain this observation. According to one mechanism, HDL-cholesterol is internalized together with *M. avium* via phagocytosis. According to the other, although approximately 30% of the engulfment of *M. avium* by THP-1 macrophages was attenuated by the coexistence of HDL (Figure 4), the HDL-cholesterol influx is enhanced by upregulation of SR-B1 in THP-1 macrophages engulfing *M. avium*. Consequently, TLR signaling and HDL-cholesterol influx via SR-B1 may promote LDs formation in the presence of both *M. avium* and HDL. Further research is needed to determine the actual mechanism.

On the basis of the present study, we believe that HDL plays a crucial role in the infection of THP-1 macrophages by *M. avium*. Identification of the lipid molecule of *M. avium* that binds to HDL via apoA-I and the analysis of the expression of the scavenger receptor, SR-B1, in an upcoming project, should shed some light on the pathogenesis of *M. avium* infections and on the relevant immune response.

5. Conclusion

HDL affects *M. avium* infection through the binding between apoA-I, the main component of HDL, and a specific lipid of

(a)

(b)

(c)

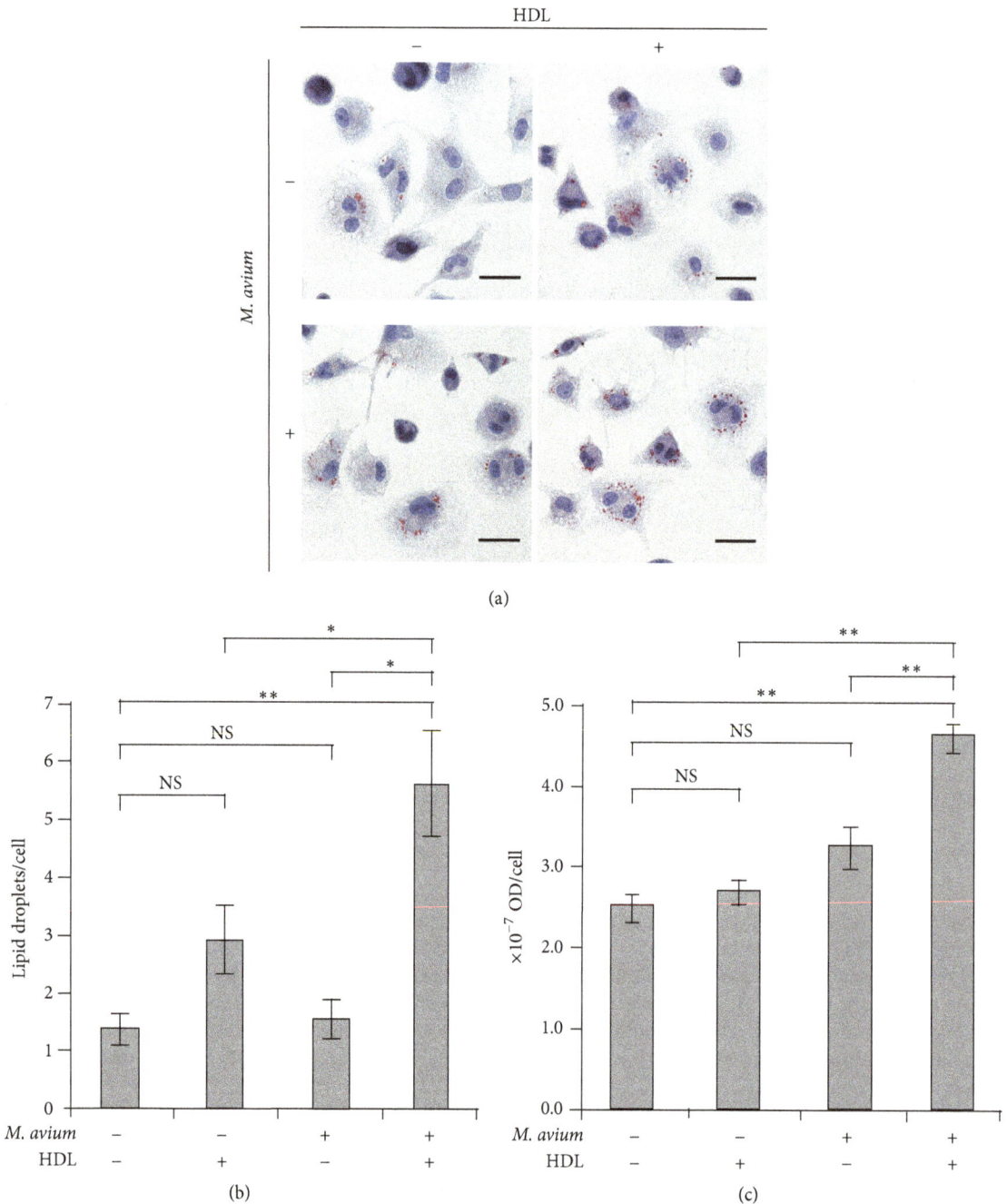

FIGURE 6: High-density lipoprotein (HDL) promotes a formation of THP-1 macrophage-derived foam cells during *M. avium* infection. THP-1 macrophages were infected by *M. avium* with (+) or without (−) HDL. (a) Representative light microscopic images of the cells stained with Oil red O (ORO) and hematoxylin are shown. The scale bar is 25 nm. (b) The numbers of lipid droplets (LDs) per cell were calculated by counting of those number for 50 cells. (c) The amount of LDs was also quantitatively analyzed by measuring absorbance (at 540 nm) of ORO extracted from the cells. The values were indicated by mean ± SEM ($n = 3$, $^*P < 0.05$, $^{**}P < 0.01$). NS: not significant.

M. avium. This interaction may enable *M. avium* to escape from innate immunity THP-1 macrophages.

Competing Interests

The authors declare that there are no competing interests regarding the publication of this paper.

Acknowledgments

The authors thank Dr. Naoko Tojo (Department of Clinical Laboratory, Medical Hospital of Tokyo Medical and Dental University) for generously helping their collaborators and coauthors. They also thank Dr. Etsuko Sawabe (Department of Clinical Laboratory, Medical Hospital of Tokyo Medical and Dental University) for generously providing assistance with handling of the microorganisms.

References

[1] K. R. Feingold and C. Grunfeld, "The role of HDL in innate immunity," *Journal of Lipid Research*, vol. 52, no. 1, pp. 1–3, 2011.

[2] L. Cai, A. Ji, F. C. de Beer, L. R. Tannock, and D. R. van der Westhuyzen, "SR-BI protects against endotoxemia in mice through its roles in glucocorticoid production and hepatic clearance," *The Journal of Clinical Investigation*, vol. 118, no. 1, pp. 364–375, 2008.

[3] J. H. M. Levels, P. R. Abraham, E. P. Van Barreveld, J. C. M. Meijers, and S. J. H. Van Deventer, "Distribution and kinetics of lipoprotein-bound lipoteichoic acid," *Infection and Immunity*, vol. 71, no. 6, pp. 3280–3284, 2003.

[4] M. Biedzka-Sarek, J. Metso, A. Kateifides et al., "Apolipoprotein A-I exerts bactericidal activity against *Yersinia enterocolitica* serotype O:3," *The Journal of Biological Chemistry*, vol. 286, no. 44, pp. 38211–38219, 2011.

[5] I. M. Orme and D. J. Ordway, "Host response to nontuberculous mycobacterial infections of current clinical importance," *Infection and Immunity*, vol. 82, no. 9, pp. 3516–3522, 2014.

[6] C. H. Weiss and J. Glassroth, "Pulmonary disease caused by nontuberculous mycobacteria," *Expert Review of Respiratory Medicine*, vol. 6, no. 6, pp. 597–613, 2012.

[7] D. A. Keown, D. A. Collings, and J. I. Keenana, "Uptake and persistence of *Mycobacterium avium* subsp. paratuberculosis in human monocytes," *Infection and Immunity*, vol. 80, no. 11, pp. 3768–3775, 2012.

[8] P. Peyron, C. Bordier, E.-N. N'Diaye, and I. Maridonneau-Parini, "Nonopsonic phagocytosis of *Mycobacterium kansasii* by human neutrophils depends on cholesterol and is mediated by CR3 associated with glycosylphosphatidylinositol-anchored proteins," *Journal of Immunology*, vol. 165, no. 9, pp. 5186–5191, 2000.

[9] J. Gatfield and J. Pieters, "Essential role for cholesterol in entry of mycobacteria into macrophages," *Science*, vol. 288, no. 5471, pp. 1647–1651, 2000.

[10] K. A. Mattos, V. C. G. Oliveira, M. Berrêdo-Pinho et al., "Mycobacterium leprae intracellular survival relies on cholesterol accumulation in infected macrophages: a potential target for new drugs for leprosy treatment," *Cellular Microbiology*, vol. 16, no. 6, pp. 797–815, 2014.

[11] C. de Chastellier and L. Thilo, "Cholesterol depletion in *Mycobacterium avium*-infected macrophages overcomes the block in phagosome maturation and leads to the reversible sequestration of viable mycobacteria in phagolysosome-derived autophagic vacuoles," *Cellular Microbiology*, vol. 8, no. 2, pp. 242–256, 2006.

[12] K. Tanigawa, K. Suzuki, H. Kimura et al., "Tryptophan aspartate-containing coat protein (CORO1A) suppresses Toll-like receptor signalling in Mycobacterium leprae infection," *Clinical and Experimental Immunology*, vol. 156, no. 3, pp. 495–501, 2009.

[13] K. A. Mattos, F. A. Lara, V. G. C. Oliveira et al., "Modulation of lipid droplets by Mycobacterium leprae in Schwann cells: a putative mechanism for host lipid acquisition and bacterial survival in phagosomes," *Cellular Microbiology*, vol. 13, no. 2, pp. 259–273, 2011.

[14] A. Brzostek, J. Pawelczyk, A. Rumijowska-Galewicz, B. Dziadek, and J. Dziadek, "Mycobacterium tuberculosis is able to accumulate and utilize cholesterol," *Journal of Bacteriology*, vol. 191, no. 21, pp. 6584–6591, 2009.

[15] S. Xu, M. Laccotripe, X. Huang, A. Rigotti, V. I. Zannis, and M. Krieger, "Apolipoproteins of HDL can directly mediate binding to the scavenger receptor SR-BI, an HDL receptor that mediates selective lipid uptake," *Journal of Lipid Research*, vol. 38, no. 7, pp. 1289–1298, 1997.

[16] J. A. Philips, E. J. Rubin, and N. Perrimon, "Drosophila RNAi screen reveals CD36 family member required for mycobacterial infection," *Science*, vol. 309, no. 5738, pp. 1251–1253, 2005.

[17] G. Schäfer, R. Guler, G. Murray, F. Brombacher, and G. D. Brown, "The role of scavenger receptor B1 in infection with mycobacterium tuberculosis in a murine model," *PLoS ONE*, vol. 4, no. 12, Article ID e8448, 2009.

[18] Y. Usami, K. Matsuda, M. Sugano et al., "Detection of chymase-digested C-terminally truncated apolipoprotein A-I in normal human serum," *Journal of Immunological Methods*, vol. 369, no. 1-2, pp. 51–58, 2011.

[19] J. M. Stevenson, I. Y. Perera, and W. F. Boss, "A phosphatidylinositol 4-kinase pleckstrin homology domain that binds phosphatidylinositol 4-monophosphate," *Journal of Biological Chemistry*, vol. 273, no. 35, pp. 22761–22767, 1998.

[20] J. Folch, M. Lees, and G. H. Sloane Stanley, "A simple method for the isolation and purification of total lipides from animal tissues," *The Journal of Biological Chemistry*, vol. 226, no. 1, pp. 497–509, 1957.

[21] T. Taki and D. Ishikawa, "TLC blotting: application to microscale analysis of lipids and as a new approach to lipid-protein interaction," *Analytical Biochemistry*, vol. 251, no. 2, pp. 135–143, 1997.

[22] M. J. Hewish, A. M. Meikle, S. D. Hunter, and S. M. Crowe, "Quantifying phagocytosis of *Mycobacterium avium* complex by human monocytes in whole blood," *Immunology & Cell Biology*, vol. 74, no. 4, pp. 306–312, 1996.

[23] J. Nuutila and E.-M. Lilius, "Flow cytometric quantitative determination of ingestion by phagocytes needs the distinguishing of overlapping populations of binding and ingesting cells," *Cytometry Part A*, vol. 65, no. 2, pp. 93–102, 2005.

[24] Q. Lin, Y.-J. Lee, and Z. Yun, "Differentiation arrest by hypoxia," *Journal of Biological Chemistry*, vol. 281, no. 41, pp. 30678–30683, 2006.

[25] K. A. Mattos, V. G. C. Oliveira, H. D'Avila et al., "TLR6-driven lipid droplets in Mycobacterium leprae-infected Schwann cells: immunoinflammatory platforms associated with bacterial persistence," *Journal of Immunology*, vol. 187, no. 5, pp. 2548–2558, 2011.

[26] H. D'Avila, R. C. N. Melo, G. G. Parreira, E. Werneck-Barroso, H. C. Castro-Faria-Neto, and P. T. Bozza, "Mycobacterium bovis bacillus Calmette-Guérin induces TLR2-mediated formation of lipid bodies: Intracellular domains for eicosanoid synthesis in vivo," *Journal of Immunology*, vol. 176, no. 5, pp. 3087–3097, 2006.

[27] A. A. Elamin, M. Stehr, and M. Singh, "Lipid droplets and *Mycobacterium leprae* infection," *Journal of Pathogens*, vol. 2012, Article ID 361374, 10 pages, 2012.

11

Influence on Adiposity and Atherogenic Lipaemia of Fatty Meals and Snacks in Daily Life

Antonio Laguna-Camacho

Medical Sciences Research Centre, Autonomous University of the State of Mexico, Toluca, MEX, Mexico

Correspondence should be addressed to Antonio Laguna-Camacho; alagunaca@uaemex.mx

Academic Editor: Zufeng Ding

The present work reviewed the connections of changes in consumption of high-fat food with changes in adiposity and lipaemia in adults with overweight or obesity. Hyperlipaemia from higher fat meals and excessive adiposity contributes to atherogenic process. Low-fat diet interventions decrease body fat, lipaemia, and atherosclerosis markers. Inaccuracy of physical estimates of dietary fat intake remains, however, a limit to establishing causal connections. To fill this gap, tracking fat-rich eating episodes at short intervals quantifies the behavioural frequency suggested to measure (by regression of changes in real time) direct effects of this eating pattern on adiposity and atherogenic lipaemia. Such evidence will provide the basis for an approach focused on a sustained decrease in frequency of fatty meals or snacks to reduce obesity, hyperlipaemia, and atherosclerosis.

1. Introduction

Atherosclerosis is globally a leading cause of death [1]. Atherogenic conditions, including dyslipidaemia, are associated with excessive adiposity [2]. The high prevalence of overweight and obesity in industrialised nations posits a serious public health concern [3]. This epidemic of obesity involves widespread patterns of unhealthy eating and physical inactivity. Although this research field is highly documented (i.e., [4]), the influence of habitual eating and exercise patterns on the atherogenic process remains to be measured as it occurs.

The current work articulates research on ingestion of dietary fat as a substantial contributor to the levels of adiposity and lipaemia. First, the physiological events are integrated from high-fat food consumption at eating occasions to infiltration of circulating atherogenic lipoprotein particles into the arterial wall. Then, evidence is presented for changes in adiposity, lipaemia, and markers of atherosclerosis by means of interventions to reduce dietary fat intake. Finally, a missing step for establishing causal links is pointed out together with a proposed approach for filling this gap.

The present work considers also the behavioural field to move current research forward. The episodes of fatty meals or snacks are here addressed as a target behaviour pattern (Figure 1). The aim is to argue that episodes of high-fat intake one by one contribute to obesity, hyperlipaemia, and atherosclerosis. Therefore, a proposed implication is that individuals with overweight or obesity would benefit from consuming fewer fatty meals or snacks on a day to day basis than usual.

2. Dietary Fat Intake and Atherogenic Process

After an eating occasion, ingested fat is transported by lipoproteins for deposition as triglycerides within large intracellular lipid droplets in adipocytes without thermogenic cost. That is, lipids are more efficiently stored compared to carbohydrates or proteins. So a pattern of frequently consuming fatty meals or snacks would facilitate the expansion of body fatness. Accordingly, many studies have found association between high-fat diet and increased weight or body mass index [5, 6]. A focus on adiposity was recently proposed as this tissue is primary affected by hypertrophy of adipocytes, which contributes significantly to the inflammatory state found in obesity [7].

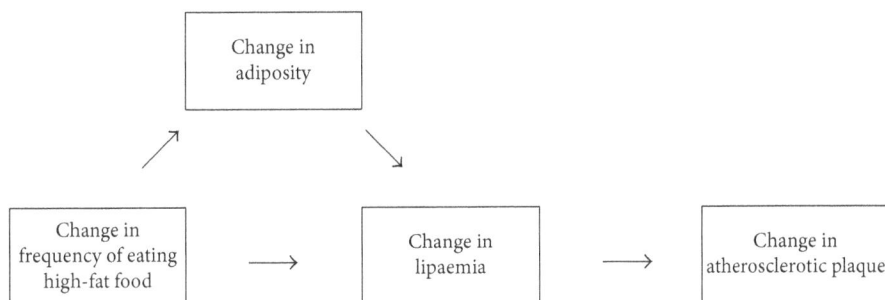

FIGURE 1: Behavioural model of influence of higher fat food consumption on adiposity and atherogenic lipaemia.

Lipolysis rate increases in response to excessive deposition of lipids in adipocytes [8–10]. Hence, hypertrophy of adipocytes contributes to higher release of free fatty acids to circulation. Also, people with overweight or obesity show an increase in plasma free fatty acids after a high-fat meal [11, 12]. The mechanisms of such postprandial abnormalities in human beings remain unclear. However, recent research shows that every episode of excessive fat intake causes adverse systemic effects by a peak of fatty acids in plasma [13, 14]. Thus, the more frequent a person consumes fatty foods, the more continuous also the physiological disturbances that contribute to atherosclerosis. In this case, the high flux of free fatty acids contributes directly to increase the synthesis in liver of lipoproteins that transport lipids in blood [15]. Thus, both high-fat meals and high adiposity cause hyperlipaemia in people with overweight or obesity. To provide an account of the process by which dyslipidaemia causes atherosclerosis, a biomechanical perspective is next considered.

There is an apolipoprotein B (apo B) molecule in the surface of each chylomicron (CM), very low-density lipoprotein (VLDL), intermediate-density lipoprotein (IDL), and low-density lipoprotein (LDL). Apo B has high molecular weight. Therefore, hyperlipaemia can elevate the blood volume or viscosity by higher number of apo B particles [16–19]. Vascular resistance compensates an increased volume of circulating blood [20]. The overstretching of the endothelium thickens the arterial wall and leads eventually to loss of compliance [21].

Endothelial failure contributes to slow blood flow as well as hematic stasis at sites of vascular resistance such as coronary arteries [22]. In state of hyperlipaemia, the blood stagnation at areas susceptible of lesion with diffuse thickening of the intima facilitates the infiltration of apo B containing lipoproteins into the arterial wall [23–25]. Lipoproteins are modified within the vascular endothelium [26]. Modified lipoproteins are a potent activator of immune cells [27]. Macrophages phagocyte modified lipoproteins in the vascular endothelium and release inflammatory cytokines such as interleukin 1, interleukin 6, and tumoral necrosis factor alpha (IL1, IL6, and TNFα) [28–30]. Macrophages release also chemokines to attract lymphocytes Th1 and Th2 that activate B cells that produce the antibodies immunoglobulin M (IgM) and G (IgG) against epitopes in the modified lipoproteins [30, 31]. Such immune response towards modified lipoproteins is a main factor of arterial inflammation [28, 32].

3. Effects of Prescribing to Decrease Dietary Fat Intake

Decreasing fat intake is a common strategy of dietetic interventions to reduce body fat and lipaemia [33–37]. Non-diabetic adults with overweight or obesity who take part in interventions based on low-fat or low-calorie diets, which usually involve strategies for decreasing dietary fat, show on average a reduction of 10% in body weight and 20% in fat mass as well as a decrease of 10% in apo B concentration [37–42]. There is also a rapid influence on apo B levels of dietetic changes. Decrease in apo B concentration of 10% was observed with a DASH diet within four weeks [43].

Prescribing dietary strategies to cut down on fat intake also decreases levels of modified lipoproteins and related immune biomarkers. For instance, reduction in body mass index by lifestyle intervention was associated with decrease in IL-6, protein-C reactive, normal T cell expressed and secreted, and other related markers of immune response and inflammation [44, 45]. People who lost weight in a long-term dietetic intervention showed also decreases in IL-1, IL-6, TNF α, and leukocytes [46].

One of the most studied modified lipoproteins has been the oxidised low-density lipoprotein (ox-LDL) [47]. Like other modified lipoproteins [48], a small fraction of the ox-LDL returns from the atherosclerotic lesion to the circulation [49]. So blood levels of ox-LDL could inform about the atherogenic process [50–55]. Interventions prescribing decrease in dietary fat intake have found parallel reduction in body weight or fat mass and in ox-LDL [56–58]. This is consistent with the mechanism that lowering fat intake and adiposity contributes to slow down the development of atherosclerosis. Indeed, there is reduction in the thickness of the atherosclerotic plaque over two years of maintained weight loss through low-fat diet [59].

4. Measurement Gap to Establish Causal Connections

Despite the well-documented positive association of changes in fat consumption with changes in fat mass and lipaemia, there is a significant bias in the core measure of dietary fat intake, the behavioural dose. Physical estimates of energy and nutrient content of diet are so inaccurate due to underreporting of food amounts that not using them in obesity research is suggested [60, 61]. Estimates are also limited by influence on intake of observing people's behaviour [62, 63]. Hence, researchers are unable to measure reliably enough in everyday life conditions if any prescribed dietary fat reduction causes the observed changes in adiposity, lipaemia, or inflammation biomarkers.

The effect of low-fat diets is taken for granted from outcome measures. However, even randomised controlled trials are not free of confounding effects from other unmeasured variables [64]. For instance, participants are liable to engage in unasked changes, which may include efforts of healthy eating, dieting, or exercising to reduce weight [65, 66]. Differences in baseline characteristics of participants can generate such behavioural responses that might account for the individual variability commonly observed in dietary interventions [67].

In any case, researchers rarely monitor the behaviour changes during dietary interventions. When change in dietary intake is monitored, it is usually at intervals of several months [68]. The effect of a decrease in dietary fat consumption could be achieved within a few weeks when such reduced rate of intake comes into balance with the lower energy expenditure from a reduced body weight [69, 70]. The outcomes of a behaviour change as they physiologically occur are still hardly considered. Identifying this prompt effect is needed for measuring the stepwise changes in adiposity and blood lipids caused by a particular pattern of eating such as eating less fat than usual at meals or between meals [71].

5. An Alternative Behavioural Measure to Narrow the Gap

Evidence from social psychology shows that people are highly accurate in their perception of what they normally do [72]. This psychological capacity helps people share efficiently information about their daily activities with other people. Giving a truthful account is cognitively less demanding than confabulating an event [73]. So researchers can collect reliable accounts of activities as they occur or elicit them without intrusion within the week period after their occurrence when the accuracy of recall is above 80–90% [74–77]. Such psychosocial measures of occurrence of everyday events include eating episodes. The count of a series of consecutive occurrences of a pattern of eating divided by the time of observation (i.e., number of episodes per week) gives a measure of its frequency, another dominant behavioural feature, instead of an estimate of energy/nutrient quantity or dose. Experimental designs that monitor episodes of any particular pattern of behaviour are well established in psychological science [78–80]. Tracking the episodes of an eating pattern informs about how often it is carried out over a time interval and how much a change in that frequency from an interval to another influences body fatness and related biomarkers. In the case of dietary fat consumption, the focus is on tracking episodes when people ate high-fat food at meals or between meals to measure the rate of its occurrence, which provides in turn a behaviour measure to regress changes in frequency of eating high-fat food onto changes in

body weight, adiposity, blood lipids, and related immune or inflammation biomarkers. The slope of the regression gives a measurement of the effect on each of the outcomes per each unit of change in behaviour frequency.

In a small study to decrease unhealthy eating in adults with overweight or obesity, episodes of fatty meals and snacks were daily tracked during four weeks [81]. Participants reported a mean baseline of 12 high-fat meals or snacks a week, a frequency level that decreased throughout the intervention by almost half. Such sustained change in frequency of high-fat eating episodes correlated positively with change in fat mass and with change in ox-LDL. Also, the change in body fat correlated positively with change in ox-LDL.

This preliminary evidence tightens previous research showing that episodes of fatty meals and snacks over time are causally connected with change in adiposity and lipaemia as well as development of atherosclerotic plaque. Moreover, this shows that the effect of eating fatty meals and snacks on the arterial wall occurs within a few weeks.

Pursuing this research avenue would contribute to elucidating the causal links of behaviour patterns with obesity, dyslipidaemia, and atherosclerosis. This research line requires, however, substantial development. For instance, future studies could combine the present measure of frequency of an eating pattern alongside amounts of food eaten on each occasion estimated also with novel approaches that overcome limitations in accuracy of conventional dietary evaluations (i.e., [82]). A characterised pattern of eating would still correspond to a constant intake dose.

6. Concluding Remarks

Broad research shows consistently a positive association of fat intake with adiposity and apo B containing lipoproteins as well as with circulating levels of immune cells, cytokines, and modified lipoproteins. However, as gaps in research remain to accurately measure fat intake, there are still no measurements of causal connection between sustained intake of either more or less dietary fat and changes in anthropometry or physiological markers. The innovative use of psychosocial measures to collect records of the timings of the fatty meals and snacks as they are carried out by individuals in their everyday life would greatly benefit research and substantiate the evidence about the impact of dietary fat consumption on body fat, blood lipids, and biomarkers of atherosclerosis.

Research is beginning to accumulate showing that a bout of high-fat ingestion rapidly initiates metabolic impairments such as acute inflammation [13, 14]. Therefore, it is reasoned that frequent episodes of fat-rich food consumption contribute in real time to the low-grade inflammatory state associated with obesity that leads to cardiovascular complications.

Reducing the habitual number of episodes of unhealthy eating may contribute to minimising the physiological disturbances found in people with overweight or obesity. So monitoring fatty meals or snacks provides an approach based on behaviour frequency that readily generates evidence to inform objective recommendations against unhealthy eating. For instance, similar to the recommendation of exercising more than three times per week [83], a recommendation could be formulated of consuming higher fat food no more than six times a week, which is about half the frequency found in people with unhealthy body mass index [81].

References

[1] C. D. Mathers and D. Loncar, "Projections of global mortality and burden of disease from 2002 to 2030," *PLoS Medicine*, vol. 3, p. e442, 2006.

[2] P. T. James, "Obesity: the worldwide epidemic," *Clinics in Dermatology*, vol. 22, no. 4, pp. 276–280, 2004.

[3] P. Kopelman, "Health risks associated with overweight and obesity," *Obesity Reviews*, vol. 8, no. 1, pp. 13–17, 2007.

[4] S. B. Heymsfield and T. A. Wadden, "Mechanisms, pathophysiology and management of obesity," *The New England Journal of Medicine*, vol. 376, pp. 254–266, 2017.

[5] W. C. Willett, "Dietary fat plays a major role in obesity: No," *Obesity Reviews*, vol. 3, no. 2, pp. 59–68, 2002.

[6] G. A. Bray and B. M. Popkin, "Dietary fat intake does affect obesity!," *American Journal of Clinical Nutrition*, vol. 68, pp. 1157–1173, 1998.

[7] J. I. Mechanick, D. L. Hurley, and W. T. Garvey, "Adiposity-based chronic disease as a new diagnostic term: american association of clinical endocrinologists and the american college of endocrinology position statement," *Endocr Pract*, vol. 23, pp. 372–378, 2017.

[8] E. Lambert and E. J. Parks, "Postprandial metabolism of meal triglyceride in humans," *Biochim et Biophys Acta*, vol. 1821, pp. 721–726, 2012.

[9] H. E. Bays, P. P. Toth, P. M. Kris-Etherton et al., "Obesity, adiposity, and dyslipidemia: a consensus statement from the National Lipid Association," *Journal of Clinical Lipidology*, vol. 7, no. 4, pp. 304–383, 2013.

[10] D. Langin, "In and out: Adipose tissue lipid turnover in obesity and dyslipidemia," *Cell Metabolism*, vol. 14, no. 5, pp. 569-570, 2011.

[11] K. G. Jackson, S. Lockyer, A. L. Carvalho-Wells, C. M. Williams, A. M. Minihane, and Lovegrove, "Dietary fat manipulation has a greater impact on lipid metabolism than the apolipoprotein E (epsilon) genotype-insights from the SAT gene study," *Molecular Nutrition and Food Research*, vol. 56, pp. 1761–1770, 2012.

[12] C. S. Katsanos, "Clinical considerations and mechanistic determinants of postprandial lipemia in older adults," *Advances in Nutrition*, vol. 5, no. 3, pp. 226–234, 2014.

[13] M. Herieka and C. Erridge, "High-fat meal induced postprandial inflammation," *Molecular Nutrition and Food Research*, vol. 58, pp. 136–146, 2014.

[14] E. Alvarez-Hernande, S. Kahl, A. Seelig, P. Begovatz, M. Irmler et al., "Acute dietary fat intake initiates alterations in energy metabolism and insulin resistance," *Journal of Clinical Investigation*, vol. 127, pp. 295–708, 2017.

[15] H. Duez, B. Lamarche, R. Valero et al., "Both intestinal and hepatic lipoprotein production are stimulated by an acute elevation of plasma free fatty acids in humans," *Circulation*, vol. 117, pp. 2369-2376, 2008.

[16] A. H. Seplowitz, S. Chien, and F. R. Smith, "Effects of lipoproteins on plasma viscosity," *Atherosclerosis*, vol. 38, pp. 89–95, 1981.

[17] S. B. Solerte, M. Fioravanti, A. L. Patti et al., "Increased plasma apolipoprotein levels and blood hyperviscosity in non-insulin-dependent diabetic patients: role of the occurrence of arterial hypertension," *Acta Diabetologica Latina*, vol. 24, pp. 341–349, 1987.

[18] W. Koenig, M. Sund, E. Ernst, W. Mraz, V. Hombach, and U. Keil, "Association between rheology and components of lipoproteins in human blood: results from the MONICA project," *Circulation*, vol. 85, pp. 2197–2204, 1992.

[19] A. Slyper, A. Le, J. Jurva, and D. Gutterman, "The influence of lipoproteins on whole-blood viscosity at multiple shear rates," *Metabolism*, vol. 54, pp. 764–768, 2005.

[20] R. B. Devereux, D. B. Case, M. H. Alderman, T. G. Pickerung, S. Chien, and J. H. Laragh, "Possible role of increased blood viscosity in the hemodynamics of systemic hypertension," *American Journal of Cardiology*, vol. 85, pp. 1265–1267, 2000.

[21] K. R. Kensey, "The mechanistic relationships between hemorheological characteristics and cardiovascular disease," *Current Medical Research and Opinion*, vol. 19, no. 7, pp. 587–596, 2003.

[22] J. M. Jung, *A study on rheological properties of blood and improvements with high-voltage plasma discharge [Ph.D. thesis]*, Drexel University, 2012.

[23] J. H. Rapp, A. Lespine, R. L. Hamilton et al., "Triglyceride-rich lipoproteins isolated by selected-affinity anti-apolipoprotein B immunosorption from human atherosclerotic plaque," *Arteriosclerosis and Thrombosis*, vol. 14, pp. 1767–1774, 1994.

[24] K. J. Williams and I. Tabas, "The response-to-retention hypothesis of early atherogenesis," *Arterioscler Thromb Vasc Biol*, vol. 15, pp. 551–561, 1995.

[25] S. D. Proctor and J. C. Mamo, "Retention of fluorescent-labelled chylomicron remnants within the intima of the arterial-wall evidence that plaque cholesterol may be derived from postprandial lipoproteins," *European Journal of Clinical Investigation*, vol. 28, pp. 497–503, 1998.

[26] T. Hevonoja, M. O. Pentikainen, M. T. Hyvonen, P. T. Kovanen, and M. Ala-Corpela, "Structure of low density lipoproteoin particles: basis for undestanding molecular changes in modified LDL," *Biochimica et Biophysica Acta*, vol. 1488, pp. 189–210, 2000.

[27] M. J. Hubler and A. J. Kennedy, "Role of lipids in the metabolism and activation of immune cells," *The Journal of Nutritional Biochemistry*, vol. 34, pp. 1–7, 2016.

[28] G. K. Hansson and P. Libby, "The immune response in atherosclerosis: a double-edged sword," *Nature Reviews Immunology*, vol. 6, pp. 508–519, 2006.

[29] S. Samson, L. Mundkur, and V. Vakkar, "Immune response to lipoproteins in atherosclerosis," *Cholesterol*, Article ID 571846, 2012.

[30] A. K. L. Robertson and G. K. Hansson, "T cells in atherogenesis: for better or for worse?" *Arteriosclerosis, Thrombosis, and Vascular Biology*, vol. 26, pp. 2421–2432, 2006.

[31] J. Frostegard, A. K. Ulfgren, P. Nyberg et al., "Cytokine expression in advanced human atherosclerotic plaques: dominance of pro-inflammatory (Th1) and macrophage-stimulating cytokines," *Atherosclerosis*, vol. 145, pp. 33–43, 1999.

[32] G. K. Hansson and A. Hermansson, "The immune system in atherosclerosis," *Nature Immunology*, vol. 12, pp. 204–212, 2011.

[33] D. C. Chan, G. F. Watts, T. W. K. Ng, S. Yamashita, and P. H. Barret, "Effect of weight loss on markers of triglyceride-rich lipoprotein metabolism in the metabolic syndrome," *European Journal of Clinical Investigation*, vol. 38, pp. 743–751, 2008.

[34] Y. Cao, D. T. Mauger, C. L. Pelkman, G. Zhao, S. M. Townsend, and P. M. Kris-Etherton, "Effects of moderate (MF) versus lower fat (LF) diets on lipids and lipoproteins: a meta-analysis of clinical trials in subjects with and without diabetes," *Journal of Clinical Lipidology*, vol. 3, no. 1, pp. 19–32, 2009.

[35] S. U. Dombrowski, A. Avenell, and F. F. Sniehott, "Behavioural interventions for obese adults with additional risk factors for morbidity: systematic review of effects on behaviour, weight and disease risk factors," *Obesity Facts*, vol. 3, pp. 377–396, 2010.

[36] L. Hooper, A. Abdelhamid, H. J. Moore, W. Douthwaite, C. M. Skeaff, and C. D. Summerbell, "Effect of reducing total fat intake on body weight: Systematic review and meta-analysis of randomised controlled trials and cohort studies," *BMJ (Online)*, vol. 345, no. 7891, Article ID e7666, 2013.

[37] A. L. Borel, J. A. Nazare, J. Smith et al., "Visceral and not subcutaneous abdominal adiposity reduction drives the benefits of 1-year lifestyle modification program," *Obesity*, vol. 20, pp. 1223–1233, 2012.

[38] P. W. Siri-Tarino, P. T. Williams, H. S. Fernstrom, R. S. Rawlings, and R. M. Krauss, "Reversal of small, dense LDL subclass phenotype by normalization of adiposity," *Obesity*, vol. 17, pp. 1768–1775, 2009.

[39] R. M. Krauss, P. J. Blanche, R. S. Rawlings, H. S. Fernstrom, and P. T. Williams, "Separate effects of reduced carbohydrate intake and weight loss on atherogenic dyslipidemia," *The American Journal of Clinical Nutrition*, vol. 83, pp. 1025–1031, 2006.

[40] G. D. Brinkworth, M. Noakes, J. D. Bucley, J. B. Keogh, and P. M. Clifton, "Long-term effects of a very-low-carbohydrate weight loss diet compared with an isocaloric low-fat diet after 12 mo," *The American Journal of Clinical Nutrition*, vol. 90, pp. 23–32, 2009.

[41] D. C. Chan, G. F. Watts, S. K. Gan, E. M. M. Ooi, and P. H. R. Barrett, "Effect of ezetimibe on hepatic fat, inflammatory markers, and apolipoprotein B-100 kinetics in insulin-resistant obese subjects on a weight loss diet," *Diabetes Care*, vol. 33, pp. 1134–1139, 2010.

[42] E. Pelletier-Beaumont, B. J. Arsenault, N. Almeras et al., "Normalization of visceral adiposity is required to normalize apolipoprotein B levels in response to a healthy eating/physical activity lifestyle modification program in viscerally obese men," *Atherosclerosis*, vol. 221, pp. 577–582, 2012.

[43] L. Hodson, K. E. Harnden, R. Roberts, A. L. Dennis, and K. N. Frayn, "Does the DASH diet lower blood pressure by altering peripheral vascular function?" *Journal of Human Hypertension*, vol. 24, pp. 312–319, 2010.

[44] C. Herder, M. Peltonen, W. Koenig et al., "Systemic immune mediators and lifestyle changes in the prevention of type 2 diabetes: Results from the Finnish Diabetes Prevention Study," *Diabetes*, vol. 55, no. 8, pp. 2340–2346, 2006.

[45] C. Herder, M. Peltonen, W. Koenig et al., "Anti-inflammatory effect of lifestyle changes in the Finnish Diabetes Prevention Study," *Diabetologia*, vol. 52, no. 3, pp. 433–442, 2009.

[46] J. S. Chae, J. K. Paik, R. Kang et al., "Mild weight loss reduces inflammatory cytokines, leucocyte count and oxidative stress in overweight and moderately obese participants treated for 3 years with dietary modification," *Nutrition Research*, vol. 33, pp. 195–203, 2013.

[47] R. Carmena, P. Duriez, and J. C. Fruchart, "Atherogenic lipoprotein particles in atherosclerosis," *Circulation*, vol. 109, pp. III2–III7, 2004.

[48] K. Nakajíma, T. Nakano, and A. Tanaka, "The oxidative modification hypothesis of atherosclerosis: the comparison of atherogenic effects on oxidized LDL and remnant lipoproteins in plasma," *Clinica Chimica Acta*, vol. 367, pp. 36–47, 2006.

[49] H. N. Hodis, D. M. Kramsch, and P. Avogaro, "Biochemical and cytotoxic Characteristics of an in vivo circulating oxidized low density lipoprotein," *Journal of Lipid Research*, vol. 35, pp. 669–677, 1994.

[50] P. Moriel, F. S. Okawabata, and D. S. P. Abdalla, "Oxidized lipoproteins in blood plasma: possible marker of atherosclerosis progression," *Life*, vol. 48, pp. 413–417, 1999.

[51] J. Hulthe and B. Fagerberg, "Circulating oxidized LDL is associated with subclinical atherosclerosis development and inflammatory cytokines (AIR Study)," *Arteriosclerosis, Thrombosis and Vascular Biology*, vol. 22, pp. 1162–1167, 2002.

[52] K. Wallenfeldt, B. Fagerberg, J. Wikstrand, and J. Hulthe, "Oxidized low-density lipoprotein in plasma is a prognostic marker of subclinical atherosclerosis development in clinically healthy men," *Journal of Internal Medicine*, vol. 256, pp. 413–420, 2004.

[53] M. Granér, T. Nakano, S. J. Sarna, M. S. Niemiene, M. Syvanne, and M. R. Taskinen, "Impact of postprandial lipaemia on low-density lipoprotein (LDL) size and oxidized LDL in patients with coronary artery disease," *European Journal of Clinical Investigation*, vol. 36, pp. 764–770, 2006.

[54] T. Weinbrenner, H. Schroder, V. Escurriol, M. Fito, R. Elosua et al., "Circulating oxidized LDL is associated with increased waist circumference independent of body mass index in men and women," *American Journal of Clinical Nutrition*, vol. 83, pp. 30–35, 2006.

[55] A. Taleb, J. L. Witztum, and S. Tsimikas, "Oxidized phospholipids on apoB-100-containing lipoproteins: a biomarker predicting cardiovascular disease and cardiovascular events," *Biomarkers in Medicine*, vol. 5, pp. 673–694, 2011.

[56] T. Vasankari, M. Fogelholm, K. Kukkonen-Harjula et al., "Reduced oxidized low-density lipoprotein after weight reduction in obese premenopausal women," *International Journal of Obesity*, vol. 25, pp. 205–211, 2001.

[57] A. Lapointe, J. Goulet, C. Couillard, B. Lamarche, and S. Lemieux, "A nutritional intervention promoting the mediterranean food pattern is associated with decrease in circulating oxidized LDL particles in healthy women from the Québec city metropolitan area," *Journal of Nutrition*, vol. 135, pp. 410–415, 2005.

[58] E. Tumova, W. Sun, P. H. Jones, M. Vrablik, C. M. Ballantyne, and R. C. Hoogeveen, "The impact of rapid weight loss on oxidative stress markers and the expression of the metabolic syndrome in obese individuals," *Journal of Obesity*, Article ID 729515, 2013.

[59] I. Shai, J. D. Spence, D. Schwarzfuchs et al., "Dietary intervention to reverse carotid atherosclerosis," *Circulation*, vol. 121, no. 10, pp. 1200–1208, 2010.

[60] N. V. Dhurandhar, D. Schoeller, A. W. Brown et al., "Energy balance measurement: when something is not better than nothing," *International Journal of Obesity*, vol. 39, pp. 1109–1113, 2015.

[61] A. F. Subar, L. S. Freedman, J. A. Tooze et al., "Addressing current criticism regarding the value of self-report dietary data," *Journal of Nutrition*, vol. 145, no. 12, pp. 2639–2645, 2015.

[62] D. A. Booth and A. Laguna-Camacho, "Physical versus psychosocial measures of influences on human obesity. Comment on Dhurandhar et al.," *International Journal of Obesity*, vol. 39, no. 7, pp. 1177-1178, 2015.

[63] D. P. French and S. Sutton, "Reactivity of measurement in health psychology: how much of a problem is it? What can be done about it?" *British Journal of Health Psychology*, vol. 15, pp. 453–468, 2010.

[64] V. W. Berger and S. Weinstein, "Ensuring the comparability of comparison groups: Is randomization enough?" *Controlled Clinical Trials*, vol. 25, no. 5, pp. 515–524, 2004.

[65] R. A. Carels, J. Harper, and K. Konrad, "Qualitative perceptions and caloric estimations of healthy and unhealthy foods by behavioral weight loss participants," *Appetite*, vol. 46, no. 2, pp. 199–206, 2006.

[66] D. Larkina and C. R. Martin, "Caloric estimation of healthy and unhealthy foods in normal-weight, overweight and obese participants," *Eating Behaviors*, vol. 23, pp. 91–96, 2016.

[67] M. R. de Boer, W. E. Waterlander, L. D. J. Kuijper, I. H. M. Steenhuis, and J. W. R. Twisk, "Testing for baseline differences in randomized controlled trials: an unhealthy research behavior that is hard to eradicate," *International Journal of Behavioral Nutrition and Physical Activity*, vol. 12, no. 4, 2015.

[68] T. Mann, A. J. Tomiyama, E. Westling, A. M. Lew, B. Samuels, and J. Chatman, "Medicare's search for effective obesity treatments: diets are not the answer," *American Psychologist*, vol. 62, pp. 220–233, 2007.

[69] J. S. Garrow, *Energy Balance and Obesity in Man*, Elsevier North Holland Pub, Amsterdam, The Netherlands, 1978.

[70] O. G. Edholm, J. M. Adam, M. J. R. Healy, H. S. Wolff, R. Goldsmith, and T. W. Best, "Food intake and energy expenditure of army recruits," *British Journal of Nutrition*, vol. 24, no. 4, pp. 1091–1107, 1970.

[71] A. Laguna-Camacho, *Patterns of eating and exercise that reduce weight [Ph.D. thesis]*, University of Birmingham, 2013.

[72] L. Jussim, "Précis of social perception and social reality: why accuracy dominates bias and self-fulfilling prophecy," *Behavioral and Brain Sciences*, vol. 16, pp. 1–66, 2015.

[73] A. Vrij, P. A. Granhag, S. Mann, and S. Leal, "Outsmarting the liars: toward a cognitive lie detection approach," *Current Directions in Psychological Science*, vol. 20, no. 1, pp. 28–32, 2011.

[74] J. J. Skowronski, A. L. Betz, C. P. Thompson, and L. Shannon, "Social Memory in Everyday Life: Recall of Self-Events and Other-Events," *Journal of Personality and Social Psychology*, vol. 60, no. 6, pp. 831–843, 1991.

[75] A. M. Armstrong, A. MacDonald, I. W. Booth, R. G. Platts, R. C. Knibb, and D. A. Booth, "Errors in memory for dietary intake and their reduction," *Applied Cognitive Psychology*, vol. 14, pp. 183–191, 2000.

[76] G. Kristo, S. M. J. Janssen, and J. M. J. Murre, "Retention of autobiographical memories: an internet-based diary study," *Memory*, vol. 17, no. 8, pp. 816–829, 2009.

[77] M. A. Conway, "Episodic memories," *Neuropsychologia*, vol. 47, no. 11, pp. 2305–2313, 2009.

[78] P. Sedlmeier and T. Betsch, *Etc. Frequency Processing and Cognition*, University Press, Oxford, UK, 2002.

[79] D. H. Barlow, M. K. Nock, and M. Hersen, *Single Case Experimental Designs: Strategies for Studying Behaviour Change*, Pearson education, Boston, USA, 3rd edition, 2009.

[80] R. L. Tate, M. Perdices, U. Rosenkoetter, S. McDonald, L. Togher, S. Shadish et al., "The single-case reporting guideline in behavioural interventions (SCRIBE) 2016: explanation and elaboration," *Archives of Scientific Psychology*, vol. 4, no. 1, pp. 10–31, 2016.

[81] A. Laguna-Camacho, A. S. Alonso-Barreto, and H. Mendieta-Zerón, "Direct effect of fatty meals and adiposity on oxidised low-density lipoprotein," *Obes Res Clin Pract*, vol. 9, pp. 298–300, 2015.

[82] F. J. Pendergast, N. D. Ridgers, A. Worsley, and S. McNaughton, "Evaluation of a smartphone food diary application using objectively measured energy expenditure," *International Journal of Behavioral Nutrition and Physical Activity*, vol. 14, no. 30, 2017.

[83] Department of Health and Human Services, *Physical Activity Guidelines for Americans*, DHHS, Washington, USA, 2008.

12

Antibacterial Activity of Free Fatty Acids from Hydrolyzed Virgin Coconut Oil using Lipase from *Candida rugosa*

Van Thi Ai Nguyen,[1,2] Truong Dang Le,[1] Hoa Ngoc Phan,[2] and Lam Bich Tran[2]

[1]*Institute of Biotechnology and Food Technology, Industrial University of Ho Chi Minh City, Ho Chi Minh City, Vietnam*
[2]*Department of Food Technology, Faculty of Chemical Engineering, Ho Chi Minh City University of Technology, Ho Chi Minh City, Vietnam*

Correspondence should be addressed to Van Thi Ai Nguyen; nguyenthiaivan@iuh.edu.vn

Academic Editor: Maurizio Averna

Free fatty acids (FFAs) were obtained from hydrolyzed virgin coconut oil (VCO) by *Candida rugosa* lipase (CRL). Four factors' influence on hydrolysis degree (HD) was examined. The best hydrolysis conditions in order to get the highest HD value were determined at VCO to buffer ratio 1 : 5 (w/w), CRL concentration 1.5% (w/w oil), pH 7, and temperature 40°C. After 16 hours' reaction, the HD value achieved 79.64%. FFAs and residual hydrolyzed virgin coconut oil (HVCO) were isolated from the hydrolysis products. They were tested for their antibacterial activity against Gram-negative and Gram-positive bacteria, which can be found in contaminated food and cause food poisoning. FFAs showed their inhibition against *Bacillus subtilis* (ATCC 11774), *Escherichia coli* (ATCC 25922), *Salmonella enteritidis* (ATCC 13076), and *Staphylococcus aureus* (ATCC 25923) at minimum inhibitory concentration (MIC) of 50%, 60%, 20%, and 40%, respectively. However, VCO and HVCO did not show their antibacterial activity against these tested bacteria.

1. Introduction

VCO is extracted from fresh kernel by using either cold press or centrifuge process. It does not go through refined, bleached, and deodorized process (RBD). Therefore, its physical properties as flavor, color, and so forth are less changed than RBD oil. VCO has many advantages in skin care, promotes the growth of hair, and enhances the beauty. Antioxidant activity and phenolic compounds in VCO also was conducted by some studies; it was suggested that the consumption of food containing phenolic compound will have a positive contribution in health [1].

Aside from benefits above, VCO is also good in health promotion and prevents some diseases because of the presence of FFAs in VCO. FFAs in VCO are rich in medium chain fatty acids (MCFAs) in which lauric acid takes the highest percentage about 46–48%. MCFAs in VCO are easily digested and absorbed, but fat is harder because it contains long chain fatty acids which need going through circulatory system

before absorbing, so VCO can be used to replace cooking oil in daily meal to improve digestion. Moreover, MCFAs are also good for obese people because they increase energy expenditure more than usual. And MCFAs are directly absorbed from the intestine and burned in the liver; this makes them have a feeling which is always early satiety, and weight is decreased [2]. And also absorbing and burning directly in liver make MCFAs not take part in biosynthesis and transport of cholesterol. Thus, MCFAs in VCO have cardioprotective ability [3]. MCFAs also showed antifungal activity; Shino and coworkers (2016) exhibited a comparison of antimicrobial activity of chlorhexidine, coconut oil, probiotics, and ketoconazole on *Candida albicans* isolated in children with early childhood caries [4]. Parfene and coworkers (2013) gave a result about antifungal activity against *Yarrowia lipolytica* of MCFAs from crude coconut oil [5]. MCFAs have effective ability to inhibit some species of virus by breaking their membranes [3]. And antibacterial activity of MCFAs was also conducted by some previous studies; Kim and Rhee (2016) presented that MCFAs were antibacterial agents against *Escherichia coli* [6]. Shilling and coworkers (2013) also studied antimicrobial effect of VCO and MCFAs against *Clostridium difficile* [7].

MCFAs are antibacterial agents; this was demonstrated by previous studies, but MCFAs used in their studies were in form of pure chemical. Therefore, the aim of this study was to use FFAs extracted from hydrolyzed VCO and evaluate their antibacterial against *Bacillus subtilis* (ATCC 11774), *Escherichia coli* (ATCC 25922), *Salmonella enteritidis* (ATCC 13076), and *Staphylococcus aureus* (ATCC 25923), which can be found in food and cause food poisoning. At the same time, the resistance of VCO and HVCO against these tested bacteria was also evaluated.

2. Materials and Methods

2.1. Materials. VCO was sponsored by Luong Quoi Coconut Co., Ltd. (Ben Tre Province, Vietnam). *Candida Rugosa* lipase (CRL) (Type VII, ≥700 unit/mg solid) was purchased from Sigma-Aldrich Co. (Canada). Chemicals used in this study were KOH, n-hexane, and iso-octane and all other chemicals from Merck (Germany) and China were analyzed with purification more than 95%. Mediums used in antibacterial test were Nutrient Broth (NB) (Italy), Mueller Hinton Agar (MHA) from HiMedia Laboratories Pvt. Ltd (India), and Mueller Hinton Broth (MHB) from Titan Biotech Ltd (India). And four types of bacteria used in this study were *Bacillus subtilis* (ATCC 11774), *Escherichia coli* (ATCC 25922), *Salmonella enteritidis* (ATCC 13076), and *Staphylococcus aureus* (ATCC 25923) provided by Microbiologics Co. (USA).

Devices used in this study were high speed homogenizer (IKA T25 digital ULTRA-TURRAX, Germany), overhead stirrer (OS20, USA), orbital shaker incubator (LM-2575RD) from Yihder Technology Co. (Taiwan), evaporator (IKA RV digital V) from Germany, and GC-FID SHIMADZU 2010 Plus (Japan).

2.2. Hydrolysis of VCO. VCO dissolved in iso-octane (VCO to solvent ratio 1 : 1 (w/w)) and phosphate buffer solution to adjust pH condition was placed in a 250 mL Erlenmeyer flask [8]. Emulsification of the mixture was carried out by using a stirrer at speed of 10000 rpm in 15 minutes; then the appropriate amount of lipase was added and dissolved by stirring at speed of 350 rpm in 5 minutes. The reaction was conducted in orbital shaker incubator at speed 150 rpm for 2 hours. To stop the reaction, add 1 ml ethanol 99.5% into the erlenmeyer flask.

The hydrolysis degree (HD) was calculated as the following formula [9]:

$$\text{HD} = \frac{V_{\text{KOH}} * M_{\text{KOH}} * M_{\text{FFAs}}}{10 * m} \, (\%), \tag{1}$$

where V_{KOH} is the volume of potassium hydroxide (KOH) titrated (mL), M_{KOH} is the molarity of KOH solution (mol/L), M_{FFAs} is the average molecular weight of free fatty acids, and m is mass of VCO (g).

2.3. Obtaining FFAs. The process was carried out according to the method of Shimada and coworkers (1998) [10]. The

excess of 0.5 N KOH was added to the hydrolyzed mixture to neutralize the released FFAs. Separatory funnel was used to extract FFAs and HVCO (tri-, di-, and monoglyceride). The upper phase containing HVCO dissolved in n-hexane was purified by using rotary evaporator. The lower phase containing FFAs was acidified with 4 N HCl solution and then was extracted by n-hexane. FFAs and HVCO were analyzed composition by Gas Chromatography with Flame Ionization Detector (GC-FID).

2.4. Antibacterial Activity Test. The test was conducted following disk diffusion method. Firstly, bacterial inoculum suspension was diluted in saline and was adjusted to get the final inoculum to 1.5×10^6 CFU/mL [11]. Mix the volume of 1 mL bacterial inoculum to 15 mL of MHA and wait until the media solidified [12]. Paper discs with diameter of 6 mm containing different solutions such as gentamicin, VCO, HVCO, and FFAs were placed in agar surface and then were incubated for 24 hours at 37°C. The antibacterial activity was determined by measuring the diameter of inhibition zone.

2.5. Determining the Minimum Inhibitory Concentration (MIC). MIC was conducted following the broth macrodilution method. Bacterial inoculum suspension was diluted directly in MHB and was adjusted to get final inoculum to 5×10^5 CFU/mL. The antibacterial agent (FFAs) was prepared by diluting FFA directly in MHB to get dilution series 100%, 90%, . . . , 10%. 1 mL of bacterial inoculum was added to each tube containing 1 mL of dilution series of FFAs (positive control tube only has MHB) and mixed the mixture. The tubes were incubated at 35°C from 16 to 20 hours before recording. The MIC is the lowest concentration of FFAs that completely inhibits growth of the bacteria in the tube. It was compared with positive control tube when determining growth end points. The results of this test are valid if growth-control well is definitely turbid or diameter of colonies ≤ 2 mm [13].

All the experiments were carried out in triplicate and the data collected were statistically analyzed by using R software.

3. Results and Discussion

3.1. Effect of VCO to Buffer Ratio. Lipase catalyzes hydrolysis reaction at interfacial area of emulsion. There are more bonds between water and oil when amount of buffer increases, so lipase has more substrates to be catalyzed leading to increase of hydrolysis degree. However, the excess of water amount will decrease hydrolysis degree because of substrate competition by binding to lipase [8]. In Figure 1(a) the VCO to buffer ratio of 1 : 5 (w/w) showed the highest hydrolysis degree. Therefore, the next experiment will be carried out at this ratio.

3.2. Effect of CRL Concentration. CRL concentration has a significant effect on hydrolysis degree. Increasing lipase concentration will get a high hydrolysis degree, but when the interface of emulsion is saturated with lipase concentration, adding more lipase will not get a higher hydrolysis degree [14]. From Figure 1(b) the hydrolysis degree cannot increase at CRL concentration of 1.5% plus anymore. As a result, the

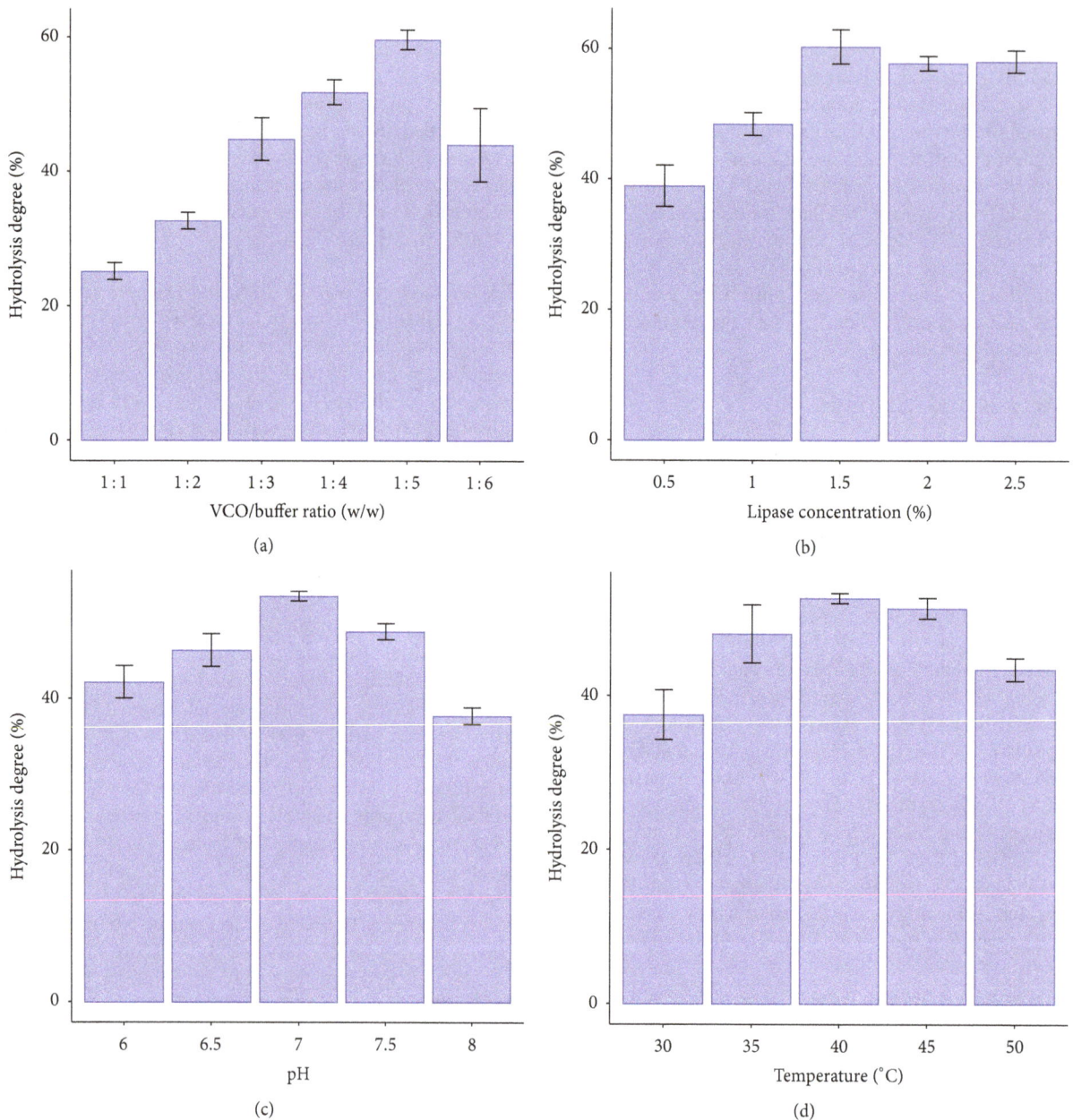

FIGURE 1: The rate of VCO to buffer (a), CRL concentration (b), pH (c), and temperature (d) effect on hydrolysis degree of VCO by CRL.

CRL concentration of 1.5% was chosen to carry out next experiments.

3.3. Effect of pH. Each lipase has an optimum pH condition. Changed pH range affects the ionization of substrate, free lipase, or lipase-substrate complex. Either high pH or low pH will make lipase denatured; substrate is broken down leading to a decrease of hydrolysis degree [15].

According to Figure 1(c), pH 7.0 was the best condition for hydrolysis reaction. And this result also was equivalent to some studies as Sharma and coworkers (2013) on hydrolysis of cod liver oil using CRL for fatty acid production. The process

was carried out at pH 7.0 [8]. Hydrolysis of soybean oil using CRL was also carried out at pH 7.0 [16]. In the research of Freitas and coworkers (2007) on hydrolysis of soybean oil by using 3 types of lipase, among them was CRL with pH for hydrolysis reaction being 7.0 [17].

3.4. Effect of Temperature. Temperature has a strong effect to hydrolysis degree of lipase. When increase temperature, lipase becomes flexible; thus the hydrolysis degree is also increased. Moreover, increasing temperature will decrease viscosity of mixture reaction. This makes lipase easy to bond to a substrate; therefore the hydrolysis degree is increased.

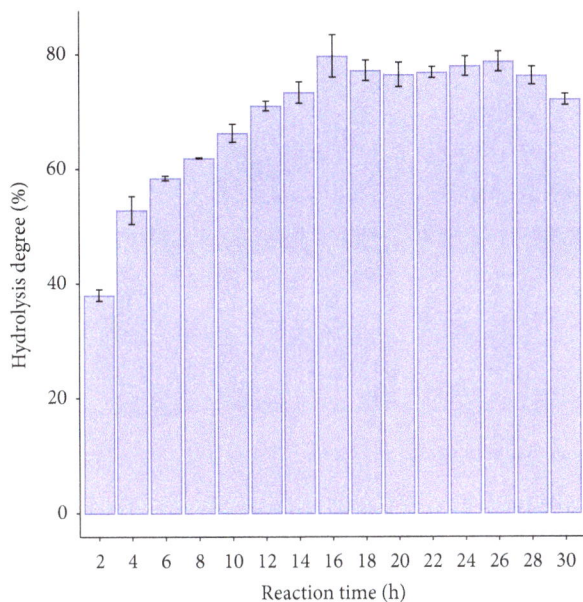

FIGURE 2: Time course of VCO hydrolysis catalyzed by CRL.

TABLE 1: Diameter of inhibition zone on 4 types of bacteria.

Bacteria	Diameter of inhibition zone ± SD (mm)		
	FFAs	HVCO	VCO
Bacillus subtilis	11.33 ± 1.15	6	6
Escherichia coli	8.67 ± 0.58	6	6
Salmonella enteritidis	10.33 ± 0.58	6	6
Staphylococcus aureus	8.33 ± 0.58	6	6

TABLE 2: Minimum inhibitory concentration of FFAs against four types of bacteria.

Bacteria	MIC (%)
Bacillus subtilis	50
Escherichia coli	60
Salmonella enteritidis	20
Staphylococcus aureus	40

However, very high temperature will denature lipase because lipase's nature is protein; as a result, hydrolysis degree is decreased [15].

As shown in Figure 1(d), the highest hydrolysis degree was obtained at 40°C. And this result was equivalent to some studies as hydrolysis of soybean oil by using CRL at 40°C of Freitas and coworkers (2007) [17]; Zhou and coworkers (2015) used CRL to hydrolyze Jatropha oil to produce biodiesel at 40°C [18]. Yiğitoğlu and Temöçin (2010) also conducted a research on hydrolysis of vegetables oil by using enzyme immobilized and free CRL at 40°C [19].

3.5. Hydrolysis Time. After 16 hours' reaction, the highest hydrolysis degree was achieved 79.64% (Figure 2). When comparing the result with other previous studies, hydrolysis degree of, for example, soybean oil by CRL was 70% after 24 hours' reaction [17]. On the other hand, tuna fish oil was hydrolyzed by using CRL, after 24 hours' reaction; the hydrolysis degree achieved was 86.5% [20]. In this research, the hydrolysis time was shorter than other studies, but it got equivalent hydrolysis degree. Hence, it can be understood that using iso-octane in this hydrolysis reaction had a positive effect. The presence of iso-octane in reaction mixture decreased VCO viscosity. VCO was emulsified easily and increased the interface of VCO and water, so this made CRL have more opportunity to contact with substrate and easily catalyze hydrolysis reaction [21]. As such, it shortened time course of VCO hydrolysis reaction.

3.6. Antibacterial Test. According to Table 1 and Figure 3, FFAs showed clearly antibacterial activity on 4 types of bacteria. Diameter of inhibition zone of FFAs on each bacteria was different. FFAs' antibacterial activity on *Salmonella enteritidis* was stronger than *Staphylococcus aureus* and *Escherichia coli*, so the result demonstrated that FFAs extracted from

hydrolyzed product were antibacterial agents. It was equivalent to some authors using FFAs in the form of pure chemicals. Shilling and coworkers (2013) showed the result of antibacterial ability of MCFAs (lauric acid in the form of pure chemicals) against *Clostridium difficile*. Moreover, hydrolyzed VCO also had an effect on the growth of *Clostridium difficile* because FFAs were released after hydrolysis of VCO [7]. Bergsson and coworkers (2002) studied antibacterial effect of fatty acids against *Helicobacter pylori* and the result exhibited that, among fatty acids, MCFAs were antibacterial agent [22]. The study of Kim and Rhee (2016) also showed that MCFAs could inhibit the growth of *Escherichia coli* [6].

VCO did not show its antibacterial activity against these tested bacteria. This was equivalent to the result of Shilling and coworkers (2013) in using VCO to inhibit *Clostridium difficile*, and VCO had a negative result [7].

HVCO in this study did not show its antibacterial ability although it might contain monoglyceride. This could be explained that, in this study, HVCO only contained a small amount of monoglyceride, so it did not have much enough to inhibit bacteria.

3.7. Minimum Inhibitory Concentration (MIC). The result of MIC was shown in Table 2 and Figures 4 and 5. The MIC of FFAs against *Salmonella enteritidis* was lowest while *Escherichia coli* needed concentration of FFAs up to 60% to inhibit it, so this showed that minimum concentration of FFAs to inhibit bacteria was completely different; it did not distinguish which is Gram-negative or Gram-positive bacteria.

4. Conclusions

This research found the best hydrolysis conditions of VCO under four parameters: VCO to buffer ratio, lipase concentration, pH, and temperature. Using lipase to hydrolyze VCO received the less changed FFAs because the hydrolysis reaction is conducted under milder condition of pH and temperature than acidic or alkaline hydrolysis. And enzymatic

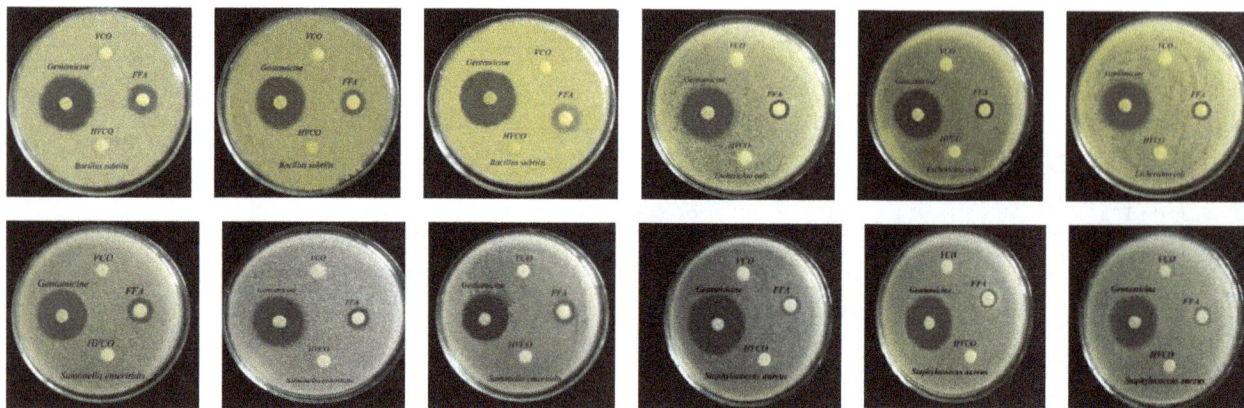

FIGURE 3: Antibacterial activity of VCO, HVCO, FFAs, and gentamicin against 4 types of tested bacteria.

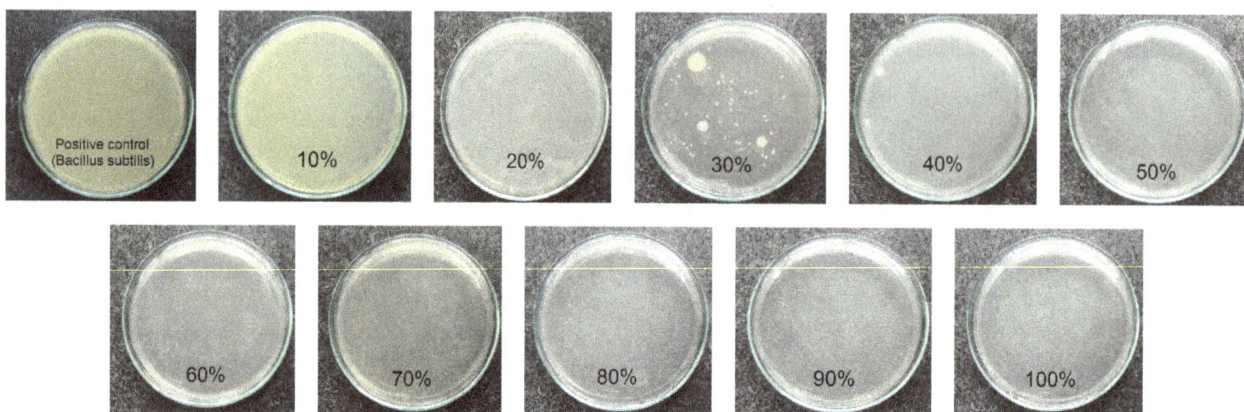

(a)

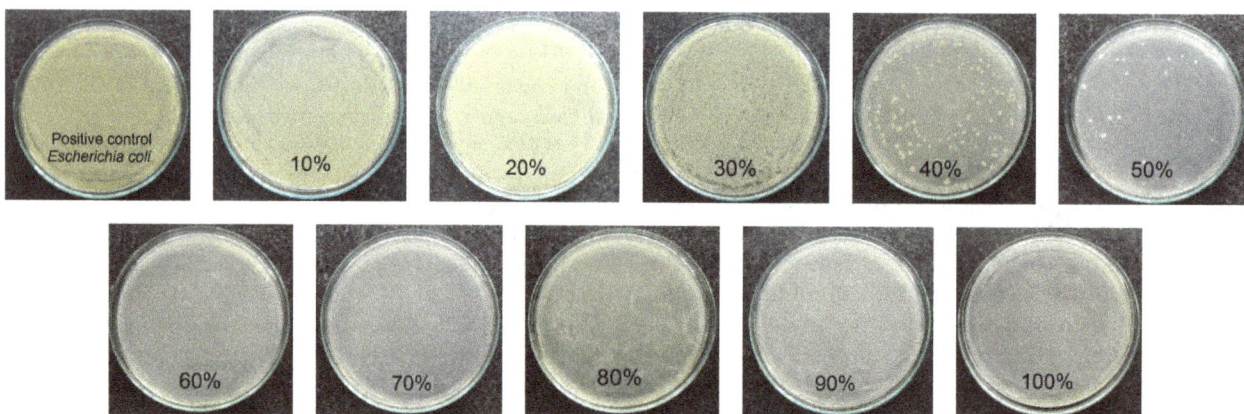

(b)

FIGURE 4: MIC of FFAs against *Bacillus subtilis* (a) and *Escherichia coli* (b).

hydrolysis can avoid some danger from undesirable side-reaction, not affecting color of products or limiting oxidized products so that FFAs can be applied in other fields as food preservation, pharmaceuticals, and so forth. Moreover, using iso-octane to dissolve VCO saved the time of hydrolysis reaction compared to other researches. FFAs were obtained,

isolated, and showed their antibacterial activity against Gram-positive and Gram-negative bacteria while VCO and HVCO did not show any positive results. MIC of FFAs was determined at different concentrations. Based on that, FFAs can be used as a preservative in food products with appropriate amount in our next researches.

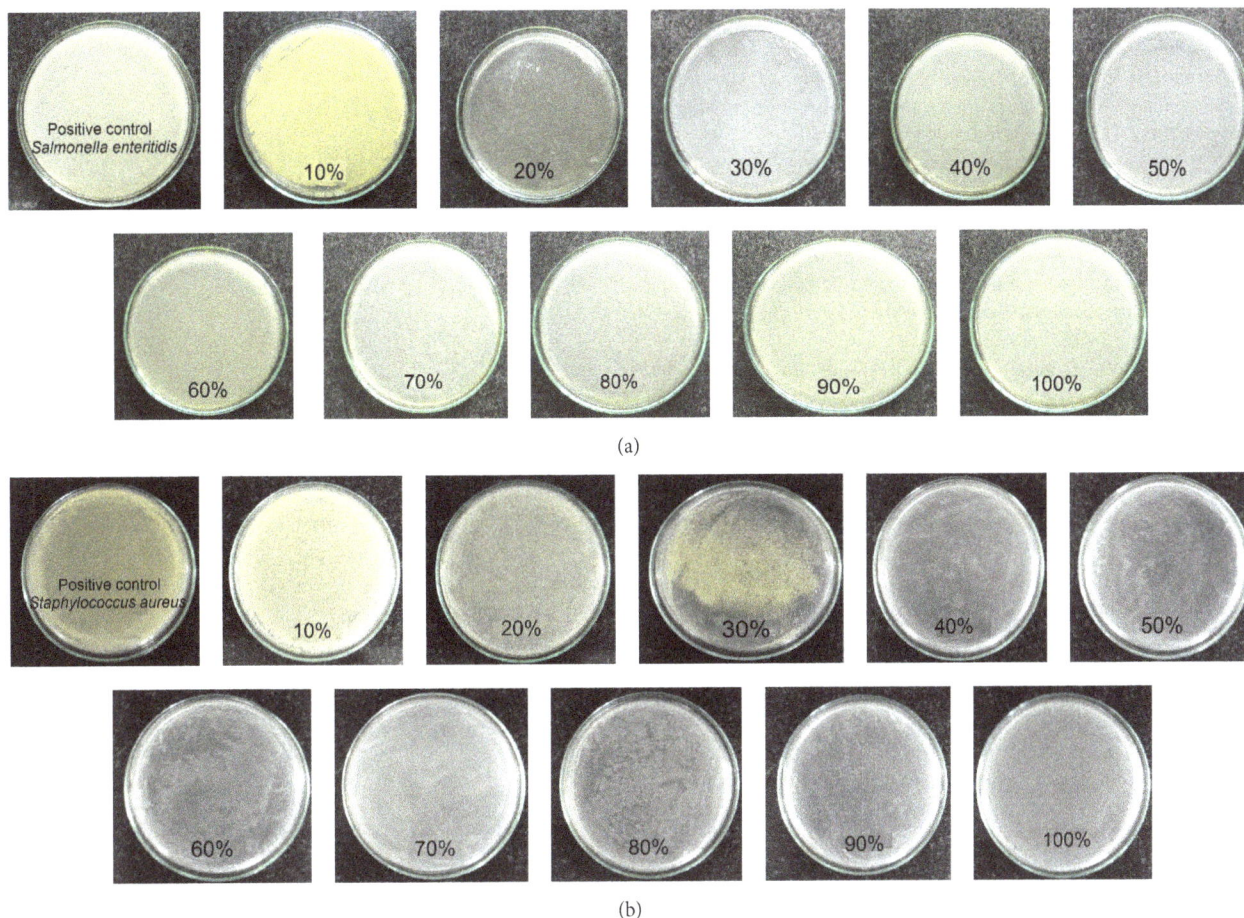

FIGURE 5: MIC of FFAs against *Salmonella enteritidis* (a) and *Staphylococcus aureus* (b).

References

[1] A. M. Marina, Y. B. Man, and I. Amin, "Virgin coconut oil: emerging functional food oil," *Trends in Food Science & Technology*, vol. 20, no. 10, pp. 481–487, 2009.

[2] T. S. T. Mansor, M. Che, M. Shuhaimi, A. Abdul, and N. Ku, "Physicochemical properties of virgin coconut oil extracted from different processing methods," *International Food Research Journal*, vol. 19, no. 3, pp. 837–845, 2012.

[3] M. DebMandal and S. Mandal, "Coconut (*Cocos nucifera* L.: Arecaceae): in health promotion and disease prevention," *Asian Pacific Journal of Tropical Medicine*, vol. 4, no. 3, pp. 241–247, 2011.

[4] B. Shino, F. C. Peedikayil, S. R. Jaiprakash, G. A. Bijapur, S. Kottayi, and D. Jose, "Comparison of antimicrobial activity of chlorhexidine, coconut oil, probiotics, and ketoconazole on *Candida albicans* isolated in children with early childhood caries: an in vitro study," *Scientifica*, vol. 2016, Article ID 7061587, 5 pages, 2016.

[5] G. Parfene, V. Horincar, A. K. Tyagi, A. Malik, and G. Bahrim, "Production of medium chain saturated fatty acids with enhanced antimicrobial activity from crude coconut fat by solid state cultivation of *Yarrowia lipolytica*," *Food Chemistry*, vol. 136, no. 3-4, pp. 1345–1349, 2013.

[6] S. A. Kim and M. S. Rhee, "Highly enhanced bactericidal effects of medium chain fatty acids (caprylic, capric, and lauric acid) combined with edible plant essential oils (carvacrol, eugenol, β-resorcylic acid, trans-cinnamaldehyde, thymol, and vanillin) against *Escherichia coli* O157: H7," *Food Control*, vol. 60, pp. 447–454, 2016.

[7] M. Shilling, L. Matt, E. Rubin et al., "Antimicrobial effects of virgin coconut oil and its medium-chain fatty acids on *Clostridium difficile*," *Journal of Medicinal Food*, vol. 16, no. 12, pp. 1079–1085, 2013.

[8] A. Sharma, S. P. Chaurasia, and A. K. Dalai, "Enzymatic hydrolysis of cod liver oil for the fatty acids production," *Catalysis Today*, vol. 207, pp. 93–100, 2013.

[9] M. C. P. Zenevicz, A. Jacques, A. F. Furigo, J. V. Oliveira, and D. de Oliveira, "Enzymatic hydrolysis of soybean and waste cooking oils under ultrasound system," *Industrial Crops and Products*, vol. 80, pp. 235–241, 2016.

[10] Y. Shimada, N. Fukushima, H. Fujita, Y. Honda, A. Sugihara, and Y. Tominaga, "Selective hydrolysis of borage oil with *Candida rugosa* lipase: Two factors affecting the reaction," *Journal of the American Oil Chemists' Society*, vol. 75, no. 11, pp. 1581–1586, 1998.

[11] CLSI, *Performance standards for antimicrobial disk susceptibility tests*, vol. 32, 2012.

[12] Vietnamese Standard No 4884:2001, *Vietnamese standard-Microbiology - General guidance for the enumeration micro-organisms - Colony count technique at 30°C*, 2001.

[13] CLSI, "Methods for Dilution Antimicrobial Susceptibility Tests for Bacteria That Grow Aerobically," *Approved Standard—Ninth Edition*, vol. 32, no. 2, p. 287, 2012.

[14] N. A. Serri, A. H. Kamarudin, and A. Rahaman, "Preliminary studies for production of fatty acids from hydrolysis of cooking palm oil using *C. rugosa* lipase," *Journal of Physical Science*, vol. 19, no. 1, pp. 79–88, 2008.

[15] D. Goswami, J. K. Basu, and S. De, "Optimization of process variables in castor oil hydrolysis by *Candida rugosa* lipase with buffer as dispersion medium," *Biotechnology and Bioprocess Engineering*, vol. 14, no. 2, pp. 220–224, 2009.

[16] W. J. Ting, K. Y. Tung, R. Giridhar, and W. T. Wu, "Application of binary immobilized *Candida rugosa* lipase for hydrolysis of soybean oil," *Journal of Molecular Catalysis B: Enzymatic*, vol. 42, no. 1-2, pp. 32–38, 2006.

[17] L. Freitas, T. Bueno, V. H. Perez, J. C. Santos, and H. F. De Castro, "Enzymatic hydrolysis of soybean oil using lipase from different sources to yield concentrated of polyunsaturated fatty acids," *World Journal of Microbiology and Biotechnology*, vol. 23, no. 12, pp. 1725–1731, 2007.

[18] G.-X. Zhou, G.-Y. Chen, and B.-B. Yan, "Two-step biocatalytic process using lipase and whole cell catalysts for biodiesel production from unrefined jatropha oil," *Biotechnology Letters*, vol. 37, no. 10, pp. 1959–1963, 2015.

[19] M. Yiğitoğlu and Z. Temoçin, "Immobilization of *Candida rugosa* lipase on glutaraldehyde-activated polyester fiber and its application for hydrolysis of some vegetable oils," *Journal of Molecular Catalysis B: Enzymatic*, vol. 66, no. 1-2, pp. 130–135, 2010.

[20] K. Bhandari, S. P. Chaurasia, and A. K. Dalai, "Hydrolysis of tuna fish oil using *Candida rugosa* lipase for producing fatty acids containing DHA," *International Journal of Applied and Natural Sciences*, vol. 2, no. 3, pp. 1–12, 2013.

[21] D. T. Raspe, L. Cardozo Filho, and C. Da Silva, "Effect of additives and process variables on enzymatic hydrolysis of macauba kernel oil (*Acrocomia aculeata*)," *International Journal of Chemical Engineering*, vol. 2013, Article ID 438270, 8 pages, 2013.

[22] G. Bergsson, Ó. Steingrímsson, and H. Thormar, "Bactericidal effects of fatty acids and monoglycerides on *Helicobacter pylori*," *International Journal of Antimicrobial Agents*, vol. 20, no. 4, pp. 258–262, 2002.

13

Effect of Supplementation with *n*-3 Fatty Acids Extracted from Microalgae on Inflammation Biomarkers from Two Different Strains of Mice

L. E. Gutiérrez-Pliego ⓘ,[1] B. E. Martínez-Carrillo ⓘ,[1]
A. A. Reséndiz-Albor ⓘ,[2] I. M. Arciniega-Martínez ⓘ,[2] J. A. Escoto-Herrera ⓘ,[1]
C. A. Rosales-Gómez ⓘ,[1] and R. Valdés-Ramos ⓘ[1]

[1]*Laboratorio de Investigación en Nutrición, Facultad de Medicina, Universidad Autónoma del Estado de México, Paseo Tollocan y Venustiano Carranza s/n, Col. Universidad, 50180 Toluca, MEX, Mexico*
[2]*Laboratorio de Inmunidad de Mucosas, Sección de Investigación y Posgrado, Escuela Superior de Medicina, Instituto Politécnico Nacional, Av. Plan de San Luis S/N, Colonia Casco de Santo Tomas, Miguel Hidalgo, 11350 Ciudad de México, Mexico*

Correspondence should be addressed to R. Valdés-Ramos; rvaldesr@uaemex.mx

Academic Editor: Gerhard M. Kostner

Background. Diabetes mellitus is considered a chronic noncommunicable disease in which inflammation plays a main role in the progression of the disease and it is known that *n*-3 fatty acids have anti-inflammatory properties. One of the most recent approaches is the study of the fatty acids of microalgae as a substitute for fish oil and a source rich in fatty acids EPA and DHA. *Objective.* To analyze the effect of supplementation with *n*-3 fatty acids extracted from microalgae on the inflammatory markers from two different strains of mice. *Methods.* Mice of two strains, db/db and CD1, were supplemented with *n*-3 fatty acids extracted from microalgae in lyophilized form and added to food; the experiment was carried out from week 8 to 16 of life. Flow cytometry was performed to determine the percentage of TCD4+ cells producing Th1 and Th2 cytokines. *Results.* Supplementation with microalgae fatty acids decreased the percentage of TCD4+ cells producing IFN-γ and TNF-α and increased the ones producing IL-17A and IL-12 in both strains; on the other hand, supplementation decreased percentage of TCD4+ cells producing IL-4 and increased the ones producing TGF-β. *Conclusions.* Microalgae *n*-3 fatty acids could be a useful tool in the treatment of diabetes as well as in the prevention of the appearance of health complications caused by inflammatory states.

1. Introduction

Diabetes mellitus is a multifactorial chronic noncommunicable disease, characterized by states of hyperglycemia resulting from defects in insulin secretion, its action, or both [1]. In Mexico, according to the National Health and Nutrition Survey Half Way 2016 (ENSANUT MC 2016 for its acronym in Spanish), 9.4% of the adult population has been diagnosed with diabetes [2]. It also represents the leading cause of negative health outcomes such as heart failure, blindness, kidney failure, amputations, and premature death [3]. The main cause of health complications in diabetes is chronic hyperglycemia, which is associated with changes in immunomodulation and inflammation [4].

The use of *n*-3 polyunsaturated fatty acids as a strategy to minimize damage caused by hyperglycemia has been deeply studied [5]. Its biological effects include benefits on the metabolism of lipoproteins [6], platelet, and endothelial and vascular function [7], as well as antioxidant and anti-inflammatory impact [8]. Evidence suggests that *n*-3 inhibits the proliferation of T lymphocytes in murine models and in humans [9, 10] and inhibits degranulation of cytotoxic T lymphocytes [11]. Thus, it suggests that polyunsaturated fatty acids have potentially immunosuppressive properties. Moreover, the supplementation with EPA (eicosapentaenoic acid) for 12 weeks has been shown to modify the fatty acids composition of the phospholipids of plasma, platelets, neutrophils, monocytes, and T and B lymphocytes [12].

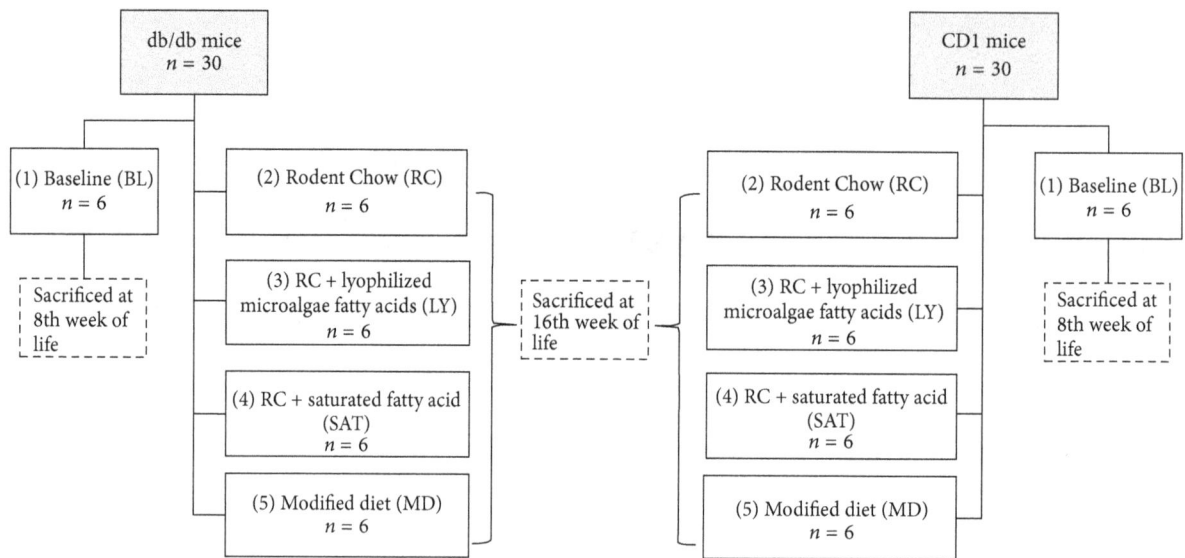

FIGURE 1: Experimental groups for both strains.

Within their anti- and proinflammatory effects, it has been shown that, in cell cultures, EPA and DHA (docosahexaenoic acid) have high anti-inflammatory and immunosuppressive effects [13–15]. Same findings have been shown on animal studies supplemented with fish oil [16–18]. It has been proven that EPA and DHA supplementation decreases proinflammatory cytokines such as tumor necrosis factor Alpha (TNF-α), interleukin-6 (IL-6), monocyte-1 chemoattractant protein (MCP-1), and plasminogen activator-1 (PAI-1) inhibitor [19].

The main dietary sources of EPA and DHA fatty acids are fish, shellfish, and marine oils [20]. However, some disadvantages of the use of these sources are undesirable nutritional and organoleptic effects, such as oxidation (due to their high polyunsaturation) and the characteristic odor of the product [21]. Another disadvantage of the use of marine oils is the risk of contamination with heavy metals and pesticides, the solution of which may be oil refining, but this process involves a higher cost of production [22]. In addition, in recent years, marine sources have been diminished because fish catch has exceeded the maximum sustainable levels [23].

Microalgae are an evolutionarily microscopic diverse eukaryotic group of unicellular and predominantly aquatic photosynthetic organisms that have recently been studied for their potential to produce compounds of high biological value and beneficial for health such as carotenoids, polyunsaturated fatty acids, and very long chain polyunsaturated fatty acids [24]. They are the primary natural producers of EPA and DHA, because they have the biosynthetic machinery to sequentially alternate between desaturation and elongation in their carbon chains [25]. Because microalgae are at the beginning of the food chain, fish are the main consumers of these fatty acids, which is why they are incorporated into the lipids of membranes and accumulated in the fats and meat of many marine species [24]. However, there are fewer studies describing their anti-inflammatory effects as those already described for fatty acids of animal origin [26–28].

For all the above and the growing study of the composition and properties of microalgae, as well as the possibility of cultivating them in artificial form, interest has aroused in the study of these microorganisms as a renewable source of n-3 fatty acids. The aim of this study was to analyze the effect of the consumption of n-3 fatty acids extracted from microalgae (Chlorophyceae and Eustigmatophyceae families) either provided as a supplement or incorporated to diet, on the inflammatory markers from two different strains of mice: db/db mice as a model of obesity and diabetes mellitus in which an inflammation state is expected and CD1 mice as a model of optimal state of health without inflammation.

2. Methods

2.1. Animals and Study Groups. The present experimental, prospective, controlled, and randomized study was conducted with sixty 8-week-old male mice from two different strains: 30 db/db mice (BKS.Cg + Leprdb + LeprdbOlaHsd Harlan®) and 30 CD1 mice (Crl: CD1 (ICR) belonging to the Faculty of Medicine from the Autonomous University of Mexico State). For each strain, five study groups were formed (n = 6): (1) a baseline (BL) group to obtain baseline values; (2) a Rodent Chow (RC) group; (3) a RC + lyophilized microalgae n-3 fatty acids (LY) group; (4) a RC + saturated fatty acid (SAT) group; (5) modified diet (MD) supplemented with microalgae n-3 fatty acids group. Supplementation was administered from 8th to 16th week of life (Figure 1). The animals were housed in acrylic cages of 19 × 29 × 13 cm, with light/dark cycles of 12/12 h with controlled temperature at 21 ± 1°C. Groups 2, 3, and 4 were fed a standard normal diet (Rodent Laboratory Chow 5001 from Purina [3.02 kcal/g]) and water ad libitum. Water consumption (mL/week) and food (g/week) were recorded weekly. Animal care and experimental procedures in rodents were carried out in accordance with the rules of the Internal Regulations for the Use of Laboratory Animals and the

TABLE 1: Nutrient composition of study groups' diet.

RC group		LY group		SAT group		MD group	
Protein, %	23.9	Protein, %	23.9	Protein, %	23.9	Protein, %	23.9
Starch, %	31.9	Starch, %	31.9	Starch, %	31.9	Starch, %	31.9
Glucose, %	0.22	Glucose, %	0.22	Glucose, %	0.22	Glucose, %	0.22
Fiber (crude), %	5.10	Fiber (crude), %	5.10	Fiber (crude), %	5.10	Fiber (crude), %	5.10
Cholesterol, ppm	200	Cholesterol, ppm	200	Cholesterol, ppm	200	Cholesterol, ppm	200
EPA + DHA, %	0.2	EPA + DHA, %	0.2	EPA + DHA, %	0.2	EPA + DHA*, %	2.0
Metabolizable energy: 3.02 kcal/g		Metabolizable energy: 3.02 kcal/g + 0.09 kcal/mg of lyophilized fatty acids*		Metabolizable energy: 3.02 kcal/g + 0.09 kcal/mg of coconut oil		Metabolizable energy: 3.07 kcal/g	

*Microalgae source.

Committee of Ethics in Research of the UAEMex, as well as the guidelines of the Ministry of Health and Agriculture of Mexico for the Production and Care of Laboratory Animals (NOM-062-ZOO-1999), Mexico City, Mexico. This protocol was approved by the Ethical Research Committee from the Faculty of Medicine of the UAEMex.

2.2. Obtaining of n-3 Fatty Acids (EPA and DHA) from Microalgae. The microalgae used in this project were native and collected and isolated by BIOMEX SA. de CV. The strains used were from Chlorophyceae and Eustigmatophyceae families which have a high content of EPA and DHA. The use of these microalgae for such purpose is of recent interest. The process for obtaining the EPA and DHA includes the cultivation of microalgae, separation of biomass, extraction of total lipids, and finally chromatographic procedures for EPA and DHA content determination (25.7% EPA + DHA). EPA and DHA were provided as free fatty acids form.

2.3. Supplementation. (a) LY group was fed with Rodent Chow and supplemented with lyophilized powder containing EPA + DHA obtained from microalgae. The supplemental dose was 1 mg/g of mouse weight, reconstituted in 100 μl of distilled water, and administered with micropipette by direct oral deposition every day at 8:00 am.

(b) SAT group was fed with Rodent Chow and supplemented with coconut oil. The daily dose of coconut oil was 1 mg/g of mouse weight administered with micropipette by oral deposition at 8:00 am.

(c) MD group was fed with a Rodent Chow added with microalgae EPA + DHA for a total content of 2.0% n-3 fatty acid which means 10x the original content; Chow was administered ad libitum (Table 1).

2.4. Determination of Body Mass Index (BMI) and Blood Glucose Concentration. The BMI and blood glucose concentrations of animals were determined at the 8th and 16th week of life. The formula used for BMI determination was BMI = [weight (g)/length (cm)2 * 100]. Weight was determined using a mouse Triple Beam 700/800 series Ohaus® brand weighing scale and length was determined by measuring the

animal from nose to anus. Blood glucose was determined with a Bayer Contour TS glucometer through tail puncture.

2.5. Collection of Biological Samples. The BL groups were sacrificed at the 8th week of life and the rest of the groups were sacrificed at the 16th week of life. Animals were anesthetized by ether camera, bled by direct cardiac puncture (using a heparinized syringe, obtaining 1mL of blood), and then sacrificed by cervical dislocation. 500 μl of the collected blood was used for lymphocyte isolation using Ficoll-Hypaque Plus (GE Healthcare Bio-Sciences AB, Sweden); lymphocytes were stored with a PBS (phosphate-buffered saline) solution to obtain a final volume of 1 ml for further flow cytometry.

2.6. Flow Cytometry Assays. Cell suspensions of peripheral blood mononuclear cell (PBMC) were adjusted to 1×10^6 cells/mL in PBS for the cytofluorometric analysis with brief modifications [29]. (i) Surface phenotype of T cells was detected by using fluorescent labeled monoclonal antibodies: anti-CD3 FITC (Cat. number 553063), anti-CD8α PE (Cat. number 553035), and anti-CD4 PerCP (Cat. number 553052) (all from BD Biosciences). Cells were incubated for 30 min at room temperature. Finally, the cells were then washed with PBS and fixed in 1% paraformaldehyde. (ii) For the detection of intracellular cytokine production, lymphocytes were stimulated with a mixture containing phorbol myristate acetate, ionomycin, and Brefeldin A (Leucocyte Activation Cocktail Kit, BD Pharmingen) and incubated for 4h at 37°C and 5% CO2. Then, antibodies to cell surface markers, anti-CD4 PerCP, were added and incubated as before. For intracellular staining of CD4+ T cells, fixation and permeabilization were performed using Cytofix/Cytoperm Kits (BD Pharmingen) according to the manufacturer's instructions. These cells were incubated with anti-IL-4 PE (Cat. number 554435), anti-IL-5 PE (Cat. number 554395), anti-IL-6 APC (Cat. number 561367), anti-IL-10 FITC (Cat. number 554466), anti-IL-17A FITC (BioLegend Cat. number 506907), anti-IFN-γ FITC (Cat. number 554411), and anti-TNF-α PE antibodies (Cat. number 554419). For all samples, the expression of CD69 was measured as an activation control. The fluorescent signal intensity was recorded and analyzed by FACS Aria Flow Cytometer (Becton Dickinson). Events were collected from

TABLE 2: Effect of supplementation with EPA and DHA extracted from microalgae on body mass index, blood glucose, food, and water consumption in db/db and CD1 mice.

	8 weeks old		16 weeks old		
	BL	RC	MD	LY	SAT
	Mean ± SD	Mean ± SD	Mean ± SD	Mean ± SD	Mean ± SD
	$n = 6$	$n = 6$	$n = 6$	$n = 6$	$n = 6$
db/db					
BMI g/cm^2	57.3 ± 4.9	61.2 ± 2.5	65 ± 1.7*	62.2 ± 1.5	59.7 ± 2.3
Glucose mg/dL	293.8 ± 131.0	551.8 ± 83.7*	505 ± 74*	525.5 ± 51*	580.7 ± 22*
Food intake, g/week	34.3 ± 2.1	32.9 ± 0.3	27.6 ± 1.0*	37.5 ± 0.8*	36.3 ± 2.4
Water intake, mL/week	64.8 ± 11.8	78.2 ± 7.2	81.4 ± 10.4*	79.6 ± 1.8	67.2 ± 8.3
CD1					
BMI g/cm^2	29.6 ± 2.0	31.4 ± 1.9	32.2 ± 3.1	34.4 ± 4.1	34.2 ± 2.4
Glucose mg/dL	126.2 ± 18.6	119.3 ± 13	127.5 ± 18.7	109.7 ± 13.3	106.8 ± 12.4
Food intake, g/week	54.6 ± 1.7	58.3 ± 5.6	42.7 ± 5	61 ± 1.9	73.3 ± 20.8*
Water intake, mL/week	63 ± 13.1	67.3 ± 3.2	53.9 ± 1.3	89.8 ± 6.1*	93.2 ± 4.8*

Data are presented as means ± standard deviations. One-way ANOVA* for comparison of differences between BL group at 8 weeks versus all the groups at 16 weeks. p value was significant at <0.05.

the lymphocyte gate on the FSC/SSC dot plot. 20,000 gated events were acquired from each sample using the CellQuest research software (Becton Dickinson). Data was analyzed using Summit software v4.3 (Dako, Colorado Inc.). Data from six mice per group are reported as the mean ± standard deviation (SD).

2.7. Statistical Analysis. One-way ANOVA was performed for comparison between groups from each strain (BL, RC, LY, SAT, and MD); Bonferroni post hoc test was applied. Differences were considered significant at $p < 0.05$. Software used to run statistical analysis was SPSS v.23 for Windows.

3. Results

3.1. BMI Was Higher in the MD Group for Diabetic Mice and Blood Glucose Was Higher in All the db/db Groups. In the diabetic db/db mice, the MD group showed a significantly higher BMI than the BL group; blood glucose concentrations were significantly higher in all groups compared to the BL group; the food intake was significantly lower in the MD group and higher in the LY group and finally the water consumption was significantly higher in the MD group, all of this compared with the BL group (Bonferroni post hoc, $p < 0.001$). In the healthy mice (CD1), there were no significant differences in BMI and blood glucose. On the other hand, the consumption of food was significantly higher in the SAT group and the water consumption was significantly higher in the LY and SAT groups, all this compared with the BL group (Bonferroni post hoc, $p < 0.001$) (Table 2).

3.2. Microalgae Fatty Acids Modified Lymphocyte Populations in db/db Mice by Lowering CD3+ and CD8+ Populations and in CD1 Mice by Lowering CD3+. In the db/db strain, the percentage of CD3+ lymphocytes was significantly higher in all the groups when compared to the BL group (Bonferroni

post hoc, $p < 0.001$). Regarding CD4+ lymphocytes, the MD group showed a significantly lower percentage and the SAT group a higher percentage (Bonferroni post hoc, $p < 0.001$). Finally, the CD8 + lymphocytes have a higher percentage in the MD, LY, and SAT groups (Bonferroni post hoc, $p < 0.001$).

For CD1 strain mice, the percentage of CD3+ lymphocytes was lower in the MD, LY, and SAT groups and higher in the RC group, all compared with the BL group (Bonferroni post hoc, $p < 0.001$). Regarding CD4 + lymphocytes, the percentage was significant in the MD and SAT groups (Bonferroni post hoc, $p < 0.001$). Finally, the percentage of CD8+ lymphocytes was significantly higher in the RC and LY groups, but lower in the MD and SAT groups, all compared with the BL group (Bonferroni post hoc, $p < 0.001$) (Table 3).

3.3. Supplementation with Microalgae Fatty Acids Increases IL-17A, IL-12, IL-4, IL-6, IL-10, and TGF-β but Decreases IFN-γ, TNF-α, and IL-5 in Diabetic Mice. In the diabetic mice, the behavior of Th1 type cytokines was the same for all study groups; the proportion of TCD4+ cells producing IFN-γ and TNF-α was significantly lower in all groups compared to the BL (Bonferroni post hoc, $p < 0.001$). On the contrary, the percentage of TCD4+ cells producing IL-12 and IL-17A was significantly lower in all groups (Bonferroni post hoc, $p < 0.001$). As for the Th2 type cytokines, the percentage of TCD4+ cells producing IL-4 was higher in the RC, MD, and SAT groups and, on the other hand, it was lower in the LY group. The percentage of TCD4+ cells producing IL-5 was significantly lower in all groups compared to the baseline group (Bonferroni post hoc, $p < 0.001$). For all groups, the percentage of TCD4+ cells producing IL-6 and IL-10 was significantly lower compared to the initial group (Bonferroni post hoc, $p < 0.001$). Finally, the percentage of TCD4+ cells producing TGF-β was significantly lower in the RC group and higher in the MD, LY, and SAT groups (Table 4).

TABLE 3: Effect of supplementation with EPA and DHA extracted from microalgae in lymphocytes populations in db/db and CD1 mice.

	8 weeks old		16 weeks old		
	BL	RC	MD	LY	SAT
	Mean ± SD	Mean ± SD	Mean ± SD	Mean ± SD	Mean ± SD
	$n = 6$	$n = 6$	$n = 6$	$n = 6$	$n = 6$
db/db					
CD3+, %	71.8 ± 0.5	70.4 ± 0.8*	69.8 ± 0.4*	70.7 ± 0.4*	68.3 ± 0.2*
CD4+, %	62.0 ± 0.2	62.1 ± 0.5	60.6 ± 0.3*	61.8 ± 0.9	65 ± 0.5*
CD8+, %	30.0 ± 1.0	29.3 ± 1.0	24.9 ± 0.3*	28.4 ± 0.2*	23 ± 0.1*
cd1					
CD3+, %	74.7 ± 0.02	77.8 ± 0.04*	70.6 ± 0.03*	72.4 ± 0.01*	69.5 ± 0.06*
CD4+, %	67.4 ± 0.2	67.5 ± 0.05	70 ± 0.04*	67.6 ± 0.03	72.3 ± 0.4*
CD8+, %	21.4 ± 0.05	24.8 ± 0.01*	17.4 ± 0.1*	24.2 ± 0.03*	19.9 ± 0.02*

Data are presented as means ± standard deviations. One-way ANOVA* for comparison of differences between BL group at 8 weeks versus all the groups at 16 weeks. p value was significant at <0.05.

TABLE 4: Effect of supplementation with EPA and DHA fatty acids extracted from microalgae on Th1 and Th2 cytokines in db/db.

	8 weeks old		16 weeks old		
	BL	RC	MD	LY	SAT
db/db mice	Mean ± SD	Mean ± SD	Mean ± SD	Mean ± SD	Mean ± SD
	$n = 6$	$n = 6$	$n = 6$	$n = 6$	$n = 6$
Th1					
% TCD4+/					
IFN-γ	22.5 ± 0.4	11.5 ± 0.4*	7.2 ± 0.5*	2.1 ± 0.4*	15.7 ± 0.4*
TNF-α	10.2 ± 0.4	7.7 ± 0.4*	8.8 ± 0.4*	2.4 ± 0.4*	1.4 ± 0.4*
IL-12	1.2 ± 0.1	12.4 ± 0.5*	10.2 ± 0.5*	6.8 ± 0.1*	3.6 ± 0.5*
IL-17A	1.4 ± 0.1	4.2 ± 0.3*	3.7 ± 0.1*	11.4 ± 0.3*	11.4 ± 0.3*
Th2					
% TCD4+/					
IL-4	1.8 ± 0.08	8.2 ± 0.3*	6.6 ± 0.3*	1.1 ± 0.1*	10.3 ± 0.2*
IL-5	19.9 ± 0.8	9.4 ± 0.4*	7.5 ± 0.6*	2.1 ± 0.2*	6.5 ± 0.2*
IL-6	1.5 ± 0.1	3.1 ± 0.4*	4.5 ± 0.2*	4.0 ± 0.1*	14.8 ± 0.2*
IL-10	1.44 ± 0.2	16.0 ± 1.3*	7.3 ± 0.2*	2.7 ± 0.08*	14.5 ± 0.2*
TGF-β	3.2 ± 0.4	2.1 ± 0.05*	5.5 ± 0.1*	4.9 ± 0.09*	7.1 ± 0.1*

Data are presented as means ± standard deviations of percentage of TCD4+ cells producing cytokines. One-way ANOVA* for comparison of differences between BL group at 8 weeks versus all the groups at 16 weeks. p value was significant at <0.05.

3.4. Supplementation with Microalgae Fatty Acids Lyophilized Increases IL-17A but Decreases IFN-γ, TNF-α, IL-12, IL-4, and IL-6 in Healthy Mice. In CD1 mice, for Th1 type cytokines, the percentage of TCD4+ cells producing IFN-γ and TNF-α was significantly lower in the RC and LY groups and significantly higher in the SAT group (Bonferroni post hoc, $p < 0.001$). The percentage of TCD4+ cells producing IL-12 was significantly lower in the RC, MD, and LY groups and higher in the SAT group, all compared with the BL group (Bonferroni post hoc, $p < 0.001$); finally, the percentage of TCD4+ producing IL-17A was significantly lower in all groups compared to the BL group (Bonferroni post hoc, $p < 0.001$). Regarding the Th2 type cytokines, the SAT group showed a significantly higher percentage of TCD4+ cells producing IL-4, IL-5, IL-6, and TGF-β compared to the BL group (Bonferroni post hoc, $p < 0.001$), and the MD group showed similar behavior for IL-5 and TGF-β. On the other hand, the percentage of TCD4+ cells producing IL-4 and IL-6 was significantly lower in the LY groups (Bonferroni post hoc, $p < 0.001$) (Table 5).

4. Discussion

The results provided in this study show evidence that supplementation with *n*-3 fatty acids obtained from microalgae improves the inflammatory profile in general by reducing the secretion of many cytokines. Therefore, these results suggest that microalgae extracts may be considered as an anti-inflammatory strategy against different diseases. These findings are summarized in Figure 2.

The BMI was significantly higher in the MD group from the diabetic mice. These results are different from

TABLE 5: Effect of supplementation with EPA and DHA fatty acids extracted from microalgae on Th1 and Th2 cytokines in CD1.

CD1 mice	8 weeks old		16 weeks old		
	BL	RC	MD	LY	SAT
	Mean ± SD	Mean ± SD	Mean ± SD	Mean ± SD	Mean ± SD
	$n = 6$	$n = 6$	$n = 6$	$n = 6$	$n = 6$
Th1					
% TCD4+/					
IFN-γ	1.8 ± 0.18	1.2 ± 0.09*	2.7 ± 0.1*	1.0 ± 0.1*	4.3 ± 0.1*
TNF-α	2.5 ± 0.13	1.5 ± 0.1*	2.6 ± 0.1	1.4 ± 0.09*	4.6 ± 0.07*
IL-12	2.8 ± 0.1	1.7 ± 0.08*	2.4 ± 0.09*	1.3 ± 0.08*	4.4 ± 0.08*
IL-17A	1.6 ± 0.08	1.8 ± 0.08*	1.8 ± 0.1*	4.1 ± 0.08*	3.6 ± 0.09*
Th2					
% TCD4+/					
IL-4	2.6 ± 0.1	1.8 ± 0.09*	2.4 ± 0.1	1.6 ± 0.1*	4.4 ± 0.1*
IL-5	2.1 ± 0.1	1.4 ± 0.1*	2.6 ± 0.1*	1.9 ± 0.1	3.4 ± 0.1*
IL-6	1.8 ± 0.1	1.7 ± 0.08	1.7 ± 0.1	1.5 ± 0.09*	2.6 ± 0.07*
IL-10	1.6 ± 0.08	1.5 ± 0.1	2.2 ± 0.1	1.2 ± 0.1	2.9 ± 1.6
TGF-β	1.5 ± 0.06	1.3 ± 0.8*	2.3 ± 0.07*	1.5 ± 0.08	5.1 ± 0.1*

Data are presented as means ± standard deviations of percentage of TCD4+ cells producing cytokines. One-way ANOVA* for comparison of differences between BL group at 8 weeks versus all the groups at 16 weeks. p value was significant at <0.05.

those reported by Zhuang in which C57Bl/6 mice were supplemented with fish oil and did not show changes in BMI [30]. There were no significant differences in plasma glucose attributable to treatment; no studies were found to match our findings; however, a study with C57B/6 mice supplemented with EPA suggests a protective effect of n-3 fatty acids on glucose metabolism [19]. Further studies on the effect of microalgae fatty acids on glucose metabolism are needed.

In this study, food consumption was lower in the MD groups for the diabetic mice. Other studies with C57Bl/6 mice fed with normal chow enriched with EPA and DHA extracted from fish oil showed no changes in food consumption [31, 32]. However, a study from Díaz- Reséndiz explains that mice regulate food intake according to the composition of the food or the presence of an extra source of energy [33].

The percentage of total T lymphocytes was lower in all study groups from both strains. In contrast to this study, Marano et al. [34] suggest that consumption of n-3 fatty acids increases CD3+ lymphocyte populations including CD4+. In agreement with our findings, several supplementation studies report that there were no changes in lymphocyte populations [35, 36].

On the other hand, in both strains the SAT groups showed a significantly lower percentage of CD8+ cells. The CD4+ populations in the SAT groups increased significantly compared to their BL group. These results are consistent with those of Baccan et al. [37] who showed that consumption of high-fat diets significantly increases lymphocyte populations.

The db/db strain is characterized by a chronic inflammatory state such as diabetes disease, which causes the pro- and anti-inflammatory cytokines to be in higher concentrations compared to the CD1 strain; however, although the strains are very different between them, the decrease in cytokine concentrations occurred in a similar way.

In this study, supplementation with n-3 fatty acids extracted from microalgae significantly decreased the percentage of TCD4+ cells producing IFN-γ and TNF-α. These cytokines play different roles in inflammatory states such as in diabetes; IFN-γ directs the differentiation of CD4+ lymphocytes into helper lymphocytes type 1 (Th1); it also intervenes in the activation of macrophages and induces a greater secretion of IL-12 [38]. However, in the diabetic mice, microalgae fatty acids were shown to increase the percentage of TCD4+ cells producing IL-12. The main functions of IL-12 are the activation of Th1 lymphocytes and to stimulate the production of IFN-γ [39]. On the other hand, TNF-α is a cytokine involved in the acute and chronic phase as well as in the activation of the production of certain anti-inflammatory cytokines such as IL-4, IL-5, and IL-6 as a form of self-regulation of the inflammatory state [40].

A study by Vigerust et al. made in transgenic TNF-α C57B/6 mice that were fed with diets enriched with either fish oil or krill oil showed no modification in this cytokine [40]. Similarly, in a study on Wistar rats supplemented with fish or soybean oil, no significant differences were found in the concentrations of IFN-γ and TNF-α [41]. Although Xavier et al. [41] showed there are no changes in TNF-α after fish oil supplementation, there are also many studies that report a significant decrease in this cytokine [42–44]. A study made by Sierra et al. [45] reports that, in Balb/c mice fed with modified diet either with EPA or with DHA for 3 weeks, spleen lymphocytes decreased their production of TNF-α only in the diet with EPA, but not in the diet with DHA. When approaching the effect of microalgae fatty acids, a study carried out in cell lines of macrophages exposed to LPS and added with extracts of different microalgae showed a significant decrease of the TNF-α compared against the control cultures. A study by Sierra et al. reported that only

Modified diet (MD)

Diabetic mice (db/db)

Lymphocyte populations:
↓ % CD3+
↓ % CD4+
↓ % CD8+

Th1 type cytokines:
↓ IFN-γ, TNF-α
↑↑ IL-12
↑ IL-17A
Th2 type cytokines:
↓ IL-5
↑↑ IL-4, IL-10
↑ IL-6, TGF-β

Healthy mice (cd1)

Lymphocyte populations:
↓ % CD3+
↑ % CD4+
↓ % CD8+

Th1 type cytokines:
↓ IL-12
↑ IFN-γ, IL-17A
Th2 type cytokines:
↓ IL-4
↑ IL-5

Lyophilized fatty acids (LY)

Diabetic mice (db/db)

Lymphocyte populations:
↓ % CD3+
↓ % CD8+

Th1 type cytokines:
↓↓ IFN-γ, TNF-α
↑ IL-12,
↑↑ IL-17A
Th2 type cytokines:
↓↓ IL-5
↓ IL-4
↑ IL-6, IL-10, TGF-β

Healthy mice (cd1)

Lymphocyte populations:
↓ % CD3+
↑ % CD8+

Th1 type cytokines:
↓ IFN-γ, TNF-α, IL-12
↑↑ IL-17A
Th2 type cytokines:
↓ IL-4, IL-6

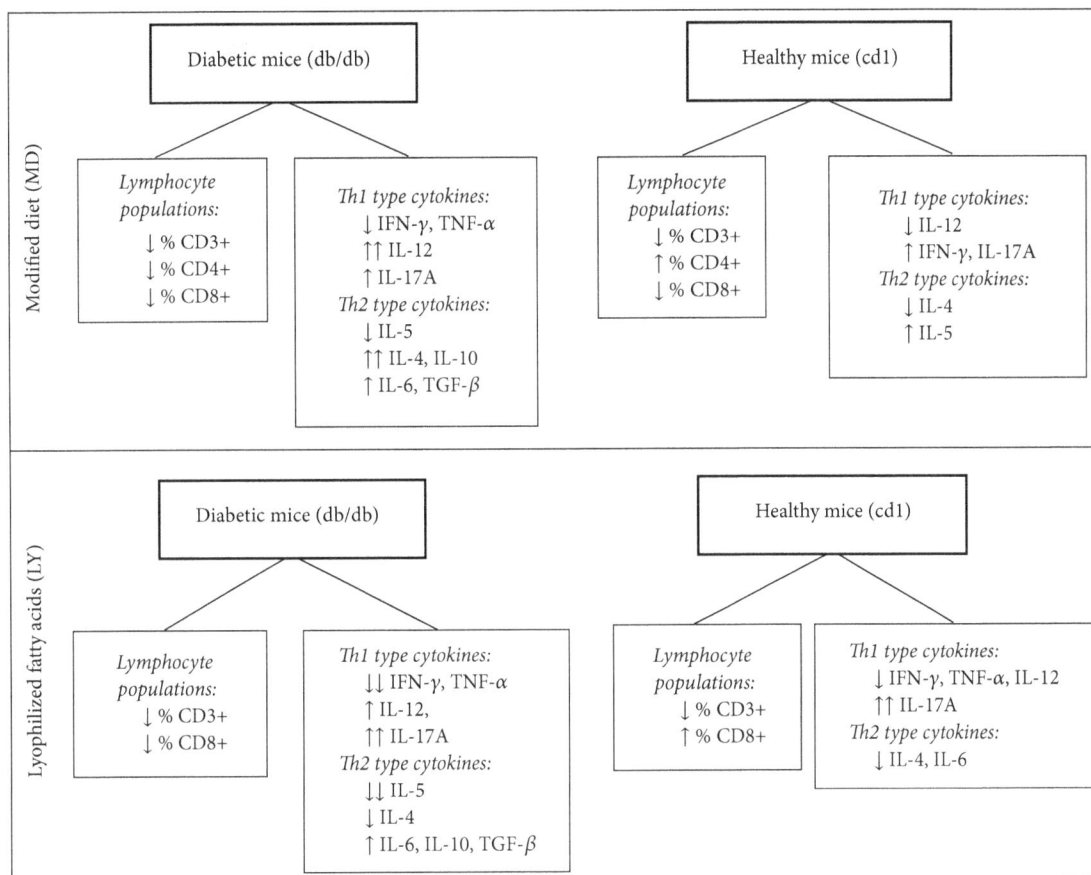

FIGURE 2: Summary of major findings about microalgae fatty acids supplementation in diabetic and healthy mice.

EPA enriched diet was able to decrease IL-12 concentrations [45].

In this study, the percentage of TCD4+ cells producing IL-17A showed a significant increase in both strains. IL-17A is known as an inflammatory cytokine whose main function is exerted on myeloid and mesenchymal cells by inducing the expression of granulocyte colony-stimulating factor (G-CSF), IL-6, and other chemokines, which increase granulopoiesis and recruit neutrophils into the site of infection [46]. Vigerust et al. also reported that, after supplementation with fish oil and krill oil enriched diets, IL-17A was shown to be increased only in the fish oil group [40], and these results agree with our findings.

The supplementation with *n*-3 fatty acids in lyophilized form and that are added in the food showed a significantly higher percentage of TCD4+ cells producing IL-10 in both strains. IL-4 is produced by type 2 T cells (Th2), basophils, and mast cells. It has anti-inflammatory function by blocking the synthesis of IL-1, TNF-α, and IL-6. In addition, it promotes the proliferation and differentiation of B lymphocytes and is considered a potent inhibitor of apoptosis [47]. IL-10 is also known as cytokine synthesis inhibiting factor (CSIF) and can inhibit the synthesis of proinflammatory cytokines by T lymphocytes and macrophages. It also regulates the growth and differentiation of B lymphocytes, NK cytotoxic and helper T cells, mast cells, granulocytes, dendritic cells,

keratinocytes, and endothelial cells [48]. A study in Wistar rats fed with a fish oil enriched diet was found to show lower concentrations of IL-4 and IL-5 at the alveolar level compared to control [41]. Also, Sierra et al. [45] showed similar results in Balb/c mice. Our findings agree with Li et al. [42] who demonstrated that fish oil supplementation decreased IL-10; however, Sierra et al. reported that only EPA supplementation was able to increase IL-10 concentrations [45]. These results suggest that the consumption of microalgae fatty acids has attenuating effects of systemic inflammation even in chronic diseases and not only in acute states of inflammation.

TGF-β is a cytokine with pleiotropic functions in hematopoiesis, angiogenesis, cell proliferation, differentiation, migration, and apoptosis. It has a strong anti-inflammatory action but may increase some immune functions. Thus, in knock-out mice for TGF-β they show defects in regulatory T lymphocytes, which generates an extensive inflammation with abundant proliferation of T lymphocytes and differentiation of CD4+ in Th1 and Th2 lymphocytes [49]. In our study both strains showed higher percentage of TCD4+ cells producing TGF-β in the LY group when compared with BL; however, they were much lower than those for the SAT group so we could agree that microalgae fatty acids have a positive effect on TGF-β expression. A study with apoE-deficient mice infused with angiotensin and treated with EPA and DHA orally for 3 weeks showed that the TGF-β gene expression was significantly decreased in the EPA group and DHA compared to untreated mice [50]. In contrast, a study on Wistar rats [51] exposed to LPS during gestation and whose offspring were supplemented with fish oil showed that TGF-β concentrations were increased as compared to controls [52].

IL-6 release is induced by IL-1 and TNF-α. It is involved in the production of immunoglobulins and in the differentiation of active B lymphocytes and plasma cells, modulates hematopoiesis, and is responsible, together with IL-1, for the synthesis of acute phase liver proteins like fibrinogen [53]. Percentage of TCD4+ cells producing IL-6 was shown to be significantly increased in LY group from both strains; however SAT group reported the highest IL-6 concentrations; we suggest that microalgae fatty acids could have a protective effect against IL-6 expression. Two studies report decreasing concentrations of IL-6 when using fish oil: one was conducted in mice [42] and the other in overweight pregnant women [43]. Additionally, a study by Robertson et al. [26] in macrophage cell line cultures treated with microalgae extracts showed a significant decrease in IL-6.

Regarding the effects of the consumption of coconut oil as a source of saturated fat, there is controversy about the properties of coconut oil, and this is because, despite being a source of saturated fat, several studies have shown that it has anti-inflammatory properties [54, 55]; in this study, the consumption of coconut oil showed a high percentage of TCD4+ cells producing Th1 and Th2 type cytokines which is consistent with other studies [56].

5. Conclusion

The results provided in this study show evidence that supplementation with n-3 fatty acids obtained from microalgae improves the inflammatory profile in general by reducing the secretion of many cytokines. Therefore, these results suggest that microalgae extracts may be considered as an anti-inflammatory strategy against different chronic diseases.

Acknowledgments

Microalgae n-3 fatty acid lyophilized and modified diet was provided by Biotecnología Mexicana de Microalgas SA de CV. GPLE is a CONACyT fellowship program member.

References

[1] Diagnosis and Classification of Diabetes Mellitus, vol. 37, supplement 1, Diabetes Care, 2014.

[2] T. Shamah-Levy, L. Cuevas-Nasu, J. Rivera-Dommarco, and M. Hernández-Ávila, Encuesta Nacional de Nutrición y Salud de Medio Camino 2016 (ENSANUT MC 2016), Inf Final Result Recuper, 2016.

[3] M. Hernández-Ávila, J. P. Gutiérrez, and N. Reynoso-Noverón, "Diabetes mellitus en México. El estado de la epidemia," Salud Pública de México, vol. 55, pp. s129–s136, 2013.

[4] K. Esposito, R. Marfella, and D. Giugliano, "Stress hyperglycemia, inflammation, and cardiovascular events [18]," Diabetes Care, vol. 26, no. 5, pp. 1650-1651, 2003.

[5] T. A. Mori, R. J. Woodman, V. Burke, I. B. Puddey, K. D. Croft, and L. J. Beilin, "Effect of eicosapentaenoic acid and docosahexaenoic acid on oxidative stress and inflammatory markers in treated-hypertensive type 2 diabetic subjects," Free Radical Biology & Medicine, vol. 35, no. 7, pp. 772–781, 2003.

[6] V. M. Montori, A. Farmer, P. C. Wollan, and S. F. Dinneen, "Fish oil supplementation in type 2 diabetes: a quantitative systematic review," Diabetes Care, vol. 23, no. 9, pp. 1407–1415, 2000.

[7] S. L. Connor and W. E. Connor, "Are fish oils beneficial in the prevention and treatment of coronary artery disease?" American Journal of Clinical Nutrition, vol. 66, no. 4, pp. 1020S–1031S, 1997.

[8] T. A. Mori and L. J. Beilin, "Long-chain omega 3 fatty acids, blood lipids and cardiovascular risk reduction," Current Opinion in Lipidology, vol. 12, no. 1, pp. 11–17, 2001.

[9] P. C. Calder, "Immunomodulation by omega-3 fatty acids," Prostaglandins, Leukotrienes and Essential Fatty Acids, vol. 77, no. 5-6, pp. 327–335, 2007.

[10] B. Liang, S. Wang, Y.-J. Ye et al., "Impact of postoperative omega-3 fatty acid-supplemented parenteral nutrition on clinical outcomes and immunomodulationsi in colorectal cancer patients," World Journal of Gastroenterology, vol. 14, no. 15, pp. 2434–2439, 2008.

[11] J. E. Teitelbaum and W. Allan Walker, "Review: The role of omega 3 fatty acids in intestinal inflammation," The Journal of Nutritional Biochemistry, vol. 12, no. 1, pp. 21–32, 2001.

[12] J. P. Schuchardt, I. Schneider, and H. Meyer, "Incorporation of EPA and DHA into plasma phospholipids in response to different omega-3 fatty acid formulations—a comparative bioavailability study of fish oil vs. krill oil," Lipids in Health and Disease, vol. 10, article 145, 2011.

[13] D. Y. Oh, S. Talukdar, E. J. Bae et al., "GPR120 is an omega-3 fatty acid receptor mediating potent anti-inflammatory and insulin-sensitizing effects," Cell, vol. 142, no. 5, pp. 687–698, 2010.

[14] M. Arita, F. Bianchini, J. Aliberti et al., "Stereochemical assignment, antiinflammatory properties, and receptor for the omega-3 lipid mediator resolvin E1," The Journal of Experimental Medicine, vol. 201, no. 5, pp. 713–722, 2005.

[15] R. Wall, R. P. Ross, G. F. Fitzgerald, and C. Stanton, "Fatty acids from fish: the anti-inflammatory potential of long-chain omega-3 fatty acids," Nutrition Reviews, vol. 68, no. 5, pp. 280–289, 2010.

[16] J. A. Hall, R. C. Wander, J. L. Gradin, S. H. Du, and D. E. Jewell, "Effect of dietary n-6-to-n-3 fatty acid ratio on complete blood and total white blood cell counts, and T-cell subpopulations in aged dogs," Am J Vet Res, vol. 60, pp. 319–327, 1999.

[17] C. A. Hudert, K. H. Weylandt, Y. Lu et al., "Transgenic mice rich in endogenous omega-3 fatty acids are protected from colitis," Proceedings of the National Acadamy of Sciences of the United States of America, vol. 103, no. 30, pp. 11276–11281, 2006.

[18] G. P. Lim, F. Calon, T. Morihara et al., "A diet enriched with the omega-3 fatty acid docosahexaenoic acid reduces amyloid burden in an aged Alzheimer mouse model," The Journal of Neuroscience, vol. 25, no. 12, pp. 3032–3040, 2005.

[19] N. S. Kalupahana, K. Claycombe, S. J. Newman et al., "Eicosapentaenoic acid prevents and reverses insulin resistance in high-fat diet-induced obese mice via modulation of adipose tissue inflammation," *Journal of Nutrition*, vol. 140, no. 11, pp. 1915–1922, 2010.

[20] J. M. Bourre, "Roles of unsaturated fatty acids (especially omega-3 fatty acids) in the brain at various ages and during ageing," *J Nutr Health Aging [Internet]*, vol. 8, no. 3, pp. 163–174, 2004.

[21] A. Valenzuela B and R. Valenzuela B, "Acidos grasos omega-3 en la nutrición ¿como aportarlos?" *Revista chilena de nutrición*, vol. 41, no. 2, pp. 205–211, 2014.

[22] I. Khozin-Goldberg, S. Leu, and S. Boussiba, "Microalgae as a source for VLC-PUFA production," *Subcellular Biochemistry*, vol. 86, pp. 471–510, 2016.

[23] C. J. Shepherd and A. J. Jackson, "Global fishmeal and fish-oil supply: Inputs, outputs and marketsa," *Journal of Fish Biology*, vol. 83, no. 4, pp. 1046–1066, 2013.

[24] R. Robertson, F. Guihéneuf, DB. Stengel, G. Fitzgerald, P. Ross, and C. Stanton, "Algae-derived Polyunsaturated fatty acids: implications for human health," in *Polyunsaturated Fatty Acifs: Sources, Antioxidant Properties and Health Benefits*, 2017.

[25] J. G. Bell and J. R. Sargent, "Arachidonic acid in aquaculture feeds: Current status and future opportunities," *Aquaculture*, vol. 218, no. 1-4, pp. 491–499, 2003.

[26] R. C. Robertson, F. Guihéneuf, B. Bahar et al., "The anti-inflammatory effect of algae-derived lipid extracts on lipopolysaccharide (LPS)-stimulated human THP-1 macrophages," *Marine Drugs*, vol. 13, no. 8, pp. 5402–5424, 2015.

[27] R. Deng and T.-J. Chow, "Hypolipidemic, antioxidant, and anti-inflammatory activities of microalgae spirulina," *Cardiovascular Therapeutics*, vol. 28, no. 4, pp. e33–e45, 2010.

[28] E. Talero, S. García-Mauriño, J. Ávila-Román, A. Rodríguez-Luna, A. Alcaide, and V. Motilva, "Bioactive compounds isolated from microalgae in chronic inflammation and cancer," *Marine Drugs*, vol. 13, no. 10, pp. 6152–6209, 2015.

[29] I. M. Arciniega-Martínez, R. Campos-Rodríguez, M. E. Drago-Serrano, L. E. Sánchez-Torres, T. R. Cruz-Hernández, and A. A. Reséndiz-Albor, "Modulatory Effects of Oral Bovine Lactoferrin on the IgA Response at Inductor and Effector Sites of Distal Small Intestine from BALB/c Mice," *Archivum Immunologiae et Therapia Experimentalis*, vol. 64, no. 1, pp. 57–63, 2016.

[30] P. Zhuang, W. Wang, Y. Zhang, and J. Jiao, "Long-term dietary EPA or DHA supplementation do not ameliorate obesity but improve glucose homeostasis via gut-adipose axis in already obese mice," *FASEB J*, vol. 31, no. 1, pp. 971-11, 2017.

[31] P. Flachs, V. Mohamed-Ali, O. Horakova et al., "Polyunsaturated fatty acids of marine origin induce adiponectin in mice fed a high-fat diet," *Diabetologia*, vol. 49, no. 2, pp. 394–397, 2006.

[32] J. M. Monk, D. M. Liddle, and A. A. De Boer, "Fish-oil-derived n-3 PUFAs reduce inflammatory and chemotactic adipokine-mediated cross-talk between co-cultured murine splenic CD8⁺ T cells and adipocytes," *Journal of Nutrition*, vol. 145, no. 4, pp. 829–838, 2015.

[33] F. de Jesús Díaz-Reséndiz, K. Franco-Paredes, A. G. Martínez-Moreno, A. López-Espinoza, and V. G. Aguilera-Cervantes, *Efectos de variables ambientales sobre la ingesta de alimento en ratas: Una revision histórico-conceptual*, vol. 8, Univ Psychol, 2009.

[34] L. Marano, R. Porfidia, M. Pezzella et al., "Clinical and immunological impact of early postoperative enteral immunonutrition after total gastrectomy in gastric cancer patients: a prospective randomized study," *Annals of Surgical Oncology*, vol. 20, no. 12, pp. 3912–3918, 2013.

[35] A. M. Rizzo, P. A. Corsetto, G. Montorfano et al., "Comparison between the AA/EPA ratio in depressed and non depressed elderly females: Omega-3 fatty acid supplementation correlates with improved symptoms but does not change immunological parameters," *Nutrition Journal*, vol. 11, no. 1, article no. 82, 2012.

[36] V. R. Mukaro, M. Costabile, K. J. Murphy, C. S. Hii, P. R. Howe, and A. Ferrante, "Leukocyte numbers and function in subjects eating n-3 enriched foods: Selective depression of natural killer cell levels," *Arthritis Research & Therapy*, vol. 10, no. 3, article no. R57, 2008.

[37] G. C. Baccan, O. Hernández, L. E. Díaz et al., "Changes in lymphocyte subsets and functions in spleen from mice with high fat diet-induced obesity," *Proceedings of the Nutrition Society*, vol. 72, no. OCE1, 2013.

[38] K. Schroder, P. J. Hertzog, T. Ravasi, and D. A. Hume, "Interferon-gamma: an overview of signals, mechanisms and functions," *J Leukoc Biol*, vol. 75, no. 2, pp. 163–189, 2004.

[39] C. L. Langrish, B. S. McKenzie, N. J. Wilson, R. De Waal Malefyt, R. A. Kastelein, and D. J. Cua, "IL-12 and IL-23: master regulators of innate and adaptive immunity," *Immunological Reviews*, vol. 202, pp. 96–105, 2004.

[40] N. F. Vigerust, B. Bjørndal, P. Bohov, T. Brattelid, A. Svardal, and R. K. Berge, "Krill oil versus fish oil in modulation of inflammation and lipid metabolism in mice transgenic for TNF-α," *European Journal of Nutrition*, vol. 52, no. 4, pp. 1315–1325, 2013.

[41] R. A. N. Xavier, K. V. de Barros, I. S. de Andrade, Z. Palomino, D. E. Casarini, and V. L. F. Silveira, "Protective effect of soybean oil- or fish oil-rich diets on allergic airway inflammation," *Journal of Inflammation Research*, vol. 9, pp. 79–89, 2016.

[42] C. C. Li, H. T. Yang, Y. C. Hou, Y. S. Chiu, and W. C. Chiu, "Dietary fish oil reduces systemic inflammation and ameliorates sepsis-induced liver injury by up-regulating the peroxisome proliferator-activated receptor gamma-mediated pathway in septic mice," *The Journal of Nutritional Biochemistry*, vol. 25, pp. 19–25, 2014.

[43] M. Ebrahimi-Mameghani, Z. Sadeghi, M. Abbasalizad Farhangi, E. Vaghef-Mehrabany, and S. Aliashrafi, "Glucose homeostasis, insulin resistance and inflammatory biomarkers in patients with non-alcoholic fatty liver disease: Beneficial effects of supplementation with microalgae Chlorella vulgaris: A double-blind placebo-controlled randomized clinical trial," *Clinical Nutrition*, vol. 36, no. 4, pp. 1001–1006, 2017.

[44] M. Haghiac, X.-H. Yang, L. Presley et al., "Dietary omega-3 fatty acid supplementation reduces inflammation in obese pregnant women: A randomized double-blind controlled clinical trial," *PLoS ONE*, vol. 10, no. 9, Article ID e0137309, 2015.

[45] S. Sierra, F. Lara-Villoslada, M. Comalada, M. Olivares, and J. Xaus, "Dietary eicosapentaenoic acid and docosahexaenoic acid equally incorporate as decosahexaenoic acid but differ in inflammatory effects," *Nutrition Journal*, vol. 24, no. 3, pp. 245–254, 2008.

[46] Y. F. Talmás-Rohana, "Interleucina 17, funciones biológicas y su receptor," *Reb*, vol. 31, no. 1, pp. 3–9, 2012.

[47] A. E. Kelly-Welch, E. M. Hanson, M. R. Boothby, and A. D. Keegan, "Interleukin-4 and interleukin-13 signaling connections maps," *Science*, vol. 300, no. 5625, pp. 1527-1528, 2003.

[48] K. W. Moore, R. de Waal Malefyt, R. L. Coffman, and A. O'Garra, "Interleukin-10 and the interleukin-10 receptor," *Annual Review of Immunology*, vol. 19, pp. 683–765, 2001.

[49] E. Gonzalo-Gil and M. Galindo-Izquierdo, "Papel del factor de crecimiento transformador-beta (TGF-β) en la fisiopatología de la artritis reumatoide," *Reumatología Clínica*, vol. 10, no. 3, pp. 174–179, 2014.

[50] T. Yoshihara, K. Shimada, K. Fukao et al., "Omega 3 polyunsaturated fatty acids suppress the development of aortic aneurysms through the inhibition of macrophage-mediated inflammation," *Circulation Journal*, vol. 79, no. 7, pp. 1470–1478, 2015.

[51] A. Rabinovitch and W. L. Suarez-Pinzon, "Cytokines and their roles in pancreatic islet β-cell destruction and insulin-dependent diabetes mellitus," *Biochemical Pharmacology*, vol. 55, no. 8, pp. 1139–1149, 1998.

[52] J. J. Fortunato, N. da Rosa, A. O. Martins Laurentino et al., "Effects of ω-3 fatty acids on stereotypical behavior and social interactions in Wistar rats prenatally exposed to lipopolysaccarides," *Nutrition Journal*, vol. 35, pp. 119–127, 2017.

[53] A. D. Pradhan, J. E. Manson, N. Rifai, J. E. Buring, and P. M. Ridker, "C-reactive protein, interleukin 6, and risk of developing type 2 diabetes mellitus," *The Journal of the American Medical Association*, vol. 286, no. 3, pp. 327–334, 2001.

[54] H. Van Der Tempel, J. E. Tulleken, P. C. Limburg, F. A. J. Muskiet, and M. H. Van Rijswijk, "Effects of fish oil supplementation in rheumatoid arthritis," *Annals of the Rheumatic Diseases*, vol. 49, no. 2, pp. 76–80, 1990.

[55] R. J. Goldberg and J. Katz, "A meta-analysis of the analgesic effects of omega-3 polyunsaturated fatty acid supplementation for inflammatory joint pain," *PAIN*, vol. 129, no. 1-2, pp. 210–223, 2007.

[56] S. Intahphuak, P. Khonsung, and A. Panthong, "Anti-inflammatory, analgesic, and antipyretic activities of virgin coconut oil," *Pharmaceutical Biology*, vol. 48, no. 2, pp. 151–157, 2010.

14

Dietary Oxidized Linoleic Acid Modulates Plasma Lipids beyond Triglycerides Metabolism

Mahdi Garelnabi,[1,2] **Gregory Ainsworth,**[3] **Halleh Mahini,**[2]
Naseeha Jamil,[1] **and Chinedu Ochin**[2]

[1]*Department of Biomedical and Nutritional Sciences, University of Massachusetts, Lowell, MA, USA*
[2]*Biomedical Engineering and Biotechnology Program, University of Massachusetts, Lowell, MA, USA*
[3]*Department of Chemistry, University of Massachusetts, Lowell, MA, USA*

Correspondence should be addressed to Mahdi Garelnabi; mahdi_garelnabi@uml.edu

Academic Editor: Gerhard M. Kostner

Introduction. Triglyceride (TG) is an independent risk factor for coronary heart disease. Previous work has shown that short-term supplementations of mouse chow with oxidized linoleic acid (OxLA) significantly reduce the level of plasma triglycerides. *Study Objective*. This study aims to determine the effects of longer-term supplementation of mouse chow with various concentrations of oxidized linoleic acid (OxLA) on plasma triglycerides. *Study Design*. The study consisted of forty C57BL/6 wildtype mice divided into four groups ($n = 10$). Two groups were kept as controls. One control group (P) was fed plain chow and the second control group (C) was fed chow supplemented with linoleic acid. The other two experimental groups (A) and (B) were fed oxidized linoleic acid supplemented chow in the following doses: 9 mg/day of oxidized linoleic acid and 18 mg/day of oxidized linoleic acid/mouse. *Results and Conclusion*. Mice that were on a diet supplemented with the higher dose of oxidized linoleic acid showed a 39% decrease in hepatic PPAR-α and a significant decrease in the plasma HDL levels compared to the mice that were fed diets of plain and linoleic acid supplemented chow. Interestingly, the longer-term consumption of oxidized linoleic acid may predispose to atheropathogenesis.

1. Introduction

Triglycerides plasma levels are major risk factor for cardiovascular diseases. It is a complex polygenic trait that is modulated by a number of major pathways in the intestine, liver, adipose, and plasma. Although triglycerides are not directly linked to the pathogenesis of atherosclerosis, they greatly influence plasma lipoproteins, especially the low-density and very low-density lipoproteins (LDL) which are known risk factors for cardiovascular disease (CVD).

Linoleic acid is the most abundant polyunsaturated fatty acid (PUFA) in many lipid rich diets. Like other polyunsaturated fatty acids (PUFA), linoleic acid is a ligand of the peroxisome proliferator-activated receptor alpha (PPAR-α) which is a regulating component of plasma triglyceride levels. We have previously shown that oxidized linoleic acid significantly lowers plasma triglyceride (TG) levels as compared to animals fed oleic acid. The changes were associated with increased APOA5 and acetyl-CoA oxidase genes expression in the mice

that were fed a diet supplemented with 9 mg/mouse/day of OxLA. Two apolipoproteins (Apo), Apo A5 and Apo CIII, were of particular interest due to their role in the level of triglycerides. These proteins modulated triglycerides by interchangeably binding to VLDL particles [1, 2].

PPAR-α induces the expression of proteins involved in the uptake, transport, and metabolism of fatty acids which results in decreasing the synthesis of triglyceride. In human ApoA5 is a direct target of PPAR-α which is consistent with the triglyceride-lowering role proposed for ApoA5 [3]. However PPAR-α link to ApoA5 in mice is not well understood. ApoA5 induces a clearance of VLDL from the plasma through electrostatic interactions with heparin sulfate proteoglycan (HSPG). This interaction aid colocalizes the VLDL to HSPG attached lipoprotein lipase (LPL), allowing for greater lipolysis [1, 4–6].

ApoCIII, like ApoA5, localized on VLDL particles but has the opposite effect as it promotes increases of plasma triglycerides. Within the liver, ApoCIII acts by utilizing the available triglyceride substrates to increase synthesis of VLDL-TG. The increased synthesis of VLDL particles results in greater plasma TG levels. ApoCIII prevented lipolysis leading to increased plasma TG [6]. As a result, these two antagonistic proteins are considered key in triglyceride modulation [1, 4, 5].

Lipases are other proteins of interest in TG metabolism. They are key enzymes that break down lipids and lipoproteins; lipoprotein lipase (LPL) is antagonized by two angiopoietin-like proteins ANGPTL3/4. The proteins cause the LPL dimers to dissociate, leading to a loss of activity and lipolysis. ANGPTL3 is promoted when its gene is bound by the Liver X Receptor (LXR). Hepatic ANGPTL4 is activated by ligands of PPAR-α, which includes PUFA [3, 7].

By investigating the hepatic gene, protein expressions, and plasma protein levels, the study determines the range of modulators for plasma triglycerides, which is known to control biosynthesis and metabolism of TG and widely affects the plasma lipoproteins. The goal is to elucidate a mechanism by which oxidized linoleic acid modulates TG and hepatic lipoproteins metabolism.

2. Experimental Design

Normal C57BL/6 male mice were cared for in an animal facility with 12-hour light/dark cycles and all the protocols pertaining to this study were approved by the University of Massachusetts, Lowell, Institutional Animal Care and the Use Committee (IACUC.) The body masses of the mice were monitored over the course of 10 weeks. Water was provided ad libitum, while set amounts of diet were measured and supplied to each study group weekly.

3. Diets

13-Hydroperoxyoctadecadienoic acid (13-HPODE) was prepared as previously described [2]. Two prepared formulas of 13-HPODE were then shipped to Harlan Laboratories, Indianapolis, Indiana, US, for preparation of the experimental mouse chow formulas. The specialized diets were kept at 2°C until used. Four different diet formulas were provided to the groups of mice: a Standard Chow as plain control (P group), a chow supplemented with linoleic acid 9 mg/mouse/day, linoleic control (C group), oxidized linoleic acid, 9 mg/mouse/day (A group), and oxidized linoleic acid 18 mg/mouse/day diet (B group). Mice were fed the dietary formulas or kept on plain chow for two months.

4. Materials and Methods

Mice were euthanized at the end of the study. Whole blood was obtained from each mouse by heart puncture and placed in heparin tubes. The containers were spun in a cold centrifuge at 3000 rpm. The samples were later aliquoted and stored at −80°C. The plasma samples were analyzed for LDL, High-Density Lipoprotein (HDL), glucose, and total cholesterol using reagents and standards from Medica Corporation, Bedford, MA 01730. ApoA5, ApoCIII, ANGPTL3, ANGPTL4, and hepatic lipase (HL) in plasma were analyzed using commercial ELISA kits.

4.1. mRNA Extraction and Analysis of ApoA5, ApoCIII, PPAR-Alpha, and SREBP1 Genes. Prior to collection of organs, chilled 1x phosphate buffered saline (PBS) was perfused through the heart, after which the liver (100 mg samples) and adipose were collected in homogenization tubes. The organs were then flash frozen in liquid nitrogen (LN2) and stored at −80°C. Single aliquots of liver and adipose were stored in 1 mL of Trizol for later RNA extraction. The RNA was extracted from liver samples with the Trizol reagent and then aliquoted. RNA was quantified using the Qubit fluorimeter (Invitrogen, Thermo Fisher Scientific, Waltham, MA, USA). Quality was assessed on a 1% agarose gel using ethidium bromide as a probe and detected on a UVP imager (UVP Biosystems, Upland, CA, USA). cDNA was prepared using iScript RT mix, which was diluted to 1 : 50 for gene expression analysis. Aliquots of 8 μL were run with a Polymerase Chain Reaction (PCR) Master mix containing EvaGreen SsoFast Supermix. ApoA5, ApoCIII, PPARa, and SREBP1 were run against GAPDH for control comparison.

4.2. Protein Extraction and Analysis of ApoCIII, ANGPTL3, and ANGPLT4. Hepatic protein was extracted through homogenization in 1 mL precipitation cocktail-10 mL radioimmunoprecipitation assay (RIPA) buffer with a complete mini ultra-protease inhibitor tablet. Homogenized samples were incubated on ice for 30 minutes and centrifuged for 15 minutes at 3000 rpm. The protein containing supernatant was removed and aliquoted. Concentrations were determined using the bicinchoninic acid (BCA) assay. All chemicals and materials used for the Western Blotting and gene expression were obtained from Bio-Rad Laboratories, Inc., Hercules, CA, USA.

For the western blot assay, 16 μL of liver protein was loaded per lane on polyacrylamide 4–15% gels and ran with Western C Protein Plus Standards Ladder. The protein was then transferred onto 0.2 um PVDF membrane for blotting. A blocking buffer was prepared by dissolving nonfat dried milk (NFDM) in 1x Tris Buffered Saline (TBS) with .1% v/v Tween 20 (T). The NFDM-TBST buffer was used to block membranes and dilute antibodies. ApoCIII, ANGPTL3, ANGPLT4, and proprotein convertase subtilisin/kexin type 9 (PCSK9) antibodies were assayed at 1 : 500. Their incubation was followed with a secondary goat anti-rabbit-HRP (1 : 10000) which was coincubated with an anti-ladder-HRP (1 : 10000). B-Actin, the reference protein, is HRP primary tagged (1 : 25000). All proteins were visualized on a UVP Biosystems Imager with Dura West ECL signaling reagent.

5. Results

5.1. Plasma Lipids, Glucose, and ELISA Measurements. The plasma was analyzed for lipids and glucose. Overall, the mice that were fed diets supplemented with fatty acids showed increasing plasma levels of glucose and lipids (Figures

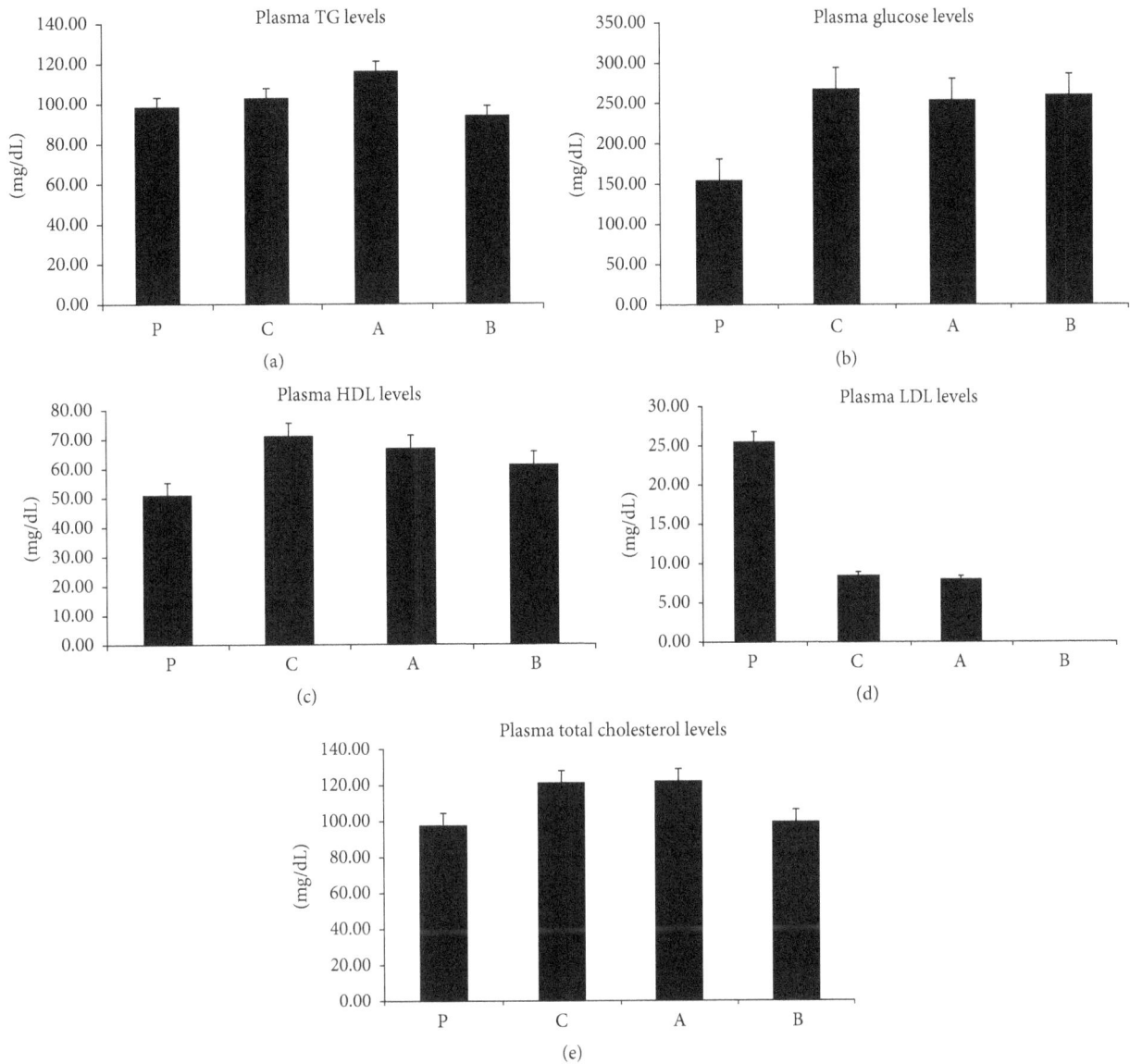

FIGURE 1: (a) The difference between the plasma levels of triglyceride in the 4 groups was not significant. (b) The difference between the plasma levels of glucose in plain control (Standard Chow) and control (chow supplemented with linoleic acid) is significant ($P < 0.01$); glucose also significantly ($P < 0.01$) decreased in higher concentration of OxLA compared to the plain chow. However the low oxidized LA has shown significantly ($P < 0.01$) increased glucose compared to the plain group. (c) Plasma HDL levels increased among all the treated groups; they were however more significant ($P < 0.05$) between the plain and control group and the LA and higher OxLA group ($P < 0.01$). (d) Higher OxLA plasma LDL levels were significantly ($P < 0.01$) reduced compared to the LA control. Plasma LDL was also greatly reduced in low OxLA compared to the plan and control groups. However the difference was not significant. (e) Total cholesterol plasma levels increased in LA and the low OxLA groups compared to the plain control and higher OxLA groups; however the difference was not significant.

1(a)–1(c) and 1(e)), with the exception of LDL (Figure 1(d)), which was decreased.

There were no substantial significant changes shown within the triglyceride measurements between the groups (Figure 1(a)) but a change in the weight gained by mice during the 10-week study was noted (data not included).

All groups that were fed a diet supplemented with fatty acids showed higher concentrations of total cholesterol (Figure 1(e)), glucose (Figure 1(b)), and HDL (Figure 1(c)) within their plasma.

Significantly greater blood glucose levels (Figure 1(b)) were seen in all three experimental groups. While the total

cholesterol (Figure 1(e)) showed an overall increase, only the groups that were fed a diet supplemented with a lower concentration of the oxidized fatty acid showed much higher levels. HDL (Figure 1(c)) levels considerably increased in the linoleic acid fed control group compared to the plain control or high oxidized fed mice. The LDL (Figure 1(d)) levels were low in all groups and they were significant only when the high oxidized fed mice are compared to the linoleic control fed group.

5.1.1. ELISA. Linoleic acid or oxidized linoleic acid supplementation has led to a decrease in plasma ApoCIII

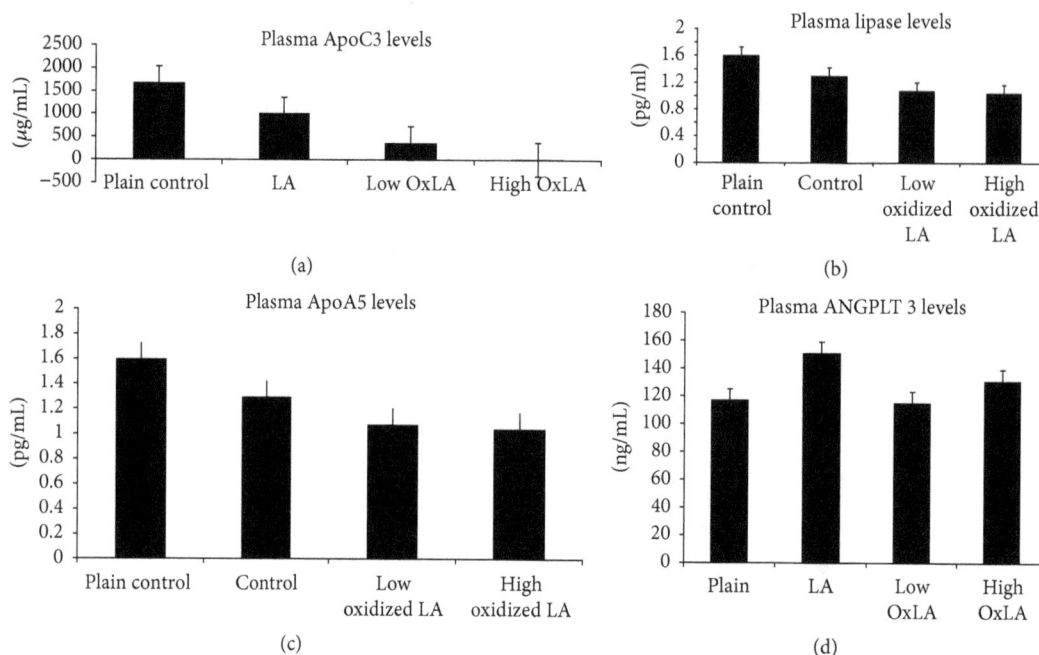

FIGURE 2: Plasma ApoC3 significantly decreased in a dose dependent manner in low OxLA ($P < 0.05$) compared to plain group. The levels were also significantly reduced ($P < 0.05$) between low OxLA and LA. Plasma ApoC3 levels were also significantly reduced ($P < 0.01$) in higher OxLA group compared to the LA. (b) Oxidized linoleic acid supplementation led to dose dependent significant ($P < 0.05$) decreases in plasma hepatic lipase when compared to the plain fed group of mice. The decreases in the hepatic lipase levels in the LA control group compared to the plain mice were not significant. (c) Plasma ApoA5 levels decreased but nonsignificantly in the treated groups compared to the plain control. (d) ANGPTL3 concentration decreased among the OxLA fed groups compared to the LA control linoleic acid group; interestingly the experimental groups had either similar or slightly elevated ANGPTL3 concentration compared to the plain groups.

(Figure 2(a)) levels. The drop was significant for both of the oxidized groups compared to the linoleic acid control group. The greatest drop was in group (B) which had the highest concentration of oxidized linoleic acid, demonstrating a dose dependent response.

Plasma Hepatic Lipase (HL). Plasma HL decreased within experimental groups (Figure 2(b)). The drop was significant for the OxLA groups compared to the plain groups; these differences were dose dependent.

Plasma ApoA5 and ANGPTL3. The group of mice that were fed the linoleic acid and 18 mg/mouse/day oxidized linoleic acid supplemented chows showed increased plasma levels of ApoA5 (Figure 2(c)) and ANGPTL3 (Figure 2(d)). The difference was not significant. The ApoA5 levels were elevated in the high oxidized linoleic acid group. However, ANGPTL3 had the highest concentration in the control linoleic acid group.

5.2. Gene Expression. ApoA5 gene was significantly upregulated (Figure 3(a)) in the linoleic acid control group. This aligns with the result for the plasma APOA5 measured by ELISA. The increase in the expression of ApoCIII noted in the linoleic acid control group was not significant (Figure 3(b)). SREBP gene expression was slightly downregulated in the linoleic acid control group. However, it was upregulated

(Figure 3(c)) in the oxidized groups. The upregulation was significant for the mice that were fed high OxLA diet.

PPAR-α expression peaked slightly for the linoleic control group and dropped for both oxidized linoleic acid groups (Figure 3(d)).

5.3. Western Blot. ApoCIII (Figure 4(a)) shows greater expression for the ApoCIII protein for all groups except the B group, which appears unchanged. The C group showed the highest protein expression followed by the P group and A group with the least upregulation.

ANGPTL3 (Figure 4(b)) shows intense protein expression for the P and C groups. The A group showed significant underexpression and the B group showed lesser expression.

ANGPTL4 (Figure 4(c)) showed significant ANGPTL4 overexpression on the linoleic acid control group. There was a slight increase in expression for the plain group. The high oxidized linoleic acid group showed no change and the low oxidized linoleic acid group showed differential expression between samples.

6. Discussion

Linoleic acid (LA) is an essential fatty acid that is required for physiological and developmental functions of mammalians, particularly humans. Like all polyunsaturated fatty acids (PUFAs), LA is susceptible to oxidation that results in

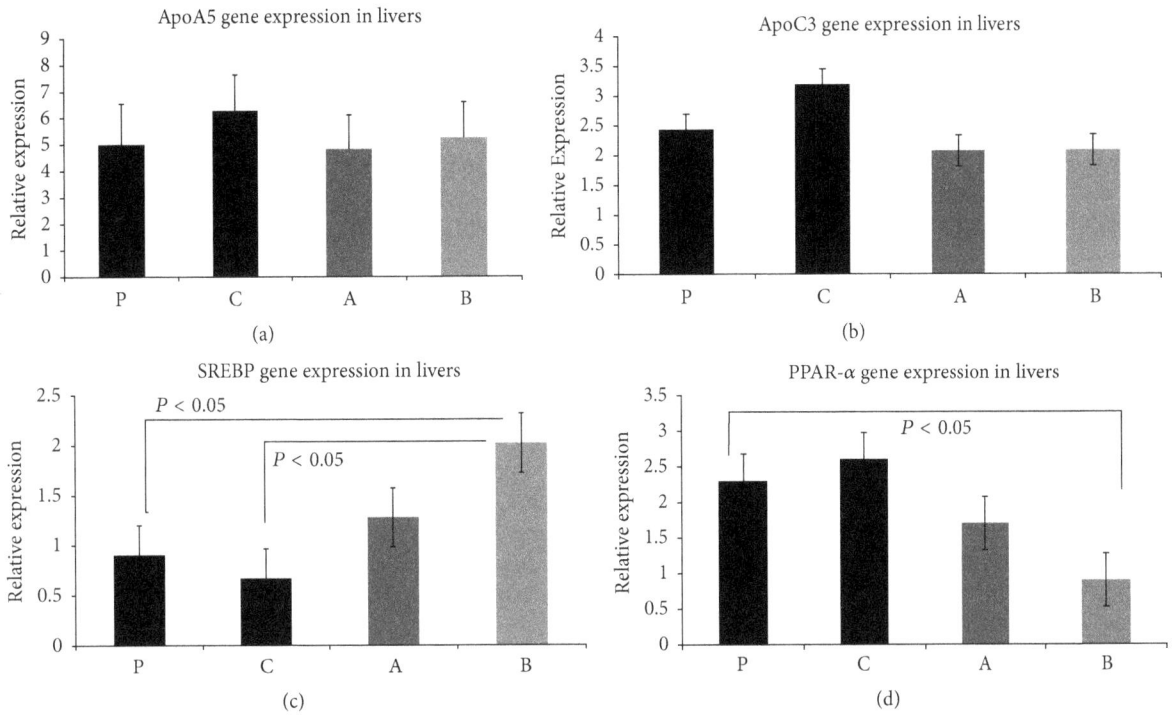

FIGURE 3: (a) ApoA5 is significantly upregulated in the linoleic acid control group (C group). P group (Standard Chow), C group (chow supplemented with linoleic acid 9 mg/mouse/day), A group (chow supplemented with oxidized linoleic acid 9 mg/mouse/day), and B group (chow supplemented with linoleic acid 18 mg/mouse/day). (b) ApoC3 is nonsignificantly upregulated in the linoleic acid control group (C group). P group (Standard Chow), C group, A group (chow supplemented with oxidized linoleic acid 9 mg/mouse/day), and B group (chow supplemented with linoleic acid 18 mg/mouse/day). (c) SREBP gene expression shows significant upregulation in the mice group fed a high concentration of oxidized linoleic acid. P group (Standard Chow), C group (chow supplemented with linoleic acid 9 mg/mouse/day), A group (chow supplemented with oxidized linoleic acid 9 mg/mouse/day), and B group (chow supplemented with linoleic acid 18 mg/mouse/day). (d) Slight peak in the PPAR-α gene expression in the linoleic acid control group (C group) and decreased expression in both oxidized linoleic acid concentration groups (A and B groups): P group (Standard Chow), C group (chow supplemented with linoleic acid 9 mg/mouse/day), A group (chow supplemented with oxidized linoleic acid 9 mg/mouse/day), and B group (chow supplemented with linoleic acid 18 mg/mouse/day).

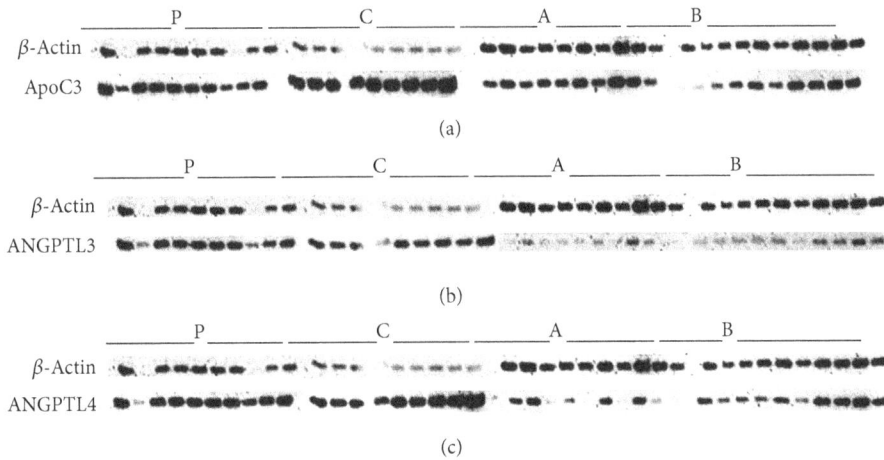

FIGURE 4: (a) Increased expression of ApoC3 over the contrast B-actin in all groups with more intense expression in group C while the B group seems unchanged. P group (Standard Chow), C group (chow supplemented with linoleic acid 9 mg/mouse/day), A group (chow supplemented with oxidized linoleic acid 9 mg/mouse/day), and B group (chow supplemented with linoleic acid 18 mg/mouse/day). (b) ANGPTL3 protein expression is increased in groups P and C, while it is underexpressed in groups A and B, with group showing much less expression. P group (Standard Chow), C group (chow supplemented with linoleic acid 9 mg/mouse/day), A group (chow supplemented with oxidized linoleic acid 9 mg/mouse/day), and B group (chow supplemented with linoleic acid 18 mg/mouse/day). (c) ANGPTL4 protein expression is increased intensely in group C, as well as the P group, which also shows a slight increase. Group A shows no change in expression levels while the B group displays much less expression between the dose dependent groups (A and B). P group (Standard Chow), C group (chow supplemented with linoleic acid 9 mg/mouse/day), A group (chow supplemented with oxidized linoleic acid 9 mg/mouse/day), and B group (chow supplemented with linoleic acid 18 mg/mouse/day).

several active metabolites that have biological relevance. 13-Hydroxyoctadecadienoic acid (13-HODE) (a common name for 13(S)-hydroxy-9Z, 11E-octadecadienoic acid (13(S)-HODE)) and 9-hydroxyoctadecadienoic acid (9-hydroxy-10(E),12(Z)-octadecadienoic acid or 9-HOPDE) are the most studied metabolites of LA. In the current study we used 9-HOPDE derivative to reassess our previous findings that showed significant reduction on TG after over two weeks of 9-HOPDE dietary intake. The present results of our study showed differential responses for how oxidized linoleic acid affects triglyceride metabolism in C57BL/6 mice. It appears that the prolonged dietary intake of the oxidized linoleic acid has a different effect than what we previously reported [2] on the short intake acute effect on TG and lipoprotein metabolism. This may suggest that extended dietary intake of oxidized fatty acids results in mixed favorable and nonfavorable liver and plasma responses. We have seen some genes and plasma lipoproteins levels changes that are dose dependent for intake of oxidized linoleic. However, it is not evident that the intake of dietary OxLA resulted in significant metabolic alterations in TG and lipoprotein.

The plasma triglyceride levels were the highest for the group that had less oxidized linoleic acid in their diet which in part incorporates our previous findings. Interestingly, plasma triglyceride levels are comparable to the gain in body weight over the 10-week period (data not shown). ApoA5 is an important lipid modulating protein that acts on TG and VLDL particles. ApoA5 itself is a protein that is not highly expressed. It is described to stabilize lipid droplets in the liver and protects LPL through an electrostatic mechanism in the bloodstream. The relative abundance of mRNA (Figure 3(a)) for ApoA5 in the liver was low, although it did have a significantly higher concentration in the linoleic acid control group. There was a lower manifestation in the low oxidized group. In comparison, the plasma level showed no significant differences for ApoA5 (Figure 2(c)). However, it demonstrated greater concentration in the bloodstream for the high oxidized and control groups compared to the low oxidized group. ApoCIII is a hepatic and plasma protein that acts to block LPL activity and causes an increase in TG levels in the blood stream. Gene expression (Figure 3(b)) data showed a greater, though insignificant, abundance of mRNA for ApoCIII in the livers of the control linoleic acid group, while there was a lower abundance seen in the mice fed oxidized linoleic acid. The western blot (Figure 4(a)) data generally agreed with the gene expression. The highest amount of active hepatic protein was found in samples from the linoleic control group followed by the plain control group. The low oxidized group showed higher expression than the higher oxidized group. Plasma (Figure 2(a)) ApoCIII concentration was lower in the groups fed fatty acids diets. However, it was more significantly lower in those fed oxidized diets. This could possibly be due to cellular degrading of ApoCIII during or after translation, or the protein may have a loss of function in these mice. Plasma glucose (Figure 1(b)), HDL (Figure 1(c)), and total cholesterol (Figure 1(e)) had some differential changes across mice samples. The glucose increased in all the experimental groups, while the plasma total cholesterol was higher in the linoleic acid control and lower oxidized linoleic

acids fed groups compared to the mice on plain chow and high oxidized linoleic acids fed groups. SREBP is known to modulate the regulation of ANGPTL3, LDLR, and PCSK9, all of which affect the plasma lipids profile [8, 9]. SREBP may have induced more expression of the LDL receptors leading to the much lower LDL concentrations (Figure 1(d)) [9].

SREBP is equally produced in the intestine and liver. It also has a secondary function in increasing triglyceride rich lipoprotein production within the intestine. By promoting the activity of MTP, PCSK9 increases lipidation of ApoB that can lead to greater plasma lipid concentrations [9]. SREBP increases activities of ANGPTL3 and SREBP also with ANGPTL4 antagonized LPL and HL which affect their activity [6, 7]. The ANGPTL family of proteins causes dissociation of many lipases. The dissociation causes a loss of activity and decreased clearance of plasma triglycerides, which is one of the reasons why VLDL particles have less lipolysis thus leading to a diminished clearance. ANGPTL3 and ANGPLT4 are essential for LPL regulation [10]. ANGPTL3 (Figure 4(b)) expression levels were regulated for P and C groups, while group A has shown significant reduction and group B slight reduction. ANGPTL3 was upregulated in group C, though not significant, in the plasma (Figure 2(d)). Although ANGPTL3 did not have any significant changes, it may have acted in preventing further clearance of triglycerides [11].

ANGPTL4 has shown significant upregulation in the linoleic control group as shown in the western blot (Figure 4(c)).

Conclusion. This study demonstrates the ambiguity of the prolonged dietary oxidized fatty acids intake. The rationale for the differences between the short and extended period intake of the oxidized linoleic fatty acid and the conflicting outcomes compared to our previous study is not very clear. However, apparently the long term intake of oxidized linoleic acid may have unfavorable effects on lipoprotein metabolism. The mechanisms of actions vary greatly between the experimental formulas and controls. These findings strongly point towards the proatherogenic roles of the oxidized fatty acids. Future studies using LDLr −/− mouse models may be necessary to establish possible linkages to the pathogenesis of atherosclerosis [12, 13].

Acknowledgments

This work was supported by a generous unrestricted fund from Genentech, a member of the Roche Group, South San Francisco, CA 94080, USA.

References

[1] M. Garelnabi, K. Lor, J. Jin, F. Chai, and N. Santanam, "The paradox of ApoA5 modulation of triglycerides: Evidence from clinical and basic research," *Clinical Biochemistry*, vol. 46, no. 1-2, pp. 12–19, 2013.

[2] M. Garelnabi, K. Selvarajan, D. Litvinov, N. Santanam, and S. Parthasarathy, "Dietary oxidized linoleic acid lowers triglycerides via APOA5/APOClll dependent mechanisms," *Atherosclerosis*, vol. 199, no. 2, pp. 304–309, 2008.

[3] M. Rakhshandehroo, B. Knoch, M. Müller, and S. Kersten, "Peroxisome proliferator-activated receptor alpha target genes," *PPAR Research*, vol. 2010, Article ID 612089, 20 pages, 2010.

[4] A. P. Jensen-Urstad and C. F. Semenkovich, "Fatty acid synthase and liver triglyceride metabolism: housekeeper or messenger?" *Biochimica et Biophysica Acta (BBA) - Molecular and Cell Biology of Lipids*, vol. 1821, no. 5, pp. 747–753, 2012.

[5] V. Sharma, R. O. Ryan, and T. M. Forte, "Apolipoprotein A-V dependent modulation of plasma triacylglycerol: a puzzlement," *Biochimica et Biophysica Acta (BBA) - Molecular and Cell Biology of Lipids*, vol. 1821, no. 5, pp. 795–799, 2012.

[6] H. C. Hassing, R. P. Surendran, H. L. Mooij, E. S. Stroes, M. Nieuwdorp, and G. M. Dallinga-Thie, "Pathophysiology of hypertriglyceridemia," *Biochimica et Biophysica Acta (BBA) - Molecular and Cell Biology of Lipids*, vol. 1821, no. 5, pp. 826–832, 2012.

[7] H. Ge, J.-Y. Cha, H. Gopal et al., "Differential regulation and properties of angiopoietin-like proteins 3 and 4," *Journal of Lipid Research*, vol. 46, no. 7, pp. 1484–1490, 2005.

[8] F. Mattijssen and S. Kersten, "Regulation of triglyceride metabolism by Angiopoietin-like proteins," *Biochimica et Biophysica Acta*, vol. 1821, no. 5, pp. 782–789, 2012.

[9] S. Rashid, H. Tavori, P. Brown et al., "PCSK9 promotes intestinal overproduction of Triglyceride-Rich Apolipoprotein-B Lipoproteins through both LDL-receptor dependent and independent mechanisms," *CirculationHello, Guest! My alerts Sign In Join Facebook Twitter Circulation AHA*, 2014.

[10] K. Nakajima, J. Kobayashi, H. Mabuchi et al., "Association of angiopoietin-like protein 3 with hepatic triglyceride lipase and lipoprotein lipase activities in human plasma," *Annals of Clinical Biochemistry*, vol. 47, no. 5, pp. 423–431, 2010.

[11] K. Schoonjans, J. Peinado-Onsurbe, A.-M. Lefebvre et al., "PPARα and PPARγ activators direct a distinct tissue-specific transcriptional response via a PPRE in the lipoprotein lipase gene," *EMBO Journal*, vol. 15, no. 19, pp. 5336–5348, 1996.

[12] S. Parthasarathy, A. Raghavamenon, M. O. Garelnabi, and N. Santanam, "Oxidized low-density lipoprotein," *Methods in Molecular Biology*, vol. 610, pp. 403–417, 2010.

[13] D. Litvinov, K. Selvarajan, M. Garelnabi, L. Brophy, and S. Parthasarathy, "Anti-atherosclerotic actions of azelaic acid, an end product of linoleic acid peroxidation, in mice," *Atherosclerosis*, vol. 209, no. 2, pp. 449–454, 2010.

15

Do Omega-3/6 Fatty Acids have a Therapeutic Role in Children and Young People with ADHD?

E. Derbyshire

Nutritional Insight Ltd, Surrey, UK

Correspondence should be addressed to E. Derbyshire; emma@nutritional-insight.co.uk

Academic Editor: Abdelgadir M. Homeida

Background. Attention deficit hyperactivity disorder (ADHD) is a debilitating behavioural disorder affecting daily ability to function, learn, and interact with peers. This publication assesses the role of omega-3/6 fatty acids in the treatment and management of ADHD. *Methods.* A systematic review of 16 randomised controlled trials was undertaken. Trials included a total of 1,514 children and young people with ADHD who were allocated to take an omega-3/6 intervention, or a placebo. *Results.* Of the studies identified, 13 reported favourable benefits on ADHD symptoms including improvements in hyperactivity, impulsivity, attention, visual learning, word reading, and working/short-term memory. Four studies used supplements containing a 9 : 3 : 1 ratio of eicosapentaenoic acid : docosahexaenoic acid : gamma linolenic acid which appeared effective at improving erythrocyte levels. Supplementation with this ratio of fatty acids also showed promise as an adjunctive therapy to traditional medications, lowering the dose and improving the compliance with medications such as methylphenidate. *Conclusion.* ADHD is a frequent and debilitating childhood condition. Given disparaging feelings towards psychostimulant medications, omega-3/6 fatty acids offer great promise as a suitable adjunctive therapy for ADHD.

1. Background

Attention deficit hyperactivity disorder (ADHD) is a common child-onset neurodevelopmental disorder occurring in children, adolescents, and adults, with an estimated prevalence of 5 to 7 per cent across cultures [1]. ADHD tends to be more common in boys than girls and is highly heritable, with pre- and perinatal factors also being implicated, although its definite cause remains unknown [2]. Although the rate of ADHD declines with age, at least half of children with the disorder will go on to have symptoms in adulthood [3]. The condition can impact heavily on mental health and education, lead to antisocial behaviour and personal dysfunction, and increase mortality risk [4]. Medications used to treat ADHD typically include methylphenidate (MPH; also, known as Ritalin), amphetamine, and atomoxetine which typically assume that there is a dopamine/norepinephrine deficit, although the aetiology of this condition is more complex [5]. Whilst MPH may ameliorate some comorbidities [6] it has been found to be ineffective in eliminating symptoms in 50 per cent of cases [7, 8]. Parents also appear to be concerned about the long-term effects of their children using medications such as MPH [9].

Long-chain polyunsaturated fatty acids (LCPUFA) and particularly omega-3 fatty acids have been under the spotlight for decades. They are key regulators of brain neurotransmission, neurogenesis, and neuroinflammation, all having an important role in the prevention and treatment of psychological and behavioural dysfunction disorders [10]. Eicosapentaenoic acid (EPA) and docosahexaenoic acid (DHA) are two fatty acids that are highly concentrated in the brain, exhibiting antioxidative, anti-inflammatory, and antiapoptotic effects, with these contributing to neuron protection [11].

The omega-6 fatty acid gamma linolenic acid (GLA) is also important in the generation of arachidonic acid (ARA) which is abundantly present in in the brain [12, 13]. A recent meta-analysis found that combinations of omega-3 and omega-6 fatty acids (EPA and GLA) helped to improve symptoms of inattention in children with ADHD [14]. Brain lipids within cell membranes also act as signalling mediums, supporting neurotransmitter function with omega-3 fatty acids thought to play a key role in this which may help in the prevention of anxiety disorders [15]. Laboratory research has also identified that omega-3 fatty acids may act in a similar way to "antipsychotics," possibly by acting on brain receptors and helping to restore oxidative balance [16].

Omega-3 deficiencies have been found to alter dopaminergic and serotonergic systems, potentially modifying cerebral receptors in specific regions of the brain [17]. EPA and DHA are regarded as "essential fatty acids (EFAs)" that need to be obtained from food or supplement sources as they cannot be made in sufficient amounts by the human body [11]. The ratio of fatty acids (omega-6 : omega-3) which complete for the same enzyme pathways can also influence neurotransmission and prostaglandin formation, both of which are crucial in the maintenance of normal brain function [18, 19]. Furthermore, as the storage of the omega-3 fatty acids is limited, a continual exogenous supply is needed to obtain suitable levels [20].

A number of studies have measured LCPUFA status in individuals with ADHD. One study conducted on young adults (22.3 to 24.3 years) found the proportion of omega-3 fatty acids was significantly lower in the plasma phospholipids and red blood cells of ADHD participants compared with controls, whilst levels of saturated fatty acids were higher [21]. Another investigation found that whilst teenagers with ADHD consume similar amounts of omega-3 and omega-6 fatty acids to controls, their DHA status was significantly lower, indicating metabolic differences in fatty acid handling in those with ADHD [22]. Similarly, another trial showed that the proportions of saturated and polyunsaturated fatty acids were higher and lower, respectively, in paediatric patients with ADHD, compared with controls again indicating differences in lipid profiles [23]. Further meta-analytical evidence has concluded that children and young people with ADHD have elevated ratios of blood omega-6/3 indicating disturbances in fatty acid metabolism in these individuals [24].

Given that the human brain is nearly 60 per cent fat and the central role that EFAs have to play in the structure, synthesis, and functions of brain neurotransmitters [25], the present article evaluates evidence on whether LCPUFAs have a therapeutic role in the management of ADHD. Particular focus will be given on their potential effects in the management of ADHD along with their role as an adjunctive therapy.

2. Methods

2.1. Approach. The National Centre for Biotechnology Information (NCBI) search engine (PubMed) was used to extract relevant publications. English-language, human, randomised controlled trials (RCTs) published between 2001 and March 2017 were included. Data files were extracted from the NCBI collection depository and imported into Covidence software used to create systematic reviews.

2.2. Exclusion/Inclusion Criteria. Publications were excluded if they were not a RCT, did not use participants with ADHD, or were conducted on older adults with ADHD. For inclusion studies needed to be conducted on children or young people (up to 18 years of age), participants were considered to

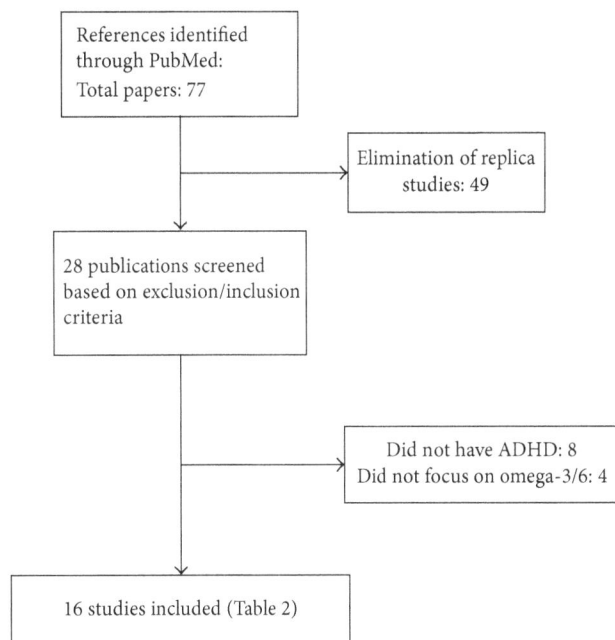

FIGURE 1: Algorithm of qualifying publications.

have ADHD at baseline and were taking an omega-3/6 supplement, including EPA, DHA, or GLA. Publications were further included if the full text was available or could be purchased.

The search terms "attention deficit hyperactivity disorder" or "ADHD" were combined with "long-chain n-3 fatty acids", "omega-3/6 fatty acids", "docosahexaenoic acid", "eicosapentaenoic acid", and "gamma linoleic acid". Data extracted from each article included (1) author(s) and country of research, (2) subjects (gender, number of participants), (3) mean age, (4) study design and methods, (5) dose of supplement, and (6) main findings.

3. Results

The NCBI search identified 77 papers. After a further adjustment for replica papers, 28 articles remained for assessment. Of these, 12 papers were discarded after reviewing the abstracts and article content as they did not meet the inclusion criteria. This left 16 RCTs for general review. Figure 1 shows the algorithm of qualifying publications. Of these, one study was conducted in the United Kingdom, five in Europe, one in the United States, one in Mexico, two in Australasia, four in Asia, and two in the Middle East.

3.1. Definitions. As shown in Table 1, all of the publications identified included children or young people with ADHD at baseline. Most studies diagnosed ADHD according to the Diagnostic and Statistical Manual of Mental Disorders, 4th Edition (DSM-IV) criteria. Others used methods such as the Conners' Parent Rating Scale (CPRS) and parent-reported learning difficulties [26–28]. Some studies focused more specifically on certain ADHD subtype. For example, Widenhorn-Müller et al. (2014) included the inattentive and

TABLE 1: Methods used to screen for ADHD.

Author	Definition used
Barragán et al. (2017)	ADHD of any subtype. Diagnosed according to the DSM-IV criteria and CGI-S scale.
Bos et al. (2015)	ADHD diagnosis confirmed by a trained researcher using the DISC-P.
Matsudaira et al. (2015)	ADHD diagnosis confirmed through a semi-structured interview based on the DSM-IV criteria.
Milte et al. (2015)	Diagnosis of ADHD or parent-rated symptoms >90th percentile on the CPRS and parent-reported learning difficulties.
Wu et al. (2015)	ADHD diagnosed according to DSM-IV and the Chinese version of CPRS. These rating scales about learning, attention, and behaviour were completed by the teachers and either parent(s) or guardians.
Widenhorn-Müller et al. (2014)	Met DSM-IV criteria for the ADHD combined subtype (hyperactive–inattentive) and the primarily inattentive or the hyperactive/impulsive subtype were included in the trial.
Manor et al. (2013)	Children were included if they had a score of at least 1.5 standard deviations above the normal for the patient's age and gender in the Teacher-Rated ADHD Rating Scale-IV School Version.
Hariri et al. (2012)	Conners' Abbreviated Questionnaires scores for hyperactivity were greater than 14.
Johnson et al. (2012)	Participants met DSM-IV criteria for a diagnosis of ADHD.
Milte et al. (2012)	Diagnosis of ADHD or parent-rated symptoms >90th percentile on the CPRS and parent-reported learning difficulties.
Perera et al. (2012)	All children in the program were clinically diagnosed using DSM-IV supported by positive scores in Swanson, Nolan, and Pelham version IV (SNAP) parent and teacher evaluation.
Gustafsson et al. (2010)	Clinical diagnosis of ADHD of combined type (fulfilling DSM-IV criteria A–E) with any neuropsychiatric comorbidity and who had been evaluated for pharmacological treatment.
Johnson et al. (2009)	Participants met DSM-IV criteria for a diagnosis of ADHD of any subtype, scoring at least 1.5 SD above the age norm for their diagnostic subtype using norms for the ADHD Rating Scale–IV–Parent Version.
Raz et al. (2009)	Parents were asked to present a formal ADHD diagnosis. The child performed a continuous performance test, while one of the parents filled in the essential fatty acids deficiency questionnaire and the DSM-IV questionnaire.
Hirayama et al. (2004)	Diagnosed or suspected as AD/HD according to DSM-IV and diagnostic interviews including behaviour observation by psychiatrists. In a strict sense, eight subjects might not be AD/HD according to the DSM-IV criteria, but two psychiatrists attending the summer camp strongly suspected them as AD/HD.
Voigt et al. (2001)	Previously been given a diagnosis of ADHD by a physician. Confirmatory diagnostic interview with a neurodevelopmental paediatrician to confirm responses to the telephone interview and to ensure that each met DSM-IV.

Key. ADHD, attention deficit hyperactivity disorder; CGI-S scale; Clinical Global Impressions-Severity scale; CPRS, Conners' Parent Rating Scale; DISC-P, Diagnostic Interview Schedule for Children-Parent Version; DSM-IV; Diagnostic and Statistical Manual of Mental Disorders.

hyperactive/impulsive subtypes within the trial whilst Voigt et al. (2001) studied those with oppositional defiant or conduct disorders. Other trials used adapted parental/researcher screening tools alongside the DSM-IV [29, 30].

3.2. Omega Fatty Acids. A total of 16 RCTs studied interrelationships between combinations of omega-3/6 fatty acids and ADHD symptoms (Table 2). Of these 13 reported beneficial effects, though the levels of effect appeared to depend on the dose of the intervention, ratio of the fatty acids, quality of the RCT, and ADHD subtype under investigation.

One of the most recent studies found that children (aged 6 to 12 years) receiving omega-3/6 fatty acids (Equazen) providing 558 mg EPA, 174 mg DHA, and 60 mg GLA in a 9 : 3 : 1 ratio over a period of 12 months did not need such a high dose of MPH to manage and reduce their ADHD symptoms (0.8 mg/kg/day versus 1.0 mg/kg/day). The completion rate was also higher in this group, whilst the

withdrawal rate and the incidence of adverse events were significantly lower. These findings indicate that omega-3/6 fatty acids may act as a useful adjunctive therapy to MPH, helping to improve tolerability, dosing, and adherence [31].

A 12-week RCT comprised of 76 male adolescents with ADHD using a similar dose of fatty acids found that supplementation improved blood levels of EPA, DHA, and total omega-3 fatty acids, though no effects on aggression, impulsivity, or anxiety were seen, possibly due to the smaller study sample size and shorter study length of this trial [32]. Two other trials have been undertaken using a similar 9 : 3 : 1 ratio of EPA, DHA, and GLA, respectively [33, 34]. This work was the first to trial omega-3/6 fatty acids finding that 1 in 8 patients benefited and experienced a reduction of more than 50% of ADHD symptoms, with strongest results seen amongst boys and those with ADHD inattentive subtype [34]. Later research by the same team of scientists found that omega-3/6 supplementation significantly improved the fatty

TABLE 2: Information extracted from trials looking at LC3PUFAs and ADHD.

Reference and country	Subjects M/F and sample size	Mean age	Study design and methods	Dose of supplement	Main findings
[31] Mexico	90 children (60 M, 30 F)	6–12 years Mean age 8.27 years	12-month trial (unblinded); MPH, omega-3/6 or a combination	Equazen: 558 mg EPA, 174 mg DHA, and 60 mg GLA (9 : 3 : 1 ratio)	Significantly better scores on ADHD. Adverse events were numerically less frequent with omega-3/6 or MPH + omega-3/6 than MPH alone.
[41] Netherlands	40 boys with ADHD and 39 matched, typically developing controls	Aged 8–14 years	16-week trial	10 g of margarine daily, enriched with either 650 mg of EPA/DHA or placebo	EPA/DHA supplementation improved parent-rated attention in both children with ADHD and typically developing children. Phospholipid DHA level at follow-up was higher for children receiving EPA/DHA supplements than placebo.
[32] United Kingdom	76 M adolescents with ADHD	12–16 years, mean = 13.7 years	12-week trial	Equazen: 558 mg EPA, 174 mg DHA, and 60 mg GLA (9 : 3 : 1 ratio)	In the treatment group, supplementation enhanced EPA, DHA, and total omega-3 fatty acid levels.
[26] Australia	90 Australian children with ADHD symptoms higher than the 90th percentile on the Conners' Rating Scales	7 to 12 years	4-month crossover study evaluating literacy and behaviour up to 12 months	Supplements rich in EPA, DHA, or LA	Increased erythrocyte EPA + DHA was associated with improved spelling ($p < 0.001$) and attention ($p < 0.001$), reduced oppositional behaviour ($p < 0.003$), hyperactivity ($p < 0.001$), cognitive problems ($p < 0.001$), DSM-IV hyperactivity ($p = 0.002$), and DSM-IV inattention ($p < 0.001$).
[27] China	179 children with lower IQs or ADHD to receive	7 to 12 years	3-month trial: evaluated effects on visual acuity	Ordinary eggs or eggs rich in EPA and DHA	Both groups of children showed a significant improvement in visual acuity ($p < 0.05$); however, visual acuity in the study group was significantly better than that of the control group ($p = 0.013$).
[37] Germany	95 children diagnosed with ADHD according to DSM-IV criteria	6–12 years	16-week trial	Omega-3 fatty acid mix	Improved working memory correlated significantly with increased EPA, DHA, and decreased ARA.
[39] Israel	200 children diagnosed with ADHD	6–13 years	15-week trial followed by an open-label extension	300 mg PS-omega-3/day	Study results demonstrate that consumption of PS-omega-3 by children with ADHD, is safe and well tolerated, without any negative effect on body weight or growth.
[38] Malaysia	103 children	6–12 years	8-week trial	635 mg EPA, 195 mg DHA	Significant reduction in levels of CRP in the omega-3 group and significant increase in SOD and glutathione reductase. Significant improvement in ASQ-P score (measure of hyperactivity).
[33] Sweden	75 children and adolescents with DSM-IV ADHD	8–18 years	3-month trial. Omega-3/6 (Equazen) or placebo, followed by 3 months of open phase	Omega-3/6 (Equazen) or placebo Equazen: 558 mg EPA, 174 mg DHA, and 60 mg gamma linoleic acid (9 : 3 : 1 ratio)	Subjects with more than 25% reduction in ADHD symptoms were classified as responders. Compared to nonresponders, the 6-month responders had significantly greater n-3 increase at 3 months and decrease in n-6/n-3 ratio at 3 and 6 months ($p < 0.05$).

TABLE 2: Continued.

Reference and country	Subjects M/F and sample size	Mean age	Study design and methods	Dose of supplement	Main findings
[28] Australia	90 Australian children with ADHD symptoms higher than the 90th percentile on the Conners' Rating Scales	7 to 12 years	4-month trial	Supplements rich in EPA, DHA, or safflower oil	Increased erythrocyte DHA was associated with improved word reading and lower parent ratings of oppositional behaviour. These effects were more evident in a subgroup of 17 children with learning difficulties.
[29] Sri Lanka	Children with ADHD $n = 48$ active group, $n = 46$ placebo	6–12 years	6-month trial	Capsule containing n3 and n6 (fish oil) and cold-pressed evening primrose oil	Statistically significant improvement was not found at 3 months of treatment between groups but was evident at 6 months of treatment ($p < 0.05$) with inattention, impulsiveness, and cooperation with parents and teachers.
[40] Norway	92 children with ADHD	7–12 years	15-week RCT	0.5 g EPA versus placebo	EPA improved CTRS, inattention/cognitive subscale ($p = 0.04$), but not Conners' total score.
[34] Sweden	75 children and adolescents with DSM-IV ADHD	8–18 years	3-month trial. Omega-3/6 (Equazen) or placebo, followed by 3 months of open phase	Equazen: 558 mg EPA, 174 mg DHA, and 60 mg GLA (9:3:1 ratio)	A subgroup of 26% responded with more than 25% reduction of ADHD symptoms and a drop of Clinical Global Impression scores to the near-normal range. After 6 months, 47% of all showed such improvement. Responders tended to have ADHD inattentive subtype and comorbid neurodevelopmental disorders.
[30] Israel	73 unmedicated children with a diagnosis of ADHD	7–13 years	7-week trial	480 mg LA, 120 mg ALA, placebo: 1000 mg of vitamin C	Both treatments ameliorated some of the symptoms, but no significant differences were found between the groups in any of the treatment effects.
[35] Japan	40 AD/HD (including eight AD/HD-suspected) children who were mostly without medication	6–12 years	2-month trial	Foods containing fish oil (fermented soybean milk, bread rolls, and steamed bread; 3.6 g DHA/week from these foods)	DHA-containing foods did not improve ADHD-related symptoms. Visual short-term memory and errors of commission (continuous performance) significantly improved in the control group compared with the changes over time in the DHA group.
[36] USA	63 children with ADHD, all receiving effective maintenance therapy with stimulant medication	6–12 years	4-month trial	345 mg DHA	No statistically significant improvement in any objective or subjective measure of ADHD symptoms.

Key. ADHD, attention deficit hyperactivity disorder; ALA, alpha-linolenic acid; ARA, arachidonic acid; CRP, C-reactive protein; CTRS, Connor Teacher Rating Scale; DHA, docosahexaenoic acid; DSM-IV; Diagnostic and Statistical Manual of Mental Disorders; EPA, eicosapentaenoic acid; F, female; GLA, gamma linoleic acid; LA, linoleic acid; M, male; MPH, methylphenidate; PS, phosphatidylserine; SOD, superoxide dismutase.

acid composition amongst study "responders," that is, those with more than a 25 per cent reduction in ADHD symptoms [33].

Three studies used functional foods providing LC3PUFA. In a double-blind RCT, the ingestion of 10 g margarine daily providing 650 mg EPA/DHA improved parent-rated attention in children with ADHD after 16 weeks in 8–14-year-olds who continued with their usual medication. Another trial using "omega eggs" providing EPA and DHA found that daily consumption by 7–12-year-olds over 3 months significantly improved visual acuity and the red blood cell fatty acid profile of children with lower intelligent quotients or ADHD, indicating that the DHA content of ordinary eggs may not be sufficient [27]. Another work giving 6- to 12-year-olds ADHD DHA-enriched foods showed that ADHD symptoms did not improve though there were some significant improvements in short-term memory and errors of continuous performance [35].

Three studies concluded that there were limited associations between omega-3/6 fatty acid supplementation and ADHD outcomes. In one study, Conners' Parent and Teacher Rating Scale was not regarded as being sensitive enough to detect small improvements in the behaviour of male adolescents [32]. Another work found that a supplement providing 480 mg of linoleic acid and 120 mg of α-linolenic acid ameliorated some ADHD symptoms amongst 7–13-year-olds, although no significant differences were found, possibly because children were unmedicated [30]. Earlier work providing 6- to 12-year-olds with 345 mg DHA over 4 months did not find this to ameliorate ADHD, indicating that a longer trial period and inclusion of arachidonic acid may have been needed [36].

Remaining studies showed general benefits. An Australian study found that children with ADHD who had increased erythrocyte EPA + DHA levels had significantly improved spelling and attention and reduced oppositional behaviour, hyperactivity, and cognitive problems [28]. An omega-3 fatty acid mix taken over 16 weeks by German children aged 6 to 12 years also increased EPA + DHA erythrocyte levels and improved working memory but had no other effects on behaviour [37]. A short 8-week trial reported significant improvements in hyperactivity scores after supplementation with EPA + DHA [38]. In a trial where children had been taking MPH, supplementation with omega-3 and omega-6 fatty acids in the ratio of 1.6 : 1 led to significant improvements in inattention and impulsiveness, along with cooperation with parents and teachers after 6 months, indicating this was a safe and effective adjunctive therapy [29].

Other trials showed findings to be more prominent in certain subgroups. For example, a large trial of 200 children found that supplementation with phosphatidylserine-omega-3 reduced ADHD symptoms in a subgroup of hyperactive-impulsive, emotionally and behaviourally dysregulated ADHD children compared with the placebo [39]. Another work found that erythrocyte DHA levels increased after 4 months of supplementation with 4 capsules daily providing either (1) 108 mg DHA and 1109 mg EPA, (2) 1032 mg DHA and 264 mg EPA, or (3) 1467 mg of linoleic acid [28]. The study also found that higher doses of DHA helped to improve the literacy and behaviour in children with ADHD, particularly in a subgroup with learning difficulties [28]. Norwegian work showed that 0.5 g EPA after 15 weeks improved symptoms in two ADHD subgroups: positional and less hyperactive/impulsive children [40].

4. Discussion

The aetiology of ADHD is complex and multifactorial though diet, nutrition, and abnormalities in the metabolism of LCP-UFA are thought to have underlying roles [21, 22]. The present review has shown that omega-3 and omega-6 fatty acids have an important role to play in the management of ADHD. Previous work has shown that the tolerability of omega-3 fatty acids given to individuals with ADHD is high with only mild side effects reported such as incidental nose bleeds and gastrointestinal discomfort [42]. Severe side effects have not been documented and these minor complaints are regarded as being less severe than methylphenidate side effects [42].

Taken together, a growing body of clinically proven evidence suggests that dietary supplementation using omega-3/6 PUFAs may help to augment conventional ADHD treatments. Research carried out in Mexico at the National Health Institute with children prescribed with MPH and taking omega-3/6 fatty acids found that they required lower doses of the prescription medicine and experienced fewer medication-related side effects [31]. Similarly, other work has shown that omega-3/6 supplementation reduced behavioural and learning difficulties in children with ADHD that was refractory to MPH treatment alone [29]. Another RCT concluded that EPA was a safe complementary treatment option in omega-3 deficient ADHD children, with scope to benefit ADHD subgroups who are less responsive to stimulant treatments [40]. A recent review of 25 clinical trials has also concluded that two patients groups, in particular, could benefit from omega-3 fatty acids. The first is those with mild ADHD where omega-3 supplements could replace stimulant medications. The second is those with severe ADHD where omega-3 supplements could reduce the amount of stimulant medication being used, in turn, potentially reducing symptoms from the medications side effects [42].

These studies are further supported by evidence from meta-analytical studies. Evidence collated from ten trials comprised of 700 children has shown that omega-3 supplementation, with higher doses of EPA had modest effects in the treatment of ADHD, indicating potential roles in augmenting traditional pharmacological treatments, whilst providing an option for families who may decline other psychopharmacologic options [43]. An earlier meta-analysis also concluded that omega-3 fatty acids offer promise as a possible supplement to traditional therapies [44]. Interestingly, a series of interviews about treatment experiences showed that over half (52%) of parents expressed initial reluctance towards psychostimulants. Once psychostimulants were used by children and adolescents with ADHD, 73% concurrently used other treatments [45]. These findings indicate that parents are concerned about their children using psychostimulants and are looking for accompanying treatment options.

In terms of study outcomes, most focused on ADHD symptoms. Whilst reduced hyperactivity and impulsivity were reported in most studies [26, 29, 31, 38, 40], other outcomes such as improved attention [41], visual acuity [27], improved word reading [28], and working/short-term memory were also observed [35, 37]. These findings indicate that LCPUFA supplementation has far-reaching effects, having additional benefits for learning. A recent 6-month 2-phase randomised trial with 154 children aged 9 and 10 years showed the omega-3/6 fatty acid supplementation improved reading ability in mainstream children and improved cognitive measures in children with attention problems, defined as those with ADHD symptom scores above the median [46].

With regards to dose a $9:3:1$ ratio of EPA (558 mg) to DHA (174 mg) and GLA (60 mg) was used in four studies [31–34]. In the largest and longest studies, this was associated with improved hyperactivity and impulsivity subscores [31] and reduced ADHD symptoms [33], especially in the inattentive ADHD subtype and those with comorbid neurodevelopment disorders [34]. Other work using 635 mg EPA and 195 mg DHA also led to significant improvements in ADHD scores [38]. Studies using lower doses (345 mg DHA) tended not to yield significant findings in terms of ADHD symptoms [36]. Taken together, it appears that higher doses of fatty acids are needed to generate measurable effects. The ratio of omega-6 to omega-3 in studies and especially the ARA/DHA ratio may also have impacted on study outcomes, as this is regarded as being important for membrane fluidity [47].

Inconsistencies of lack of findings in some trials may have been attributed to interventions being too short. As erythrocytes survive in the body for 120 days, supplementation trials shorter than 12 weeks (84 days) may not be sufficient enough to detect changes in LCPUFA compositions [32]. Equally, as the turnover of fatty acids in the brain is thought to be rather low in 6- to 12-year-olds, longer periods of supplementation and/or higher doses may be needed to modify the fatty acid content of the central nervous system [36]. It should also be considered that some studies used different tools to assess ADHD symptoms, measures of attention, and scales of hyperactivity, some of which may be more sensitive than others. On a final note, it should be considered that the Diagnostic and Statistical Manual of Mental Disorders, 5th Edition (DSM-5) is now out which includes new diagnostic groups such as "disruptive mood regulation" which have potential to be applied in future studies [48]. This could possibly increase prevalence rates of mental health disorders in future trials [48]. Continued research using the latest DSM-5 criteria, along with larger and longer interventions (more than 12 weeks), is now needed.

5. Conclusion

In conclusion, ADHD is a debilitating neurodevelopmental disorder that can impact heavily on children and young people's behaviour, mental health, education, and social/family lives. Whilst conventional medications have a role to play in the management of ADHD symptoms, new clinically trialled evidence indicates that omega-3/6 supplementation programmes can provide a promising adjunctive therapy, lowering the dose of psychopharmacologic medications needed and subsequently improving compliance with these. It also appears that parents are looking for complementary treatments for their children to use alongside traditional treatments.

Abbreviations

ARA:	Arachidonic acid
ADHD:	Attention deficit hyperactivity disorder
CPRS:	Conners' Parent Rating Scale
CTRS:	Connors' Teacher Rating Scale
DHA:	Docosahexaenoic acid
DSM:	Diagnostic and Statistical Manual of Mental Disorders
EFAs:	Essential fatty acids
EPA:	Eicosapentaenoic acid
GLA:	Gamma linoleic acid
LCPUFA:	Long-chain polyunsaturated fatty acids
MPH:	Methylphenidate
RCT:	Randomized controlled trial.

Acknowledgments

The author received funding from Equazen.

References

[1] D. Antai-Otong and M. L. Zimmerman, "Treatment Approaches to Attention Deficit Hyperactivity Disorder," *Nursing Clinics of North America*, vol. 51, no. 2, pp. 199–211, 2016.

[2] A. Thapar and M. Cooper, "Attention deficit hyperactivity disorder," *The Lancet*, vol. 387, no. 10024, pp. 1240–1250, 2016.

[3] J. Biederman and S. V. Faraone, "Attention-deficit hyperactivity disorder," *The Lancet*, vol. 366, no. 9481, pp. 237–248, 2005.

[4] E. Taylor, "Attention deficit hyperactivity disorder: Overdiagnosed or diagnoses missed?" *Archives of Disease in Childhood*, vol. 102, no. 4, pp. 376–379, 2016.

[5] A. Sharma and J. Couture, "A Review of the Pathophysiology, Etiology, and Treatment of Attention-Deficit Hyperactivity Disorder (ADHD)," *Annals of Pharmacotherapy*, vol. 48, no. 2, pp. 209–225, 2014.

[6] H. Abikoff, L. Hechtman, R. G. Klein et al., "Symptomatic improvement in children with ADHD treated with long-term methylphenidate and multimodal psychosocial treatment," *Journal of the American Academy of Child and Adolescent Psychiatry*, vol. 43, no. 7, pp. 802–811, 2004.

[7] J. Biederman, M. C. Monuteaux, E. Mick et al., "Young adult outcome of attention deficit hyperactivity disorder: a controlled 10-year follow-up study," *Psychological Medicine*, vol. 36, no. 2, pp. 167–179, 2006.

[8] J. C. Blader, S. R. Pliszka, P. S. Jensen, N. R. Schooler, and V. Kafantaris, "Stimulant-responsive and stimulant-refractory aggressive behavior among children with ADHD," *Pediatrics*, vol. 126, no. 4, pp. e796–e806, 2010.

[9] I. Berger, T. Dor, Y. Nevo, and G. Goldzweig, "Attitudes toward attention-deficit hyperactivity disorder (ADHD) treatment: Parents' and children's perspectives," *Journal of Child Neurology*, vol. 23, no. 9, pp. 1036–1042, 2008.

[10] M. M. Pusceddu, P. Kelly, C. Stanton, J. F. Cryan, and T. G. Dinan, "N-3 Polyunsaturated Fatty Acids through the Lifespan: Implication for Psychopathology," *International Journal of Neuropsychopharmacology*, vol. 19, no. 12, 2016.

[11] R. Crupi, A. Marino, and S. Cuzzocrea, "n-3 fatty acids: role in neurogenesis and neuroplasticity," *Current Medicinal Chemistry*, vol. 20, no. 24, pp. 2953–2963, 2013.

[12] J. Dobryniewski, S. D. Szajda, N. Waszkiewicz, and K. Zwierz, "Biology of essential fatty acids (EFA)," *Przeglad Lekarski*, vol. 64, no. 2, pp. 91–99, 2007.

[13] S. M. Alashmali, K. E. Hopperton, and R. P. Bazinet, "Lowering dietary n-6 polyunsaturated fatty acids: Interaction with brain arachidonic and docosahexaenoic acids," *Current Opinion in Lipidology*, vol. 27, no. 1, pp. 54–66, 2016.

[14] B. K. Puri and J. G. Martins, "Which polyunsaturated fatty acids are active in children with attention-deficit hyperactivity disorder receiving PUFA supplementation? A fatty acid validated meta-regression analysis of randomized controlled trials," *Prostaglandins Leukotrienes and Essential Fatty Acids*, vol. 90, no. 5, pp. 179–189, 2014.

[15] C. P. Müller, M. Reichel, C. Mühle, C. Rhein, E. Gulbins, and J. Kornhuber, "Brain membrane lipids in major depression and anxiety disorders," *Biochimica et Biophysica Acta (BBA)—Molecular and Cell Biology of Lipids*, vol. 1851, no. 8, pp. 1052–1065, 2015.

[16] M. H. Kokacya, S. Inanir, U. S. Copoglu, R. Dokuyucu, and O. Erbas, "The antipsychotic effects of omega-3 fatty acids in rats," *American Journal of the Medical Sciences*, vol. 350, no. 3, pp. 212–217, 2015.

[17] S. Chalon, "Omega-3 fatty acids and monoamine neurotransmission," *Prostaglandins Leukotrienes and Essential Fatty Acids*, vol. 75, no. 4-5, pp. 259–269, 2006.

[18] M. Haag, "Essential fatty acids and the brain," *Canadian Journal of Psychiatry*, vol. 48, no. 3, pp. 195–203, 2003.

[19] C. H. S. Ruxton, P. C. Calder, S. C. Reed, and M. J. A. Simpson, "The impact of long-chain n-3 polyunsaturated fatty acids on human health," *Nutrition Research Reviews*, vol. 18, no. 1, pp. 113–129, 2005.

[20] L. M. Arterburn, E. B. Hall, and H. Oken, "Distribution, interconversion, and dose response of n-3 fatty acids in humans," *The American Journal of Clinical Nutrition*, vol. 83, no. 6 Suppl, pp. 1467S–1476S, 2006.

[21] C. J. Antalis, L. J. Stevens, M. Campbell, R. Pazdro, K. Ericson, and J. R. Burgess, "Omega-3 fatty acid status in attention-deficit/hyperactivity disorder," *Prostaglandins Leukotrienes and Essential Fatty Acids*, vol. 75, no. 4-5, pp. 299–308, 2006.

[22] A. L. Colter, C. Cutler, and K. A. Meckling, "Fatty acid status and behavioural symptoms of attention deficit hyperactivity disorder in adolescents: A case-control study," *Nutrition Journal*, vol. 7, no. 1, article no. 8, 2008.

[23] S. Spahis, M. Vanasse, S. A. Bélanger, P. Ghadirian, E. Grenier, and E. Levy, "Lipid profile, fatty acid composition and pro- and anti-oxidant status in pediatric patients with attention-deficit/hyperactivity disorder," *Prostaglandins Leukotrienes and Essential Fatty Acids*, vol. 79, no. 1-2, pp. 47–53, 2008.

[24] L. LaChance, K. McKenzie, V. H. Taylor, and S. N. Vigod, "Omega-6 to Omega-3 Fatty Acid Ratio in Patients with ADHD: A Meta-Analysis," *Journal of the Canadian Academy of Child and Adolescent Psychiatry*, vol. 25, no. 2, pp. 87–96, 2016.

[25] C. Y. Chang, D. S. Ke, and J. Y. Chen, "Essential fatty acids and human brain," *Acta Neurologica Taiwanica*, vol. 18, no. 4, pp. 231–241, 2009.

[26] C. M. Milte, N. Parletta, J. D. Buckley, A. M. Coates, R. M. Young, and P. R. C. Howe, "Increased erythrocyte eicosapentaenoic acid and docosahexaenoic acid are associated with improved attention and behavior in children with adhd in a randomized controlled three-way crossover trial," *Journal of Attention Disorders*, vol. 19, no. 11, pp. 954–964, 2015.

[27] Q. Wu, T. Zhou, L. Ma, D. Yuan, and Y. Peng, "Protective effects of dietary supplementation with natural ω-3 polyunsaturated fatty acids on the visual acuity of school-age children with lower IQ or attention-deficit hyperactivity disorder," *Nutrition*, vol. 31, no. 7-8, pp. 935–940, 2015.

[28] C. M. Milte, N. Parletta, J. D. Buckley, A. M. Coates, R. M. Young, and P. R. C. Howe, "Eicosapentaenoic and docosahexaenoic acids, cognition, and behavior in children with attention-deficit/hyperactivity disorder: a randomized controlled trial," *Nutrition*, vol. 28, no. 6, pp. 670–677, 2012.

[29] H. Perera, K. C. Jeewandara, S. Seneviratne, and C. Guruge, "Combined ω3 and ω6 supplementation in children with attention-deficit hyperactivity disorder (ADHD) refractory to methylphenidate treatment: A double-blind, placebo-controlled study," *Journal of Child Neurology*, vol. 27, no. 6, pp. 747–753, 2012.

[30] R. Raz, R. L. Carasso, and S. Yehuda, "The influence of short-chain essential fatty acids on children with attention-deficit/hyperactivity disorder: a double-blind placebo-controlled study," *Journal of Child and Adolescent Psychopharmacology*, vol. 19, no. 2, pp. 167–177, 2009.

[31] E. Barragán, D. Breuer, and M. Döpfner, "Efficacy and Safety of Omega-3/6 Fatty Acids, Methylphenidate, and a Combined Treatment in Children With ADHD," *Journal of Attention Disorders*, vol. 21, no. 5, pp. 433–441, 2017.

[32] T. Matsudaira, R. V. Gow, J. Kelly et al., "Biochemical and Psychological Effects of Omega-3/6 Supplements in Male Adolescents with Attention-Deficit/Hyperactivity Disorder: A Randomized, Placebo-Controlled, Clinical Trial," *Journal of Child and Adolescent Psychopharmacology*, vol. 25, no. 10, pp. 775–782, 2015.

[33] M. Johnson, J.-E. Månsson, S. Östlund et al., "Fatty acids in ADHD: Plasma profiles in a placebo-controlled study of Omega 3/6 fatty acids in children and adolescents," *ADHD Attention Deficit and Hyperactivity Disorders*, vol. 4, no. 4, pp. 199–204, 2012.

[34] M. Johnson, S. Östlund, G. Fransson, B. Kadesjö, and C. Gillberg, "Omega-3/omega-6 fatty acids for attention deficit hyperactivity disorder: A randomized placebo-controlled trial in children and adolescents," *Journal of Attention Disorders*, vol. 12, no. 5, pp. 394–401, 2009.

[35] S. Hirayama, T. Hamazaki, and K. Terasawa, "Effect of docosahexaenoic acid-containing food administration on symptoms of attention-deficit/hyperactivity disorder—a placebo-controlled

double-blind study," *European Journal of Clinical Nutrition*, vol. 58, no. 3, pp. 467–473, 2004.

[36] R. G. Voigt, A. M. Llorente, C. L. Jensen, J. K. Fraley, M. C. Berretta, and W. C. Heird, "A randomized, double-blind, placebo-controlled trial of docosahexaenoic acid supplementation in children with attention-deficit/hyperactivity disorder," *The Journal of Pediatrics*, vol. 139, no. 2, pp. 189–196, 2001.

[37] K. Widenhorn-Müller, S. Schwanda, E. Scholz, M. Spitzer, and H. Bode, "Effect of supplementation with long-chain ω-3 polyunsaturated fatty acids on behavior and cognition in children with attention deficit/hyperactivity disorder (ADHD): a randomized placebo-controlled intervention trial," *Prostaglandins Leukotrienes and Essential Fatty Acids*, vol. 91, no. 1-2, pp. 49–60, 2014.

[38] M. Hariri, A. Djazayery, M. Djalali, A. Saedisomeolia, A. Rahimi, and E. Abdolahian, "Effect of n-3 supplementation on hyperactivity, oxidative stress and inflammatory mediators in children with attention-deficit-hyperactivity disorder," *Malaysian Journal of Nutrition*, vol. 18, no. 3, pp. 329–335, 2012.

[39] I. Manor, A. Magen, D. Keidar et al., "Safety of phosphatidylserine containing omega3 fatty acids in ADHD children: A double-blind placebo-controlled trial followed by an open-label extension," *European Psychiatry*, vol. 28, no. 6, pp. 386–391, 2013.

[40] P. A. Gustafsson, U. Birberg-Thornberg, K. Duchén et al., "EPA supplementation improves teacher-rated behaviour and oppositional symptoms in children with ADHD," *Acta Paediatrica*, vol. 99, no. 10, pp. 1540–1549, 2010.

[41] D. J. Bos, B. Oranje, E. S. Veerhoek et al., "Reduced Symptoms of Inattention after Dietary Omega-3 Fatty Acid Supplementation in Boys with and without Attention Deficit/Hyperactivity Disorder," *Neuropsychopharmacology*, vol. 40, no. 10, pp. 2298–2306, 2015.

[42] A. Königs and A. J. Kiliaan, "Critical appraisal of omega-3 fatty acids in attention-deficit/hyperactivity disorder treatment," *Neuropsychiatric Disease and Treatment*, vol. 12, pp. 1869–1882, 2016.

[43] M. H. Bloch and A. Qawasmi, "Omega-3 fatty acid supplementation for the treatment of children with attention-deficit/hyperactivity disorder symptomatology: systematic review and meta-analysis," *Journal of the American Academy of Child and Adolescent Psychiatry*, vol. 50, no. 10, pp. 991–1000, 2011.

[44] E. Hawkey and J. T. Nigg, "Omega-3 fatty acid and ADHD: Blood level analysis and meta-analytic extension of supplementation trials," *Clinical Psychology Review*, vol. 34, no. 6, pp. 496–505, 2014.

[45] C. Leggett and E. Hotham, "Treatment experiences of children and adolescents with attention-deficit/ hyperactivity disorder," *Journal of Paediatrics and Child Health*, vol. 47, no. 8, pp. 512–517, 2011.

[46] M. Johnson, G. Fransson, S. Östlund, B. Areskoug, and C. Gillberg, "Omega 3/6 fatty acids for reading in children: a randomized, double-blind, placebo-controlled trial in 9-year-old mainstream schoolchildren in Sweden," *Journal of Child Psychology and Psychiatry*, vol. 58, no. 1, pp. 83–93, 2017.

[47] A. J. Hulbert, "Life, death and membrane bilayers," *Journal of Experimental Biology*, vol. 206, no. 14, pp. 2303–2311, 2003.

[48] M. Zulauf Logoz, "The Revision and 5th Edition of the Diagnostic and Statistical Manual of Mental Disorders (DSM-5): Consequences for the Diagnostic Work with Children and Adolescents," *Praxis der Kinderpsychologie und Kinderpsychiatrie*, vol. 63, no. 7, pp. 562–576, 2014.

16

Cardiovascular Outcomes of PCSK9 Inhibitors: With Special Emphasis on its Effect beyond LDL-Cholesterol Lowering

Dhrubajyoti Bandyopadhyay (iD),[1] **Kumar Ashish,**[2] **Adrija Hajra,**[3] **Arshna Qureshi,**[4] **and Raktim K. Ghosh**[5]

[1]*Internal Medicine, Mount Sinai St Luke's Roosevelt Hospital Center, New York, NY, USA*
[2]*University of Texas MD Anderson Cancer Center, Houston, TX, USA*
[3]*IPGMER, Kolkata, India*
[4]*Department of Medicine, Lady Hardinge Medical College, New Delhi, India*
[5]*Metro Health, Case Western Reserve University, Cleveland, OH, USA*

Correspondence should be addressed to Dhrubajyoti Bandyopadhyay; drdhrubajyoti87@gmail.com

Academic Editor: Maurizio Averna

PCSK9 inhibitors, monoclonal antibodies, are novel antihypercholesterolemic drugs. FDA first approved them in July 2015. PCSK9 protein (692-amino acids) was discovered in 2003. It plays a major role in LDL receptor degradation and is a prominent modulator in low-density lipoprotein cholesterol (LDL-C) metabolism. PCSK9 inhibitors are monoclonal antibodies that target PCSK9 protein in liver and inhibiting this protein leads to drastically lowering harmful LDL-C level in the bloodstream. Despite widespread use of the statin, not all the high-risk patients were able to achieve targeted level of LDL-C. Using PCSK9 inhibitors could lead to a substantial decrement in LDL-C plasma level ranging from 50% to 70%, either as a monotherapy or on top of statins. A large number of trials have shown robust reduction of LDL-C plasma level with the use of PCSK9 inhibitors as a monotherapy or in combination with statins in familial and nonfamilial forms of hypercholesterolemia. Moreover, PCSK9 inhibitors do not appear to increase the risk of hepatic and muscle-related side effects. PCSK9 inhibitors proved to be a highly potent and promising antihypercholesterolemic drug by decreasing LDL-R lysosomal degradation by PCSK9 protein. Statin drugs are known to have some pleiotropic effects. In this article, we are also focusing on the effects of PCSK9 inhibitor beyond LDL-C reduction like endothelial inflammation, atherosclerosis, its safety in patients with diabetes, obesity, and chronic kidney disease, and its influence on neurocognition and stroke.

1. Introduction

Heart disease is the leading cause of death in the US (23.7% of total deaths in 2011) [1]. Approximately one out of three Americans died of heart disease and stroke [2]. People with high cholesterol level are twice more likely to be suffering from heart disease than normal adults. 73.7 million or 31.7% of US adults are found to have high LDL-C. Currently, near about half of the adults (48.1%) with elevated LDL-C is getting treatment. Less than one-third (29.5%) of the population with high LDL-C is under control [1]. Familial hypercholesterolemia (FH) which is due to the mutation of specific LDL receptor gene has been found in 1 in 299 population in the US [3]. In the case of homozygous FH, the cholesterol level can be elevated even up to 1000 mg/dl (with LDL-C > 600 mg/dL) and in heterozygous FH this level may reach up to 350–550 mg/dl (with LDL-C = 200–400 mg/dL). Patients with untreated FH are prone to develop widespread atherosclerosis from their early life. Most of the untreated homozygous FH patients usually develop heart attack in their late teens and about half of the heterozygous FH suffer from heart disease at around 45 years for men and 55 to 60 years for females [4, 5]. According to 2013 AHA/ACC guidelines individuals with LDL-C level more than 190 mg/dl require high-intensity statin therapy to achieve 50% reduction. It is noteworthy that maximally tolerated dose of statin even with the combination of other nonstatin cholesterol-lowering medications is not sufficient to attain this goal, particularly in the case of FH [6]. In a study only 21% of patients achieved the target LDL-C level with the use of statin as a single agent [7] and a data from the UK showed among patients using combination therapy (statin and ezetimibe) only 44% patients achieved the target LDL-C level [8].

2. Existing Lipid-Lowering Agents

The primary lipid-lowering agents include the statin, ezetimibe, bile acid sequestrants, nicotinic acid, and fibrates. Among them, statin, ezetimibe, and bile acid sequestrants are mainly used to lower LDL-C level. Statin acts by inhibition of HMG-CoA reductase, thereby increasing LDL receptor activity. Ezetimibe inhibits cholesterol absorption by inhibiting Niemann-Pick C1-like 1 protein. Nicotinic acid and fibrates are popularly known for their triglyceride reducing property [5]. Statin is widely used to lower LDL-C and thus for primary and secondary prevention of cardiovascular disease. But this effect does not come without any side effect. Hepatic dysfunction (seen in 0.5 to 3.0% of patients) [9], myopathy (approximately 0.1% of patients develop myopathy) [10], myositis and rhabdomyolysis (near about 5% patients develop statin-associated muscle symptoms) [11], proteinuria, acute kidney injury [12], cognitive changes [13], induction of diabetes mellitus, rare cases of neuropathy [14], and drug-induced lupus have been reported [9]. In the US, the statin is considered as category X in pregnancy [9]. Overall statin intolerance is seen approximately in 10–15% of patients in clinical practice [15]. Statin is not sufficiently useful in patients with very high plasma levels of LDL-C including FH patients and patients with elevated plasma levels of lipoprotein(a) even with combination with ezetimibe. Most of the cases are due to statin intolerance or their LDL-C levels are too high to control with statin-dependent therapy. So there is a pressing need to think beyond statin in such patients.

3. Newer Hypolipidemic Drugs Either Approved Recently Or in Late Stage Development

Recently several new classes of lipid-lowering drugs have been evolved.

Lomitapide, approved by the FDA in December 2012, is an inhibitor of microsomal triglyceride transfer protein (MTP). It is used orally and indicated mainly in homozygous FH or severe heterozygous FH [16].

Mipomersen (inhibitor of apolipoprotein B-100), an antisense oligonucleotide complementary to the coding region of human apo-B mRNA, was approved by the FDA in January 2013. It is used subcutaneously in FH patients mainly [16].

Inhibitors of cholesteryl-ester transfer protein (CETP) causes increase HDL and decrease LDL-C by 40–45% [17]. According to REVEAL trial on anacetrapib, use of CETP

inhibitors in the patient with atherosclerotic vascular disease along with intensive statin regime resulted in lower incidence of major coronary events compared to the placebo arm [18]. Bempedoic acid (ETC-1002), a novel small molecule, is known to affect carbohydrate and lipid metabolism and reduce the LDL-C level near about 27% [19].

Proprotein convertase subtilisin/kexin type-9 (PCSK9) inhibitors are monoclonal antibodies which bind to the PCSK9 protein and regulate LDL-C level in blood. Alirocumab (Praluent, marketed by Sanofi-Aventis) and evolocumab (Repatha, sold by Amgen Inc.) have been approved by US-FDA in July and August 2015, respectively. Bococizumab (RN316, Pfizer) was undergoing cardiovascular safety trial, and, after showing inadequate results, the trials have been stopped, and Pfizer also discontinued its production [20].

4. PCSK9 and Cholesterol Pathway in the Body

The discovery of proprotein convertase subtilisin kexin 9 (PCSK9) by Abifadel et al. in the year 2003 has revolutionized the management of FH and subject not responding to statins regime [21]. PCSK9 gene is located on chromosome 1p32.3 [22]. PCSK9, a serine protease, is mainly produced by the liver, intestine, and kidney. PCSK9 is synthesized as 692-amino acid protein (73-kDa zymogen). After intramolecular autocatalytic processing in the endoplasmic reticulum (ER) 73-kDa zymogen gives rise to a 14-kDa prodomain and a 63 kDa mature PCSK9 [23]. Under normal circumstances, the binding of LDL to its receptor (LDL-R) is followed by the endocytosis of the complex by endosomes. At the plasma membrane, PCSK9 interacts with the LDL-R, but the neutral pH negatively modulates this interaction. On the contrary, the acidic pH of the endosome increases the affinity of the two by manifolds. Consequently, the positively charged C-terminal domain of PCSK9 binds to the negatively charged ligand-binding domain of the LDL-R [24–26] and thereby locks the LDL-R in an open conformation. The failure to attain a closed conformation in the endosome prevents normal recycling of the LDL-R to the plasma membrane. The LDL-R is then routed to lysosomes for degradation [24, 27]. As a consequence of decreased recycling, LDL-R at the cell surface is attenuated, and so does the LDL-C clearance. This normal physiology is magnified by gain-of-function mutations of PCSK9 leading to elevated LDL-C level and cardiovascular disease (CVD). Loss-of-function mutations of the PCSK9 result in increased surface LDL-R and improved LDL-C clearance [28].

The inverse relation between PCSK9 activity levels and LDL-R suggests that PCSK9 inhibition could have a synergistic effect with statins on LDL-C.

PCSK9 gene mutation is implicated in approximately 1-2% of patients with FH. PCSK9 gene mutation is the third commonest cause of FH, after LDL receptor or apolipoprotein B (ApoB) genes mutation [29]. The loss-of-function mutation of PCSK9 gene exhibited mitigation of CVD risk by 88% in the black population [30]. This observation fueled the concept that PCSK9 inhibitor might be beneficial in cases of FH and CVD.

5. Why PCSK9 Inhibitors Have Great Potential?

Patient population with FH is relatively small. The prevalence of heterozygous FH and homozygous FH due to loss of function of various gene are estimated as 1 in 500 population and 1 in 1 million, respectively [5].

However, some patients who are intolerant of statin treatment as high as 3 million (or up to 15% of patients taking statins) [31]. Till now vitamin and minerals like coenzyme-Q10 supplementation do not appear to prevent statin-induced muscular problems [32].

The number of the patients on statins but not achieving the target LDL-C levels would be even higher (only about one-third patient will achieve target LDL-C < 70 mg/dl even with high dose statin) [33]. In statin-treated patients there is upregulation of PCSK9 which attenuates the efficacy of statin. This observation led to the development of PCSK9 inhibitor [34]. On the other hand, PCSK9 inhibitor lowers LDL level in a dose-dependent manner. It reduces LDL-C by 70 percent and 60 percent in statin naïve patients and patients currently on statin therapy, respectively [35]. This reduction of LDL-C has been proven to have significant benefits in clinical studies irrespective of baseline cardiovascular risks.

FOURIER trial on evolocumab, a randomized, double-blind, placebo-controlled trial enrolling 27,564 participants with history of atherosclerotic disease, has shown a substantial reduction of all cause of mortality, cardiovascular mortality, and myocardial infarction with use of evolocumab on a background of statin therapy [36]. Heightened risk of major adverse cardiac (MACE) and limb events (MALE) typically present in patient with symptomatic peripheral arterial disease. As per the results of subanalysis study of FOURIER trial, evolocumab also mitigates the risk of MALE and the relation between achieved LDL-C and lower limb events are directly proportional. Thus, reduction of LDL-C to extremely low level should be considered in a subjects with PAD, regardless of history of MI or stroke, to diminish the chance of MACE and MALE [37].

6. A Brief Description of PCSK9 Inhibitors

Alirocumab (approximate molecular weight of 146 kDa), a human monoclonal antibody (IgG1), consists of two disulfide-linked heavy chains which are disulfide-linked to a light chain [38, 39]. It is used at a dose of 75–150 mg subcutaneously once every two weeks. The onset of action is 4–6 hours, and elimination half-life is usually 17–20 days. It undergoes proteolysis in many tissues to form polypeptides and amino acids [40].

Evolocumab (approximate molecular weight of 141.8 kDa), a human monoclonal antibody (IgG2) lambda with gamma 2 heavy chain linked by a disulfide bond to lambda light chain. It is administered subcutaneously at a dose of 140 mg every two weeks or 420 mg once monthly. The onset of action is within 4 hours and half-life of elimination is 11–17 days. It is metabolized by nonsaturable proteolysis [41, 42].

Bococizumab (approximate molecular weight of 145.1 kDa) is a humanized monoclonal antibody IgG2-Kappa with gamma 2 heavy chain linked by disulfide bond with kappa light chain. It is administered at a dose of 150 mg every two weeks or 300 mg once monthly by subcutaneous injection [43].

7. Brief Preclinical Studies

In mice lack of PCSK9 is protective against atherosclerosis and overexpression of it causes increased accumulation of cholesteryl-esters in aorta leading to accelerated atherosclerosis [44]. Alirocumab reduced atherosclerosis lesion size, monocyte and T-cell recruitment, smooth muscle cells proliferation, collagen, and macrophage content, thus improving plaque morphology in mice [45]. Infection and inflammation play a key role in the expression of PCSK9 in mice model. Clearance of LPS requires LDL or HDL binding as transport protein, and PCSK9 inhibition leads to increase expression of LDL-R causing more clearance of LDL along with LPS. This finding highlights the anti-inflammatory action of PCSK9 inhibitors [46]. Preclinical studies by Walley et al. and Dwivedi et al. also supported the anti-inflammatory PCSK9 inhibitors in mice animal model [47, 48]. Alirocumab administration in mice has shown reduced circulating neutrophil, monocytes, and decreased expression of endothelial ICAM-1. Thus it attenuates monocytes attachment to vascular endothelium and dampens vascular inflammation. Food intake, body weight, and weight of the liver were unaltered with alirocumab therapy in mice [45]. Reversible liver parenchymal hypertrophy and nonsignificant adrenal cortex hypertrophy have been reported in animals. There are no increased risks of hepatitis-C virus infection, immunosuppression, neurocognitive dysfunction, type 2 diabetes mellitus, and no increased bile acid concentration in intestine, thus possessing no significant increased risk of the intestinal tumor in the animal model [49]. In rats, bococizumab administration demonstrates no adverse effects on embryo-fetal development even in dose greater than usual clinical dosage [50].

8. Clinical Development of PCSK9 Inhibitors

8.1. Alirocumab. Alirocumab is a novel PCSK9 inhibitor. Many studies have been done and still going on to find out its efficacy and safety profile which have been summarized in Table 1. Its LDL lowering effect is also independent of the site of injection [51].

Also, a pooled data from 3 double-blind, randomized, placebo-controlled, phase 2 studies showed alirocumab to reduce LDL-apo® and Lp(a) significantly from baseline in comparison to placebo [52]. As we know that lipoprotein is an

TABLE 1: Trials on alirocumab. A-mAb: alirocumab; ATV: atorvastatin; EZE: ezetimibe; RSV: rosuvastatin; ASCVD: atherosclerotic cardiovascular disease; CHD: coronary heart disease; heFH: heterozygous familial hypercholesterolemia.

S. No	Trial [reference]	Participants	Comparison	LDL-C reduction	Comments
(1)	ODYSSEY MONO Completion date: July 2013 [50].	Patients with hypercholesterolemia	A-mAb versus EZE	47.2% versus 15.6%	-
(2)	ODYSSEY COMBO I Completion date: April 2014 [46].	Hypercholesterolemia + CHD or CHD equivalents, on treatment with maximal tolerated statin dose	A-mAb versus Placebo	48.2% versus 2.3%	-
(3)	ODYSSEY OPTIONS I Completion date: May 2014 [44].	Hyperlipidemia + risk of ASCVD, on baseline treatment with ATV	ATV + A-mAb versus ATV + EZE versus ATV (double dose) versus RSV	44.1% versus 20.5% versus 5.0% versus 21.4 %	-
(4)	ODYSSEY OPTIONS II Completion date: May 2014 [45].	Hyperlipidemia + risk of ASCVD, on baseline treatment with RSV	A-mAb versus EZE versus RSV	50.6% versus 14.4% versus 16.3%	-
(5)	ODYSSEY LONG TERM TRIAL Completion date: November 2014 [47].	Hypercholesterolemia + risk of ASCVD, on treatment with maximally tolerated statin dose	A-mAb versus placebo	61% versus 0.8%	-
(6)	ODYSSEY FH I Completion date: December 2014 [49]	Familial heterozygous hypercholesterolemia on maximally tolerated statin dose	A-mAb versus placebo	57.9% reduction in A-Mab group	-
(7)	ODYSSEY FH II Completion date: January 2015 [49]	Familial heterozygous hypercholesterolemia on maximally tolerated statin dose	A-mAb versus placebo	51.4% reduction in A-Mab group	-
(8)	ODYSSEY COMBO II Completion date: July 2015 [48].	Hypercholesterolemia + risk of ASCVD, on treatment with maximally tolerated statin dose	A-mAb versus EZE	50.6% versus 20.7%	-
(9)	ODYSSEY HIGH FH Completion date: 2016 Sep	Patients having heFH and LDL-C ≥ 160 mg/dl even after maximum tolerated dose of statin	A-mAb versus Placebo	45.7% versus 6.6%	

TABLE 1: Continued.

S. No	Trial [reference]	Participants	Comparison	LDL-C reduction	Comments
(9)	Phase 2 pooled analysis [51]	Primary hypercholesterolemia on lipid lowering therapy	A-mAb versus placebo	68.4% versus 10.5%	-
(10)	Randomized controlled trial [54]	Hypercholesterolemia on treatment with ATV	A-mAb versus placebo	40% to 70% versus 5%	-
(11)	Pooled analysis of 14 randomized controlled trials	-	A-mAb versus control (placebo or EZE)	LDL-C reduced to as low as 15 mg/dl in A-mAb group	Rates of adverse events in those achieving LDL-C < 25 mg/dl (72.7%) and <15 mg/dl (71.7%) were similar to those who did not (76.7%)

independent risk factor for CAD, this finding holds promise [53].

8.2. Evolocumab. Several clinical studies have established the efficacy and safety of evolocumab [Table 2]. GAUSS-3 Randomized Clinical Trial also proved its tolerability in patients with muscle-related statin intolerance [54].

8.3. Bococizumab. Despite the fact that the initial studies showed promising result [Table 3], recently SPIRE trials showed attenuation of the effect of bococizumab in 15–20% of patients due to the formation of antibody against its murine component. This led to the interruption of further development of this drug [55].

9. The Role of PCSK9 beyond LDL-C Lowering

PCSK9 inhibitors after being established as a valid option, now its effects on inflammation, endothelial function, atherosclerosis, diabetes, and obesity are now actively investigated.

(i) The PCSK9 Level in CKD Patients and HD Patients. In a study, it was shown that serum PCSK9 level was decreased in chronic kidney disease patients who were on hemodialysis (CLD-HD) and PCSK9 had a positive correlation with LDL-C level. This signifies that PCSK9 plays a major role in regulating LDL-C even in CKD-HD patients. PCSK9 also is involved in the metabolism of triglyceride-rich lipoproteins in CKD-HD patients [56]. PCSK9 level tends to rise in patients with nephrotic syndrome, and it has a positive correlation with proteinuria. The PCSK9 level is also higher in patients on peritoneal dialysis in comparison to hemodialysis or renal transplant patients [57].

(ii) PCSK9 and Lipoprotein A. Lp (a) is a widely accepted cardiovascular risk factor, except for regular extracorporeal lipoprotein apheresis which is the only available modality to reduce Lp (a). The possible mechanism of Lp (a) reduction with use of PCSK9 inhibitors is because of the enormous expression of LDL-R due to PCSK9 inhibition, which unmasks clearance mechanism of Lp (a) by abundant hepatic LDL-R [58]. Additionally, Canuel et al. have reported that PCSK9 degrades LDL related protein-1, which catabolizes Lp(a). Thus, the inhibition of PCSK9 increases the catabolism of Lp(a) [59].

Alirocumab also reduces non-HDL cholesterol and fasting triglycerides significantly. Alirocumab also increases the HDL level and apolipoprotein A-1. Though the possible mechanism which increases HDL-C is not entirely clear, one likely hypothesis is that reduction of LDL-C causes reduction in cholesteryl-ester transfer protein activity, as less LDL-C is available to transfer out cholesterol from HDL particles. Eventually, it causes a relatively high level of HDL-C [60].

(iii) PCSK9 Inhibitors on Inflammation and Atherosclerosis. Atherosclerosis is a chronic inflammatory process within the arterial wall. Proinflammatory effect of PCSK9 has been shown in different experimental models. This effect is supposed to be responsible for promoting atherosclerosis independent of LDL-C level. Vascular smooth muscle cells (VSMC), oxidized LDL-C, have shown a high level of PCSK9 expression [61].

LOX-1, a receptor for oxidized LDL in VSMC, is upregulated in inflammation. It has been reported that PCSK9 stimulates transcription of LOX-1 and LOX-1, in turn, and stimulates PCSK9 expression, which facilitates atherogenesis [62]. The interaction between PCSK9 and LDL-R favors the entry of inflammatory monocytes into the arterial wall and thus promotes atherosclerosis [63]. Though it had been shown in a study that there is no effect of PCSK9 inhibitors on hs-CRP concentration level [64], the relationship of PCSK9 with systemic inflammation cannot be denied. Walley et al. revealed that loss of PCSK9 function in both mice model and human enhances pathogen lipid clearance with the help of LDL receptor and regulates inflammatory response in septic shock with better survival [48].

ATHEROREMO-IVUS study conducted by Cheng et al. revealed a linear correlation between PCSK9 level and the amount of necrotic tissue in atherosclerotic plaque [65]. Lowering of LDL-C by PCSK9 inhibitors reduces inflammation, endothelial apoptosis, and the concentration of oxidized LDL-C within the plaque. This alters the composition of the plaque more favorably. GLAGOV Randomized Clinical Trial (n = 968), which was concluded last year, revealed that addition of evolocumab to statin-treated patient causes decrease in percent atheroma plaque volume assessed by sequential intravascular ultrasound after 76 weeks of therapy [66]. Macrophage recycles the cell membrane lipid from the dead cells including RBC. When recycling capacity of macrophages exceeds, those cell membrane lipids accumulate as atheroma. Macrophage fat catabolism capacity is associated with underlying atherosclerosis, and this could be quantified by accumulation acyl-carnitine intermediates in ECF which is the direct parameter of the adequacy of beta-oxidation to recycle membrane fatty acid. A study conducted by Blair et al. revealed that minimizing macrophage fat overload by reducing fat metabolism rate is favorable, which could be achieved by using statin and the newer PCSK9 inhibitors [67].

(iv) PCSK9 in Diabetic Patients. Diabetes is the major well-established risk factor for cardiovascular diseases. Diabetes increases the risk for atherosclerosis due to endothelial inflammation. Inflammation in blood vessels is one of the primary drivers for atherosclerosis and diabetes makes it much worse. A study conducted by Sattar et al. revealed that PCSK9 inhibitor markedly decreases atherogenic lipoproteins in diabetic patients and results are similar as seen in nondiabetic patients [68]. They did not alter the normal glucose homeostasis. Arsenault and colleagues have put the light on the fact that serum PCSK9 protein level is higher in insulin-resistant subjects, which also supports its temporal association of hyperlipidemia in diabetic patients [69]. The results from FOURIER trial demonstrated no significant difference in new-onset diabetes and neurocognitive events between evolocumab and placebo arm [36]. They reported

TABLE 2: Studies on evolocumab. E-mAb: evolocumab; EZE: ezetimibe; ATV: atorvastatin.

S. No.	Trial	Participants	Comparison	LDL-C reduction	Adverse events (AE)
(1)	MENDEL-2 Completion date: October 2013	Hypercholesterolemia	E-mAb versus Placebo versus EZE	Reduction in E-mab group; 55–57% more than placebo & 38–40% more than EZE	44% versus 44% versus 46%
(2)	DESCARATES Completion date: November 2013	Hypercholesterolemia, on background therapy with diet or ATV (10 mg or 80 mg) or EZE singly or in combination.	E-mAb versus Placebo	50.1% versus 6.8%	74.8% versus 74.2%
(3)	GAUSS-2 Completion date: November 2013	Hypercholesterolemia	E-mAb versus EZE	53–56% versus 37–39%	Muscle AE's; more frequent in EZE group (23%) versus E-mab group (12%)
(4)	RUTHERFORD-2 Primary completion date: November 2013	Familial Hypercholesterolemia	E-mAb versus placebo	60% in E-mAb group	Similar adverse events profile
(5)	LAPLACE-2 Completion date: Dec. 2013	Hypercholesterolemia	E-mAb versus EZE versus Placebo	−66% to 75% in E-mAb group	36% versus 40% versus 39%
(6)	FOURIER Completion date: 2017	Patient with h/o CVD on maximum tolerated statin therapy but LDL is more than 70 mg/dl	E-mAb versus placebo in statin treated patients	59% reduction of LDL in comparison to placebo	Primary end point. that is, Cv events was 9.8% versus 11.3%; Injection site reaction was more in E-mAb 2.1% versus 1.6%
(7)	OSLER I & OSLER II Expected completion date: June 2018 & August 2018, respectively	Patients who completed "parent trials" of evolocumab and eligible patients were randomly assigned in 2:1 ratio to receive either evolocumab plus standard therapy or standard therapy alone	E-mAb versus standard therapy	61% in E-mAb group.	69.2% versus 64.8%

TABLE 3: Studies on bococizumab.

S. number	Trial	Participants	Comparison	LDL reduction	Adverse events	Comments
(1)	Dose ranging trial (NCT01592240)	Those with LDL-C > 80 mg/dl on stable statin therapy	Bococizumab versus placebo	54.2% versus 2.8%	Similar adverse events profile	Despite dose reduction in many subjects, bococizumab significantly reduced LDL-C across all the doses
(2)	SPIRE-1 and SPIRE-2	Those with background lipid-lowering treatment and have an LDL-C of > / = 70 mg/dl (SPIRE-1) or LDL-C > / = 100 mg/dl (SPIRE-2)	Bococizumab versus placebo	This study has been terminated. Completion date: Jan. 2017		Bococizumab being a humanized monoclonal antibody, a strong immune response was seen against it which mitigate the LDL-C lowering effect. Anti-drug Ab was seen in 48% patients and neutralizing Ab developed in 29% patients

similar levels of HbA1c and FPG between both the groups in patients with diabetes, prediabetes, and normoglycemia [70].

(v) PCSK9 and Its Relation with Vitamin E, Cortisol, Adreno-corticotropic Hormone, and Gonadal Hormones. Vitamin E is one of the important antioxidants which prevent oxidative damage of long-chain PUFA and cell membrane disruption. The function of vitamin E transport and steroidogenesis are intricately related to LDL-C metabolism. In a recent study, 901 patients with LDL-C ≥ 2.0 mmol/L were randomly assigned to monthly subcutaneous evolocumab for 52 weeks and placebo. In evolocumab-treated substudy group, the level of vitamin E in LDL-C was decreased substantially, and vitamin E level in HDL was increased significantly. Cortisol level was elevated in evolocumab-treated patients, but there were no changes in ACTH, cortisol:ACTH ratio, and gonadal hormone levels. So from this above data, we can conclude that in spite of lowering LDL-C level, evolocumab does not reduce the cholesterol normalized level of vitamin E, ACTH, and gonadal hormones [71].

(vi) PCSK9 in Ischemic Stroke. PCSK9 and LDL receptor are involved in mouse brain development. After 24 to 72 hrs. of reperfusion in ischemic stroke, PCSK9 is upregulated in the dentate gyrus of the mouse model but without affecting de novo neurogenesis. PCSK9 degrades the LDL receptor in the brain (telencephalon and cerebellum) both during development and ischemia/reperfusion. PCSK9 is expressed only in the olfactory peduncle in adult mice, but it does not degrade LDL receptor. In adult mice, blood-brain barrier is impermeable to PCSK9 which explains the absence of LDL receptor lowering effect of PCSK9 in the brain of adult mice. The effect of LDL receptor in the brain is still not clear. LDL receptor-negative mice show impaired learning and memory. Though our understanding of the effects of LDL receptor excess in the brain is still evolving, ablation of the PCSK9 gene in mice did not reveal any significant effect of decreased PCSK9 level on brain recovery after an ischemic stroke. So we can hope that PCSK9 inhibition by monoclonal antibody should not hamper brain recovery after an ischemic insult [72]. A meta-analysis involving 11 studies showed there is no increased incidence of stroke with its use [73].

(vii) Effect of PCSK9 beyond Liver. PCSK9 effect beyond LDL receptor degradation in the liver is largely unknown. PCSK9 is expressed by extrahepatic tissues such as intestine, pancreas, kidney, smooth vascular cells, endothelial cells, goblet cells, and brain. In insulin-resistant diabetic patient, PCSK9 level is significantly low in duodenum compared to insulin sensitive obese patient undergoing bariatric surgery. But it remains to conclude whether insulin induces expression of PCSK9 in the intestine as it does in the liver. PCSK9-deficient mice had sevenfold increased LDL receptor expression in intestine. PCSK9-deficient mice showed reduced postprandial hypertriglyceridemia due to reduced level of ApoB. A route of cholesterol excretion that is upregulated in the PCSK9-deficient mice leads to fecal cholesterol excretion. It is also involved in nephrogenesis and binds

with amiloride-sensitive epithelial sodium channel (ENaC) and mediates their degradation by proteasome pathways. PCSK9 downregulates the LDL receptor expression on the surface of the isolated human pancreatic beta cell and in PCSK9 (−/−) mice; there were increased cell surface LDL receptors. But there is an inconsistency about the deleterious effect of LDL-C accumulation in beta cells insulin secretion of the PCSK9 (−/−) mice model. This discrepancy can be explained by the PCSK9 inhibitor-mediated reduction of LDL-C level which counterbalances the deleterious effect of LDL-C accumulation inside the beta cells [74].

(viii) PCSK9 and Neurocognitive Effect. In 2012, FDA issued a warning for all statin drugs: "ill-defined memory loss or impairment" [75]. The PCSK9 inhibitor is one of the most potent and promising new therapies to lower LDL-C level nowadays. So it is imperative to discuss any plausible role of it in causing neurocognitive impairment. LDL-R also is expressed in the brain and helps in clearing apolipoprotein-E, which is responsible for the formation of amyloid-β which accumulates in the brain of Alzheimer's disease patient. PCSK9 gene deleted mice have shown reduced apolipoprotein-E and amyloid-β formation. PCSK9-inhibitors cannot cross blood-brain barrier in human. Long-term LDL-C lowering by PCSK9 improves arterial health which in turn protects against the development of dementia. Though there are some studies which signal to unfavorable effects on neurocognition, recently Robinson et al. showed no increased neurocognitive risk in a pooled analysis of 14 trials on PCSK9 inhibitors even after attaining an extremely low LDL-C level [76]. Loss of function of PCSK9 is not associated with any symptoms of mental retardation as well.

Currently, few trials are ongoing to search any adverse neurocognitive effect of PCSK9 inhibitors, and we have to wait until the end of 2017-2018 for a final opinion regarding this aspect.

(ix) Available Safety Data on PCSK9 Inhibitors. The most common adverse events occurring in alirocumab treated patients were gastrointestinal disorders, infections and infestations, musculoskeletal disorders, and skin and subcutaneous tissue disorders [77]. One patient with the history of atrial fibrillation and chronic obstructive pulmonary disorder had a pulmonary embolism after alirocumab therapy in ODYSSEY MONO study [78].

Nasopharyngitis, upper respiratory tract infection, influenza, and back pain are commonly found adverse effects regarding the use of evolocumab [79]. The increment of creatine kinase levels to more than five times the ULN occurred in 1.2% of patients in the evolocumab group in DESCARTES Trial [80]. Acute pancreatitis has been reported in MENDEL 2 trial as well [81].

Upper respiratory tract infection, nasopharyngitis, diarrhea, urinary tract infection, arthralgia, bronchitis, injection site erythema, gastroesophageal reflux disease, and cough are commonly found adverse effects after bococizumab use [82].

Interestingly, PCSK9 inhibitors do not lead to an increased rate of new-onset diabetes [83]. Though high blood glucose during the treatment period and baseline

HbA1c of more than 6.5% have been found in alirocumab treated patients, there was no pattern in changes in either blood glucose or HbA1c from screening to week 24 [84]. Evolocumab treatment did not show any adverse effect on glycemic measures in a 52-week placebo-controlled trial [80].

A meta-analysis of 25 randomized controlled trials has shown no significant difference regarding the occurrence of adverse events between PCSK9 inhibitor group and placebo group (or ezetimibe group) [85].

Cholesterol is an important component of myelin protein. So, lipid-lowering therapies may play a role to hamper the neural structure and function. Along with that reduced serum cholesterol may also enhance the blood-brain barrier permeability. It may result in increased exposure of the central nervous system to the toxins in the blood. So there is a concern whether PCSK9 inhibitors can cause cognitive dysfunction by lowering cholesterol level [75]. A meta-analysis has shown a significant increase in neurocognitive events with PCSK9 inhibitor therapy in comparison with placebo therapy. But the analysis has some limitations due to lack of uniform data, heterogeneity of the studies, and lack of uniform definitions of the cardiovascular events [86]. On the other hand, LDL receptor causes clearance of apolipoprotein-E, a protein responsible for Alzheimer's disease. So PCSK9 inhibitors may have the protective role for this disease. Long-term PCSK9 inhibitor therapy can prevent vascular dementia by improving arterial health as well. So there may be a positive effect of PCSK9 inhibitors on neurocognitive functions. Cognitive side effects have been found as an uncommon finding in the OSLER study. Less than 1% patients showed amnesia and less than 1% patients showed mental impairment, but no precise data is available till now [75]. Apart from neurocognitive impairment, increased incidences of hemorrhagic stroke, hormonal insufficiency, and hemolytic anemia have been found to be associated with the very low LDL-C level [79]. There was only statistically insignificant increased risk of cataract in a recent analysis [76].

Pooled analysis from 10 ODYSSEY Trials established no increase adverse events in patients on alirocumab therapy [87].

Naturally more time-tested trials are required for conclusive data. To summarize, published evidence from the trials suggests that PCSK9 inhibitors are well tolerated and with good safety profile [84].

10. Conclusion and Future Direction

The discovery of PCSK9 proteins has changed the dynamics of lipid control in hypercholesterolemic patients. PCSK9 inhibitors pave the path of achieving an extremely low plasma level of LDL-C and have shown to reduce lifetime risk for CVD events. They are also involved in decreasing endothelial inflammation which is the key factor for atherosclerosis. Very aggressive lowering of LDL-C by PCSK9 inhibitors leads to plaque stabilization and regression. Large phase II and III trials for these monoclonal Abs have shown

its safety, efficacy, and effectiveness in patients who are at risk for cardiovascular diseases due to dyslipidemia. Many trials have revealed that PCSK9 inhibitors have reduced all causes of mortality including cardiovascular mortality with less adverse effects like myopathy and hepatotoxicity. Though SPIRE trials were terminated early for the concern of immunogenicity, the immunogenicity of evolocumab is extremely low and neutralizing anti-drug antibody was seen only in 1.3% patients on alirocumab. Newer drug Inclisiran, a PCSK9-specific small interfering RNA, is also being studied. ODYSSEY OUTCOME study and other trials are also in progress to evaluate the whole spectrum of PCSK9 inhibitors, and the results are scheduled to come by the end of 2017 to 2018. It is to be evaluated by subsequent trials whether these pleiotropic effects would confer substantial morbidity and mortality benefits.

References

[1] Cholesterol Facts High, http://www.cdc.gov/cholesterol/facts.htm.March.

[2] A. S. Go, D. Mozaffarian, V. L. Roger et al., "Heart disease and stroke statistics—2014 update: a report from the American heart association," *Circulation*, vol. 129, no. 3, pp. e28–e292, 2014.

[3] S. D. D. Ferranti, A. M. Rodday, M. Mendelson, J. B. Wong, L. K. Leslie, and R. C. Sheldrick, "6: What is the Prevalence of Familial Hypercholesterolemia in the US? Session Title: Lipid-Lowering Trials. Circulation.2014; 130: A19656," in *Core 2. Epidemiology and Prevention of CV Disease: Physiology, Pharmacology and Lifestyle Abstract*, A19656, 130, 1965.

[4] Bob Carlson, "Familial Hypercholesterolemia Captures Gene Test Controversies.Biotechnol Healthc," in *Familial Hypercholesterolemia Captures Gene Test Controversies*, vol. 7, p. 89, Spring, Biotechnol Healthc, 2010.

[5] G. K. Hovingh, M. H. Davidson, J. J. P. Kastelein, and A. M. O'Connor, "Diagnosis and treatment of familial hypercholesterolaemia," *European Heart Journal*, vol. 34, no. 13, pp. 962–971, 2013.

[6] N. J. Stone, "ACC/AHA Guideline on the Treatment of Blood Cholesterol to Reduce Atherosclerotic Cardiovascular Risk in Adults," *A Report of the American College of Cardiology/American Heart Association*, vol. 129, supplement 2, pp. S1–S45, 2014.

[7] A. H. Pijlman, R. Huijgen, S. N. Verhagen et al., "Evaluation of cholesterol lowering treatment of patients with familial hypercholesterolemia: a large cross-sectional study in The Netherlands," *Atherosclerosis*, vol. 209, no. 1, pp. 189–194, 2010.

[8] Audit of the Management of Familial Hypercholesterolaemia 2010: Full Report. Royal College of Physicians website 2010, http://www.rcplondon.ac.uk/resources/audits/FH.

[9] Rosenson RS, "statins: Actions, side effects, and administration," http://www.uptodate.com/contents/statins-actions-side-effects-andadministrationsource=search_result&search=statin&selectedTitle=1~150.

[10] D. J. Graham, J. A. Staffa, D. Shatin et al., "Incidence of hospitalized rhabdomyolysis in patients treated with lipid-lowering drugs," *The Journal of the American Medical Association*, vol. 292, no. 21, pp. 2585–2590, 2004.

[11] U. Laufs, H. Scharnagl, M. Halle, E. Windler, M. Endres, and W. März, "Treatment options for statin-associated muscle symptoms," *Deutsches Ärzteblatt International*, vol. 112, no. 44, pp. 748-55, 2015.

[12] A. A. Alsheikh-Ali, M. S. Ambrose, J. T. Kuvin, and R. H. Karas, "The safety of rosuvastatin as used in common clinical practice: A postmarketing analysis," *Circulation*, vol. 111, no. 23, pp. 3051–3057, 2005.

[13] D. A. Redelmeier, D. Thiruchelvam, and N. Daneman, "Delirium after elective surgery among elderly patients taking statins," *Canadian Medical Association Journal*, vol. 179, no. 7, pp. 645–652, 2008.

[14] D. Gaist, U. Jeppesen, M. Andersen, L. A. García Rodríguez, J. Hallas, and S. H. Sindrup, "Statins and risk of polyneuropathy: A case-control study," *Neurology*, vol. 58, no. 9, pp. 1333–1337, 2002.

[15] M. Banach, M. Rizzo, and P. P. Toth, "Statin intolerance – an attempt at a unified definition. Position paper from an international lipid expert panel," *Archives of Medical Science*, vol. 11, no. 1, pp. 1–23, 2015.

[16] Elena Citkowitz, "Familial Hypercholesterolemia Treatment & Management," 2015, http://emedicine.medscape.com/article/121298-treatment#d7.

[17] J. J. P. Kastelein, J. Besseling, S. Shah et al., "Anacetrapib as lipid-modifying therapy in patients with heterozygous familial hypercholesterolaemia (REALIZE): A randomised, double-blind, placebo-controlled, phase 3 study," *The Lancet*, vol. 385, no. 9983, pp. 2153–2161, 2015.

[18] "Effects of Anacetrapib in Patients with Atherosclerotic Vascular Disease," *The New England Journal of Medicine*, vol. 377, no. 13, pp. 1217–1227, 2017.

[19] C. M. Ballantyne, M. H. Davidson, D. E. MacDougall et al., "Efficacy and safety of a novel dual modulator of adenosine triphosphate-citrate lyase and adenosine monophosphate-activated protein kinase in patients with hypercholesterolemia: Results of a multicenter, randomized, double-blind, placebo-controlled, parallel-group trial," *Journal of the American College of Cardiology*, vol. 62, no. 13, pp. 1154–1162, 2013.

[20] A. M. Sible, J. J. Nawarskas, and J. R. Anderson, "PCSK9 inhibitors: An innovative approach to treating hyperlipidemia," *Cardiology in Review*, vol. 24, no. 3, pp. 141–152, 2016.

[21] M. Abifadel, M. Varret, J. Rabès et al., "Mutations in *PCSK9* cause autosomal dominant hypercholesterolemia," *Nature Genetics*, vol. 34, no. 2, pp. 154–156, 2003.

[22] D. Urban, J. Pöss, M. Böhm, and U. Laufs, "Targeting the proprotein convertase subtilisin/kexin type 9 for the treatment of dyslipidemia and atherosclerosis," *Journal of the American College of Cardiology*, vol. 62, no. 16, pp. 1401–1408, 2013.

[23] N. G. Seidah and A. Prat, "The proprotein convertases are potential targets in the treatment of dyslipidemia," *Journal of Molecular Medicine*, vol. 85, no. 7, pp. 685–696, 2007.

[24] P. L. Surdo, M. J. Bottomley, A. Calzetta et al., "Mechanistic implications for LDL receptor degradation from the PCSK9/LDLR structure at neutral pH," *EMBO Reports*, vol. 12, no. 12, pp. 1300–1305, 2011.

[25] T. Yamamoto, C. Lu, and R. O. Ryan, "A two-step binding model of PCSK9 interaction with the low density lipoprotein receptor," *The Journal of Biological Chemistry*, vol. 286, no. 7, pp. 5464–5470, 2011.

[26] K. Tveten, Ø. L. Holla, J. Cameron et al., "Interaction between the ligand-binding domain of the LDL receptor and the C-terminal domain of PCSK9 is required for PCSK9 to remain bound to the LDL receptor during endosomal acidification," *Human Molecular Genetics*, vol. 21, no. 6, Article ID ddr578, pp. 1402–1409, 2012.

[27] S. C. Blacklow, "Versatility in ligand recognition by LDL receptor family proteins: advances and frontiers," *Current Opinion in Structural Biology*, vol. 17, no. 4, pp. 419–426, 2007.

[28] V. Bittner, "Pleiotropic Effects of PCSK9 (Proprotein Convertase Subtilisin/Kexin Type 9) Inhibitors?" *Circulation*, vol. 134, no. 22, pp. 1695-1696, 2016.

[29] J. Cohen, A. Pertsemlidis, I. K. Kotowski, R. Graham, C. K. Garcia, and H. H. Hobbs, "Low LDL cholesterol in individuals of African descent resulting from frequent nonsense mutations in PCSK9," *Nature Genetics*, vol. 37, no. 2, pp. 161–165, 2005.

[30] N. G. Seidah, Z. Awan, M. Chrétien, and M. Mbikay, "PCSK9: A key modulator of cardiovascular health," *Circulation Research*, vol. 114, no. 6, pp. 1022–1036, 2014.

[31] D. H. Fitchett, R. A. Hegele, and S. Verma, "Statin intolerance," *Circulation*, vol. 131, no. 13, pp. e389–e391, 2015.

[32] J. M. Young, C. M. Florkowski, S. L. Molyneux et al., "Effect of Coenzyme Q10 Supplementation on Simvastatin-Induced Myalgia," *American Journal of Cardiology*, vol. 100, no. 9, pp. 1400–1403, 2007.

[33] D. G. Karalis, R. D. Subramanya, S. E. Hessen, L. Liu, and M. F. Victor, "Achieving optimal lipid goals in patients with coronary artery disease," *American Journal of Cardiology*, vol. 107, no. 6, pp. 886–890, 2011.

[34] G. D. Norata, G. Tibolla, and A. L. Catapano, "PCSK9 inhibition for the treatment of hypercholesterolemia: Promises and emerging challenges," *Vascular Pharmacology*, vol. 62, no. 2, pp. 103–111, 2014.

[35] Rosenson RS, "Lipid lowering with drugs other than statins and fibrates," http://www.uptodate.com/contents/lipid-lowering-with-drugs-other-than-statins-andfibrates?source=machine-Learning&search=pcsk9&selectedTitle=1~26&anchor=H370516502&sectionRank=1#H370516502.

[36] M. S. Sabatine, R. P. Giugliano, A. C. Keech et al., "Evolocumab and clinical outcomes in patients with cardiovascular disease," *The New England Journal of Medicine*, vol. 376, no. 18, pp. 1713–1722, 2017.

[37] M. P. Bonaca, P. Nault, R. P. Giugliano et al., "Low-Density Lipoprotein Cholesterol Lowering With Evolocumab and Outcomes in Patients With Peripheral Artery Disease," *Circulation*, vol. 137, no. 4, pp. 338–350, 2018.

[38] Alirocumab label Label revised,.

[39] International Nonproprietary Names for Pharmaceutical Substances (INN), "WHO Drug Information," Vol. 27, No. 1, 2013.

[40] "Alirocumab: Drug information," http://www.uptodate.com/contents/alirocumab-drug-information?source=see_link.

[41] A. F. G. Cicero, A. Colletti, and C. Borghi, "Profile of evolocumab and its potential in the treatment of hyperlipidemia," *Drug Design, Development and Therapy*, vol. 9, pp. 3073–3082, 2015.

[42] "Evolocumab: Drug information," http://www.uptodate.com/contents/evolocumab-drug-information?source=see_link.

[43] "Bococizumab," http://medcheminternational.blogspot.in/2015/09/bococizumab.html.

[44] M. Denis, J. Marcinkiewicz, A. Zaid et al., "Gene inactivation of proprotein convertase subtilisin/kexin type 9 reduces atherosclerosis in mice," *Circulation*, vol. 125, no. 7, pp. 894–901, 2012.

[45] S. Kühnast, J. W. A. Van Der Hoorn, E. J. Pieterman et al., "Alirocumab inhibits atherosclerosis, improves the plaque morphology, and enhances the effects of a statin," *Journal of Lipid Research*, vol. 55, no. 10, pp. 2103–2112, 2014.

[46] K. R. Feingold, A. H. Moser, J. K. Shigenaga, S. M. Patzek, and C. Grunfeld, "Inflammation stimulates the expression of PCSK9," *Biochemical and Biophysical Research Communications*, vol. 374, no. 2, pp. 341–344, 2008.

[47] D. J. Dwivedi, P. M. Grin, M. Khan et al., "Differential expression of PCSK9 modulates infection, inflammation, and coagulation in a murine model of sepsis," *Shock*, vol. 46, no. 6, pp. 672–680, 2016.

[48] K. R. Walley, K. R. Thain, J. A. Russell et al., "PCSK9 is a critical regulator of the innate immune response and septic shock outcome," *Science Translational Medicine*, vol. 6, no. 258, pp. 258–ra143, 2014.

[49] "The endocrinologic and metabolic drugs advisory committee meeting. Briefing document," Praluent(alirocumab) injection, 2015, http://www.fda.gov/downloads/AdvisoryCommittees/CommitteesMeetingMaterials/Drugs/EndocrinologicandMetabolicDrugsAdvisoryCommittee/UCM449865.pdf.

[50] S. N. Campion, B. Han, G. D. Cappon et al., "Decreased maternal and fetal cholesterol following maternal bococizumab (anti-PCSK9 monoclonal antibody) administration does not affect rat embryo-fetal development," *Regulatory Toxicology and Pharmacology*, vol. 73, no. 2, pp. 562–570, 2015.

[51] C. Lunven, T. Paehler, F. Poitiers et al., "A randomized study of the relative pharmacokinetics, pharmacodynamics, and safety of alirocumab, a fully human monoclonal antibody to PCSK9, after single subcutaneous administration at three different injection sites in healthy subjects," *Cardiovascular Therapeutics*, vol. 32, no. 6, pp. 297–301, 2014.

[52] D. Gaudet, D. J. Kereiakes, J. M. McKenney et al., "Effect of alirocumab, a monoclonal proprotein convertase subtilisin/kexin 9 antibody, on lipoprotein(a) concentrations (a pooled analysis of 150 mg every two weeks dosing from phase 2 trials)," *American Journal of Cardiology*, vol. 114, no. 5, pp. 711–715, 2014.

[53] D. Gaudet, G. F. Watts, J. G. Robinson et al., "Effect of Alirocumab on Lipoprotein(a) Over ≥1.5 Years (from the Phase 3 ODYSSEY Program)," *American Journal of Cardiology*, vol. 119, no. 1, pp. 40–46, 2017.

[54] S. E. Nissen, E. Stroes, R. E. Dent-Acosta et al., "Efficacy and tolerability of evolocumab vs ezetimibe in patients with muscle-related statin intolerance: The GAUSS-3 randomized clinical trial," *Journal of the American Medical Association*, vol. 315, no. 15, pp. 1580–1590, 2016.

[55] P. M. Ridker, J. Revkin, P. Amarenco et al., "Cardiovascular efficacy and safety of bococizumab in high-risk patients," *The New England Journal of Medicine*, vol. 376, no. 16, pp. 1527–1539, 2017.

[56] H. Abujrad, J. Mayne, M. Ruzicka et al., "Chronic kidney disease on hemodialysis is associated with decreased serum PCSK9 levels," *Atherosclerosis*, vol. 233, no. 1, pp. 123–129, 2014.

[57] P. Pavlakou, E. Liberopoulos, E. Dounousi, and M. Elisaf, "PCSK9 in chronic kidney disease," *International Urology and Nephrology*, vol. 49, no. 6, pp. 1015–1024, 2017.

[58] S. Tsimikas, "Lipoprotein(a): Novel target and emergence of novel therapies to lower cardiovascular disease risk," *Current Opinion in Endocrinology, Diabetes and Obesity*, vol. 23, no. 2, pp. 157–164, 2016.

[59] M. Canuel, X. Sun, M.-C. Asselin, E. Paramithiotis, A. Prat, and N. G. Seidah, "Proprotein Convertase Subtilisin/Kexin Type 9 (PCSK9) Can Mediate Degradation of the Low Density Lipoprotein Receptor-Related Protein 1 (LRP-1)," *PLoS ONE*, vol. 8, no. 5, Article ID e64145, 2013.

[60] T. Teramoto, M. Kobayashi, H. Tasaki et al., "Efficacy and safety of alirocumab in Japanese patients with heterozygous familial hypercholesterolemia or at high cardiovascular risk with hypercholesterolemia not adequately controlled with statins – ODYSSEY JAPAN randomized controlled trial," *Circulation Journal*, vol. 80, no. 9, pp. 1980–1987, 2016.

[61] L. Liberale, F. Montecucco, G. G. Camici et al., "Treatment with proprotein convertase subtilisin/kexin type 9 (PCSK9) inhibitors to reduce cardiovascular inflammation and outcomes," *Current Medicinal Chemistry*, vol. 24, no. 14, pp. 1403–1416, 2017.

[62] Z. Ding, S. Liu, X. Wang et al., "Cross-Talk between LOX-1 and PCSK9 in vascular tissues," *Cardiovascular Research*, vol. 107, no. 4, pp. 556–567, 2015.

[63] M. D. Shapiro and S. Fazio, "PCSK9 and atherosclerosis - lipids and beyond," *Journal of Atherosclerosis and Thrombosis*, vol. 24, no. 5, pp. 462–472, 2017.

[64] A. Sahebkar, P. Di Giosia, C. A. Stamerra et al., "Effect of monoclonal antibodies to PCSK9 on high-sensitivity C-reactive protein levels: A meta-analysis of 16 randomized controlled treatment arms," *British Journal of Clinical Pharmacology*, vol. 81, no. 6, pp. 1175–1190, 2016.

[65] J. M. Cheng, R. M. Oemrawsingh, H. M. Garcia-Garcia et al., "PCSK9 in relation to coronary plaque inflammation: Results of the ATHEROREMO-IVUS study," *Atherosclerosis*, vol. 248, pp. 117–122, 2016.

[66] S. J. Nicholls, R. Puri, T. Anderson et al., "Effect of evolocumab on progression of coronary disease in statin-treated patients: The GLAGOV randomized clinical trial," *Journal of the American Medical Association*, vol. 316, no. 22, pp. 2373–2384, 2016.

[67] H. C. Blair, J. Sepulveda, and D. J. Papachristou, "Nature and nurture in atherosclerosis: The roles of acylcarnitine and cell membrane-fatty acid intermediates," *Vascular Pharmacology*, vol. 78, pp. 17–23, 2016.

[68] N. Sattar, D. Preiss, J. G. Robinson et al., "Lipid-lowering efficacy of the PCSK9 inhibitor evolocumab (AMG 145) in patients with type 2 diabetes: A meta-analysis of individual patient data," *The Lancet Diabetes & Endocrinology*, vol. 4, no. 5, pp. 403–410, 2016.

[69] B. J. Arsenault, E. Pelletier-Beaumont, N. Alméras et al., "PCSK9 levels in abdominally obese men: Association with cardiometabolic risk profile and effects of a one-year lifestyle modification program," *Atherosclerosis*, vol. 236, no. 2, pp. 321–326, 2014.

[70] M. S. Sabatine, L. A. Leiter, S. D. Wiviott et al., "Cardiovascular safety and efficacy of the PCSK9 inhibitor evolocumab in patients with and without diabetes and the effect of evolocumab on glycaemia and risk of new-onset diabetes: A prespecified analysis of the FOURIER randomised controlled trial," *The Lancet Diabetes & Endocrinology*, 2017.

[71] D. J. Blom, C. S. Djedjos, M. L. Monsalvo et al., "Effects of evolocumab on Vitamin E and steroid hormone levels: Results from the 52-week, phase 3, double-blind, randomized,placebo-controlled DESCARTES study," *Circulation Research*, vol. 117, no. 8, pp. 731–741, 2015.

[72] E. Rousselet, J. Marcinkiewicz, J. Kriz et al., "PCSK9 reduces the protein levels of the LDL receptor in mouse brain during development and after ischemic stroke," *Journal of Lipid Research*, vol. 52, no. 7, pp. 1383–1391, 2011.

[73] A. R. Khan, C. Bavishi, H. Riaz et al., "Increased Risk of Adverse Neurocognitive Outcomes with Proprotein Convertase Subtilisin-Kexin Type 9 Inhibitors," *Circulation: Cardiovascular Quality and Outcomes*, vol. 10, no. 1, Article ID e003153, 2017.

[74] B. Cariou, K. Si-Tayeba, and C. Le Maya, "Role of PCSK9 beyond liver involvement," *Current Opinion in Pediatrics*, vol. 26, no. 3, pp. 155–161, 2015.

[75] K. J. Swiger and S. S. Martin, "PCSK9 inhibitors and neurocognitive adverse events: Exploring the FDA directive and a proposal for N-of-1 trials," *Drug Safety*, vol. 38, no. 6, pp. 519–526, 2015.

[76] J. G. Robinson, R. S. Rosenson, M. Farnier et al., "Safety of Very Low Low-Density Lipoprotein Cholesterol Levels With Alirocumab: Pooled Data From Randomized Trials," *Journal of the American College of Cardiology*, vol. 69, no. 5, pp. 471–482, 2017.

[77] M. J. Koren, E. M. Roth, J. M. McKenney et al., "Safety and efficacy of alirocumab 150 mg every 2 weeks, a fully human proprotein convertase subtilisin/kexin type 9 monoclonal antibody: A phase II pooled analysis," *Postgraduate Medical Journal*, vol. 127, no. 2, pp. 125–132, 2015.

[78] E. M. Roth and J. M. McKenney, "ODYSSEY MONO: Effect of alirocumab 75 mg subcutaneously every 2 weeks as monotherapy versus ezetimibe over 24 weeks," *Future Cardiology*, vol. 11, no. 1, pp. 27–37, 2015.

[79] M. Hassan and M. Yacoub, "GAUSS-2, RUTHERFORD-2, LAPLACE-2, DESCARTES, and TESLA Part B: PCSK9 inhibitors gain momentum," *Global Cardiology Science and Practice*, vol. 2014, no. 4, p. 49, 2014.

[80] D. J. Blom, T. Hala, M. Bolognese et al., "A 52-week placebo-controlled trial of evolocumab in hyperlipidemia," *The New England Journal of Medicine*, vol. 370, no. 19, pp. 1809–1819, 2014.

[81] M. J. Koren, P. Lundqvist, M. Bolognese et al., "Anti-PCSK9 monotherapy for hypercholesterolemia: The MENDEL-2 randomized, controlled phase III clinical trial of evolocumab," *Journal of the American College of Cardiology*, vol. 63, no. 23, pp. 2531–2540, 2014.

[82] C. M. Ballantyne, J. Neutel, A. Cropp et al., "Results of bococizumab, a monoclonal antibody against proprotein convertase subtilisin/kexin type 9, from a randomized, placebo-controlled, dose-ranging study in statin-treated subjects with hypercholesterolemia," *American Journal of Cardiology*, vol. 115, no. 9, pp. 1212–1221, 2015.

[83] K. G. Parhofer, "PCSK9 inhibitors in hypercholesterolemia: New hope for patients with diabetes mellitus?" *Herz*, vol. 41, no. 3, pp. 217–223, 2016.

[84] I. Gouni-Berthold and H. K. Berthold, "PCSK9 antibodies for the treatment of hypercholesterolemia," *Nutrients*, vol. 6, no. 12, pp. 5517–5533, 2014.

[85] X.-L. Zhang, Q.-Q. Zhu, L. Zhu et al., "Safety and efficacy of anti-PCSK9 antibodies: A meta-analysis of 25 randomized, controlled trials," *BMC Medicine*, vol. 13, no. 1, article no. 123, 2015.

[86] M. J. Lipinski, U. Benedetto, R. O. Escarcega et al., "The impact of proprotein convertase subtilisin-kexin type 9 serine protease inhibitors on lipid levels and outcomes in patients with primary hypercholesterolaemia: A network meta-analysis," *European Heart Journal*, vol. 37, no. 6, pp. 536–545, 2016.

[87] K. K. Ray, H. N. Ginsberg, M. H. Davidson et al., "Reductions in Atherogenic Lipids and Major Cardiovascular Events: A Pooled Analysis of 10 ODYSSEY Trials Comparing Alirocumab with Control," *Circulation*, vol. 134, no. 24, pp. 1931–1943, 2016.

17

8-Hydroxyeicosapentaenoic Acid Decreases Plasma and Hepatic Triglycerides via Activation of Peroxisome Proliferator-Activated Receptor Alpha in High-Fat Diet-Induced Obese Mice

Hidetoshi Yamada,[1] Sayaka Kikuchi,[1] Mayuka Hakozaki,[1] Kaori Motodate,[1] Nozomi Nagahora,[1] and Masamichi Hirose[2]

[1]Iwate Biotechnology Research Center, 22-174-4 Narita, Kitakami, Iwate 024-0003, Japan
[2]Department of Molecular and Cellular Pharmacology, Iwate Medical University School of
 Pharmaceutical Sciences, Shiwa, Iwate 028-3694, Japan

Correspondence should be addressed to Hidetoshi Yamada; hyamada@ibrc.or.jp

Academic Editor: Rosemary Lee Walzem

PPARs regulate the expression of genes involved in lipid homeostasis. PPARs serve as molecular sensors of fatty acids, and their activation can act against obesity and metabolic syndromes. 8-Hydroxyeicosapentaenoic acid (8-HEPE) acts as a PPAR ligand and has higher activity than EPA. However, to date, the PPAR ligand activity of 8-HEPE has only been demonstrated *in vitro*. Here, we investigated its ligand activity *in vivo* by examining the effect of 8-HEPE treatment on high fat diet-induced obesity in mice. After the 4-week treatment period, the levels of plasma and hepatic triglycerides in the 8-HEPE-fed mice were significantly lower than those in the HFD-fed mice. The expression of genes regulated by PPARα was significantly increased in 8-HEPE-fed mice compared to those that received only HFD. Additionally, the level of hepatic palmitic acid in 8-HEPE-fed mice was significantly lower than in HFD-fed mice. These results suggested that intake of 8-HEPE induced PPARα activation and increased catabolism of lipids in the liver. We found no significant differences between EPA-fed mice and HFD-fed mice. We demonstrated that 8-HEPE has a larger positive effect on metabolic syndrome than EPA and that 8-HEPE acts by inducing PPARα activation in the liver.

1. Background

An imbalance between energy intake and expenditure can result in the accumulation of excess triglycerides in adipose tissues. Long-term continuation of this imbalance can result in chronic obesity, which is associated with hyperlipidemia, fatty liver, and adipocyte hypertrophy. Hyperlipidemia is a risk factor for atherosclerosis [1] and is also a predisposing factor for cardiovascular diseases. Fatty liver is another risk factor for cardiovascular disease and is thought to have a negative influence on the regulation of insulin signaling [2]. Adipocyte hypertrophy contributes to dysfunction in the adipose tissues and is associated with insulin resistance and increased risk of developing diabetes [3]. As a chronic high level of triglycerides is a common factor for hyperlipidemia, fatty liver, and adipocyte hypertrophy, any increase in triglyceride expenditure should potentially alleviate the risk of cardiovascular diseases and diabetes.

PPARs are members of a nuclear receptor superfamily and play critical roles in the regulation of storage and catabolism of lipids [4]. They contribute to these regulation processes by activating gene expression in a ligand-dependent manner, which involves recognition of and binding to peroxisome proliferator response elements (PPREs) that are composed of TGACCT-related direct repeats separated by one nucleotide [5, 6]. PPARs form heterodimers on PPREs via the retinoid-X receptor, the receptor for 9-cis-retinoic acid [7, 8]. Three types of PPAR have been identified, namely, α, γ, and δ. PPARα is expressed at high levels in the liver where it promotes fatty acid oxidation, ketogenesis, lipid transport, and gluconeogenesis [9, 10]. PPARα responds to the concentration of fatty acids in the liver and enhances fatty acid breakdown by upregulating genes encoding β-oxidation enzymes [11–13]. Phenoxyalkyl-carboxylic acid derivatives (fibrates) have been used to treat hypertriglyceridemia through activation of PPARα [14–16]. Activation of PPAR by fibrates results in a substantially reduced level of serum triglycerides [16]. However, these drugs have the adverse side effects of hepatic toxicity, myopathy, and cholelithiasis. Thus, other PPAR activators are being investigated to determine whether they show fewer adverse effects than fibrates.

The identification of unsaturated fatty acids as PPAR ligands provided firm evidence that the direct interaction of nuclear receptors with these fatty acids is required for some PPAR-dependent transcription activity [7, 11, 17–20]. Unsaturated fatty acids can bind to all three types of PPAR, with PPARα exhibiting the highest affinity for concentrations equivalent to circulating blood levels [11, 21]. In contrast, the long-chain fatty acid erucic acid (C22:1) is a weak ligand that appears to have more affinity for PPARδ [22]. Overall, saturated fatty acids are poor PPAR ligands compared to unsaturated fatty acids [7, 11, 19]. Hydroxyeicosapentaenoic acids (HEPEs) are unsaturated fatty acids and are the oxylipin products of the lipoxygenase pathway. In a previous study, we showed that dried pacific krill is a source of HEPEs (5-HEPE, 8-HEPE, 9-HEPE, 12-HEPE, and 18-HEPE) and that 8-HEPE has high ligand activity for PPARs [23]. 8-HEPE increases the levels of expression of genes regulated by PPARs in Fao (rat hepatoma cell line), 3T3-F442A (mouse preadipocyte cell line), and C2C12 (mouse myoblast cell line) cells. Furthermore, 8-HEPE enhances adipogenesis and glucose uptake. By contrast, at the same concentrations, eicosapentaenoic acid (EPA) shows only a weak effect, indicating that 8-HEPE is a more potent inducer of physiological effects.

8-HEPE has a greater affinity for PPAR activation than EPA *in vitro*; it is possible that 8-HEPE might be of value in the treatment of obesity and metabolic syndrome. Pacific krill contain about 20 mg of 8-HEPE per 100 g [23] and therefore could potentially be used as a food supplement. However, no animal experiments on the *in vivo* effectiveness of 8-HEPE have been reported. Here, we treated high-fat diet-induced obese mice with 8-HEPE to investigate its antiobesity effects.

2. Material and Methods

2.1. 8-HEPE and EPA Purification from Pacific Krill. 8-HEPE and EPA were purified as described previously [23]. Dried krill (Kawashu, Iwate, Japan) were powdered and then extracted with methanol. The extract was subjected to column chromatography using Diaion HP-20 (Mitsubishi Chemical, Tokyo, Japan). The 8-HEPE- and EPA-containing fraction was eluted with methanol from the HP-20. 8-HEPE and EPA were separated on an InertSustain ODS-3 column (20.0 mm dia. × 250 mm; GL Science Inc.).

2.2. Animal Experiments. Four-week-old, male C57BL/6J mice (Charles River Laboratories, Tokyo, Japan) were housed singly in a temperature-controlled environment (23 ± 1°C) with a 12 h light/dark cycle. The animal experiments reported here were approved by the institutional animal care and use committee at Iwate Biotechnology Research Center (IBRC-ARC-2014-01). The low-fat diet AIN-93G (protein 13.9% calorie, fat 9.7% calorie, and carbohydrate 77.0% calorie) (total 377 kcal/100 g diet) and the high-fat diet HFD-60 (protein 18.2% calorie, fat 62.2% calorie, and carbohydrate 19.6% calorie) (total 506 kcal/100 g diet) were purchased from Charles River Laboratories. The animals were maintained on HFD for seven weeks. Then, they were randomly assigned to six groups, with 9 mice in each group: HFD; HFD with EPA (10 mg/kg); HFD with 8-HEPE (10 mg/kg); low-fat diet (LFD); LFD with EPA (10 mg/kg); and LFD with 8-HEPE (10 mg/kg). EPA and 8-HEPE were added to diet without exchange of other dietary components. The animals were maintained on these diets for 4 weeks prior to all analyses. Food intake was measured twice a week, and body weight was measured weekly. Blood samples were collected using heparin sodium (Wako) coated tubes, and plasma was separated from the blood by centrifugation at 14000 g for 10 min.

2.3. Measurement of Triglycerides, Glucose, Cholesterol, Transaminase, Leptin, and Adiponectin. Hepatic lipids were extracted from mouse livers as described by Folch et al. [32]. In brief, the liver (~50 mg) was homogenized in 700 μL of 0.1 M acetic acid solution-methanol (2 : 5, v/v). Chloroform (500 μL) was added to the mixture and, after 10 min, 250 μL of 0.1 M acetic acid was added. The mixture was centrifuged at 2400 g for 10 min at room temperature. The chloroform layer was allowed to evaporate and the pellet was suspended in 200 μL of isopropanol. Triglycerides in blood plasma and liver extracts were measured using the Triglyceride E-Test (Wako). Total cholesterol, glucose, and transaminase in plasma were measured using the Total Cholesterol E-Test (Wako), Glucose C2-Test (Wako), and Transaminase C2-Test (Wako), respectively. Plasma leptin was measured using a mouse leptin measurement kit (Morinaga Institute of Biological Science, Inc., Kanagawa, Japan). Plasma adiponectin was measured using an Adiponectin ELISA Kit (Otsuka Pharmaceutical Co., Ltd., Tokushima, Japan). All kits were used according to the manufacturers' recommendations.

2.4. Measurement of Palmitic Acid and Stearic Acid in Liver. A 100 mg piece of the liver was homogenized in 1 mL of cold methanol/water (4 : 1, v/v). After sonication, the sample was placed on ice for 20 minutes and then deproteinized by centrifugation at 21,000 g for 10 min. The supernatant (800 μL) was freeze-dried and dissolved in 100 μL of methanol/water (4 : 1, v/v) before analysis. Palmitic acid and stearic acid were separated on an InertSustain ODS-3 column (2.0 mm dia. × 250 mm; GL Science Inc.) with gradient elution (acetonitrile/water/formic acid, 30/70/0.1 to 90/10/0.1 in 30 min) at a flow rate of 0.2 mL/min. The compounds were identified and quantified by LC-TOFMS using Agilent Mass Hunter

TABLE 1: Body weights, liver weights, the relative amount of gonadal WAT per g body weight in male C57BL/6J-DIO mice fed HFD, HFD with EPA, HFD with 8-HEPE, LFD, LFD with EPA, or LFD with 8-HEPE for 4 wk. Values are means ± SDs, $n = 9$. Labeled means without a common letter differ; $P < 0.05$. HFD: high-fat diet; LFD: low-fat diet; HEPE: hydroxyeicosapentaenoic acid; EPA: eicosapentaenoic acid.

	Body weight (g)	Liver weight (g)	Gonadal WAT per g body weight (%)
HFD	35.96 ± 2.19[a]	1.28 ± 0.13[a]	5.94 ± 0.61[a]
HFD with EPA	35.92 ± 2.95[a]	1.26 ± 0.13[a]	5.85 ± 1.13[ab]
HFD with 8-HEPE	35.09 ± 1.27[a]	1.29 ± 0.10[a]	5.74 ± 0.39[b]
LFD	26.02 ± 1.17[b]	0.96 ± 0.22[b]	2.16 ± 0.40[c]
LFD with EPA	25.42 ± 3.64[b]	1.02 ± 0.14[b]	2.71 ± 0.66[c]
LFD with 8-HEPE	26.10 ± 1.46[b]	1.00 ± 0.12[b]	2.36 ± 0.48[c]

Workstation Software. Analytes were quantified with internal standard methods and 5-point calibration curves. Assay variability was assessed by analyzing sample replicate.

2.5. Quantitative Real-Time PCR.

Total RNAs were extracted with an RNeasy Lipid Tissue Kit (QIAGEN, Tokyo, Japan) and used to synthesize cDNAs using a PrimeScript RT Reagent Kit (Takara, Shiga, Japan); all kits were used according to the manufacturers' recommendations. Quantitative real-time PCR was performed with the gene specific primers listed in Supplemental Table 1 in Supplementary Material available online at http://dx.doi.org/10.1155/2016/7498508 and Fast SYBR Green Master Mix (Applied Biosystems, Foster City, CA, USA).

2.6. Histology.

Gonadal fat depots were fixed in 10% formalin (Wako) for 3 hours and embedded in paraffin (Wako). Paraffin embedded samples were sectioned at 6 μm thickness with a microtome (Leica, Tokyo, Japan). Sections were subjected to standard hematoxylin and eosin staining. Adipocyte area was measured with ImageJ software (http://rsbweb.nih.gov/ij/).

2.7. Statistical Analysis.

Statistically significant differences between the experimental groups were identified using one-way ANOVA and Tukey's post hoc tests. Data are shown as means ± SD.

3. Results

3.1. Food Intake and Weight Gain.

We did not find any significant differences among the groups of mice for food intake per day or body weight (see Figures S1 and S2 and Table 1). Based on food intake, the mice ingested 23.98 ± 1.99 μg of EPA or 23.67 ± 1.69 μg of 8-HEPE per day, respectively.

3.2. Plasma Triglyceride Levels Are Suppressed in Mice Fed 8-HEPE.

Plasma triglyceride levels in mice fed HFD with 8-HEPE were significantly lower than in the HFD group (Table 2). There was no significant difference in plasma triglyceride levels between mice fed HFD with EPA and mice fed HFD or among the mice fed LFD, LFD with EPA, and LFD with 8-HEPE (Table 2). 8-HEPE consumption did not affect plasma glucose or total cholesterol levels.

3.3. 8-HEPE Reduces Triglyceride Accumulation in Liver.

Next, we examined the effect of 8-HEPE on the liver. Triglyceride levels in mice fed HFD with 8-HEPE were significantly lower than the HFD group (Table 2). As fatty liver is the most common hepatic disorder in Western countries [24], we examined liver function in the three groups of mice by measuring aspartate aminotransferase (AST) and alanine aminotransferase (ALT) activities in the plasma. The levels of activity of these enzymes in the plasma are used as a marker to indicate hepatic disorders. ALT activity in mice fed 8-HEPE was lower than the HFD group (Table 2).

In our previous study, we showed that 8-HEPE acts as a PPARα ligand in the rat hepatoma cell line FaO [23]. Moreover, Wy14,643, a PPARα agonist, induces expression of Cpt1a, Cpt2, Ehhadh, Fabp1, and Cyp4a in the liver [25, 26]. Here, we used real-time PCR to determine whether 8-HEPE consumption affects gene expression levels in the mouse liver. We found that the level of Cpt1a, Ehhadh, and Cyp4a10 expression in mice fed HFD with 8-HEPE was higher than in mice fed HFD (Figure 1(a)). However, there was no significant difference in hepatic gene expression levels between mice fed LFD and mice fed LFD with 8-HEPE (Figure S3). We also investigated expression of sterol regulatory element binding transcription factor 1 (Srebf1) and fatty acids synthase (Fasn) as it has been reported that EPA can suppress hepatic lipogenesis and steatosis by reducing transcription of Srebf1 [27]. However, we could not detect any change in Srebf1 or Fasn expression in the mice fed 8-HEPE (Figure S4). The mice fed HFD or HFD with EPA showed no significant differences for liver weight, liver triglyceride levels, plasma AST or ALT activities, and gene expression levels (Table 2 and Figure 1). In the mice fed HFD with 8-HEPE, the palmitic acid level was significantly lower than in the HFD group (Figure 1(b)). These results indicate that 8-HEPE activates PPARα and increases fatty acid oxidation in liver.

3.4. 8-HEPE Reduces Adipocyte Hypertrophy.

Obesity is associated with adipocyte hypertrophy and functional disorder of the adipose tissue [28]. To determine whether 8-HEPE ameliorated these effects, we compared gonadal white adipose tissue (gonadal WAT) in the three groups of mice. We found that relative gonadal WAT was slightly reduced in the mice fed HFD with 8-HEPE compared to the HFD group (Table 1). Moreover, the gonadal adipocyte cells were smaller in the

TABLE 2: Plasma triglycerides, glucose, total cholesterol, AST, ALT, and hepatic triglyceride in male C57BL/6J-DIO mice fed HFD, HFD with EPA, HFD with 8-HEPE, LFD, LFD with EPA, or LFD with 8-HEPE for 4 wk. Values are means ± SDs, $n = 9$. Labeled means without a common letter differ; $P < 0.05$. HFD: high-fat diet; LFD: low-fat diet; HEPE: hydroxyeicosapentaenoic acid; EPA: eicosapentaenoic acid.

	Triglyceride (mg/dL)	Glucose (mg/dL)	Plasma Cholesterol (mg/dL)	AST (Karmen)	ALT (Karmen)	Hepatic Triglyceride (mg/g)
HFD	91.36 ± 22.14[a]	335.23 ± 59.78	172.04 ± 16.35[a]	30.85 ± 3.59	10.88 ± 2.15[a]	50.18 ± 10.87[a]
HFD with EPA	82.10 ± 23.53[ab]	327.58 ± 37.75	168.27 ± 21.07[a]	32.45 ± 6.87	10.86 ± 3.53[ab]	44.31 ± 13.79[ab]
HFD with 8-HEPE	78.12 ± 18.44[b]	327.05 ± 47.14	166.13 ± 14.38[a]	30.35 ± 3.41	9.30 ± 1.39[b]	40.53 ± 7.65[b]
LFD	74.76 ± 25.10[ab]	308.01 ± 28.80	103.07 ± 9.12[b]	31.76 ± 4.24	5.93 ± 1.13[c]	29.27 ± 5.32[c]
LFD with EPA	70.99 ± 44.21[ab]	279.73 ± 25.72	109.13 ± 17.21[b]	32.33 ± 6.19	6.73 ± 1.39[c]	28.47 ± 4.92[c]
LFD with 8-HEPE	68.14 ± 23.10[b]	319.66 ± 52.42	107.02 ± 11.98[b]	35.71 ± 7.15	7.72 ± 1.43[c]	30.85 ± 8.52[c]

(a)

(b)

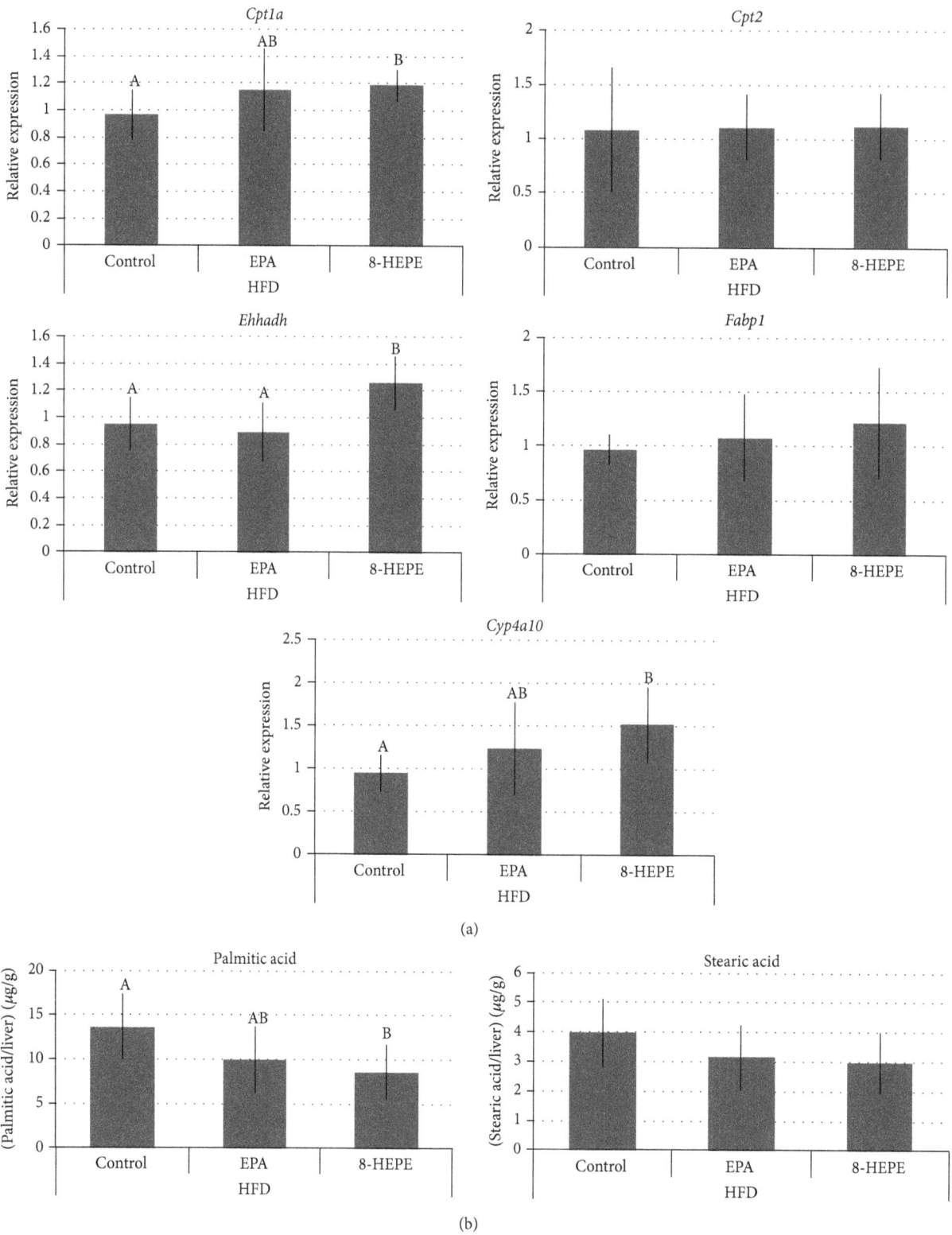

FIGURE 1: Changes in hepatic gene expression (a) and in concentrations of long-chain saturated fatty acids (b) in male C57BL/6J-DIO mice fed HFD, HFD with EPA, or HFD with 8-HEPE for 4 wk. Gene expression levels were measured by real-time PCR and normalized against expression of *Actb*. Values are means ± SDs, $n = 9$. Labeled means without a common letter differ; $P < 0.05$. HFD: high-fat diet; HEPE: hydroxyeicosapentaenoic acid; EPA: eicosapentaenoic acid.

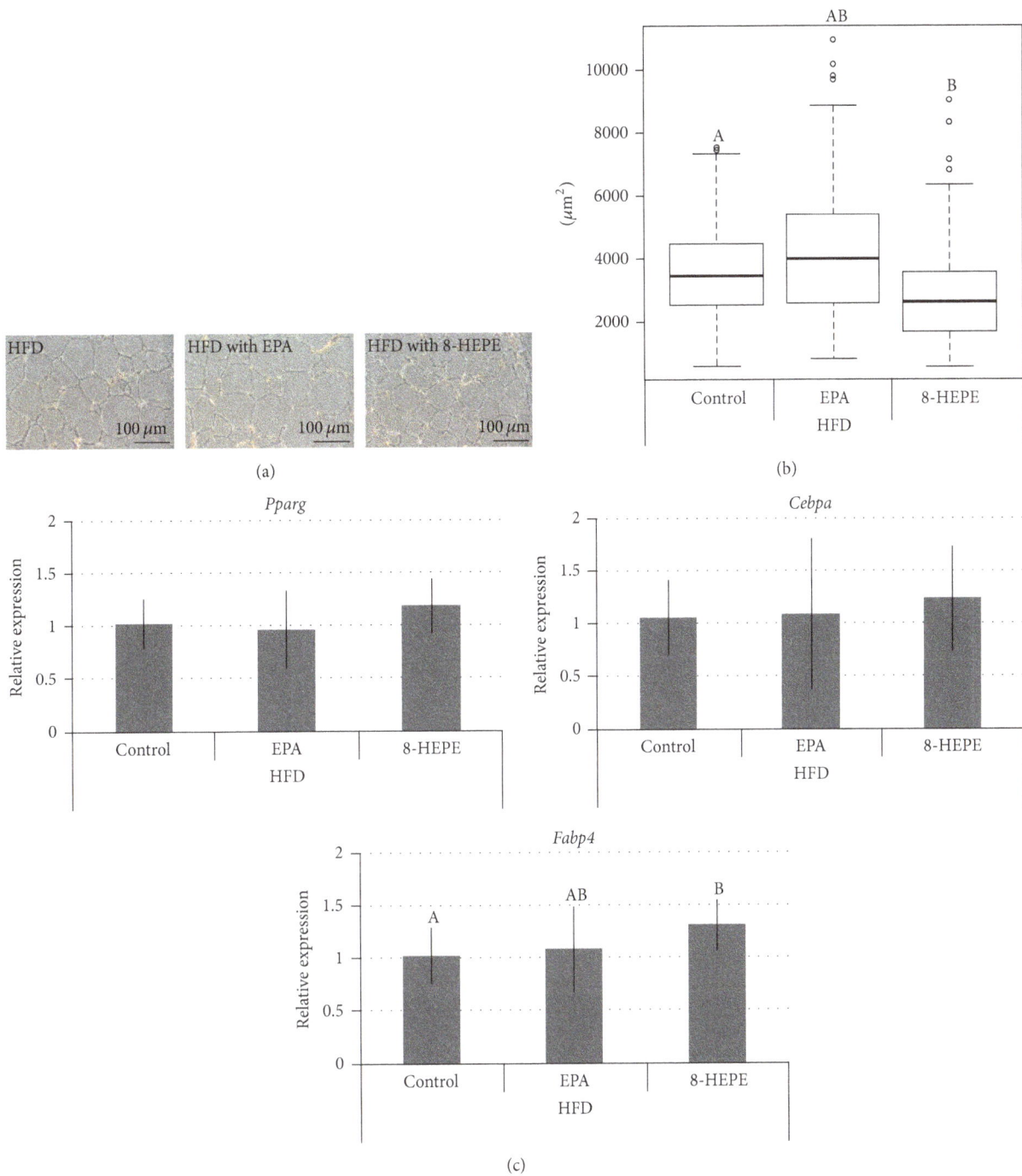

FIGURE 2: Hematoxylin and eosin staining of gonadal WAT (a), quantification of adipocyte cross sections (b), and expression of *Pparg*, *Cebpa*, and *Fabp4* in gonadal WAT (c) in male C57BL/6J-DIO mice fed HFD, HFD with EPA, or HFD with 8-HEPE for 4 wk. Data were collected from H&E-stained sections from five mice; two fields of view were analyzed per mouse, and 10–15 cells per field were analyzed using ImageJ software. Gene expression levels were measured by real-time PCR and normalized against expression of *Actb*. Values are means ± SDs, $n = 9$. Labeled means without a common letter differ; $P < 0.05$. HFD: high-fat diet; HEPE: hydroxyeicosapentaenoic acid; EPA: eicosapentaenoic acid.

mice fed 8-HEPE compared to those fed HFD (Figures 2(a) and 2(b)). As adipose tissue regulates energy homeostasis and insulin sensitivity via the secretion of leptin and adiponectin [29], we measured the concentrations of adiponectin and leptin in the plasmas of the mice. No significant differences were present between mice fed HFD and HFD with 8-HEPE (Table S4). Our previous study showed that 8-HEPE can activate PPARγ in mice preadipocyte cells [23]. In line with this finding, mice fed 8-HEPE showed increased expression of *Fabp4* in gonadal WAT (Figure 2(c)); however, they did not show any difference in expression of *Pparg* and *Cebpa*. The activation of PPARδ induces *Angptl4* expression in muscle [23] and *Lpin2* and *St3gal5* in the liver [30]. To estimate PPARδ activation here, we measured *Angptl4*, *Lpin2*, and *St3gal5* expression; however, *Angptl4*, *Lpin2*, and *St3gal5* did not show any increase in expression level in mice fed 8-HEPE (Figures S4 and S5).

4. Discussion

Our analyses here demonstrate that 8-HEPE intake reduces the levels of triglycerides in the blood and the liver and also lessens adipocyte hypertrophy in mice. Mice fed HFD with 8-HEPE consumed $23.67 \pm 1.69\,\mu g$/day of 8-HEPE and had an average body weight of 32.3 g. Thus, the relative 8-HEPE consumption was approximately 0.73 mg/kg/day. Compared to the control HFD diet, 8-HEPE was associated with a decrease in plasma triglyceride levels (Cohen's $d = 0.64$), in liver triglycerides (Cohen's $d = 1.01$), and in gonadal WAT (Cohen's $d = 0.49$) (Tables 1 and 2). However, supplementation of the diet with 8-HEPE did not alter daily food intake or plasma glucose and cholesterol levels (Tables 2, S1, and S2). Compared to the LFD diet, 8-HEPE did not affect plasma triglycerides, hepatic triglycerides, or hepatic gene expression. These results indicate that the major effect of 8-HEPE was to reduce triglyceride levels in mice fed a high-fat diet intake. The consequent reduction in plasma ALT activity in the 8-HEPE mice (Table 2) suggests that this supplement might aid in the prevention of fatty liver induced hepatic disorders in the mice. Previous studies have shown that EPA administration is effective in decreasing plasma triglyceride levels in experimental animals and humans [31]. However, in the present study, there were no significant differences in plasma triglyceride levels between mice fed HFD and HFD with EPA (Table 2). The contrasting outcomes of EPA on plasma triglycerides may be due to differences in the dose of EPA.

Both Forman et al. and our group have reported that 8-HEPE can act as a PPAR ligand [11, 23]. However, this is the first study to report on the *in vivo* effects of 8-HEPE on PPARs. As detailed above, the expression of genes regulated by PPARα, such as *Cpt1a*, *Ehhadh*, and *Cyp4a10*, was increased in the livers of mice fed 8-HEPE (Figure 1(a)). These results indicate that 8-HEPE taken orally can activate PPARα and cause a decrease in triglyceride levels in the liver. Moreover, 8-HEPE was detected in the plasma of mice fed 8-HEPE (Table S3). We also found that *Fabp4* expression was increased in the gonadal WAT of mice fed 8-HEPE (Figure 2(c)). However, plasma concentrations of adiponectin and leptin were not altered in these mice (Table S4). In view of our results, the daily intake of 8-HEPE used here appeared to be too low to activate PPARγ. PPARδ activation was not observed either (Figures S4 and S5). Overall, our results indicate that a 0.73 mg/kg/day intake of 8-HEPE in mice is sufficient to activate PPARα in the liver, but that activation of PPARγ and PPARδ may require higher levels of 8-HEPE intake.

5. Conclusions

This study is the first demonstration that supplementation of the diet with 8-HEPE can have a beneficial effect on various characteristics of obesity through the activation of PPARα. When we compared mice fed 8-HEPE to those fed unmodified HFD, statistically significant changes were observed in plasma and liver triglyceride levels and the amount of gonadal WAT. Statistically significant effects were not observed, if the mice were fed a diet supplemented with EPA. Thus, 8-HEPE induced a higher level of lipid catabolism in liver than EPA. Notably, 8-HEPE can easily be obtained from krill for use as a food supplement. Our results here indicate that use of 8-HEPE as a supplement may offer protective effects against hyperlipidemia, fatty liver, and adipose tissue dysfunction in patients with chronic obesity.

Abbreviations

HEPE: Hydroxyeicosapentaenoic acid
PPARs: Peroxisome proliferator-activated receptors
EPA: Eicosapentaenoic acid
Cyp4a10:   Cytochrome P450 CYP4A enzymes
Fabp: Fatty acid-binding protein
Cpt: Carnitine palmitoyltransferase
Ehhadh: Enoyl-CoA hydratase/3-hydroxyacyl CoA.

Competing Interests

Hidetoshi Yamada, Sayaka Kikuchi, Mayuka Hakozaki, Kaori Motodate, Nozomi Nagahora, and Masamichi Hirose declare that there is no conflict of interests.

Authors' Contributions

Hidetoshi Yamada and Mayuka Hakozaki designed the research; Hidetoshi Yamada, Sayaka Kikuchi, Mayuka Hakozaki, Kaori Motodate, Nozomi Nagahora, and Masamichi Hirose conducted the research; Hidetoshi Yamada, Sayaka Kikuchi, and Mayuka Hakozaki analyzed the data; Hidetoshi Yamada and Nozomi Nagahora wrote the paper; and Hidetoshi Yamada had primary responsibility for the final content. All authors read and approved the final paper.

Acknowledgments

This work was supported by funds from the Basic Biotechnology Project of Iwate Prefecture, Japan, and grants-in-aid from the Sanriku Foundation and the Center for Revitalization Promotion, Japan Science and Technology Agency.

References

[1] D. H. Blankenhorn and H. N. Hodis, "Atherosclerosis—reversal with therapy," *Western Journal of Medicine*, vol. 159, no. 2, pp. 172–179, 1993.

[2] G. Targher, C. P. Day, and E. Bonora, "Risk of cardiovascular disease in patients with nonalcoholic fatty liver disease," *The*

New England Journal of Medicine, vol. 363, no. 14, pp. 1341–1350, 2010.

[3] C. Weyer, J. E. Foley, C. Bogardus, P. A. Tataranni, and R. E. Pratley, "Enlarged subcutaneous abdominal adipocyte size, but not obesity itself, predicts type II diabetes independent of insulin resistance," *Diabetologia*, vol. 43, no. 12, pp. 1498–1506, 2000.

[4] K. Schoonjans, B. Staels, and J. Auwerx, "The peroxisome proliferator activated receptors (PPARs) and their effects on lipid metabolism and adipocyte differentiation," *Biochimica et Biophysica Acta—Lipids and Lipid Metabolism*, vol. 1302, no. 2, pp. 93–109, 1996.

[5] J. D. Tugwood, I. Issemann, R. G. Anderson, K. R. Bundell, W. L. McPheat, and S. Green, "The mouse peroxisome proliferator activated receptor recognizes a response element in the 5' flanking sequence of the rat acyl CoA oxidase gene," *The EMBO Journal*, vol. 11, no. 2, pp. 433–439, 1992.

[6] B. Zhang, S. L. Marcus, F. G. Sajjadi et al., "Identification of a peroxisome proliferator-responsive element upstream of the gene encoding rat peroxisomal enoyl-CoA hydratase/3-hydroxyacyl-CoA dehydrogenase," *Proceedings of the National Academy of Sciences of the United States of America*, vol. 89, no. 16, pp. 7541–7545, 1992.

[7] S. A. Kliewer, S. S. Sundseth, S. A. Jones et al., "Fatty acids and eicosanoids regulate gene expression through direct interactions with peroxisome proliferator-activated receptors α and γ," *Proceedings of the National Academy of Sciences*, vol. 94, no. 9, pp. 4318–4323, 1997.

[8] S. L. Marcus, K. S. Miyata, B. Zhang, S. Subramani, R. A. Rachubinski, and J. P. Capone, "Diverse peroxisome proliferator-activated receptors bind to the peroxisome proliferator-responsive elements of the rat hydratase/dehydrogenase and fatty acyl-CoA oxidase genes but differentially induce expression," *Proceedings of the National Academy of Sciences of the United States of America*, vol. 90, no. 12, pp. 5723–5727, 1993.

[9] C. Bernal-Mizrachi, S. Weng, C. Feng et al., "Dexamethasone induction of hypertension and diabetes is PPAR-alpha dependent in LDL receptor-null mice," *Nature Medicine*, vol. 9, no. 8, pp. 1069–1075, 2003.

[10] J. K. Reddy and T. Hashimoto, "Peroxisomal β-oxidation and peroxisome proliferator-activated receptor α: an adaptive metabolic system," *Annual Review of Nutrition*, vol. 21, pp. 193–230, 2001.

[11] B. M. Forman, J. Chen, and R. M. Evans, "Hypolipidemic drugs, polyunsaturated fatty acids, and eicosanoids are ligands for peroxisome proliferator-activated receptors alpha and delta," *Proceedings of the National Academy of Sciences of the United States of America*, vol. 94, no. 9, pp. 4312–4317, 1997.

[12] B. Desvergne and W. Wahli, "Peroxisome proliferator-activated receptors: nuclear control of metabolism," *Endocrine Reviews*, vol. 20, no. 5, pp. 649–688, 1999.

[13] S. S. Lee, T. Pineau, J. Drago et al., "Targeted disruption of the alpha isoform of the peroxisome proliferator-activated receptor gene in mice results in abolishment of the pleiotropic effects of peroxisome proliferators," *Molecular and Cellular Biology*, vol. 15, no. 6, pp. 3012–3022, 1995.

[14] A. Tenenbaum and E. Z. Fisman, "Fibrates are an essential part of modern anti-dyslipidemic arsenal: spotlight on atherogenic dyslipidemia and residual risk reduction," *Cardiovascular Diabetology*, vol. 11, article 125, 2012.

[15] L. Berglund, J. D. Brunzell, A. C. Goldberg et al., "Evaluation and treatment of hypertriglyceridemia: an Endocrine Society clinical practice guideline," *The Journal of Clinical Endocrinology & Metabolism*, vol. 97, no. 9, pp. 2969–2989, 2012.

[16] M. Miller, N. J. Stone, C. Ballantyne et al., "Triglycerides and cardiovascular disease: a scientific statement from the American Heart Association," *Circulation*, vol. 123, no. 20, pp. 2292–2333, 2011.

[17] B. M. Forman, P. Tontonoz, J. Chen, R. P. Brun, B. M. Spiegelman, and R. M. Evans, "15-Deoxy-delta 12, 14-prostaglandin J2 is a ligand for the adipocyte determination factor PPAR gamma," *Cell*, vol. 83, no. 5, pp. 803–812, 1995.

[18] P. Dowell, V. J. Peterson, T. Mark Zabriskie, and M. Leid, "Ligand-induced peroxisome proliferator-activated receptor α conformational change," *The Journal of Biological Chemistry*, vol. 272, no. 3, pp. 2013–2020, 1997.

[19] G. Krey, O. Braissant, F. L'Horset et al., "Fatty acids, eicosanoids, and hypolipidemic agents identified as ligands of peroxisome proliferator-activated receptors by coactivator-dependent receptor ligand assay," *Molecular Endocrinology*, vol. 11, no. 6, pp. 779–791, 1997.

[20] M. Göttlicher, E. Widmark, Q. Li, and J. A. Gustafsson, "Fatty acids activate a chimera of the clofibric acid-activated receptor and the glucocorticoid receptor," *Proceedings of the National Academy of Sciences of the United States of America*, vol. 89, no. 10, pp. 4653–4657, 1992.

[21] K. Yu, W. Bayona, C. B. Kallen et al., "Differential activation of peroxisome proliferator-activated receptors by eicosanoids," *The Journal of Biological Chemistry*, vol. 270, no. 41, pp. 23975–23983, 1995.

[22] T. E. Johnson, M. Holloway, R. Vogel et al., "Structural requirements and cell-type specificity for ligand activation of peroxisome proliferator-activated receptors," *The Journal of Steroid Biochemistry and Molecular Biology*, vol. 63, no. 1–3, pp. 1–8, 1997.

[23] H. Yamada, E. Oshiro, S. Kikuchi, M. Hakozaki, H. Takahashi, and K.-I. Kimura, "Hydroxyeicosapentaenoic acids from the Pacific krill show high ligand activities for PPARs," *Journal of Lipid Research*, vol. 55, no. 5, pp. 895–904, 2014.

[24] B. Palmentieri, I. de Sio, V. La Mura et al., "The role of bright liver echo pattern on ultrasound B-mode examination in the diagnosis of liver steatosis," *Digestive and Liver Disease*, vol. 38, no. 7, pp. 485–489, 2006.

[25] J. P. Vanden Heuvel, D. Kreder, B. Belda et al., "Comprehensive analysis of gene expression in rat and human hepatoma cells exposed to the peroxisome proliferator WY14,643," *Toxicology and Applied Pharmacology*, vol. 188, no. 3, pp. 185–198, 2003.

[26] E. Ip, G. C. Farrell, G. Robertson, P. Hall, R. Kirsch, and I. Leclercq, "Central role of PPARalpha-dependent hepatic lipid turnover in dietary steatohepatitis in mice," *Hepatology*, vol. 38, pp. 123–132, 2003.

[27] A. Sato, H. Kawano, T. Notsu et al., "Antiobesity effect of eicosapentaenoic acid in high-fat/high-sucrose diet-induced obesity: importance of hepatic lipogenesis," *Diabetes*, vol. 59, no. 10, pp. 2495–2504, 2010.

[28] Y.-H. Yu and H. N. Ginsberg, "Adipocyte signaling and lipid homeostasis: sequelae of insulin-resistant adipose tissue," *Circulation Research*, vol. 96, no. 10, pp. 1042–1052, 2005.

[29] S. Klaus, "Adipose tissue as a regulator of energy balance," *Current Drug Targets*, vol. 5, no. 3, pp. 241–250, 2004.

18

Safety and Efficacy of Extremely Low LDL-Cholesterol Levels and its Prospects in Hyperlipidemia Management

Dhrubajyoti Bandyopadhyay [ORCID],[1] **Arshna Qureshi,**[2] **Sudeshna Ghosh,**[3] **Kumar Ashish,**[4] **Lyndsey R. Heise,**[5] **Adrija Hajra,**[6] and **Raktim K. Ghosh**[7]

[1]*Department of Internal Medicine, Mount Sinai St Luke's Roosevelt, New York, NY, USA*
[2]*Department of Medicine, Lady Hardinge Medical College, New Delhi, India*
[3]*IPGMER, Kolkata, India*
[4]*The University of Texas MD Anderson Cancer Center, Houston, TX, USA*
[5]*Department of Internal Medicine, University of Nebraska Medical Center, Omaha, NE, USA*
[6]*Department of Internal Medicine, IPGMER, Kolkata, India*
[7]*Division of Cardiovascular Diseases, Metrohealth Medical Center, Case Western Reserve University, Cleveland, OH, USA*

Correspondence should be addressed to Dhrubajyoti Bandyopadhyay; drdhrubajyoti87@gmail.com

Academic Editor: Gerhard M. Kostner

The risk of cardiovascular disease has been reported to have a linear relationship with LDL levels. Additionally, the currently recommended LDL target goal of 70 mg/dl does not diminish the CV risk entirely leaving behind some residual risk. Previous attempts to maximally lower the LDL levels with statin monotherapy have met dejection due to the increased side effects associated with the treatment. Nevertheless, with the new advancements in clinical medicine, it has now become possible to bring down the LDL levels to as low as 15 mg/dl using PCSK9 monoclonal antibodies alone or in combination with statins. The development of inclisiran, siRNA silencer targeting PCSK9 gene, is a one step forward in these endeavors. Moreover, various studies aiming to lower the CV risk and mortality by lowering LDL levels have demonstrated encouraging results. The current challenge is to explore this arena to redefine the target LDL levels, if required, to avoid any suboptimal treatment. After thorough literature search in the PubMed, Embase, Scopus, and Google Scholar, we present this article to provide a brief overview of the safety and efficacy of lowering LDL below the current goal.

1. Introduction

Hyperlipidemia has always been a topic of interest owing to the concomitant increased risk of adverse cardiovascular events. Coronary artery disease, a leading cause of death in the United States with almost 400,000 deaths/year, is found to be strongly associated with hyperlipidemia [1]. Moreover, increased LDL levels are found to be positively correlated with the increased CV risk. Thus, the treatment of hyperlipidemia plays a crucial role in the management of patients with CAD or those at increased risk of CAD all around the world. About 73.5 million adults in the USA have elevated LDL-cholesterol [2]. The American College of Cardiology/American Heart Association (NCEP IV) guidelines recommend prescription of evidence-based doses of statins independent of the LDL level [3].

Interestingly, most physicians prefer treating to an LDL goal and consider 70 mg/dl to be an appropriate target goal for people at the highest risk for cardiovascular disease [4]. However, despite achieving the target level of 70 mg/dl with high-intensity statin therapy, there is residual CV risk. Furthermore, targeting HDL and TG levels to reduce this residual risk has been proved futile [5]. Meanwhile, the recent availability of PCSK9 inhibitors has revalidated the discussion on further lowering of LDL and has brought back the age-old question: how low is in fact low enough to bring the CV risk to the minimum?

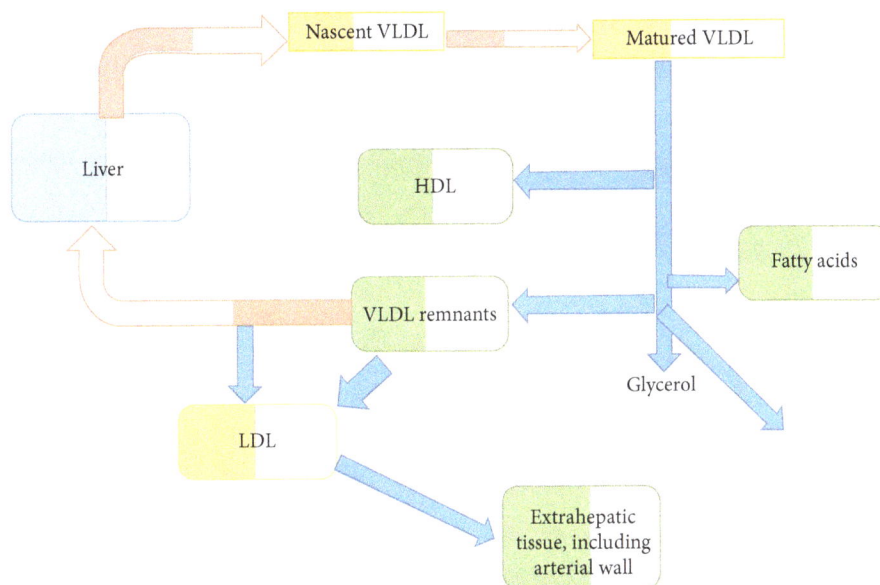

FIGURE 1: LDL metabolism.

2. LDL Metabolism and Pathophysiology of Atherosclerosis

The level of LDL is the single most important marker of atherosclerosis (Figure 1). Deranged LDL metabolism leads to coronary artery disease that is often fatal, especially in patients with diabetes. It has been found that not only elevated levels of LDL lead to coronary heart disease, but changes in composition can also result in the same. As we all know, cholesterol is an integral part of the plasma membrane, and a minimum level of LDL needs to be present to maintain structural integrity and sustain normal function of cells.

The development of atherosclerosis is indeed a complicated process where LDL plays a pivotal role. LDL causes endothelial damage which helps in the progression and formation of fatty streaks. Atherosclerosis, the most important factor behind the coronary vascular disease, affecting mostly medium- and large-sized arteries is characterized by the presence of modified smooth muscles, foam cells, endothelial cells, WBCs, and lipid in the center. With the growing comprehension of inflammatory process and mediators, studies have revealed that lipid-related inflammation could be cornerstone mediator for atherosclerosis [6] (Figure 2). The most likely site for plaque formation is the regions that experience low endothelial stress rather than area experiencing high stress. The plaques continue to grow into the lumen, and they experience increasingly high stress as the lumen diameter becomes narrower which ultimately contributes to the destabilization of the plaque [7]. Atherosclerosis can be prevented by implementing lifestyle modifications, controlling the risk factors of which controlling high LDL is of paramount importance.

3. Commonly Used LDL-Lowering Drugs

Statins. Statins inhibit the HMG-CoA reductase enzyme, the rate-controlling enzyme in the biosynthesis of cholesterol [8].

Statins lower LDL-C and triglycerides while slightly raising HDL. It is the standard of care for dyslipidemia management. Liver damage, muscle pain, and increased risk of type 2 diabetes mellitus are some of the side effects of statins, but benefits outweigh the risks [9]. They also prevent SMC migration and proliferation and impede the activation of TNF-alpha, IL-1 beta, and other interleukins which play an active role in inflammation [8].

Ezetimibe. This medication prevents the absorption of bile acid in the small intestine, lowers LDL, increases HDL slightly, and, to a little extent, lowers triglycerides. However, ezetimibe can cause myalgia and abdominal pain [10].

PCSK-9 Inhibitors. Proprotein Convertase Subtilisin/Kexin Type 9 (PCSK9) causes degradation of LDL receptors in the liver. Alirocumab and evolocumab are the two monoclonal antibodies directed against PCSK-9, and thus it prevents degradation of LDL receptors in the liver. Nasopharyngitis, reactions at the injection sites, flu-like symptoms, and muscle soreness are a few of the side effects that have been reported in the patients treated with alirocumab [11].

Fibrates, bile acid binding resins, and niacin are also used for lowering LDL-cholesterol.

4. The Historical Perspective of Extremely Low LDL Level

Individuals with hypobetalipoproteinemia and PCSK9 mutation have inherited natural protection from CAD. It is because of low LDL and consequently lower incidence of atherosclerosis and associated events. Patients with a total deficiency of PCSK9 have been reported to have LDL-C levels in the range of 15 mg/dl without having any adverse effects from these extremely low LDL levels [12].

Anthropological and historical evidence showed that, nearly 10,000 years ago, our ancestor hunter-gatherers, who

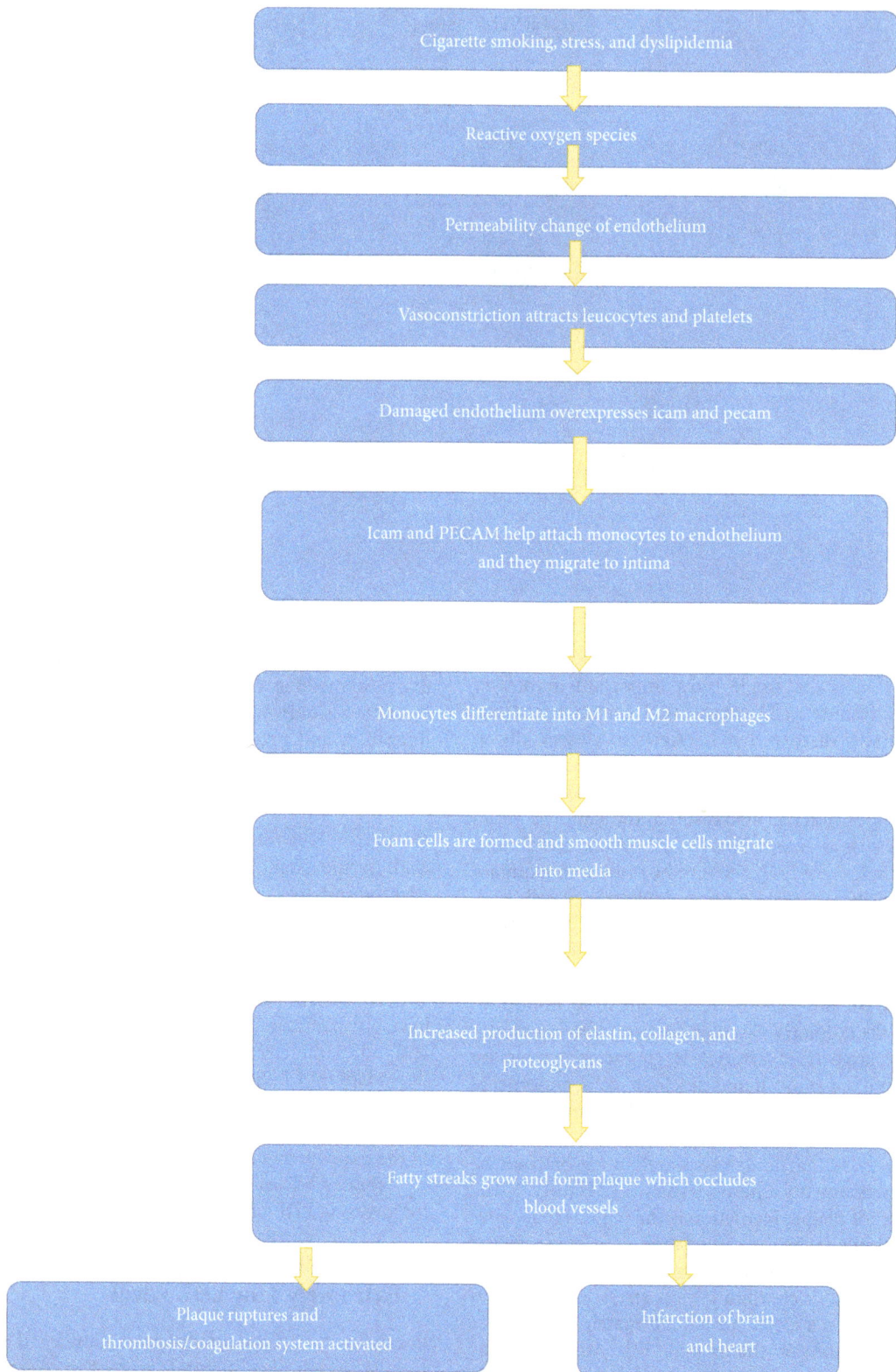

FIGURE 2: Mechanism of atherosclerosis.

were primarily dependent on wildlife diet which was mostly nuts, fruits, vegetables, and flesh of wild animals, were free from atherosclerosis with an average cholesterol level 50–75 mg/dl. Only after 500 generations, after the agricultural revolution, the modern day evolved human beings are mostly reliant on processed food, refined sugars, and carbohydrates. Even the meat that we consume today is obtained from animals which are fed processed grains and corns that make the meat deficient in omega-3 fatty acids. In this short period, the massive changes in our dietary habits took place, which is not long enough for the genetic adaptations to happen to handle this excess load of cholesterol and this causes the rise of average serum cholesterol level to somewhat around 220–230 mg/dl. These findings suggest drastic changes in our diet in comparison to genetic adaptations which are somewhat responsible for the rise of serum LDL level and increased incidence of atherosclerotic diseases [13]. Also, we know about South Asian paradox which denotes that South Asian people are more prone to develop CAD despite having within-target LDL level. So, we have to consider whether further lowering of LDL below the existing target level would help in reduction of atherosclerosis burden and CV events in those cases [14].

5. What Defines Low and Extremely Low LDL and Its Proposed Benefits and Adverse Effects

The LDL-C level of less than 50 mg/dl is considered low while a level of less than 20 mg/dl is considered extremely low. Intensive lipid lowering treatment has been found to halt the progression of atherosclerosis as compared to moderate lipid lowering treatment. It also regresses the atheroma plaque volume as reported by the REVERSAL (Reversal of Atherosclerosis with Aggressive Lipid Lowering) trial and ASTEROID (A Study to Evaluate the Effect of Rosuvastatin on Intravascular Ultrasound-Derived Coronary Atheroma Burden) trial. SATURN (Study of Coronary Atheroma by Intravascular Ultrasound: Effect of Rosuvastatin Versus Atorvastatin) trial also supports this [15]. An LDL level below 2.5 mmol/l can cause an atherosclerotic plaque to regress. Similarly, GLAGOV trial reported that patients who received evolocumab on a baseline treatment with statins demonstrated plaque regression in a larger number of patients as compared to placebo (64.3% versus 47.3%) after 76 weeks of therapy [16]. In a retrospective analysis, coronary calcium score was reduced with the aggressive lowering of LDL [17].

Though the intensive lowering of LDL reduces the plaque size, there is an ongoing debate regarding its side effects. Few previous clinical trials had reported increased incidences of adverse events such as hemorrhagic strokes, dementia, depression, hematuria, and cancers with extremely low LDL-C. The Dallas Heart Study ($n = 12887$), a population-based study, stretched over a period of 15 years found that a PCSK9 mutation is associated with significantly low LDL level. People with PCSK9 mutation exhibited a low incidence of CAD (a reduction of 88 percent in black and 47 percent of whites) with no increase in the hemorrhagic stroke or cancer. A person with a complete absence of PCSK9 has LDL level of about 15 mg/dl, and there has not been any report of any adverse incidents [18]. The brain itself contains 25% of total cholesterol, and it is needed for maintaining its complex neuronal circuit. Blood-brain barrier is impermeable to circulatory cholesterol. This fact implies that the cholesterol regulation in the brain is not similar to that of extracerebral cholesterol. So, cholesterol level outside of the brain should

not affect the brain functioning as these two cholesterol pools are different. On the surface, a target LDL level of less than 70 mg/dl may appear markedly low, but its cogency can be supported by a physiological rationale that we are born with an LDL level of 30–40 mg/dl, and, at that time, the development of the brain is at its peak. The safety of low LDL level has been supported by Ray et al. They reported that the reduction of LDL levels to as low as that of a neonate is safe as well as beneficial in reducing the risk of angina, MI, or cerebrovascular disorder and total mortality [19]. These might be the levels to which humans are inherently adapted, and the levels ventured to be achieved [14]. Human brain is the most cholesterol-enriched organ but, unlike other peripheral organs, human brain is primarily dependent on de novo cholesterol synthesis rather than peripheral plasma cholesterol [20]. These above pieces of evidence lead to the hypothesis that the lowering of plasma LDL would not affect the normal brain function.

6. Why Extremely Low or Low LDL Is Now Discussed

Most of the statin trials showed an average of 31% of relative risk reduction which means that 69% of relative risk is still present. Despite the widespread use of statins, cardiovascular diseases and strokes are responsible for 25% of deaths worldwide. There is certainly need to address this residual risk. A meta-analysis by the Cholesterol Treatment Trialists (CTT) contributors reported that a reduction of 1 mmol per liter in LDL-cholesterol levels results in a consistent 20% to 25% decrease in the risk of the major cardiovascular events as well as the total mortality decreasing by 12 percent [21]. PROVE IT-TIMI study noted a residual CV risk of 22.4% despite reducing LDL-C to 62 mg/dl. This residual risk was targeted in various studies by modulating HDL and TG levels but showed disappointing results. However, recently PCSK9 inhibitors are emerging as a promising alternative to achieve LDL levels even below the target. Statin monotherapy upregulates the PCSK9 by 25–35% on average, along with LDL receptors in hepatocytes, which counterbalances the beneficial effects of statin. Thus, PCSK9 inhibitors would also mitigate the intrinsic counterbalancing effect of statins when given in combination [22]. However, the dilemma that continues to trouble physicians is determining how aggressively LDL needs to be treated. After the recent pooled analysis of 14 trials by Robinson et al. which showed the safety and efficacy of alirocumab in attaining low LDL level even below 15 mg/dl, this topic gains momentum [23].

7. Studies Showed Promising Results of Extremely Low LDL/Low LDL Level and Safety

Many trials piloted to ascertain the effects of lower-than-recommended LDL levels have reported promising results (Table 1). The TNT Trial was conducted to investigate the impact of very low LDL-C levels on major cardiovascular events compared with relatively higher LDL-C levels. The

TABLE 1: Summary of the trials that attained lower-than-recommended LDL level.

Trial, year of publication	N	Comparison	LDL reduction	CV event reduction	Adverse events	Comment
Cannon et al., 2004	4162	PVS versus AVS	-	26.3% versus 22.4%	21.4% versus 22.8%	Median LDL level - 62 mg/dl
TNT, 2005	18,003	AVS 10 mg versus 80 mg	-	20–30% fewer CV events in AVS 80 mg group	5.8% versus 8.1%	-
SPARCL, 2006	4731	AVS versus placebo	-	-	93% versus 91%	16% relative reduction in the risk of stroke
JUPITER, 2008	17,802	RSV versus placebo	50% lower LDL in RSV group	20% reduction for each 1 mmol/L decline in LDL level	15.2% versus 15.5%	Median LDL in RSV group of 55 g/dl at 12 months
IMPROVE-IT, 2015	18,144	EZE + statin versus statin	EZE lowered LDL-C further by 24%	7.2% lower risk of major vascular events	No significant difference	Mean LDL level I EZE group of 53.2 mg/dl
YUKAWA-2, 2016	404	E-mAb versus placebo	75.9% reduction in E-mAb group		46.5% versus 51%	Median LDL in E-mAb group of 28 g/dl
Post hoc analysis of 10 ODYSSEY trials, 2016	4974	A-mAb versus placebo, A-mAb versus EZE	−55.4% versus +2.7% −48.1% versus −18.0%	Every 39 mg/dl fall in LDL was translated into 24% lower risk of MACE	79.9% versus 81.3%, 76% versus 73.9%	33.1% of the pooled cohort achieved LDL < 50 mg/dl
Pooled analysis of 14 randomized trials, 2017	5234	A-mAb versus control (placebo or EZE)	LDL levels reported to be as low as 15 mg/dl in A-mAb group with an increase in adverse event rates	-	Low LDL levels (even <15 mg/dl) were not associated	-
FOURIER, 2017	27,564	E-mAb versus placebo	59% reduction as compared to placebo	-	No significant differences except injection site reactions were more common with E-mAb	Rate of CV events was 9.8% versus 11.3%
ORION-1 (dose ranging trial), 2017	501	Inclisiran versus placebo	51% reduction with 2 dose regimen	-	11% versus 8%	48% subjects attained LDL levels below 50 mg/dl

PVS: pravastatin, RSV: rosuvastatin, AVS: atorvastatin, EZE: ezetimibe, A-mAb: alirocumab, E-mAb: evolocumab, MACE: major adverse cardiac event.

study revealed a highly significant reduction in the rate of major cardiovascular events with descending levels of LDL-cholesterol ($p < 0.0001$) with a decrease of 22 percent in combined cardiovascular end point (including coronary artery disease, nonfatal MI, and resuscitated cardiac arrest and a reduction of 20 percent in cardiac deaths with lower LDL levels) [24]. Additionally, the dreaded side effects of a very low LDL-C level such as muscle pain, hemorrhagic stroke, and death due to cancer were not increased. In the IMPROVE-IT (The Improved Reduction of Outcomes: Vytorin Efficacy International Trial), 18,144 participants with acute coronary syndrome were assigned to either simvastatin (40 mg) plus ezetimibe (10 mg) or simvastatin (40 mg) plus placebo randomly. At seven years, the rate of combined cardiovascular death, major coronary event (nonfatal myocardial infarction, unstable angina, or coronary revascularization), or nonfatal stroke was significantly lower in the simvastatin-plus-ezetimibe group (32.7% versus 34.7%). It was observed in those who had baseline LDL level well below the current LDL goal [25]. In a study in 2007, a statin was prescribed to a group of patients with LDL-C less than 60 mg/dl who also had other comorbid conditions such as diabetes mellitus or ischemic heart disease. After a follow-up period of 2.0 +/− 1.4 years, it was found that statin improved survival not only in patients taking them at the baseline level (HR, 0.58; 95% CI, 0.38 to 0.88) but also in those who have LDL-C below 40 mg/dl. Even patients without ischemic heart disease showed improved survival. However, there was no increased risk of elevated transaminases, malignancy, or rhabdomyolysis [26].

The JUPITER trial compared the clinical outcomes and adverse events in patients treated with rosuvastatin who attained LDL-C less than 50 mg/dl and those who did not. The study revealed reduced major cardiovascular events by 65% among those attaining LDL-C < 50 mg/dl and by 44% for the rest of the cohort. Similarly, all-cause mortality was decreased by 46% among patients achieving LDL-C < 50 mg/dl and by 20% for the remaining cohort. However, there was also a higher rate of adverse events including diabetes, hepatobiliary disorders, and insomnia in patients with LDL-C < 30 mg/dl [27]. The SPARCL (Stroke Prevention by Aggressive Reduction in Cholesterol Levels) study conducted on patients with stroke or transient ischemic attack with atorvastatin 80 mg found that the statin reduced the chances of stroke in these groups of patients but increased the incidence of hemorrhagic stroke [28]. On the contrary, another study carried out among subjects with a history of myocardial infarction who were treated with either 80 mg simvastatin as a part of intensive statin therapy or 20 mg simvastatin reported no difference regarding hemorrhagic stroke after a mean follow-up period of 6.7 years (SD of 1.5 years). However, myopathy cases were reported in a higher number, among 80 mg simvastatin users [29]. Robinson et al. evaluated the safety of alirocumab. They described LDL-C levels to be as low as 15 mg/dl and did not report any adverse neurocognitive event, although a nonsignificant increase in cataract incidence seemed to be more in the group achieving LDL-C levels < 25 mg/dl [23]. In the same vein, Sabatine et al. did not report any significant increase in adverse reactions with very low LDL-C. Also, Prostate Cancer Prevention Trial indicated that low

cholesterol is associated with reduced risk of high grade of prostate cancer [30]. A Retrospective Observational Study conducted in Quebec, Canada, on patients admitted with acute myocardial infarction concluded that high dose statin use might be associated with significant reduction in the cancer incidence [31]. The substudy of PROVE IT-TIMI 22 investigating 80 mg atorvastatin versus 40 mg pravastatin also proved that achieving LDL-C level below the expected level (80 to 100 mg/dl) is not associated with increased adverse events [32].

A meta-analysis by Boekholdt et al. reported an increased risk of hemorrhagic stroke in those with very low levels of LDL as compared to moderately low levels. However, the absolute number was low, and the statistical power was therefore insufficient to draw a definite conclusion. Furthermore, they believed that significantly lower risk of cerebrovascular events outweighed the potential for hemorrhagic stroke [33]. A Phase 3 Study of Evolocumab showed no increase in the adverse events despite a median LDL level of 26–36 mg/dl over 12 weeks [34]. The most recent data about safety and efficacy of low LDL comes from the FOURIER trial. They have shown a significant reduction of LDL from a baseline value of 92 mg/dl to 30 mg/dl with evolocumab. Most importantly, there was a significant decrease in the risk of major cardiovascular events without any major rise in the adverse events. They reported a 17% decrease in the cardiovascular death, myocardial infarction, and stroke on lowering the LDL to 43 mg/dl while reducing the LDL levels further to 22 mg/dl decreased the risk to 20%. Additionally, they reported consistent clinical improvements per unit reduction in LDL [35]. A post hoc analysis of 10 ODYSSEY trials comparing alirocumab with the control indicated that low LDL-C was associated with a lower incidence of major adverse cardiovascular events with no significant increment in the treatment-emergent adverse reactions [19].

Very recently, a prespecified safety analysis of IMPROVE-IT involving 15281 patients showed that patients ($n = 971$) with LDL level below 30 mg/dl had no increased adverse events over six years' follow-up [36].

8. Future Directions and Ongoing Studies

Low and extremely low LDL-C levels are being supported widely; however, many have raised concerns about their long-term effects which still stands unexplored. The mystery behind the advantages and disadvantages of prolonged exposure to pharmacologically induced low LDL levels needs to be unveiled.

A common finding among the LDL-lowering trials was the time lag between the onset of LDL lowering and the appearance of full clinical benefits regarding risk reduction presenting itself as another lacuna in the better understanding of the link between LDL-C lowering and CV risk reduction.

Bringing LDL to very low levels with statin monotherapy poses safety concerns in some patients. While the emergence of PCSK9 inhibitors has appeared to solve the problem, the cost effectiveness of treatment with PCSK9 inhibitors remains questionable. The immunogenic effects of the PCSK9 inhibitors varying from mild injection site reactions to anaphylaxis and loss of drug efficacy need to be scrutinized further. Recently SPIRE trial showed antibodies against the murine component of bococizumab in 15–20% of patients and a neutralizing antidrug antibody was seen only in 1.3% patients on alirocumab [37]. Poor adherence to the treatment due to multiple injections is another issue with the PCSK9 monoclonal antibodies.

A therapeutic strategy involving small (21–25 bp) interfering RNA (siRNA) targeting PCSK9 has gained our attraction recent past. Inclisiran, a novel therapeutic drug that inhibits PCSK9 through RNA interference, has shown encouraging results in an average reduction of LDL by 51% with only a 2-dose regime over a period of 9 months. This was investigated in a clinical phase 2 trial, ORION-1. The result of this trial is encouraging as ease of using this drug will be impactful as it needs only one or two injections over six-month to 1-year period [38]. Nevertheless, the impact on cardiovascular outcomes is yet to be studied in ORION-4. None of the studies have thus far mentioned the duration of treatment with PCSK9 inhibitors needed to maintain the risk reduction.

Peptide-based anti-PCSK9 vaccines are being tested on mice which supposedly have the potential to control LDL level for a longer duration [39].

Several ongoing studies are aiming to enhance our knowledge regarding the safety of low LDL and reduction in cardiovascular risk. In ODYSSEY OUTCOMES, a placebo-controlled phase 3 trial, 18600 post-MI patients are being randomized to alirocumab or placebo arm. It intends to compare the effects of alirocumab with placebo on the occurrence of cardiovascular events over a period of 64 months. The ODYSSEY APPRISE is a multicountry, multicenter phase 3 study aimed at investigating the safety of alirocumab in patients with severe hypercholesterolemia over a period of 30 months. TAUSSIG is another ongoing study designed to assess the long-term safety, tolerability, and efficacy of evolocumab in patients with severe hypercholesterolemia. Meanwhile, PACMAN-AMI is evaluating the effects of PCSK9 inhibition on the morphology of coronary plaque in patients with acute myocardial infarction.

9. Conclusion

In summary, the residual risk despite achieving the current target LDL levels needs to be addressed. Although the clinical benefits of lowering LDL have been well stated, their long-term consequences are still under investigation. Many trials conducted in the past were successful in reducing the LDL levels well below the target with a consequent reduction in CV risk. Though there is ample evidence that low LDL does protect from residual CV risk, there have also been a few studies claiming an increasing number of adverse events with low and extremely low LDL levels. Nevertheless, we have to wait for the result of ongoing trials to have a conclusive answer on the long-term effect of lowering the current LDL goal.

References

[1] D. Mozaffarian, E. J. Benjamin, A. S. Go et al., "Heart disease and stroke statistics—2016 update," in *Circulation*, vol. 132, American Heart Association, 2015.

[2] D. Mozaffarian, E. J. Benjamin, A. S. Go et al., "Heart Disease and Stroke Statistics—2015 Update," in *Circulation*, vol. 131, American Heart Association, 2014.

[3] N. J. Stone, J. Robinson, and A. H. Lichtenstein, "2013 ACC/AHA guideline on the treatment of blood cholesterol to reduce atherosclerotic cardiovascular risk in adults: a report of the American College of Cardiology/American Heart Association Task Force on Practice Guidelines," *Circulation*, 2013.

[4] M. Nayor and R. S. Vasan, "Recent update to the us cholesterol treatment guidelines: A comparison with international guidelines," *Circulation*, vol. 133, no. 18, pp. 1795–1806, 2016.

[5] W. E. Boden, J. L. Probstfield, T. Anderson et al., "Niacin in patients with low HDL cholesterol levels receiving intensive statin therapy," *The New England Journal of Medicine*, vol. 365, no. 24, pp. 2255–2267, 2011.

[6] C. Weber and H. Noels, "Atherosclerosis: current pathogenesis and therapeutic options," *Nature Medicine*, vol. 17, no. 11, pp. 1410–1422, 2011.

[7] J. J. Wentzel, Y. S. Chatzizisis, F. J. H. Gijsen, G. D. Giannoglou, C. L. Feldman, and P. H. Stone, "Endothelial shear stress in the evolution of coronary atherosclerotic plaque and vascular remodelling: Current understanding and remaining questions," *Cardiovascular Research*, vol. 96, no. 2, pp. 234–243, 2012.

[8] K. Pahan, "Lipid-lowering drugs," *Cellular and Molecular Life Sciences*, vol. 63, no. 10, pp. 1165–1178, 2006.

[9] WebMD, "Which Medicines Lower "Bad" (LDL) Cholesterol," http://www.webmd.com/cholesterol-management/guide/cholesterol-lowering-medication#1, 2017.

[10] *Cholesterol Medications: Consider The Options*, Mayo Clinic, 2015, http://www.mayoclinic.org/diseases-conditions/high-blood-cholesterol/in-depth/cholesterol-medications/art-20050958.

[11] https://www.mayoclinic.org/drugs-supplements/alirocumab-subcutaneous-route/side-effects/drg-20151256.

[12] J. D. Horton, J. C. Cohen, and H. H. Hobbs, "PCSK9: a convertase that coordinates LDL catabolism," *Journal of Lipid Research*, vol. 50, pp. S172–S177, 2009.

[13] J. H. O'Keefe Jr. and L. Cordain, "Cardiovascular Disease Resulting from a Diet and Lifestyle at Odds with Our Paleolithic Genome: How to Become a 21st-Century Hunter-Gatherer," *Mayo Clinic Proceedings*, vol. 79, no. 1, pp. 101–108, 2004.

[14] J. H. O'Keefe Jr., L. Cordain, W. H. Harris, R. M. Moe, and R. Vogel, "Optimal low-density lipoprotein is 50 to 70 mg/dl: Lower is better and physiologically normal," *Journal of the American College of Cardiology*, vol. 43, no. 11, pp. 2142–2146, 2004.

[15] T. Dave, J. Ezhilan, H. Vasnawala, and V. Somani, "Plaque regression and plaque stabilisation in cardiovascular diseases," *Indian Journal of Endocrinology and Metabolism*, vol. 17, no. 6, p. 983, 2013.

[16] S. J. Nicholls, R. Puri, T. Anderson et al., "Effect of evolocumab on progression of coronary disease in statin-treated patients: The GLAGOV randomized clinical trial," *Journal of the American Medical Association*, vol. 316, no. 22, pp. 2373–2384, 2016.

[17] B. Ibanez, G. Vilahur, and J. J. Badimon, "Plaque progression and regression in atherothrombosis," *Journal of Thrombosis and Haemostasis*, vol. 5, no. 1, pp. 292–299, 2007.

[18] T. McCormack, R. Dent, and M. Blagden, "Very low LDL-C levels may safely provide additional clinical cardiovascular benefit: the evidence to date," *International Journal of Clinical Practice*, vol. 70, no. 11, pp. 886–897, 2016.

[19] K. K. Ray, H. N. Ginsberg, M. H. Davidson et al., "Reductions in Atherogenic Lipids and Major Cardiovascular Events: A Pooled Analysis of 10 ODYSSEY Trials Comparing Alirocumab with Control," *Circulation*, vol. 134, no. 24, pp. 1931–1943, 2016.

[20] M. Orth and S. Bellosta, "Cholesterol: its regulation and role in central nervous system disorders," *Cholesterol*, vol. 2012, Article ID 292598, 19 pages, 2012.

[21] C. Baigent, A. Keech, and P. M. Kearney, "Efficacy and safety of cholesterol-lowering treatment: prospective meta-analysis of data from 90,056 participants in 14 randomised trials of statins," *The Lancet*, vol. 366, no. 9493, pp. 1267–1278, 2005.

[22] N. E. Lepor and D. J. Kereiakes, "The PCSK9 inhibitors: A novel therapeutic target enters clinical practice," *American Health and Drug Benefits*, vol. 8, no. 9, pp. 483–488, 2015.

[23] J. G. Robinson, R. S. Rosenson, M. Farnier et al., "Safety of Very Low Low-Density Lipoprotein Cholesterol Levels With Alirocumab: Pooled Data From Randomized Trials," *Journal of the American College of Cardiology*, vol. 69, no. 5, pp. 471–482, 2017.

[24] J. C. LaRosa, S. M. Grundy, J. J. P. Kastelein et al., "Safety and efficacy of Atorvastatin-induced very low-density lipoprotein cholesterol levels in patients with coronary heart disease (a post hoc analysis of the treating to new targets [TNT] study)," *American Journal of Cardiology*, vol. 100, no. 5, pp. 747–752, 2007.

[25] S. A. Murphy, C. P. Cannon, M. A. Blazing et al., "Reduction in Total Cardiovascular Events with Ezetimibe/Simvastatin Post-Acute Coronary Syndrome the IMPROVE-IT Trial," *Journal of the American College of Cardiology*, vol. 67, no. 4, pp. 353–361, 2016.

[26] N. J. Leeper, R. Ardehali, E. M. DeGoma, and P. A. Heidenreich, "Statin use in patients with extremely low low-density lipoprotein levels is associated with improved survival," *Circulation*, vol. 116, no. 6, pp. 613–618, 2007.

[27] J. Hsia, J. G. MacFadyen, J. Monyak, and P. M. Ridker, "Cardiovascular event reduction and adverse events among subjects attaining low-density lipoprotein cholesterol <50 mg/dl with rosuvastatin: The JUPITER trial (justification for the use of statins in prevention: An intervention trial evaluating rosuvastatin)," *Journal of the American College of Cardiology*, vol. 57, no. 16, pp. 1666–1675, 2011.

[28] L. B. Goldstein, P. Amarenco, M. Szarek et al., "Hemorrhagic stroke in the Stroke Prevention by Aggressive Reduction in Cholesterol Levels study.," *Neurology*, vol. 70, no. 24, pp. 2364–2370, 2008.

[29] J. Armitage, L. Bowman, K. Wallendszus et al., "Intensive lowering of LDL cholesterol with 80 mg versus 20 mg simvastatin daily in 12,064 survivors of myocardial infarction: a double-blind randomised trial," *Lancet*, vol. 376, no. 9753, pp. 1658–1669, 2010.

[30] E. A. Platz, C. Till, P. J. Goodman et al., "Men with low serum cholesterol have a lower risk of high-grade prostate cancer in the placebo arm of the prostate cancer prevention trial," *Cancer Epidemiology, Biomarkers & Prevention*, vol. 18, no. 11, pp. 2807–2813, 2009.

[31] I. Karp, H. Behlouli, J. LeLorier, and L. Pilote, "Statins and Cancer Risk," *American Journal of Medicine*, vol. 121, no. 4, pp. 302–309, 2008.

[32] S. D. Wiviott, C. P. Cannon, D. A. Morrow, K. K. Ray, M. A. Pfeffer, and E. Braunwald, "Can low-density lipoprotein be too low? The safety and efficacy of achieving very low low-density lipoprotein with intensive statin therapy: A PROVE IT-TIMI 22 substudy," *Journal of the American College of Cardiology*, vol. 46, no. 8, pp. 1411–1416, 2005.

[33] S. M. Boekholdt, G. K. Hovingh, S. Mora et al., "Very low levels of atherogenic lipoproteins and the risk for cardiovascular events: a meta-analysis of statin trials," *Journal of the American College of Cardiology*, vol. 64, pp. 485–494, 2014.

[34] A. Kiyosue, N. Honarpour, C. Kurtz, A. Xue, S. M. Wasserman, and A. Hirayama, "A Phase 3 Study of Evolocumab (AMG 145) in Statin-Treated Japanese Patients at High Cardiovascular Risk," *American Journal of Cardiology*, vol. 117, no. 1, pp. 40–47, 2016.

[35] M. S. Sabatine, R. P. Giugliano, A. C. Keech et al., "Evolocumab and clinical outcomes in patients with cardiovascular disease," *The New England Journal of Medicine*, vol. 376, no. 18, pp. 1713–1722, 2017.

[36] R. P. Giugliano, S. D. Wiviott, M. A. Blazing et al., "Long-term safety and efficacy of achieving very low levels of low-density lipoprotein cholesterol: A prespecified analysis of the IMPROVE-IT trial," *JAMA Cardiology*, vol. 2, no. 5, pp. 547–555, 2017.

[37] P. M. Ridker, J. Revkin, P. Amarenco et al., "Cardiovascular efficacy and safety of bococizumab in high-risk patients," *The New England Journal of Medicine*, vol. 376, no. 16, pp. 1527–1539, 2017.

[38] K. K. Ray, U. Landmesser, L. A. Leiter et al., "Inclisiran in patients at high cardiovascular risk with elevated LDL cholesterol," *The New England Journal of Medicine*, vol. 376, no. 15, pp. 1430–1440, 2017.

[39] G. Galabova, S. Brunner, G. Winsauer et al., "Peptide-based anti-PCSK9 vaccines-an approach for long-term LDLc management," *PLoS ONE*, vol. 9, no. 12, Article ID e114469, 2014.

19

Effect of Locally Manufactured Niger Seed Oil on Lipid Profile Compared to Imported Palm and Sunflower Oils on Rat Models

Zewdie Mekonnen ⓘ,[1,2] **Abrha Gebreselema,**[1] **and Yohannes Abere**[1]

[1]*Department of Biochemistry, College of Medicine and Health Sciences, Bahir Dar University, Bahir Dar, Ethiopia*
[2]*Department of Biomedical Research, Biotechnology Research Institute, Bahir Dar University, Bahir Dar, Ethiopia*

Correspondence should be addressed to Zewdie Mekonnen; zewdmek2007@gmail.com

Academic Editor: Gerhard M. Kostner

Background. Different types of dietary lipids have been shown to affect lipid metabolism and lipid profile differently. *Objective.* This study aims to assess the effect of local niger seed oil on serum lipid profile compared to palm oil and sunflower oil in rats. *Methods.* The effect of the 15% plant oils on serum lipid profile, body weight gain percentage, and feed efficiency ratio was assessed after 8 weeks of experimental period. *Results and Conclusion.* The 15% niger seed oil showed decrease and increase in the level of lipid profile as compared to rats fed with 15% palm oil and sunflower oil (except Triacylglycerol), respectively. The 15% niger seed oil showed significant decrease and increase in body weight gain percentage as compared to the 15% palm oil and 15% sunflower oil, respectively. The feed efficiency ratio was significantly higher and lower in the 15% niger seed oil compared to rats fed with 15% sunflower oil and control group and the palm oil fed rats, respectively. The current study concluded that consumption of locally manufactured niger seed oil decreased the blood lipid profiles, body weight gain percentage, and feed efficiency ratio as compared to palm oil. Utilization of oils containing more unsaturated fatty acids like niger seed oil is recommended to reduce the risk of developing cardiovascular disease.

1. Introduction

Lipoprotein disorder is among the most common metabolic diseases occurring in human. It may lead to coronary heart disease [1–3]. Cardiovascular diseases are a group of diseases of the heart and blood vessels which include coronary heart disease, cerebrovascular disease, peripheral arterial disease, rheumatic heart disease, congenital heart disease, deep vein thrombosis, and also pulmonary embolism. They are reported to be responsible for 31% of global deaths [4]. Coronary heart disease develops from the occlusion of coronary vessels by atherosclerotic plaques [5]. Excess levels of blood cholesterol accelerate atherogenesis. Controlling the blood cholesterol level reduces the incidence of coronary heart disease [6, 7]. Knowledge about the levels of cholesterol subfractions is reported to be more meaningful than simple plasma cholesterol level. The higher the level of LDL-c, the greater the risk of atherosclerotic heart disease. Conversely, the higher the level of HDL-c, the lower the risk of coronary

heart disease [8]. Different types of dietary lipids have been reported to affect lipid metabolism and serum lipid profile differently [9]. Plasma cholesterol levels are moderately decreased when low-cholesterol diets are used [10, 11]. It is now generally believed that vegetable oils decrease plasma cholesterol levels, although they differ in their cholesterol-lowering capacity. Furthermore, the effect of dietary cholesterol on plasma cholesterol levels may be influenced by the types of fatty acid consumed [12–15].

Oilseeds are reported as mainstay of rural and national economy in Ethiopia. Niger is an oilseed crop mainly cultivated in different parts of Ethiopia and India [16]. Niger seed oil provides about 50–60% of the oil for domestic consumption in Ethiopia [17]. It is also used as an oilseed crop in India, where it provides about 3% of the edible oil requirement of the country [18]. Palm oil, obtained from the fruits of the palm trees, is the most widely produced edible vegetable oil in the world, used in food preparation for over 5,000 years, and its nutritional and health attributes have been well documented [19, 20]. The oil is consumed in its fresh state or at various levels of oxidation. Feeding experiments in various animal species and humans have highlighted controversial evidences on the beneficial and harmful effects of fresh palm oil on health. The reported benefits include reduction in the risk of arterial thrombosis and atherosclerosis, inhibition of cholesterol biosynthesis, platelet aggregation, and reduction in blood pressure [21]. On the other hand, when used in the oxidized state, it possesses potential danger to the physiological and biochemical functions of the body. The reduction of the dietetic level of oxidized oil or the level of oxidation may reduce the health risk [21]. Sunflower oil is one of the major vegetable oils produced worldwide. Sunflower plants can only be grown in limited geographical locations because of the soil and climatic conditions required [22]. The oil of sunflower contains more polyunsaturated fatty acids and many other molecules that are responsible for the different health benefits of sunflower oil. The health effect assessment of sunflower oil was carried out in the 1970s and 1980s and was essential to our understanding of the specific effects of different fatty acids on heart health. Due to its high omega-6 fatty acid, sunflower oil is reported to improve blood lipid profile and reduce the risk of cardiovascular disease. The oil is used in spread manufacture, in cooking, and for dressing salads [23]. From many years of research, it has been established that the primary cholesterol-elevating fatty acids are the saturated fatty acids with 12 (lauric acid), 14 (myristic acid), and 16 carbon atoms (palmitic acid) with a concomitant increase in the risk of coronary heart disease. World Health Organization in its report in 2005 stated that there is a convincing evidence that palmitic oil consumption contributes to an increased risk of developing cardiovascular diseases [2, 3, 19, 24].

The total cholesterol, LDL-c, HDL-c, and TAG are collectively called blood lipids. Their levels could be modified by the type and amount of fat in the diet [25, 26]. The comparison of oils from different origin could be useful to establish the quality of oil and give direction or clue about the impact of their consumption on the nutritional state, cardiovascular disease, and related health of consumers. As to the researchers' knowledge, there was no literature or studies conducted in Ethiopia which can indicate the effect of niger seed oil on serum lipid profile compared to the imported palm oil and sunflower oil. Therefore, this study attempts to compare the effect of niger seed oil manufactured in Bahir Dar to the imported palm and sunflower oils on serum lipid profile in rat model.

2. Materials and Methods

2.1. Oil Sample Collection. Niger seed oil was purchased from niger seed oil extracting local factory in Bahir Dar, Ethiopia, and imported palm and sunflower oils were purchased from a supermarket in Bahir Dar, Ethiopia.

2.2. Experimental Diets Preparation. Test diets were prepared by mixing sample oils with normal commercial rat pellet to contain 15% of the oils. Each of the 15% test diets was prepared

by adding 15 g niger seed oil, 15 g palm oil, and 15 g sunflower oil to 85 g rat pellet in a separate container. The oil and the rat pellet were mixed manually and left to absorb the vegetable oils at room temperature overnight and stored at 20°C before the feeding trial is conducted.

2.3. Experimental Animals. Male Wistar albino rats ($n = 12$) were purchased from Animal Science Department, Gondar University, Gondar, Ethiopia. All animals were given one week of acclimatization in animal housing conditions before being used for the study. All the animals were fed with standard animal feed and had access to water and were handled in accordance with the National Institutes of Health (NIH) Guidelines for Care and Use of Laboratory Animals. The rats were divided into 4 groups, each containing 3 rats. Each group was housed in a separate cage. Group 1 served as the normal control and received standard animal feed and water; group 2 was fed with 15% niger seed oil; group 3 was fed with 15% of sunflower oil; and group 4 was fed with 15% palm oil. The experiment lasted for 8 weeks.

2.4. Biological Evaluation. During the experimental period, food intake (FI) was recorded every second day for each group, and the animals were weighted twice weekly in all groups. The biological values of different diets were assessed by the determination of body weight gain percentage (BWG %) which was calculated at the end of the experimental period, using the following equation:

$$\text{body weight gain percentage (BWG\%)}$$
$$= \frac{\text{Final b.wt} - \text{Initial b.wt} \times 100}{\text{Initial b.wt}}, \tag{1}$$

where, b.wt is body weight of rat.

The feed efficiency ratio (FER) was calculated using the following equation [27]:

$$\text{feed efficiency ratio (FER)}$$
$$= \frac{\text{gain in body weight (g)}}{\text{feed consumed (g)}}. \tag{2}$$

2.5. Blood Collection and Serum Separation. At the end of the experiment (8 weeks), the feeding of rats was stopped for 12 hours before they were scarified by cervical dislocation and blood was taken through cardiac puncture by trained individual. The serum was separated through centrifugation with speed of 3000 revolutions per minute at room temperature for 10 minutes. Then, test tubes were placed in ice-box and transported to the GAMBY General Teaching Hospital for the analysis.

2.6. Laboratory Analysis. The serum levels of total cholesterol, triglyceride, HDL-c, and LDL-c were analyzed by ABX Pentra 400 clinical chemistry autoanalyzer (manufactured by HORIBA ABX SAS) as per the manufacturer's instructions. To ensure the accuracy and precision of the test results, all preanalytical, analytical, and postanalytical precautions

TABLE 1: The effect of niger seed oil, sunflower oil, and palm oil on blood serum total cholesterol, TAG, HDL-c, and LDL-c at the end of study period.

Group	HDL-c (mmol/l)	LDL-c (mmol/l)	Total cholesterol (mmol/l)	TAG (mmol/l)
Control (feed with normal pellet)	0.31 ± 0.01^a	0.17 ± 0.01^a	1.16 ± 0.02^a	$0.74 \pm 0.0.01^a$
Feed with 15% niger seed oil	0.29 ± 0.03^a	0.19 ± 0.02^a	1.04 ± 0.03^b	0.34 ± 0.01^b
Feed with 15% sunflower oil	0.26 ± 0.02^a	0.12 ± 0.01^{ab}	0.82 ± 0.02^{ab}	0.36 ± 0.02^b
Feed with 15% palm oil	0.42 ± 0.03^b	0.26 ± 0.03^c	1.4 ± 0.003^c	0.81 ± 0.02^c

Data were expressed as means ± SD ($p < 0.05$). Values with different letters within the same column are statistically significant.

were taken into consideration. The results obtained from the laboratory staff were validated and verified by trained personnel before release. In addition, to maintain internal quality control, known standards were run and the equipment was calibrated prior to analysis. As external quality control, the laboratory also participates in the international digital Proficiency Testing (PT) program.

2.7. Statistical Analysis. All statistical calculations were performed on the Statistical Package for Social Sciences (SPSS) version 20.0 software (IBM Corp., released in 2011, Armonk, NY). All data were expressed as the mean ± SD. Independent sample t-test and one-way ANOVA were used. In all the statistical tests, a confidence level of 95% and $p < 0.05$ were considered significant.

3. Result

The current study showed decrease in the level of total cholesterol, triglyceride, and HDL-c and slight increase in the level of LDL-c in rats fed with 15% niger seed oil as compared to the control group. It also showed increase in the level of total cholesterol, HDL-c, and LDL-c and slight decrease in the level of triglyceride in rats fed with 15% niger seed oil compared to rats fed with 15% sunflower oil. High levels of total cholesterol, HDL-c, LDL-c, and TAG were observed in rats fed with 15% palm oil compared to all other groups (Table 1 and Figure 1).

As shown in Table 2 and Figure 2, significant ($p < 0.05$) decrease in body weight gain percentage was observed in rats fed with 15% niger seed oil compared with rats fed with 15% palm oil and significant ($p < 0.05$), as well as increase in body weight gain percentage compared to control group. A significant ($p < 0.05$) decrease in the body weight gain percentage in rats fed with 15% sunflower oil compared to rats fed with 15% niger seed oil and 15% palm oil and control group was shown.

The study also showed a significant ($p < 0.05$) increase in the feed efficiency ratio (FER) of rats fed with 15% niger seed oil compared to the control group throughout the study period. It also indicated nonsignificant ($p > 0.05$) increase in feed efficiency ratio (FER) of rats fed with 15% niger seed oil compared to rats fed with 15% sunflower oil in the first week of the experimental period and significant ($p < 0.05$) increase was observed after the first week of the experimental

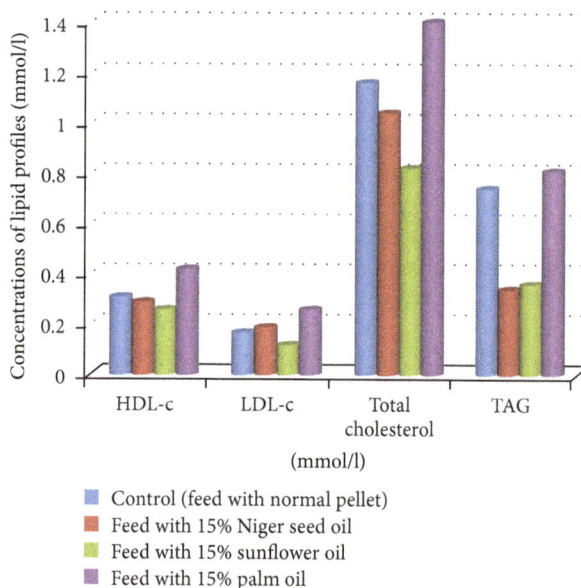

FIGURE 1: The comparisons of the mean values of total cholesterol, TAG, HDL-c, and LDL-c of blood serum at the end of study period.

period. Rats fed with palm oil had a significant ($p < 0.05$) increase in feed efficiency ratio (FER) as compared to other groups (Table 3 and Figure 3).

4. Discussion

The oil content of niger is reported to be 42–44% [28–30]. It has been reported to contain up to 20.9% carbohydrate and 27.8% protein [16]. Niger seed oil has a fatty acid composition typical for seed oils of the Asteraceae plant family. The oil extracted from niger seed cultivated in Ethiopia was reported to contain more than 70% linoleic acid [16, 28–31], while the one cultivated in India contains approximately 55% linoleic acid [16]. In all other works done so far on the fatty acid composition of niger, linoleic acid is unequivocally the dominant fatty acid present in niger seed oil followed by palmitic, oleic, and stearic acids [28–32]. Regarding the fatty acid profile, niger seed oil resembles that of sunflower oil with its high content of linoleic acid, which may be up to 85% depending on the origin [17]. Due to its high linoleic acid content, the oil of niger seed is reported to be

TABLE 2: The characteristic mean level of the body weight gain percentage (BWG %) of experimental animals (Wistar albino rats) through different weeks of the experimental period.

Feed	Body weight gain % of rat in the first one week	Body weight gain % of rat in the 2nd two weeks	Body weight gain % of rat in the 3rd two weeks	Body weight gain % of rat in the 4th two weeks	Body weight gain % of rat in the final 8th weeks
Control (feed with normal pellet)	53.96 ± 0.94^a	93.45 ± 0.96^a	98.74 ± 0.42^a	99.83 ± 0.43^a	101.34 ± 0.87^a
Feed with 15% niger seed oil	37.15 ± 0.97^b	190.13 ± 0.70^b	220.93 ± 0.65^b	221.77 ± 0.67^b	217.65 ± 0.51^b
Feed with 15% sunflower oil	33.62 ± 1.4^c	48.68 ± 1.15^c	65.04 ± 0.87^c	65.11 ± 0.90^c	71.23 ± 0.96^c
Feed with 15% palm oil	209.57 ± 0.51^d	248.20 ± 1.0^d	273.73 ± 0.38^d	271.97 ± 0.23^d	265.47 ± 0.47^d

Data were expressed as means $\pm$ SD ($p < 0.05$). Values with different letters within the same column are statistically significant.

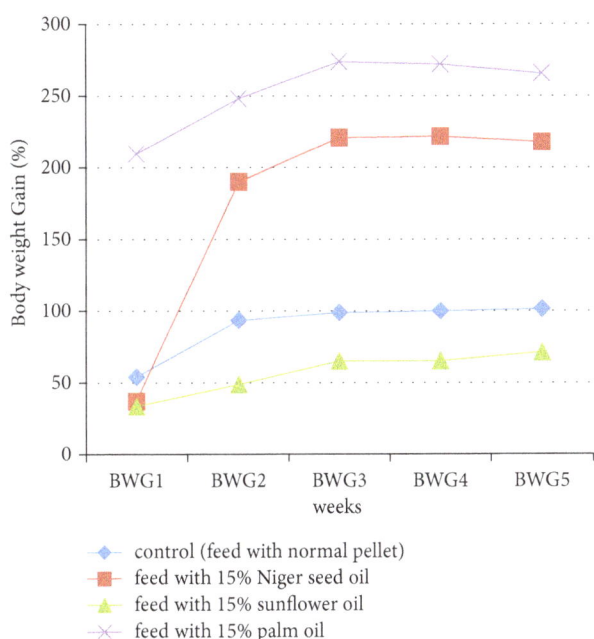

FIGURE 2: Characteristics of the body weight gain percentage (BWG %) of experimental animals (Wistar albino rats) (BWG1: body weight gain percentage of rats in the first one week; BWG2: body weight gain percentage of rats in the 2nd two weeks; BWG3: body weight gain percentage of rats in the 3rd two weeks; BWG4: body weight gain percentage of rats in the 4th two weeks; BWG5: body weight gain percentage of rats in the final 8th weeks).

nutritionally important. The niger seed oil is also reported to have significant antioxidant activity due to its content of sterols and tocopherols [16].

The palm oil has been reported to have 50% saturated fatty acids esterified with glycerol. The saturated fatty acid content of palm oil included palmitic acid (C-16:0), and the unsaturated fatty acid content included 10% linoleic acid (C-18:2), which is omega-6 fatty acid. Palm oil also contains vitamin K, magnesium, tocotrienols, and small amounts of squalene and ubiquinone [33]. In addition to the polyunsaturated fatty acids, sunflower oil also contains a number of other compounds including tocopherols, plant sterol and

stanol esters, phospholipids, carotenoids, and trace elements. It contains alpha-tocopherol, which makes it resistant to photooxidation, and it is low in gamma-tocopherol, which is required to provide stability against oxidation [23].

The levels of total cholesterol and LDL-c were significantly ($p < 0.05$) higher in rats fed with niger seed oil as compared to rats fed with sunflower oil. The mean values of total cholesterol and TAG were significantly ($p < 0.05$) higher in rats feed with niger seed oil compared to control group. The difference in the mean value of TAG between niger seed oil and sunflower oil was not significant ($p > 0.05$). There was also nonsignificant ($p > 0.05$) increase in the level of HDL-c in rats fed with niger seed oil as compared to rats fed with sunflower oil. These observed results might be due to the high percentage of polyunsaturated fatty acid in sunflower oil compared to niger seed oil [21, 34]. Polyunsaturated fatty acids have been reported to facilitate the transportation and utilization of lipids [34]. Studies reported that sunflower oil is very high in polyunsaturated fatty acid and low in saturated fatty acid contents. They reported that standard sunflower oil is predominantly composed of linoleic acid (C-18:2) and oleic acid (C-18:1). These two acids account for about 90% of the total fatty acid content of sunflower oil. The remaining 8–10% is comprised of palmitic acid (C-16:0) and stearic acid (C-18:0) [23]. Slight increase in the level of triglyceride has been shown in rats fed with 15% sunflower oil as compared to rats fed with niger seed oil. This result agreed with the finding of a study that reported that increasing the polyunsaturated fatty acid content in food increases the Triacylglycerol level [35, 36]. The levels of total cholesterol, HDL-c, LDL-c, and TAG were significantly ($p < 0.05$) higher in rats fed with 15% palm oil compared to control group, rats fed with 15% niger seed oil, and rats fed with 15% sunflower oil.

The increase in the level of lipid profiles in rats fed with palm oil compared to the other groups might be due to the percentage of saturated fatty acids. The reported percentage of saturation and unsaturation in the fatty acid composition of palm oil was one to one, where as in niger seed and sunflower oil, the percentage of saturation was relatively lower compared to palm oil. The existence of more unsaturated fatty acid percentage in niger seed and sunflower oil might help the experimental animal to process the feed more

TABLE 3: The characteristic mean level of feed efficiency ratio (FER) of experimental animals (Wistar albino rats) through different weeks of the experimental period.

Feed	FER of rat in the first one week	FER of rat in the 2nd two weeks	FER of rat in the 3rd two weeks	FER of rat in the 4th two weeks	FER of rat in the final 8th weeks
Control (feed with normal pellet)	0.26 ± 0.01^{a}	0.45 ± 0.01^{a}	0.47 ± 0.006^{a}	0.48 ± 0.01^{a}	0.49 ± 0.006^{a}
Feed with 15% niger seed oil	0.16 ± 0.02^{b}	0.82 ± 0.044^{b}	0.95 ± 0.064^{b}	0.96 ± 0.067^{b}	0.94 ± 0.067^{b}
Feed with 15% sunflower oil	0.16 ± 0.035^{b}	0.23 ± 0.055^{c}	0.31 ± 0.056^{c}	0.31 ± 0.056^{c}	0.34 ± 0.069^{c}
Feed with 15% palm oil	0.97 ± 0.015^{c}	1.15 ± 0.012^{d}	1.27 ± 0.006^{d}	1.26 ± 0.012^{d}	1.24 ± 0.032^{d}

Data were expressed as means $\pm$ SD ($p < 0.05$). Values with different letters within the same column are statistically significant. In all comparisons the p-value is less than 0.05 except the comparison between 15% Sunflower oil and 15% Niger seed oil on FER of Rat in the first one week of the experiment in which the $p > 0.05$.

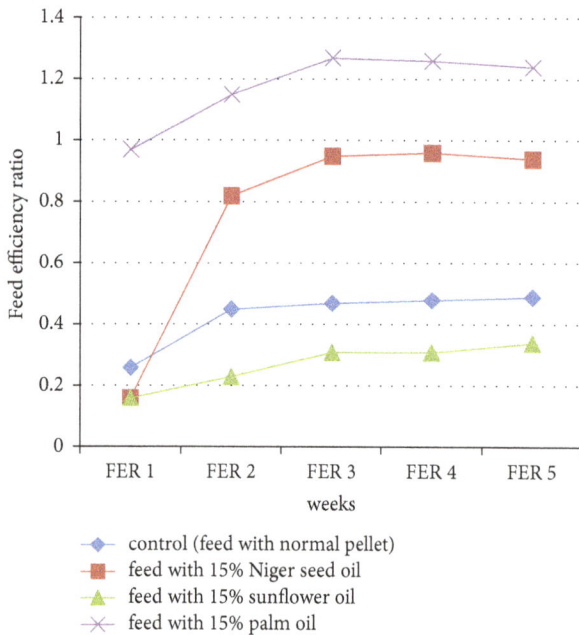

FIGURE 3: The characteristics of feed efficiency ratio (FER) of experimental animals (Wistar albino rats) (FER1: FER of rats in the first one week; FER2: FER of rats in the 2nd two weeks; FER3: FER of rats in the 3rd two weeks; FER4: FER of rats in the 4th two weeks; FER5: FER of rats in the final 8th weeks).

efficiently compared to the palm oil-fed rats [21]. High intake of saturated fatty acid has been linked to increased cholesterol levels which can lead to cardiovascular disease [37]. In line with this, dietary recommendations have limited the intake of saturated fatty acids for the prevention of cardiovascular disease. World Health Organization's 2005 report stated that there is a convincing evidence that palm oil consumption contributes to an increased risk of developing cardiovascular disease [24]. In line with World Health Organization's 2005 report, a number of studies suggested the association of high contents of saturated fats in palm oil with the detrimental atherogenic profile [23, 38, 39]. Other studies done on the effect of palm oil on plasma lipoprotein profile reported

positive correlation with the risk of developing cardiovascular disease [15, 40]. A study carried out for determination of the effect of lipid profile by supplementing polyunsaturated fatty acids like linoleic acid and saturated fatty acids like palmitic acid and stearic acid also reported similar results [41]. Another research done by supplementation of dietary fat containing more unsaturated fatty acid (saturated fatty acids replaced with polyunsaturated fatty acids) reported more LDL fractional catabolic rate that might contribute to variability of plasma cholesterol levels [42].

The significant decrease in body weight gain percentage of rats fed with 15% sunflower oil as compared to rats fed with 15% niger seed oil, rats fed with 15% palm oil, and control group and the significant ($p < 0.05$) decrease in body weight gain percentage of rats fed with 15% niger seed oil compared with rats fed with 15% palm oil might be because of the fatty acid composition difference of sunflower oil, niger seed oil, and palm oil [28–32]. Researchers indicated that the utilization of polyunsaturated fatty acids increases the utilization of proteins and cholesterol and might assist in losing weight [35, 36]. Research done on dietary fatty acid with utilization of polyunsaturated fatty acids reported that the polyunsaturated fatty acids facilitated the lipid metabolism and resulted in decrease in obesity [43, 44]. The rats fed with palm oil have shown high body weight gain percentage compared to other groups. This might be because of fatty acid composition differences in which the palm oil contains more saturated fatty acid compared to the fatty acid composition of sunflower oil and niger seed oil; particularly the palm oil contains more palmitic acid [40, 45]. Different studies reported that saturated fatty acids (SFAs) have been shown to produce higher rates of weight gain compared with other types of fatty acids [36, 42, 43, 46].

The significant increase in feed efficiency ratio (FER) in rats fed with niger seed oil compared to rats fed with sunflower oil and control group and significant decrease compared to rats fed with palm oil (Table 3 and Figure 3) might be due to the fatty acid composition difference of the oils which precipitated to either accumulation of the body lipid or more processing of the body lipid. In the current study, the rats fed with palm oil gained more progressive body weight compared to the other groups. This might contribute to the significant increase in body weight gain percentage and feed efficiency ratio in rats fed with palm oil as compared to the other groups. Different research findings indicated that more lipid accumulation affects the body weight gain percentage and the feed efficiency ratio [35, 36, 43, 44]. Other research findings also reported that the utilization of unsaturated fatty acids such as monounsaturated fatty acids reduced fat gain, which agreed with the current study, in which the utilization of more unsaturated fatty acid containing niger seed oil and sunflower oil showed significant decrease in body weight gain percentage and feed efficiency ratio compared to palm oil [9, 36, 46].

5. Conclusion

The current study concluded that consumption of locally manufactured niger seed oil decreased the blood lipid profiles, body weight gain percentage, and feed efficiency ratio as compared to palm oil. It also indicated decease in blood lipid profile, body weight gain percentage, and feed efficiency ratio in rats fed with sunflower oil as compared to rats fed with palm oil and niger seed oil. Niger seed oil lowers the plasma lipid profile that precipitated to low risk of cardiovascular disease. Because the polyunsaturated fatty acids have been found to facilitate lipid transportation and metabolism, utilization of oils containing more unsaturated fatty acids like niger seed oil and sunflower oil is recommended to reduce the risk of developing cardiovascular diseases.

Abbreviations

BWG %: Body weight gain percentage
CHD: Coronary heart disease
CVD: Cardiovascular disease
FER: Feed efficiency ratio
FI: Food intake
T-C: Total cholesterol
HDL-c: High-density lipoprotein cholesterol
LDL-c: Low-density lipoprotein cholesterol
TAG: Triacylglycerol
NSO: Niger seed oil
SFO: Sunflower oil
PO: Palm oil
PUSFA: Polyunsaturated fatty acid
LA: Linoleic acid
SD: Standard deviation
SFA: Saturated fatty acids
mmol/l: Millimoles per liter.

Acknowledgments

This research work was fully funded by the College of Medicine and Health Sciences of Bahir Dar University. The authors acknowledge College of Medicine and Health Sciences of Bahir Dar University for provision of the fund to conduct the research. The authors would like to thank Mr. Habtie Getahun who is the laboratory technician in the Department of Biochemistry, College of Medicine and Health Sciences, Bahir Dar University. The authors also acknowledge GAMBY General Teaching Hospital for the analysis of serum sample.

References

[1] D. McNamara, "Dietary cholesterol and atherosclerosis," *Biochimica et Biophysica Acta (BBA) - Molecular and Cell Biology of Lipids*, vol. 1529, no. 1-3, pp. 310–320.

[2] R. P. Mensink, P. L. Zock, A. D. Kester, and M. B. Katan, "Effects of dietary fatty acids and carbohydrates on the ratio of serum total to HDL cholesterol and on serum lipids and apolipoproteins: a meta-analysis of 60 controlled trials," *American Journal of Clinical Nutrition*, vol. 77, no. 5, pp. 1146–1155, 2003.

[3] P. W. Siri-Tarino, Q. Sun, F. B. Hu, and R. M. Krauss, "Saturated fat, carbohydrate, and cardiovascular disease," *American Journal of Clinical Nutrition*, vol. 91, no. 3, pp. 502–509, 2010.

[4] K. N. Jeejeebhoy, "Short bowel syndrome: a nutritional and medical approach," *Canadian Medical Association Journal*, vol. 166, no. 10, pp. 1297–1302, 2002.

[5] E. D. Grech, "ABC of interventional cardiology: Pathophysiology and investigation of coronary artery disease," *BMJ*, vol. 326, no. 7397, pp. 1027–1030.

[6] A. A. Al-Othman, "Growth and lipid metabolism responses in rats fed different dietary fat sources," *International Journal of Food Sciences and Nutrition*, vol. 51, no. 3, pp. 159–167, 2000.

[7] J. M. R. Gill, J. C. Brown, M. J. Caslake et al., "Effects of dietary monounsaturated fatty acids on lipoprotein concentrations, compositions, and subfraction distributions and on VLDL apolipoprotein B kinetics: Dose-dependent effects on LDL," *American Journal of Clinical Nutrition*, vol. 78, no. 1, pp. 47–56, 2003.

[8] O. B. Ajayi and D. D. Ajayi, "Effect of oilseed diets on plasma lipid profile in albino rats," *Pakistan Journal of Nutrition*, vol. 8, no. 2, pp. 116–118, 2009.

[9] Y. S. Diniz, A. C. Cicogna, C. R. Padovani, L. S. Santana, L. A. Faine, and E. L. B. Novelli, "Diets Rich in Saturated and Polyunsaturated Fatty Acids: Metabolic Shifting and Cardiac Health," *Nutrition Journal*, vol. 20, no. 2, pp. 230–234, 2004.

[10] S. Sahu, R. Chawla, and B. Uppal, "Comparison of two methods of estimation of low density lipoprotein cholesterol, the direct versus Friedewald estimation," *Indian Journal of Clinical Biochemistry*, vol. 20, no. 2, pp. 54–61, 2005.

[11] P. P. Toth, "High-Density Lipoprotein and Cardiovascular Risk," *Circulation*, vol. 109, no. 15, pp. 1809–1812, 2004.

[12] S. Yu-Poth, D. Yin, P. M. Kris-Etherton, G. Zhao, and T. D. Etherton, "Long-chain polyunsaturated fatty acids upregulate LDL receptor protein expression in fibroblasts and HepG2 cells," *Journal of Nutrition*, vol. 135, no. 11, pp. 2541–2545, 2005.

[13] C. Xie, L. A. Woollett, S. D. Turley, and J. M. Dietschy, "Fatty acids differentially regulate hepatic cholesteryl ester formation and incorporation into lipoproteins in the liver of the mouse," *Journal of Lipid Research*, vol. 43, no. 9, pp. 1508–1519, 2002.

[14] R. Mushtaq, R. Mushtaq, and Z. T. Khan, "Effect of walnut on lipid profile in obese female in different ethnic groups of Quetta, Pakistan," *Pakistan Journal of Nutrition*, vol. 8, no. 10, pp. 1617–1622, 2009.

[15] C. A. Idris and K. Sundram, "Effect of dietary cholesterol, trans and saturated fatty acids on serum lipoproteins in non-human primates," *Asia Pacific Journal of Clinical Nutrition*, vol. 11, no. s7, pp. S408–S415, 2002.

[16] M. Syume and B. S. Chandravanshi, "Nutrient composition of Niger seed (Guizotia abyssinica (L. f.) Cass.) cultivated in different parts of Ethiopia," *Bulletin of the Chemical Society of Ethiopia*, vol. 29, no. 3, pp. 341–355, 2015.

[17] W. Riley and H. Belayneh, *NigerIn:Oil crops of the world*, G. Robbelen, R. K. Downey, and A. Ashri, Eds., McGrow-Hill, New York, NY, USA, 1989.

[18] A. Getinet and S. Sharma, *Niger Guizotia Abyssinica (L.f.) Cass. Promoting the conservat-ion and use of underutilized and neglected crops*, Institute of Plant Genetics and Crop Plant Research, Gatersleben/International Plant Genetic Resources Institute, Rome, Italy, 1996, http://www.bioversityinternational.org/uploads/tx_news/Niger_Guizotia_abyssinica_L.f._Cass._136.

[19] D. O. Edem, "Palm oil: Biochemical, physiological, nutritional, hematological, and toxicological aspects: A review," *Plant Foods for Human Nutrition*, vol. 57, no. 3-4, pp. 319–341, 2002.

[20] N. Chandrasekharan, K. Sundram, and Y. Basiron, "Changing nutritional and health perspectives on palm oil," *Brunei International Medical Journal*, vol. 2, pp. 417–427, 2002.

[21] M. Karaji-Bani, F. Montazeni, and M. Hashemi, "Effect of palm oil and serum lipid profile in rats," *Pakistan Journal of Nutrition*, vol. 5, no. 3, pp. 234–236, 2006.

[22] M. K. Gupta, "Sunflower oil. Vegetable oils in food technology," in *Vegetable Oils in Food Technology Composition, Properties and Uses*, pp. 128–156, Blackwell Publishing: Oxford, 2002.

[23] R. Foster, C. S. Williamson, and J. Lunn, "Culinary oils and their health effects," *Nutrition Bulletin*, vol. 34, no. 1, pp. 4–47, 2009.

[24] Diet, *Nutrition and the Prevention of Chonic Diseases*, World Health Organization (WHO), 2005.

[25] M. L. Fernandez and K. L. West, "Mechanisms by which Dietary Fatty Acids Modulate Plasma Lipids," *Journal of Nutrition*, vol. 135, no. 9, pp. 2075–2078, 2005.

[26] W. Puavilai and D. Laoragpongse, "Is calculated LDL-C by using the new modified Friedewald equation better than the standard Friedewald equation?" *Journal of the Medical Association of Thailand*, vol. 87, no. 6, pp. 589–593, 2004.

[27] E. Bethke, M. Bernreuther, and R. F. Tallman, "Feed Efficiency Versus Feed Conversion Ratio – Demonstrated on Feeding Experiments with Juvenile Cod (Gadus Morhua)," *SSRN Electronic Journal*.

[28] M. Ramadan and J. Mörsel, "Proximate neutral lipid composition of niger (Guizotia abyssinica Cass.) seed," *Czech Journal of Food Sciences*, vol. 20, no. No. 3, pp. 98–104, 2018.

[29] M. F. Ramadan and J.-T. Mörsel, "Determination of the lipid classes and fatty acid profile of niger (Guizotia abyssinica Cass.) seed oil," *Phytochemical Analysis*, vol. 14, no. 6, pp. 366–370, 2003.

[30] M. Fawzy Ramadan and J.-T. Mörsel, "Phospholipid composition of niger (Guizotia abyssinica cass.) seed oil," *LWT- Food Science and Technology*, vol. 36, no. 2, pp. 273–276, 2003.

[31] Z. Hussain, S. Yadav, S. Kumar et al., "Molecular characterization of niger [Guizotia abyssinica (L.f.) Cass.] germplasms diverse for oil parameters," *Indian Journal of Biochemistry and Biophysics*, vol. 14, no. 3, pp. 344–350, 2015.

[32] M. F. Ramadan and J.-T. Mörsel, "Analysis of glycolipids from black cumin (Nigella sativa L.), coriander (Coriandrum sativum L.) and niger (Guizotia abyssinica Cass.) oilseeds," *Food Chemistry*, vol. 80, no. 2, pp. 197–204, 2003.

[33] S. Mukherjee and A. Mitra, "Health Effects of Palm Oil," *Journal of Human Ecology*, vol. 26, no. 3, pp. 197–203, 2017.

[34] R.-E. Go, K.-A. Hwang, Y.-S. Kim, S.-H. Kim, K.-H. Nam, and K.-C. Choi, "Effects of palm and sunflower oils on serum cholesterol and fatty liver in rats," *Journal of Medicinal Food*, vol. 18, no. 3, pp. 363–369, 2015.

[35] C. L. Pelkman, V. K. Fishell, D. H. Maddox, T. A. Pearson, D. T. Mauger, and P. M. Kris-Etherton, "Effects of moderate-fat (from monounsaturated fat) and low-fat weight-loss diets on the serum lipid profile in overweight and obese men and women," *American Journal of Clinical Nutrition*, vol. 79, no. 2, pp. 204–212, 2004.

[36] S.-C. Yang, S.-H. Lin, J.-S. Chang, and Y.-W. Chien, "High fat diet with a high monounsaturated fatty acid and polyunsaturated/saturated fatty acid ratio suppresses body fat accumulation and weight gain in obese hamsters," *Nutrients*, vol. 9, no. 10, article no. 1148, 2017.

[37] S. R. Ismail, S. K. Maarof, S. Siedar Ali, A. Ali, and S. L. Atkin, "Systematic review of palm oil consumption and the risk of cardiovascular disease," *PLoS ONE*, vol. 13, no. 2, p. e0193533, 2018.

[38] F. B. Hu, J. E. Manson, and W. C. Willett, "Types of Dietary Fat and Risk of Coronary Heart Disease: A Critical Review," *Journal of the American College of Nutrition*, vol. 20, no. 1, pp. 5–19, 2001.

[39] E. K. Kabagambe, A. Baylin, A. Ascherio, and H. Campos, "The type of oil used for cooking is associated with the risk of nonfatal acute myocardial infarction in Costa Rica," *Journal of Nutrition*, vol. 135, no. 11, pp. 2674–2679, 2005.

[40] P. J. van Jaarsveld and A. J. S. Benadé, "Effect of palm olein oil in a moderate-fat diet on low-density lipoprotein composition in non-human primates," *Asia Pacific Journal of Clinical Nutrition*, vol. 11, pp. S416–423, 2002.

[41] S. V. Gupta and P. Khosla, "Palmitic and stearic acids similarly affect plasma lipoprotein metabolism in cynomolgus monkeys fed diets with adequate levels of linoleic acid," *Journal of Nutrition*, vol. 131, no. 8, pp. 2115–2120, 2001.

[42] J. W. Stewart, M. L. Kaplan, and D. C. Beitz, "Pork with a high content of polyunsaturated fatty acids lowers LDL cholesterol in women," *American Journal of Clinical Nutrition*, vol. 74, no. 2, pp. 179–187, 2001.

[43] M. Pellizzon, A. Buison, F. Ordiz Jr., L. Santa Ana, and K. L. Jen, "Effects of dietary fatty acids and exercise on body-weight regulation and metabolism in rats," *Obesity Research*, vol. 10, no. 9, pp. 947–955, 2002.

[44] K.-L. C. Jen, A. Buison, M. Pellizzon, F. Ordiz Jr., L. Santa Ana, and J. Brown, "Differential effects of fatty acids and exercise on body weight regulation and metabolism in female wistar rats," *Experimental Biology and Medicine*, vol. 228, no. 7, pp. 843–849, 2003.

[45] P. J. van Jaarsveld, C. M. Smuts, and A. S. Benade, "Effect of palm olein oil in a moderate-fat diet on plasma lipoprotein profile and aortic atherosclerosis in non-human primates," *Asia Pacific Journal of Clinical Nutrition*, vol. 11, no. s7, pp. S424–S432, 2002.

[46] N. Crespo and E. Esteve-Garcia, "Dietary polyunsaturated fatty acids decrease fat deposition in separable fat depots but not in the remainder carcass," *Poultry Science*, vol. 81, no. 4, pp. 512–518, 2002.

20

Correlation between Cholesterol, Triglycerides, Calculated, and Measured Lipoproteins: Whether Calculated Small Density Lipoprotein Fraction Predicts Cardiovascular Risks

Sikandar Hayat Khan,[1] Nadeem Fazal,[2] Athar Abbas Gilani Shah,[3] Syed Mohsin Manzoor,[1] Naveed Asif,[4] Aamir Ijaz,[4] Najmusaqib Khan Niazi,[5] and Muhammad Yasir[2]

[1]*Department of Pathology, PNS Hafeez, Islamabad, Pakistan*
[2]*Department of Medicine, PNS Hafeez, Islamabad, Pakistan*
[3]*Department of Surgery, PNS Hafeez, Islamabad, Pakistan*
[4]*Department of Chemical Pathology & Clinical Endocrinology, AFIP, Rawalpindi, Pakistan*
[5]*Administration Department, PNS Hafeez, Islamabad, Pakistan*

Correspondence should be addressed to Sikandar Hayat Khan; sik_cpsp@yahoo.com

Academic Editor: Akihiro Inazu

Background. Recent literature in lipidology has identified LDL-fractions to be more atherogenic. In this regard, small density LDL-cholesterol (sdLDLc) has been considered to possess more atherogenicity than other LDL-fractions like large buoyant LDL-cholesterol (lbLDLc). Recently, Srisawasdi et al. have developed a method for calculating sdLDLc and lbLDLc based upon a regression equation. Using that in developing world may provide us with a valuable tool for ASCVD risk prediction. *Objective.* (1) To correlate directly measured and calculated lipid indices with insulin resistance, UACR, glycated hemoglobin, anthropometric indices, and blood pressure. (2) To evaluate these lipid parameters in subjects with or without metabolic syndrome, nephropathy, and hypertension and among various groups based upon glycated hemoglobin results. *Design.* Cross-sectional study. *Place and Duration of Study.* From Jan 2016 to 15 April 2017. *Subjects and Methods.* Finally enrolled subjects (male: 110, female: 122) were evaluated for differences in various lipid parameters, including measured LDL-cholesterol (mLDLc), HDLc and calculated LDL-cholesterol (cLDLc), non-HDLc, sdLDLC, lbLDLC, and their ratio among subjects with or without metabolic syndrome, nephropathy, glycation index, anthropometric indices, and hypertension. *Results.* Significant but weak correlation was mainly observed between anthropometric indices, insulin resistance, blood pressure, and nephropathy for non-HDLc, sdLDLc, and sdLDLc/lbLDLc. Generally lipid indices were higher among subjects with metabolic syndrome [{sdLDLc: 0.92 + 0.33 versus 0.70 + 0.29 ($p < 0.001$)}, {sdLDLc/lbLDLc: 0.55 + 0.51 versus 0.40 + 0.38 ($p = 0.010$)}, {non-HDLc: 3,63 + 0.60 versus 3.36 + 0.65 ($p = 0.002$)}]. The fact that the sdLDLc levels provided were insignificant in Kruskall Wallis Test indicated a sharp increase in subjects with HbA1c > 7.0%. Subjects having nephropathy (UACR > 2.4 mg/g) had higher concentration of non-HDLc levels in comparison to sdLDLc [{non-HDLc: 3.68 + 0.59 versus 3.36 + 0.43 ($p = 0.007$), {sdLDLc: 0.83 + 0.27 versus 0.75 + 0.35 ($p =$ NS)}]. *Conclusion.* Lipid markers including cLDLc and mLDLc are less associated with traditional ASCVD markers than non-HDLc, sdLDLc, and sdLDLc/lbLDLc in predicting metabolic syndrome, nephropathy, glycation status, and hypertension.

1. Introduction

Atherosclerotic cardiovascular diseases (ASCVD) have emerged as the leading cause of human morbidity and mortality across all races and ethnicities. Literature review strongly signifies the increasing frequency of stroke, IHD, peripheral vascular disease (PVD), and diabetes in subcontinental countries and countries with emerging economies [1]. In the developing world the concept of adipocytes having "thrifty genotype" and "starvation genes" has been associated with higher prevalence of diseases resulting from ASCVD [2].

Genetics, lifestyles, and environmental triggers can all help in accelerating cholesterol deposition to cause ASCVD. Traditionally the ultimate villain in this interplay had always been the (low density lipoprotein cholesterol) LDLc [3]. The convention to date had seen the plight of lipoproteins classification as good and evil, that is, HDLc and LDLc, with most literature guidelines relying upon them as diagnostic and clinical intervention markers in managing various categories of ASCVD [4, 5]. However, various evolving technologies have now allowed the researchers to measure and study the role of different subclasses of lipoproteins [6]. An insight into defining these lipoproteins is technically based upon their particular size, which vary from less than 1.06 (LDL) and greater than 1.06 nm to 1.23 nm as HDL after segregation through ultracentrifugation. [7]. These lipoproteins are actually mixtures of various proportions of esterified and nonesterified cholesterol, phospholipids, proteins, triglycerides, and surface apolipoproteins [8]. Kinetic studies have identified a lot of variability in terms of shape, size, and lipid composition which are difficult to measure as perfection in clinical laboratories provided improvement in laboratory science and calibration practices [9]. The recent data has subcategorized LDL particles based upon their size and density into small and dense LDL-cholesterol particles termed small density LDLc (sdLDLc) and large dense LDL-cholesterol, which has been proven to be more predictive to highlight underlying cardiovascular risks [10]. The former category of lipoproteins is now considered to easily penetrate vessel wall to become oxidized and thus causing nondesirable ASCVD outcomes [11]. Thus current evolution in lipidology is now converging to recognize the importance of sdLDLc in causation of ASCVD risks; however, the technologies measuring LDL particle number are yet not available in most developing healthcare markets along with cost-effectiveness being another consideration. Srisawasdi et al. have recommended a surrogate for measuring sdLDLc and lbLDLc by utilizing mathematical modeling incorporating step wise multivariate regression equation and recommended its use for worldwide clinical practice [12]. Koba et al. have also observed that LDL mass rather than size is more significant as LDL particle concentration in IHD progresses [10]. Moreover, the same authors have also felt that the risk predicting capability of sdLDLc is superior to that of non-HDL cholesterol and LDL-cholesterol.

With this background information the authors have decided to study the correlation of calculated small dense LDL-cholesterol (sdLDLc) and calculated large buoyant LDL-cholesterol (lbLDLc) and traditional lipid markers with varying ASCVD associated risk factors based upon glycemic status, insulin resistance (IR) status, nephropathy status, metabolic syndrome, and blood pressure.

2. Materials and Methods

After formal approval by hospital's ethical review committee, this cross-sectional study was conducted at department of pathology and medicine, PNS Hafeez (Islamabad), and department of chemical pathology and endocrinology, Armed Forces Institute of Pathology (AFIP), Rawalpindi. The study duration was 1 year starting from Jan 2016 to Jan 2017. From a target population of referrals from medical and surgical OPD subjects to laboratory for estimation of lipid profile and fasting plasma glucose, 232 OPD subjects were finally enrolled after complete explanation of study concept, probable outcomes, and nature of clinical interventions involved with formally signing the consent form. Subjects who had some chronic or acute disorder, pregnancy, children, and admitted cases on medication known to alter lipid/related parameters were excluded from the study. Few samples were excluded later due to hemolysis and related technical reasons. The OPD patients were interviewed according to predesigned clinical Performa and were clinically evaluated using various anthropometric indices as per WHO criteria [13]. 10 ml of blood was drawn in EDTA, plain bottles, and Na-Fluoride tubes for measuring various biochemical parameters. Fasting plasma glucose, cholesterol, and triglycerides were measured using GOD-PAP, CHOD-PAP, and GPO-PAP method on Selectra-ProM, while (measured LDLc) mLDLc and HDLc were measured by cholesterol esterase method on ADVIA 1800 Chemistry System, respectively. Calculated LDLc (cLDLc) was measured using Friedewald's formula and sdLDLc and lbLDLc were calculated as per the regression equation recommended by Srisawasdi et al. [12] as follows:

$$\text{sdLDL-c mmol/L} = 0.580\,(\text{non-HDL-c})$$
$$+ 0.407\,(\text{mLDL-c}) \quad (1)$$
$$- 0.719\,(\text{cLDL-c}) - 0.312.$$

Glycated hemoglobin was measured using fast ion-exchange resin separation method; serum insulin by chemiluminescence's technique on Immulite® 1000 and spot urine specimen in 174 subjects for measuring urine albumin creatinine ratio (UACR) were evaluated by immunoturbidimetric method on ADVIA 1800. Homeostasis Model Assessment for insulin resistance (HOMA-IR) was calculated as per the method of Matthews' et al. [14]. Metabolic syndrome was diagnosed using (National Cholesterol Education Program) NCEP and International Diabetic Federation (IDF) criteria [15, 16]. Based upon glycated hemoglobin results, four groups were made, namely, Group-1: HbA1c levels < 5.5%, Group-2: HbA1c levels = 5.6–6.5%, Group-3: HbA1c levels = 6.6–7.0%, and Group-4: HbA1c levels > 7.0%. Two groups for nephropathy related impact were made based upon patient's UACR results as Group-1 with UACR < 2.5 mg/g and Group-2 with UACR > 2.4 mg/g.

2.1. Data Analysis. All data were entered into Excel program (Microsoft Office-2007) and later transferred into SPSS version-15. Descriptive statistics in terms of mean ± SD were calculated for age. All lipid indices were compared between gender groups through independent sample t-statistics. Pearson's correlation was calculated between various lipid parameters with anthropometric indices, blood pressure, and biochemical risk factors. Nonparametric "Kruskal Wallis Test" was employed to compare various groups formulated based upon the presence or absence of metabolic syndrome components (as per the IDF criteria) to compare

TABLE 1: Gender-wise comparison of various lipid indices.

Parameter	Gender	N	Mean	Std. deviation	Sig. (2-tailed)*
Total cholesterol (mmol/L)	Male	110	4.54	0.59	0.171
	Female	122	4.43	0.62	
Fasting triglycerides (mmol/L)	Male	110	1.69	0.82	0.112
	Female	122	1.53	0.67	
HDLc (mmol/L)	Male	109	0.91	0.21	0.000
	Female	121	1.04	0.28	
mLDLc (mmol/L)**	Male	108	2.71	0.68	0.583
	Female	122	2.66	0.76	
Non-HDLc (mmol/L)	Male	110	3.63	0.58	0.008
	Female	122	3.41	0.68	
cLDLc (mmol/L)***	Male	110	2.88	0.51	0.017
	Female	122	2.70	0.57	
sdLDLc (mmol/L)****	Male	110	0.82	0.35	0.676
	Female	122	0.80	0.35	
lbLDLc (mmol/L)*****	Male	110	1.84	0.56	0.818
	Female	122	1.86	0.50	
sdLDLc/lbLDLc	Male	110	0.50	0.55	0.486
	Female	122	0.45	0.37	
LDL-c/HDLc	Male	109	3.08	0.94	0.002
	Female	121	2.70	0.91	
VLDL-cholesterol (mmol/L)	Male	110	0.34	0.16	0.112
	Female	122	0.31	0.13	

*Measured using independent sample t-test (SPSS); **measured LDL-cholesterol (mLDLc) by cholesterol esterase method; ***calculated LDL-cholesterol (cLDLc) by Friedewald's formula; ****small density LDL-cholesterol (sdLDLc) by Srisawasdi et al. regression equation; *****large buoyant LDL-cholesterol (lbLDLc) by Srisawasdi et al. regression equation.

lipid parameters and later the same test was employed to compare various groups formulated upon the glycated hemoglobin results for the ratio between small density and large buoyant LDL-cholesterol. Independent sample t-test was employed to compare lipid indices between subjects with or without metabolic syndrome and subjects with or without nephropathy based upon UACR results. Hypertensive and nonhypertensive groups were compared for various lipid indices by employing Mann–Whitney U test.

3. Results

The study population constituted 122 females with age 45.27 + 12.42 years and 110 males with 47.98 + 11.30 years. Gender-wise comparison for various lipid parameters is depicted in Table 1 where differences were significant for HDLc, non-HDLc, and LDLc. Table 2 demonstrates Pearson's correlation for lipid parameter with anthropometric, blood pressure, and biochemical risk factors, where non-HDLc, sdLDLc, and sdLDLc/lbLDLc were found to be better correlated with aforementioned designated risk factors. The differences for non-HDLc and sdLDLc were found to be most significant among subjects with or without metabolic syndrome (Table 3). Assessing metabolic cluster-wise increment (as per metabolic syndrome definition) we observed that (excluding criteria inclusive markers like triglycerides and HDLc), serum non-HDLc, sdLDLc, and sdLDLc/lbLDLc increased gradually among subjects with no component to subjects

having all components of metabolic syndrome (Table 4). The results for various glycated hemoglobin based groups for sdLDLc/lbLDLc were not found to be significant which may be due to noninclusion of known diabetics. However, Figure 1 suggests a rapid increase in the number of sdLDLc in comparison to lbLDLc (sdLDLc/lbLDLc) with patient HbA1c group having HbA1c > 7.0%; however, the results were not significant but authors feel that type-2 statistical error due to small size of group-4 ($n = 12$) could be one reason behind this nonsignificance. There were no differences among any of the lipid markers between subjects with or without hypertension (Table 5). Based upon urine albumin creatinine ratio (UACR) we only observed significant differences for non-HDLc and cLDLc (Table 6).

4. Discussion

Calculated sdLDLc and its ratio with lbLDLc have provided marginally improved risk prediction by being better and significantly correlated with multiple traditional and established ASCVD markers. In this regard it is important to appreciate that sdLDLc levels were clearly found to be increased in subjects having metabolic syndrome and insulin resistance and these levels increase in a staircase manner from no risk factors to acquiring all five components of metabolic syndrome as also demonstrated by other researchers [17, 18]. However, it appears that other lipid markers especially non-HDLc, VLDL, triglycerides, and HDLc also worsened

TABLE 2: Correlation between various lipid parameters with anthropometric, blood pressure, and biochemical risk factors with lipid parameters.

		Total cholesterol	Fasting triglyceride	HDLc	mLDLc	Non-HDLc	cLDLc	sdLDLc	lbLDLc	sdLDLc/lbLDLc	mLDLc/HDLc
Body Mass Index (BMI)	Pearson Correlation	0.197**	0.115	0.126	0.032	0.139*	0.080	0.099	0.018	0.093	-0.045
	Sig. (2-tailed)	0.003	0.081	0.056	0.626	0.035	0.224	0.132	0.783	0.160	0.500
	N	232	232	230	230	232	232	232	232	232	230
Waist to hip ratio (WhpR)	Pearson Correlation	0.205**	0.173**	-0.004	0.169*	0.191**	0.123	0.202**	0.079	0.122	0.095
	Sig. (2-tailed)	0.002	0.008	0.957	0.010	0.004	0.062	0.002	0.231	0.063	0.150
	N	232	232	230	230	232	232	232	232	232	230
Glycated hemoglobin (HbA1c %)	Pearson Correlation	-0.050	0.101	0.032	-0.011	-0.040	-0.132*	0.074	-0.068	0.149*	-0.028
	Sig. (2-tailed)	0.456	0.129	0.632	0.864	0.546	0.046	0.268	0.309	0.025	0.671
	N	228	228	227	226	228	228	228	228	228	227
Serum insulin (mIU/L)	Pearson Correlation	0.091	0.169*	-0.068	0.001	0.109	0.026	0.090	-0.062	0.135*	0.011
	Sig. (2-tailed)	0.169	0.010	0.310	0.989	0.102	0.696	0.178	0.348	0.041	0.867
	N	228	228	227	227	228	228	228	228	228	227
HOMA-IR***	Pearson Correlation	0.097	.290**	-0.085	-0.035	0.125	-0.032	0.143*	-0.143*	0.0322**	0.001
	Sig. (2-tailed)	0.146	0.000	0.199	0.598	0.060	0.627	0.031	0.031	0.000	0.989
	N	228	228	227	227	228	228	228	228	228	227
HOMA % B****	Pearson Correlation	0.041	0.022	-0.029	0.022	0.041	0.042	0.021	0.014	-0.029	0.007
	Sig. (2-tailed)	0.536	0.744	0.659	0.743	0.535	0.530	0.751	0.830	0.666	0.911
	N	228	228	227	227	228	228	228	228	228	227
Urine albumin creatinine ratio (UACR)	Pearson Correlation	0.107	0.114	-0.079	0.098	0.154*	0.083	0.130	0.059	0.049	0.153*
	Sig. (2-tailed)	0.162	0.135	0.304	0.197	0.042	0.276	0.088	0.443	0.523	0.044
	N	174	174	173	173	174	174	174	174	174	173
Systolic BP (SBP) mm of Hg	Pearson Correlation	0.125	0.143*	0.112	0.020	0.078	-0.001	0.087	-0.048	0.122	-0.073
	Sig. (2-tailed)	0.058	0.029	0.090	0.758	0.238	0.984	0.185	0.468	0.065	0.271
	N	232	232	230	230	232	232	232	232	232	230
Diastolic BP (DBP) mm of Hg	Pearson Correlation	0.145*	0.160*	0.056	0.006	0.110	0.031	0.096	-0.043	0.130*	-0.018
	Sig. (2-tailed)	0.028	0.015	0.401	0.934	0.095	0.640	0.144	0.512	0.047	0.781
	N	232	232	230	230	232	232	232	232	232	230

*Correlation is significant at the 0.05 level (2-tailed); ** correlation is significant at the 0.01 level (2-tailed); *** Homeostasis Model Assessment for Insulin Resistance (HOMA-IR); **** Homeostasis Model Assessment for insulin sensitivity (HOMA % B).

TABLE 3: Comparison of lipid indices among subjects with and without metabolic syndrome as per IDF criteria.

Lipid parameter	Metabolic syndrome (as per IDF criteria)	N	Mean	Std. dev	Sig. (2-tailed)[**]
HDLc (mmol/L)	Present	121	0.94	0.25	_0.028_
	Not present	108	1.02	0.26	
mLDLc[*] (mmol/L)	Present	121	2.80	0.76	_0.013_
	Not present	107	2.56	0.66	
Non-HDLc (mmol/L)	Present	121	3.63	0.60	_0.002_
	Not present	108	3.36	0.65	
cLDLc[**] (mmol/L)	Present	121	2.79	0.52	0.569
	Not present	108	2.75	0.54	
sdLDLc[***] (mmol/L)	Present	121	0.92	0.33	_0.000_
	Not present	108	0.70	0.29	
lbLDLc (mmol/L)[****]	Present	121	1.87	0.54	0.575
	Not present	108	1.83	0.51	
sdLDLc/lbLDLc	Present	121	0.55	0.51	_0.010_
	Not present	108	0.40	0.38	

[*]Measured LDL-cholesterol (mLDLc); [**]calculated LDL-cholesterol (cLDLc) as per Friedewald's equation; [***]small dense LDL-cholesterol (sdLDLc); [****]large buoyant LDL-cholesterol (lbLDLc).

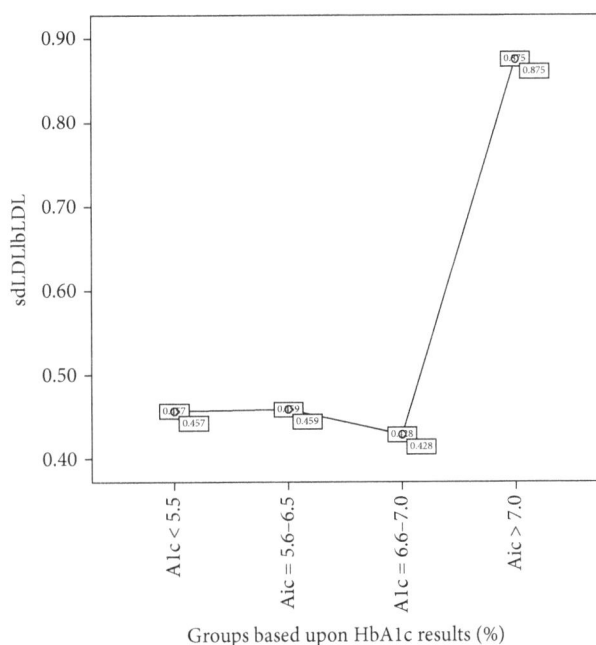

FIGURE 1: Comparison between groups based upon HbA1c values for sdLDLc/lbLDLc by Kruskal Wallis Test ($p = 0.430$).

with accumulation of various metabolic cluster which brings us to the reality that these lipoprotein bound and free lipids are constantly modifying and contributing to each other. Therefore the previously used entity of "atherogenic dyslipidemia" being low HDLc and high triglycerides can be broadened to also include increases in sdLDLc, non-HDLc, and VLDLc [18–20].

Non-HDLc showed more correlation with BMI and WhpR than other lipid markers including sdLDLc and its ratio with lbLDLc; however, the latter seem to be better associated with WhpR. Recent studies have also highlighted

WhpR to be more predictive of ASCVD risk than BMI which seems to be more representative of muscle mass [21, 22].

Glycation rates have been associated with enhanced atherosclerosis and morbidity and mortality liked to CVD [23]. In this regard our study which did not include any known diabetics has only demonstrated sdLDLc/lbLDLc ratios to have mild weak correlation with glycated hemoglobin and slightly higher results group of diagnosed diabetics. This strengthens our viewpoint that some degree of lipid derangements does start with increasing glycation in the shape of increased numbers of small-sized LDL in comparison to large LDL particles in the plasma as highlighted by some researchers [23–26].

While both diastolic and systolic blood pressure are included in metabolic syndrome, still we could not observe significant differences for various lipid markers among hypertensive and nonhypertensive patients which is in line with the findings of Esteghamati et al. [27]. However, we found the ratio between sdLDLc/lbLDLc to have weak correlation with systolic and diastolic blood pressures, which indicates that slight derangements in lipid metabolism do develop in subjects having raised blood pressures [28, 29]. sdLDL/lbLDLc along with non-HDLc and mLDLc/HDLc did show some weak correlation with UACR but it was only non-HDLc that demonstrated significant differences between subjects with and without nephropathy. These findings are consistent with the results of Palazhy et al. [30, 31].

Certain _limitations_ to the study must be acknowledged. We have utilized Srisawasdi et al.'s regression equation for measuring sdLDLc and lbLDLc, which still needs to be validated by epidemiological studies. Moreover, our study has small sample size and cross-sectional design where type-2 statistical errors could have confounded our findings so large clinical randomized clinical trials may be carried out to augment or disapprove our observations.

The is a _clinically important_ study as it not only has highlighted association between lipid parameters with various

TABLE 4: Comparison of lipid indices among various groups formulated based upon the number of metabolic syndrome components present in subjects from 0 implying absence of any metabolic syndrome component to 5 implying all 5 components present in a subject.

Metabolic syndrome groups		Total cholesterol (mmol/L)	Fasting triglycerides (mmol/L)	HDLc (mmol/L)	mLDLc (mmol/L)	Non-HDLc (mmol/L)	cLDLc (mmol/L)	sdLDLc (mmol/L)**	lbLDLc (mmol/L)***	sdLDL/lbLDL	mLDLc/HDLc	VLDLc (mmol/L)
0	Mean	4.33	1.07	1.15	2.63	3.18	2.69	0.67	1.97	0.34	2.36	0.21
	N	18	18	18	18	18	18	18	18	18	18	18
	Std. dev	0.49	0.31	0.18	0.65	0.53	0.50	0.24	0.44	0.08	0.74	0.06
1	Mean	4.31	1.28	0.99	2.54	3.31	2.73	0.65	1.82	0.37	2.66	0.26
	N	42	42	42	41	42	42	42	42	42	42	42
	Std. dev	0.69	0.73	0.27	0.70	0.68	0.52	0.33	0.56	0.53	0.99	0.15
2	Mean	4.43	1.54	0.97	2.67	3.45	2.76	0.80	1.87	0.45	2.88	0.31
	N	62	62	62	62	62	62	62	62	62	62	62
	Std. dev	0.63	0.62	0.26	0.64	0.62	0.56	.26	0.48	0.18	0.87	0.12
3	Mean	4.49	1.66	0.93	2.77	3.59	2.80	0.86	1.90	0.45	3.03	0.33
	N	46	46	46	46	46	46	46	46	46	46	46
	Std. dev	0.57	0.56	0.19	0.80	0.60	0.54	0.37	0.50	0.17	0.88	0.11
4	Mean	4.62	1.92	0.91	2.79	3.68	2.84	0.93	1.86	0.54	3.16	0.38
	N	36	36	36	36	36	36	36	36	36	36	36
	Std. dev	0.53	0.71	0.19	0.82	0.54	0.45	0.35	0.57	0.27	1.02	0.14
5	Mean	4.78	2.26	1.00	2.68	3.78	2.76	0.99	1.69	.8413	2.9269	0.45
	N	25	25	25	25	25	25	25	25	25	25	25
	Std. dev	0.56	0.99	0.39	0.74	0.69	0.59	0.29	0.63	1.03	1.03	0.19
p value*		0.015	<0.001	0.001	0.725	0.001	0.896	0.000	0.910	<0.001	0.055	<0.001

*Kruskal Wallis Test. **Small density LDL-cholesterol (sdLDL-c) by Srisawasdi et al. regression equation. ***Large buoyant LDL-cholesterol (lbLDL-c) by Srisawasdi et al. regression equation.

TABLE 5: Comparison of lipid indices among subjects with or without hypertension.

Lipid parameter	Hypertension	N	Mean rank	Asymp. sig.
Total cholesterol (mmol/L)	Absent	205	116.43	0.966
	Present	27	117.02	
Fasting triglycerides (mmol/L)	Absent	205	115.18	0.409
	Present	27	126.52	
HDLc (mmol/L)	Absent	203	115.47	0.985
	Present	27	115.72	
mLDLc (mmol/L)[*]	Absent	204	115.98	0.760
	Present	26	111.75	
Non-HDLc (mmol/L)	Absent	205	116.85	0.825
	Present	27	113.81	
cLDLc (mmol/L)[**]	Absent	205	118.01	0.345
	Present	27	105.04	
sdLDLc (mmol/L)[***]	Absent	205	116.06	0.783
	Present	27	119.85	
lbLDLc (mmol/L)[****]	Absent	205	117.85	0.399
	Present	27	106.26	
sdLDLc/lbLDLc	Absent	205	116.70	0.903
	Present	27	115.02	
mLDLc/HDLc	Absent	203	116.43	0.561
	Present	27	108.50	
VLDL-cholesterol (mmol/L)	Absent	205	115.18	0.409
	Present	27	126.52	

[*]As per Friedewald's equation; [**]measured using nonparametric test (SPSS); [***]small density LDL-cholesterol (sdLDLc) by Srisawasdi et al. regression equation; [****]large buoyant LDL-cholesterol (lbLDLc) by Srisawasdi et al. regression equation.

TABLE 6: Comparison of lipid parameters among subjects with and without nephropathic changes as measured by urine albumin creatinine ratio (UACR).

Lipid parameter	Urine albumin creatinine ratio (UACR)	N	Mean	Std. deviation	Sig. (2-tailed)
Total cholesterol (mmol/L)	<2.5 mg/g	135	4.36	0.54	0.006
	>2.4 mg/g	39	4.6	0.45	
Fasting triglycerides (mmol/L)	<2.5 mg/g	135	1.47	0.669	0.082
	>2.4 mg/g	39	1.68	0.63	
HDLc (mmol/L)	<2.5 mg/g	134	.9951	0.27	0.084
	>2.4 mg/g	39	.9254	0.20	
mLDLc (mmol/L)[*]	<2.5 mg/g	134	2.6166	0.73662	0.413
	>2.4 mg/g	39	2.7156	.63762	
Non-HDLc (mmol/L)	<2.5 mg/g	135	3.3656	.59011	0.000
	>2.4 mg/g	39	3.6797	.43173	
cLDLc (mmol/L)[**]	<2.5 mg/g	135	2.6997	.52434	0.007
	>2.4 mg/g	39	2.9110	.38825	
sdLDLc (mmol/L)[***]	<2.5 mg/g	135	.7586	.35181	0.172
	>2.4 mg/g	39	.8315	.27045	
lbLDLc (mmol/L)[****]	<2.5 mg/g	135	0.4277	.32988	0.411
	>2.4 mg/g	39	0.4599	.16809	
sdLDLc/lbLDLc	<2.5 mg/g	134	2.7685	.88169	0.103
	>2.4 mg/g	39	3.0651	1.01300	
mLDLc/HDLc	<2.5 mg/g	135	.2950	.13228	0.082
	>2.4 mg/g	39	.3360	.12625	

[*]As per Friedewald's equation; [**]measured using independent sample t-test (SPSS); [***]small density LDL-cholesterol (sdLDLc) by Srisawasdi et al. regression equation; [****]large buoyant LDL-cholesterol (lbLDLc) by Srisawasdi et al. regression equation.

traditional risk factors but also has allowed us to understand how different lipid indices vary across various anthropometric and biochemical groups. The study has also opened up some new avenues for research on LDL-fractions so as to learn in detail the risk association between lipoprotein indices and cardiovascular diseases. Moreover, the study was also able to highlight the superiority of non-HDLc over available lipid indices in measuring ASCVD risk.

5. Conclusion

Calculated sdLDLc and its ratio with lbLDLc were not able to augment any ASCVD risk prediction over and above non-HDLc. However, it becomes apparent that other lipid markers including calculated LDLc and measured LDLc are less associated with traditional ASCVD markers than non-HDLc, sdLDLc, and sdLDLc/lbLDLc in predicting metabolic syndrome, nephropathy, glycation status, and hypertension. However, the results need to be validated by methods which directly measure sdLDLc or LDL-fractions.

References

[1] A. E. Moran, M. H. Forouzanfar, G. A. Roth et al., "Temporal trends in ischemic heart disease mortality in 21 world regions, 1980 to 2010: The global burden of disease 2010 study," *Circulation*, vol. 129, no. 14, pp. 1483–1492, 2014.

[2] A. M. Prentice, P. Rayco-Solon, and S. E. Moore, "Insights from the developing world: Thrifty genotypes and thrifty phenotypes," *Proceedings of the Nutrition Society*, vol. 64, no. 2, pp. 153–161, 2005.

[3] M.-R. Taskinen, "LDL-cholesterol, HDL-cholesterol or triglycerides - Which is the culprit?" *Diabetes Research and Clinical Practice*, vol. 61, no. 1, pp. S19–S26, 2003.

[4] B. B. Adhyaru and T. A. Jacobson, "New cholesterol guidelines for the management of atherosclerotic cardiovascular disease risk: a comparison of the 2013 american college of cardiology/american heart association cholesterol guidelines with the 2014 national lipid association recommendations for patient-centered management of dyslipidemia," *Cardiology Clinics*, vol. 33, no. 2, pp. 181–196, 2015.

[5] J. Yeboah, T. S. Polonsky, R. Young et al., "Utility of Nontraditional Risk Markers in Individuals Ineligible for Statin Therapy According to the 2013 American College of Cardiology/American Heart Association Cholesterol Guidelines," *Circulation*, vol. 132, no. 10, pp. 916–922, 2015.

[6] K. Hübner, T. Schwager, K. Winkler, J.-G. Reich, and H.-G. Holzhütter, "Computational lipidology: Predicting lipoprotein density profiles in human blood plasma," *PLoS Computational Biology*, vol. 4, no. 5, Article ID e1000079, 2008.

[7] J. R. Patsch, R. L. Jackson, and A. M. Gotto Jr., "Evaluation of the classical methods for the diagnosis of type III hyperlipoproteinemia," *Klinische Wochenschrift*, vol. 55, no. 21, pp. 1025–1030, 1977.

[8] C. J. Packard and J. Shepherd, "Lipoprotein heterogeneity and apolipoprotein B metabolism," *Arteriosclerosis, Thrombosis, and Vascular Biology*, vol. 17, no. 12, pp. 3542–3556, 1997.

[9] E. Multia, H. Sirén, K. Andersson et al., "Thermodynamic and kinetic approaches for evaluation of monoclonal antibody - Lipoprotein interactions," *Analytical Biochemistry*, vol. 518, pp. 25–34, 2017.

[10] S. Koba, T. Hirano, Y. Ito et al., "Significance of small dense low-density lipoprotein-cholesterol concentrations in relation to the severity of coronary heart diseases," *Atherosclerosis*, vol. 189, no. 1, pp. 206–214, 2006.

[11] M. Rizzo and K. Berneis, "Small, dense low-density-lipoproteins and the metabolic syndorme," *Diabetes/Metabolism Research and Reviews*, vol. 23, no. 1, pp. 14–20, 2007.

[12] P. Srisawasdi, S. Chaloeysup, Y. Teerajetgul et al., "Estimation of plasma small dense LDL cholesterol from classic lipid measures," *American Journal of Clinical Pathology*, vol. 136, no. 1, pp. 20–29, 2011.

[13] *Waist circumference and waist-hip ratio: Report of a WHO expert consultation*, Geneva, 2008, http://citeseerx.ist.psu.edu/viewdoc/download.

[14] D. R. Matthews, J. P. Hosker, A. S. Rudenski, B. A. Naylor, D. F. Treacher, and R. C. Turner, "Homeostasis model assessment: insulin resistance and β-cell function from fasting plasma glucose and insulin concentrations in man," *Diabetologia*, vol. 28, no. 7, pp. 412–419, 1985.

[15] M. González-Ortiz, E. Martínez-Abundis, O. Jacques-Camarena, S. O. Hernández-González, I. G. Valera-González, and M. G. Ramos-Zavala, "Prevalence of metabolic syndrome in adults with excess of adiposity: comparison of the adult treatment panel III criteria with the international diabetes federation definition," *Acta Diabetologica*, vol. 43, no. 3, pp. 84–86, 2006.

[16] Y. T. Bee Jr., K. K. Haresh, and S. Rajibans, "Prevalence of metabolic syndrome among malaysians using the international diabetes federation, national cholesterol education program and modified world health organization definitions," *Malays J Nutr*, vol. 14, no. 1, pp. 65–77, 2008.

[17] K. Kikkawa, K. Nakajima, Y. Shimomura et al., "Small dense LDL cholesterol measured by homogeneous assay in Japanese healthy controls, metabolic syndrome and diabetes patients with or without a fatty liver," *Clinica Chimica Acta*, vol. 438, pp. 70–79, 2015.

[18] P. P. Toth, "Insulin resistance, small LDL particles, and risk for atherosclerotic disease," *Current Vascular Pharmacology*, vol. 12, no. 4, pp. 653–657, 2014.

[19] D. Nikolic, N. Katsiki, G. Montalto, E. R. Isenovic, D. P. Mikhailidis, and M. Rizzo, "Lipoprotein subfractisons in metabolic syndrome and obesity: clinical significance and therapeutic approaches," *Nutrients*, vol. 5, no. 3, pp. 928–948, 2013.

[20] S. W. Kim, J. H. Jee, H. J. Kim et al., "Non-HDL-cholesterol/HDL-cholesterol is a better predictor of metabolic syndrome and insulin resistance than apolipoprotein B/apolipoprotein A1," *International Journal of Cardiology*, vol. 168, no. 3, pp. 2678–2683, 2013.

[21] N. Motamed, D. Perumal, F. Zamani et al., "Conicity index and waist-to-hip ratio are superior obesity indices in predicting 10-year cardiovascular risk among men and women," *Clinical Cardiology*, vol. 38, no. 9, pp. 527–534, 2015.

[22] X. Bi, S. L. Tey, C. Leong, R. Quek, Y. T. Loo, and C. J. Henry, "Correlation of adiposity indices with cardiovascular disease

risk factors in healthy adults of Singapore: a cross-sectional study," *BMC Obesity*, vol. 3, no. 1, 2016.

[23] S.-I. Yamagishi, K. Nakamura, and T. Matsui, "Advanced glycation end products (AGEs) and their receptor (RAGE) system in diabetic retinopathy," *Current Drug Discovery Technologies*, vol. 3, no. 1, pp. 83–88, 2006.

[24] S. Suh, H.-D. Park, S. W. Kim et al., "Smaller Mean LDL particle size and higher proportion of small dense LDL in Korean type 2 diabetic patients," *Diabetes & Metabolism*, vol. 35, no. 5, pp. 536–542, 2011.

[25] Y. Yoon, J. Song, H. D. Park, K.-U. Park, and J. Q. Kim, "Significance of small dense low-density lipoproteins as coronary risk factor in diabetic and non-diabetic Korean populations," *Clinical Chemistry and Laboratory Medicine*, vol. 43, no. 4, pp. 431–437, 2005.

[26] T. Hayashi, T. Hirano, T. Yamamoto, Y. Ito, and M. Adachi, "Intensive insulin therapy reduces small dense low-density lipoprotein particles in patients with type 2 diabetes mellitus: relationship to triglyceride-rich lipoprotein subspecies," *Metabolism - Clinical and Experimental*, vol. 55, no. 7, pp. 879–884, 2006.

[27] A. Esteghamati, S. Asnafi, M. Eslamian, S. Noshad, and M. Nakhjavani, "Associations of small dense low-density lipoprotein and adiponectin with complications of type 2 diabetes," *Endocrine Research*, vol. 40, no. 1, pp. 14–19, 2015.

[28] P. Sharma, P. Purohit, and R. Gupta, "Cardiac risk factors in descendants of parents with history of coronary artery disease (CAD): an evaluation focusing on small dense low density lipoprotein cholesterol (sdLDLc) and high density lipoprotein cholesterol (HDLc)," *Indian Journal of Biochemistry and Biophysics*, vol. 50, no. 5, pp. 453–461, 2013.

[29] H. Shen, J. Zhou, G. Shen, H. Yang, Z. Lu, and H. Wang, "Correlation between serum levels of small, dense low-density lipoprotein cholesterol and carotid stenosis in cerebral infarction patients >65 years of age," *Ann Vasc Surg*, vol. 28, no. 2, pp. 375–380, 2014.

[30] S. Palazhy and V. Viswanathan, "Lipid abnormalities in type 2 diabetes mellitus patients with overt nephropathy," *Diabetes & Metabolism Journal*, vol. 41, no. 2, p. 128, 2017.

[31] E. Abd-Allha, B. Hassan, M. Abduo, S. Omar, and H. Sliem, "Small dense low-density lipoprotein as a potential risk factor of nephropathy in type 2 diabetes mellitus," *Indian Journal of Endocrinology and Metabolism*, vol. 18, no. 1, pp. 94–98, 2014.

21

Elucidation of the Role of Lectin-Like oxLDL Receptor-1 in the Metabolic Responses of Macrophages to Human oxLDL

Danielle W. Kimmel,[1,2] William P. Dole,[3] and David E. Cliffel[1,2]

[1]*Department of Chemistry, Vanderbilt University, VU Station B, Nashville, TN 37235-1822, USA*
[2]*Vanderbilt Institute for Integrative Biosystems Research and Education, Vanderbilt University, Nashville, TN 37235-1809, USA*
[3]*Novartis Institutes for Biomedical Research, 220 Massachusetts Ave. 360C, Cambridge, MA 02139, USA*

Correspondence should be addressed to David E. Cliffel; d.cliffel@vanderbilt.edu

Academic Editor: Zufeng Ding

Atherogenesis is the narrowing of arteries due to plaque build-up that results in cardiovascular disease that can lead to death. The macrophage lectin-like oxidized LDL receptor-1 (LOX-1), also called the oxidized low-density lipoprotein receptor 1 (OLR1), is currently thought to aid in atherosclerotic disease progression; therefore metabolic studies have potential to both provide mechanistic validation for the role of LOX-1 in disease progression and provide valuable information regarding biomarker strategies and clinical imaging. One such mechanistic study is the upregulation of LOX-1 by methylated bacterial DNA and deoxy-cytidylate-phosphate-deoxy-guanylate-DNA (CpG)-DNA exposure. CpG-DNA is known to promote oxidative burst responses in macrophages, due to its direct binding to toll-like receptor 9 (TLR9) leading to the initiation of an NF-κB mediated immune response. In addition to the upregulation of macrophage LOX-1 expression, these studies have also examined the macrophage metabolic response to murine LOX-1/OLR1 antibody exposure. Our data suggests the antibody exposure effectively blocks LOX-1 dependent oxLDL metabolic activation of the macrophage, which was quantified using the multianalyte microphysiometer (MAMP). Using the MAMP to examine metabolic fluctuations during various types of oxLDL exposure, LOX-1 upregulation and inhibition provide valuable information regarding the role of LOX-1 in macrophage activation of oxidative burst.

1. Introduction

The uptake of oxidized low-density lipoprotein (oxLDL) is thought to promote atherogenesis and lead to clinically relevant cardiovascular diseases [1–4]. Our previous work using the MAMP showed that there is a correlation between oxLDL uptake by macrophages and increases in glucose and oxygen consumption and increases in lactate production and extracellular acidification rates [5]. Research suggests that macrophage uptake of oxLDL is regulated by cell surface atherogenic oxLDL receptors including class A macrophage scavenger receptors, class B type I scavenger receptors, macrosialin (CD68) [1, 3]. Previous research suggests the importance of the lectin-like oxidized low-density lipoprotein receptor-1 (LOX-1) [1, 3].

LOX-1 is a 50 kDa type II membrane glycoprotein composed of 273 amino acids [1, 3], which can selectively bind and internalize oxLDL [6]. Studies have shown that LOX-1 is expressed in macrophages among other cells but can also be stimulated in a variety of cell types using TNF-α, PMA, and shear stress [3]. The signaling pathway activated by LOX-1 promotes superoxide radical formation through the NADPH oxidase complex as well as promoting transcription of VCAM-1 through NF-κB activation, both of which induce inflammatory responses leading to foam cell formation and atherogenesis [2, 4, 6]. Researchers have also studied LOX-1 deficient mice and subsequent knockout hybridization to find resistance toward oxLDL-induced arterial impairment, suggesting a vital role for LOX-1 in atherogenesis and arterial remodeling [6].

In the present study, our focus is to elucidate the role of LOX-1 in the biological response of murine macrophages to human oxLDL. Previous work has demonstrated that macrophages are a key cell type in atherosclerotic lesions

[1, 3, 4, 6–8]. Their activity and responses to oxLDL particles play a major role in the pathogenesis of the disease. Understanding the metabolic responses of macrophages to oxLDL mediated by LOX-1 has the potential to both provide additional mechanistic validation of this receptor in disease processes and support clinical imaging and biomarker strategies for future investigation. Here, we utilize the MAMP to simultaneously measure glucose and oxygen consumption, lactate production, and extracellular acidification before, during, and after macrophage assault by oxLDL. This study focuses on the characterization of macrophage metabolic responses to various oxidation levels of LDL and the characterization of LOX-1 mediated effects of oxLDL on macrophage metabolism.

2. Materials and Methods

2.1. Materials. Unless otherwise noted, all materials used were previously reported in prior work [5]. In order to examine the differences in macrophage metabolic responses from various oxidation levels of LDL, an oxLDL kit was purchased by Novartis International (Boston, MA) from Kalen Biomedical (Montgomery Village, MD). The kit contained LDL, low oxLDL, medium oxLDL, and high oxLDL. Additional oxLDL was purchased through Intracel (Frederick, MD) for comparison with Kalen high oxLDL. Murine deoxy-cytidylate-phosphate-deoxy-guanylate- (CpG-) DNA was purchased through Hycult Biotech (Plymouth Meeting, PA). Mouse LOX-1/OLR1 antibody (MAB1564) and rat IgG2A isotype control (MAB006) were purchased through R&D Systems (Minneapolis, MN). Trehalose was purchased through Sigma-Aldrich (St. Louis, MO).

2.2. Methods. Cell culturing methods and experimental protocols were previously reported in prior work [5], all exceptions noted below. Initial oxLDL experiments were performed using each of the four types of LDL: native, low, medium, and highly oxidized. Throughout the course of multiple experiments, macrophages were exposed to each type of LDL at four concentrations: $1 \mu g/mL$, $10 \mu g/mL$, $50 \mu g/mL$, and $100 \mu g/mL$. Additionally, after an hour of stabilization time, macrophages were exposed to $25 \mu g/mL$ mouse LOX-1/OLR1 antibody for 20 min, followed by 60 min of recovery to establish a baseline for the metabolic profile of the antibody. Macrophages were also exposed to $3 \mu mol/L$ CpG-DNA for 30 min. For the studies examining the effects of mouse LOX-1/OLR1 antibody prior to oxLDL exposure, macrophages were exposed to $25 \mu g/mL$ mouse LOX-1/OLR1 antibody for 10 min followed directly by a 6 min exposure to the 4 previously mentioned dosages of Kalen high oxLDL. Treatment and analysis of data were performed as previously reported [5].

3. Results

The present study is designed to investigate the relationship between the dynamic metabolic profiles of oxLDL uptake by macrophages with and without murine LOX-1/OLR1

antibodies present. To this end, it was vital to study the metabolic response of macrophages exposed to variations in oxidation of LDL as well as concentrations to ensure optimization of response before antibody exposure. To do this, experiments were executed using 6 min dosages of native LDL, low oxLDL, medium oxLDL, or high oxLDL at four concentrations: $1 \mu g/mL$, $10 \mu g/mL$, $50 \mu g/mL$, and $100 \mu g/mL$. During each experiment glucose and oxygen consumption, lactate production, and extracellular acidification were simultaneously measured to produce the dynamic response seen. Each data set was then compared with the baseline to assess significance. Additionally, a comparison of two commercially available oxLDL sources was performed to gauge which type incited a substantial and reproducible oxidative burst phenomenon. A p value of less than 0.05 was taken as statistically significant.

Native LDL was used as a baseline to determine if there were any significant variations in response during exposure or recovery due to the introduction of lipoproteins into the chamber. LDL exposure was performed at the four separate concentrations previously mentioned. Maximum changes in peak heights were taken during LDL exposure and compared with the baseline, prior to exposure. Using LDL response as a control, various forms of oxLDL were then studied and the resulting data was treated in the same manner. OxLDL from the Kalen Biomedical kit contained various forms of three different oxidation levels: low, medium, and high. In addition to the Kalen oxLDL, the Intracel version, with only one oxidation level, was also studied. For ease of comparison between LDL and each type of oxLDL each analyte is compiled into separate graphs.

Glucose consumption by the macrophages during exposure showed great variation between native and oxidized LDL, as detailed in Figure 1 and Table 1. For native LDL, there were notable increases in glucose consumption for the $10 \mu g/mL$ and $100 \mu g/mL$ concentrations. Low oxLDL exposure did not reveal any significant increases in glucose consumption; however the medium oxLDL at $1 \mu g/mL$ and $50 \mu g/mL$ did show significance. While this does not show a clear trend in peak height increases for medium oxLDL exposures, it does showcase the increase in glucose consumption, which we have found to be a pivotal signature of oxLDL initiated oxidative burst. The moderately oxidized LDL is not providing the onset of foam cell formation but is inhibiting basal metabolism. High oxLDL and the Intracel oxLDL provided the clearest peak height increases, but with great biological variation causing an increase in standard error. The data suggests a rough trend of increasing glucose consumption based on an increase in oxidation of LDL. The decrease of glucose consumption with increasing concentrations of LDL exposure indicates that there might be receptor saturation preventing exponential update of glucose, thus limiting the onset of oxidative burst.

A second analyte, lactate production, is also indicative of cellular metabolic health so it was simultaneously measured while macrophages underwent varying types of LDL exposures. The resulting data can be seen in Figure 2 and Table 2. Native LDL produced a statistically significant increase in response for the $1 \mu g/mL$ and $100 \mu g/mL$ concentration but

TABLE 1: Metabolic glucose consumption of various LDL exposure to macrophages. Mean peak MAMP height changes and standard errors for each experimental condition during exposure. All p values are calculated based on peak height versus 10 min average basal value prior to exposure. Terms in bold are representative of $p < 0.05$, indicating statistical significance.

| | Glucose | | | | |
	Native LDL (%)	Low oxLDL (%)	Medium oxLDL (%)	High oxLDL (%)	Intracel oxLDL (%)
1 μg/ml	0.14 ± 18.3	31.2 ± 32.3	**3.67 ± 1.4**	7.5 ± 4.4	7.63 ± 12.4
	$p = 0.9944$	$p = 0.3793$	**p = 0.0348**	$p = 0.5992$	$p = 0.582$
10 μg/ml	**8.2 ± 2.0**	2.5 ± 3.9	20.8 ± 22.0	56.4 ± 59.3	0.8 ± 3.2
	p = 0.0263	$p = 0.5498$	$p = 0.377$	$p = 0.3792$	$p = 0.8164$
50 μg/ml	5.9 ± 10.0	1.3 ± 5.5	**3.19 ± 0.41**	30.6 ± 22.3	34.12 ± 29.8
	$p = 0.5968$	$p = 0.8451$	**p = 0.0046**	$p = 0.2964$	$p = 0.3309$
100 μg/ml	**6.6 ± 1.4**	19.1 ± 17.8	8.3 ± 2.6	3.93 ± 4.07	**5.91 ± 1.5**
	p = 0.0023	$p = 0.4201$	$p = 0.0506$	$p = 0.3785$	**p = 0.0397**

TABLE 2: Metabolic lactate production of various LDL exposure to macrophages. Mean peak MAMP height changes and standard errors for each experimental condition during exposure. All p values are calculated based on peak height versus 10 min average basal value prior to exposure. Terms in bold are representative of $p < 0.05$, indicating statistical significance.

| | Lactate | | | | |
	Native LDL (%)	Low oxLDL (%)	Medium oxLDL (%)	High oxLDL (%)	Intracel oxLDL (%)
1 μg/ml	**3.2 ± 0.68**	1.42 ± 1.58	1.81 ± 1.3	1.54 ± 2.8	2.0 ± 2.1
	p = 0.0022	$p = 0.3987$	$p = 0.2047$	$p = 0.6001$	$p = 0.3875$
10 μg/ml	1.0 ± 1.0	0.39 ± 1.5	17.2 ± 20.0	0.91 ± 0.63	**2.7 ± 0.66**
	$p = 0.3699$	$p = 0.8039$	$p = 0.4134$	$p = 0.2198$	**p = 0.0143**
50 μg/ml	0.47 ± 1.1	**0.84 ± 0.16**	0.38 ± 3.8	2.0 ± 1.8	**8.1 ± 2.35**
	$p = 0.698$	**p = 0.0040**	$p = 0.9534$	$p = 0.3012$	**p = 0.0117**
100 μg/ml	**4.0 ± 1.5**	**2.7 ± 1.1**	1.21 ± 1.2	**15.2 ± 4.29**	**7.03 ± 3.0**
	p = 0.0445	**p = 0.0441**	$p = 0.379$	**p = 0.0172**	**p = 0.0438**

TABLE 3: Metabolic oxygen consumption of various LDL exposure to macrophages. Mean peak MAMP height changes and standard errors for each experimental condition during exposure. All p values are calculated based on peak height versus 10 min average basal value prior to exposure. Terms in bold are representative of $p < 0.05$, indicating statistical significance.

| | Oxygen | | | | |
	Native LDL (%)	Low oxLDL (%)	Medium oxLDL (%)	High oxLDL (%)	Intracel oxLDL (%)
1 μg/ml	**2.9 ± 0.36**	11.8 ± 6.3	0.39 ± 0.72	21.7 ± 20.3	1.6 ± 4.2
	p = 0.0013	$p = 0.1053$	$p = 0.6596$	$p = 0.3639$	$p = 0.7151$
10 μg/ml	0.3 ± 1.2	3.47 ± 5.9	17.1 ± 14.1	6.27 ± 6.55	1.0 ± 1.65
	$p = 0.8103$	$p = 0.8338$	$p = 0.2794$	$p = 0.3822$	$p = 0.5915$
50 μg/ml	1.3 ± 2.2	12.8 ± 11.2	22.6 ± 20.7	**2.8 ± 0.78**	**8.88 ± 1.47**
	$p = 0.583$	$p = 0.3748$	$p = 0.3114$	**p = 0.0161**	**p = 0.0028**
100 μg/ml	**1.7 ± 0.6**	12.1 ± 15.9	20.7 ± 18.2	2.36 ± 1.21	3.14 ± 1.7
	p = 0.0371	$p = 0.5028$	$p = 0.3084$	$p = 0.1594$	$p = 0.1085$

exhibited no other trends toward oxidative burst. Using lowly oxidized LDL, a response was elicited for the 50 μg/mL and 100 μg/mL concentrations. This shows an increase in anaerobic metabolism, indicating that the macrophages have not significantly undergone oxidative burst during low oxLDL exposure, likely due to the lack of oxidation of the LDL. Both medium and highly oxidized LDL show limited responses, none of which are significant until the 100 μg/ml concentration of medium oxLDL. This lack of lactate production increases shows that the cell is utilizing aerobic respiration normally, not necessarily indicative of undergoing oxidative burst. Finally, the Intracel oxLDL elicited responses at 10 μg/mL, 50 μg/mL, and 100 μg/mL showing that it was effective at increasing aerobic respiration and thus promoting oxidative burst.

During oxLDL exposure, increase in oxygen consumption coupled with increases in glucose consumption is indicative of oxidative burst onset. Data was collected from exposure to various forms of LDL and is presented in Figure 3 and Table 3. Native LDL exposure revealed statistically significant

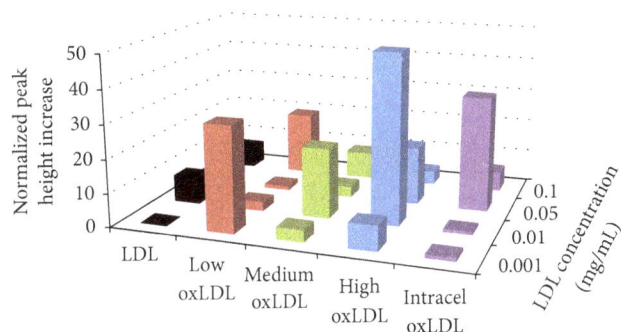

FIGURE 1: Average metabolic glucose consumption response of RAW 264.7 cells to a 6 min exposure of various LDL. Percent increases were calculated based on the variation from the maximum peak height chance during exposure to the baseline average, prior to exposure. For native LDL, the glucose consumption at $100\,\mu g/mL$ the number of averaged experiments (n) is 4 and the standard error of the mean for the basal metabolic rate of the baseline is $\pm 0.20\%$. For $50\,\mu g/mL$ $n = 2$; $\pm 0.27\%$, $10\,\mu g/mL$ $n = 2$; $\pm 0.01\%$, and $1\,\mu g/mL$ $n = 4$; $\pm 6.1\%$. For low oxLDL, $100\,\mu g/mL$ $n = 3$; $\pm 12.5\%$, $50\,\mu g/mL$ $n = 5$; $\pm 3.4\%$, $10\,\mu g/mL$ $n = 3$; $\pm 0.36\%$, and $1\,\mu g/mL$ $n = 2$; $\pm 0.38\%$. For medium oxLDL at $100\,\mu g/mL$ $n = 2$; $\pm 0.33\%$, $50\,\mu g/mL$ $n = 3$; $\pm 0.51\%$, $10\,\mu g/mL$ $n = 4$; $\pm 0.16\%$, and $1\,\mu g/mL$ $n = 5$; $\pm 0.32\%$. For high oxLDL at $100\,\mu g/mL$ $n = 3$; $\pm 0.5\%$, $50\,\mu g/mL$ $n = 3$; $\pm 9.7\%$, $10\,\mu g/mL$ $n = 3$; $\pm 0.44\%$, and $1\,\mu g/mL$ $n = 3$; $\pm 8.6\%$. For Intracel oxLDL at $100\,\mu g/mL$ $n = 4$; $\pm 1.8\%$, $50\,\mu g/mL$ $n = 6$; $\pm 15.4\%$, $10\,\mu g/mL$ $n = 3$; $\pm 0.67\%$, and $1\,\mu g/mL$ $n = 2$; $\pm 0.31\%$.

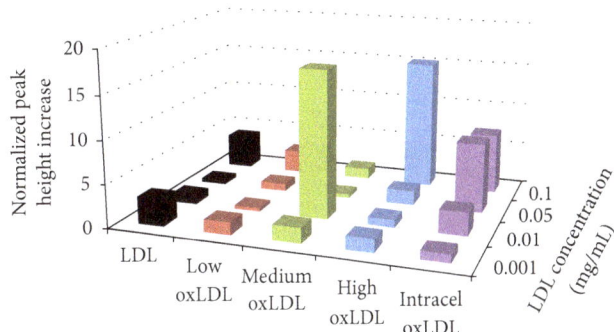

FIGURE 2: Average metabolic lactate production response of RAW 264.7 cells to a 6 min exposure of various LDL. Percent increases were calculated based on the variation from the maximum peak height chance during exposure to the baseline average, prior to exposure. For native LDL, the lactate production at $100\,\mu g/mL$ $n = 3$; $\pm 0.01\%$, $50\,\mu g/mL$ $n = 2$; $\pm 0.005\%$, $10\,\mu g/mL$ $n = 5$; $\pm 0.35\%$, and $1\,\mu g/mL$ $n = 4$; $\pm 0.04\%$. For low oxLDL at $100\,\mu g/ml$ $n = 4$; $\pm 0.6\%$, $50\,\mu g/ml$ $n = 4$; $\pm 0.12\%$, $10\,\mu g/ml$ $n = 5$; $\pm 0.14\%$, and $1\,\mu g/ml$ $n = 4$; $\pm 0.02\%$. For medium oxLDL at $100\,\mu g/ml$ $n = 4$; $\pm 0.47\%$, $50\,\mu g/mL$ $n = 4$; $\pm 4.6\%$, $10\,\mu g/mL$ $n = 4$; $\pm 0.26\%$, and $1\,\mu g/mL$ $n = 5$; $\pm 0.25\%$. For high oxLDL at $100\,\mu g/mL$ $n = 3$; $\pm 0.67\%$, $50\,\mu g/mL$ $n = 5$; $\pm 0.26\%$, $10\,\mu g/mL$ $n = 3$; $\pm 0.12\%$, and $1\,\mu g/mL$ $n = 4$; $\pm 0.9\%$. For Intracel oxLDL at $100\,\mu g/mL$ $n = 5$; $\pm 0.03\%$, $50\,\mu g/mL$ $n = 4$; $\pm 0.46\%$, $10\,\mu g/mL$ $n = 4$; $\pm 0.51\%$, and $1\,\mu g/mL$ $n = 4$; $\pm 0.55\%$.

increases in oxygen consumption at $1\,\mu g/ml$ and $100\,\mu g/ml$. The lack of an increasing trend of oxygen consumption during LDL exposure provides little evidence for oxidative burst onset. Kalen low and medium oxLDL exposure provided no statistically significant increases in oxygen consumption. However, both Kalen high oxLDL and the Intracel oxLDL provided a significant increase in oxygen consumption during the $50\,\mu g/ml$. While there are limited statistically significant increases in oxygen consumption when compared with the baseline, biological samples promote wide variations in standard error. That said, there is an overall trend of increase in oxygen consumption with increase in oxidation level and concentration, up to a point. The higher levels of oxidation of LDL do not show this trend, suggesting a saturation of available receptors allowing for macrophage uptake.

Extracellular acidification provides an overall view of the health of the cell during exposure to various forms of LDL. The data obtained during exposure is presented in Figure 4 and Table 4. Native LDL and low oxLDL provided statistically significant increases in extracellular acidification at the highest dosage concentration, $100\,\mu g/ml$. The medium and high oxLDLs showed no dramatically significant increases in extracellular acidification; however the Intracel oxLDL showed an insignificant yet clear increasing trend with increasing dosages up to $50\,\mu g/ml$. The extracellular acidification data does not provide conclusive evidence of oxidative burst onset; however extracellular acidification is the sum result of both anaerobic and aerobic metabolism. Therefore, this series does not adequately reflect the complexities of the macrophage metabolic response.

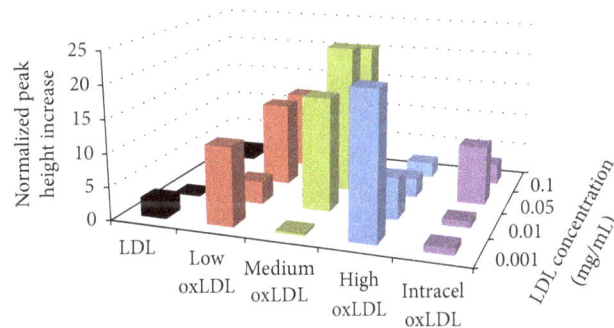

FIGURE 3: Average metabolic oxygen consumption response of RAW 264.7 cells to a 6 min exposure of various LDL. Percent increases were calculated based on the variation from the maximum peak height chance during exposure to the baseline average, prior to exposure. For native LDL at $100\,\mu g/mL$ $n = 4$; $\pm 0.28\%$, $50\,\mu g/mL$ $n = 3$; $\pm 0.27\%$, $10\,\mu g/mL$ $n = 4$; $\pm 0.09\%$, and $1\,\mu g/mL$ $n = 3$; $\pm 0.27\%$. For low oxLDL at $100\,\mu g/ml$ $n = 2$; $\pm 1.03\%$, $50\,\mu g/ml$ $n = 2$; $\pm 5.1\%$, $10\,\mu g/ml$ $n = 3$; $\pm 0.56\%$, and $1\,\mu g/ml$ $n = 3$; $\pm 0.33\%$. For medium oxLDL at $100\,\mu g/mL$ $n = 3$; $\pm 1.51\%$, $50\,\mu g/mL$ $n = 4$; $\pm 0.74\%$, $10\,\mu g/mL$ $n = 3$; $\pm 0.004\%$, and $1\,\mu g/mL$ $n = 3$; $\pm 0.42\%$. For high oxLDL at $100\,\mu g/mL$ $n = 2$; $\pm 0.37\%$, $50\,\mu g/mL$ $n = 3$; $\pm 0.09\%$, $10\,\mu g/mL$ $n = 3$; $\pm 0.11\%$, and $1\,\mu g/mL$ $n = 2$; $\pm 0.94\%$. For Intracel oxLDL at $100\,\mu g/mL$ $n = 4$; $\pm 0.16\%$, $50\,\mu g/mL$ $n = 3$; $\pm 0.69\%$, $10\,\mu g/mL$ $n = 3$; $\pm 0.57\%$, and $1\,\mu g/mL$ $n = 4$; $\pm 0.28\%$.

From the native LDL studies we see that LDL exposure induces a slight response seen primarily in glucose consumption and lactate production. This illustrates that the macrophage is increasing glucose uptake to ensure proper breakdown of the consumed LDL, which is highlighted by the increase in lactate production. These experiments allowed for

TABLE 4: Extracellular acidification rates of various LDL exposure to macrophages. Mean peak MAMP height changes and standard errors for each experimental condition during exposure. All p values are calculated based on peak height versus 10 min average basal value prior to exposure. Terms in bold are representative of $p < 0.05$, indicating statistical significance.

| | Acidification | | | | |
	Native LDL (%)	Low oxLDL (%)	Medium oxLDL (%)	High oxLDL (%)	Intracel oxLDL (%)
1 μg/ml	0.73 ± 1.6	8.1 ± 3.3	1.03 ± 4.7	2.13 ± 1.7	2.13 ± 1.8
	$p = 0.7001$	$p = 0.0918$	$p = 0.8452$	$p = 0.268$	$p = 0.3317$
10 μg/ml	1.2 ± 2.4	8.28 ± 6.6	4.53 ± 8.4	3.61 ± 4.8	28.5 ± 22.4
	$p = 0.6481$	$p = 0.4708$	$p = 0.6254$	$p = 0.4864$	$p = 0.3044$
50 μg/ml	5.8 ± 5.2	1.12 ± 1.96	**2.8 ± 0.60**	13.1 ± 11.3	46.8 ± 23.6
	$p = 0.3156$	$p = 0.6205$	**p = 0.025**	$p = 0.3280$	$p = 0.1044$
100 μg/ml	**25.6 ± 2.4**	**1.3 ± 0.1**	3.1 ± 5.2	21.6 ± 10.6	**22.0 ± 2.7**
	p = 0.0018	**p < 0.0001**	$p = 0.8183$	$p = 0.2628$	**p = 0.0005**

TABLE 5: Metabolic effects of CpG-DNA exposure on macrophages. Mean peak MAMP height changes and standard errors for each experimental condition during exposure and recovery are shown. All p values are calculated based on peak height versus 10 min average basal value prior to exposure. Terms in bold are representative of $p < 0.05$, indicating statistical significance.

| | CpG-DNA | | | |
	Glucose (%)	Lactate (%)	Oxygen (%)	Acidification (%)
During Exposure	**4.0 ± 1.0**	1.4 ± 1.0	2.3 ± 2.5	**15.6 ± 5.6**
	p = 0.0108	$p = 0.2107$	$p = 0.3999$	**p = 0.039**
Recovery	27.0 ± 5.3	**5.6 ± 1.8**	1.6 ± 9.5	11.3 ± 4.8
	p = 0.0038	**p = 0.0174**	$p = 0.8729$	$p = 0.066$

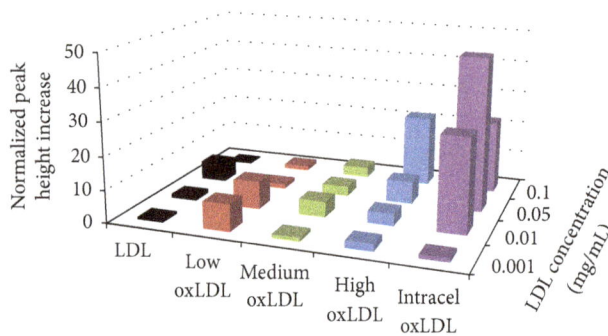

FIGURE 4: Average extracellular acidification response of RAW 264.7 cells to a 6 min exposure of various LDL. Percent increases were calculated based on the variation from the maximum peak height chance during exposure to the baseline average, prior to exposure. For native LDL at 100 μg/mL $n = 2$; $\pm 0.03\%$, 50 μg/mL $n = 3$; $\pm 0.15\%$, 10 μg/mL $n = 4$; $\pm 0.76\%$, and 1 μg/mL $n = 3$; $\pm 0.8\%$. For low oxLDL at 100 μg/ml $n = 3$; $\pm 0.9\%$, 50 μg/mL $n = 2$; $\pm 0.55\%$, 10 μg/mL $n = 5$; $\pm 8.8\%$, and 1 μg/ml $n = 2$; $\pm 0.24\%$. For medium oxLDL at 100 μg/mL $n = 2$; $\pm 0.22\%$, 50 μg/mL $n = 2$; $\pm 0.3\%$, 10 μg/mL $n = 2$; $\pm 0.003\%$, and 1 μg/mL $n = 3$; $\pm 0.78\%$. For high oxLDL at 100 μg/mL $n = 2$; $\pm 4.45\%$, 50 μg/mL $n = 3$; $\pm 4.3\%$, 10 μg/mL $n = 3$; $\pm 0.38\%$, and 1 μg/mL $n = 2$; $\pm 0.04\%$. For Intracel oxLDL at 100 μg/mL $n = 3$; $\pm 0.17\%$, 50 μg/mL $n = 3$; $\pm 1.0\%$, 10 μg/mL $n = 3$; $\pm 10.9\%$, and 1 μg/mL $n = 3$; $\pm 0.83\%$.

a baseline metabolic response from the introduction of native LDL to compare with further experiments on macrophage exposure to various forms of oxLDL. The lowly oxidized LDL is not providing the onset of foam cell formation but is inhibiting basal metabolism. High oxLDL studies exhibited

increases in metabolism signifying an increase in aerobic respiration, which is a hallmark of oxidative burst. This level of LDL oxidation showcases the ability of oxLDL to induce oxidative burst; however the lack of response for the highest dosage concentration also suggests an oversaturation of oxLDL receptors, inhibiting oxidative burst occurrence. There was also a marked increase seen with Intracel oxLDL, particularly at the 50 μg/mL dosage level, indicating oxidative burst induction. The 100 μg/mL dosage also has a response but is less intense than the 50 μg/mL dosage, again suggesting oversaturation of oxLDL receptors.

Virulent components of pathogens can induce inflammatory responses that can be observed and measured by the MAMP. One known inducer and LOX-1 receptor upregulator is CpG-DNA, which directly binds to toll-like receptor 9 (TLR9) to promote an oxidative burst signaling cascade [9]. To incite a metabolic inflammatory response, macrophages were exposed to 3 μmol/L CpG-DNA for 30 min (Figure 5, Table 5). During the exposure time, a significant peak height increase in glucose consumption and extracellular acidification was seen, indicating an increase in aerobic respiration to promote oxidative burst. During recovery significant peak height increases were also seen in lactate production. This suggests that after CpG-DNA is removed from the system, routine use of anaerobic and aerobic metabolism takes effect, thus leaving no lasting metabolic effects from the CpG-DNA exposure.

The murine LOX-1/OLR1 antibody is thought to completely block LOX-1 dependent oxLDL from binding to cells and will be utilized in concert with oxLDL exposure in later experiments. To ensure there are no lasting negative

TABLE 6: Metabolic effects of mouse LOX-1/OLR1 antibody exposure on macrophages. Mean peak MAMP height changes and standard errors for each experimental condition during exposure and recovery are shown. All p values are calculated based on peak height versus 10 min average basal value prior to exposure. Terms in bold are representative of $p < 0.05$, indicating statistical significance.

| | Mouse LOX-1/OLR1 antibody | | | |
	Glucose (%)	Lactate (%)	Oxygen (%)	Acidification (%)
During exposure	**−24.7 ± −9.2**	**7.4 ± 2.3**	**−12.1 ± −4.4**	**−26.7 ± −1.6**
	p = 0.0314	**p = 0.0105**	**p = 0.0105**	**p = 6.9E^{-7}**
Recovery	**−46.8 ± −12.5**	6.2 ± 3.5	−11.6 ± −5.4	−14.8 ± −6.5
	p = 0.0072	$p = 0.1103$	$p = 0.0845$	$p = 0.0569$

TABLE 7: Metabolic effects of trehalose exposure on macrophages. Mean peak MAMP height changes and standard errors for each experimental condition during exposure and recovery are shown. All p values are calculated based on peak height versus 10 min average basal value prior to exposure. Terms in bold are representative of $p < 0.05$, indicating statistical significance.

| | Trehalose exposure | | | |
	Glucose (%)	Lactate (%)	Oxygen (%)	Acidification (%)
During exposure	**53.9 ± 3.79**	**51.9 ± 3.53**	9.72 ± 6.48	**36.4 ± 7.08**
	p < 0.0001	**p < 0.0001**	$p = 0.1815$	**p = 0.0037**

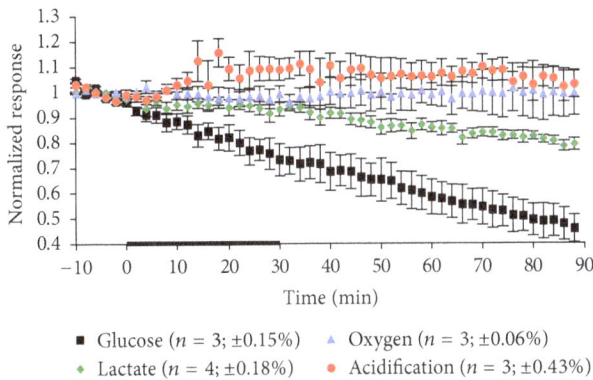

FIGURE 5: Average metabolic response of RAW 264.7 cells to a 30 min exposure of 3 μmol/L CpG-DNA. The black bar indicates the exposure time. The number of experiments and standard error of the mean for the basal metabolic rate 10 min before exposure to CpG-DNA are given in parentheses.

■ Glucose ($n = 3$; ±0.15%) ▲ Oxygen ($n = 3$; ±0.06%)
♦ Lactate ($n = 4$; ±0.18%) ● Acidification ($n = 3$; ±0.43%)

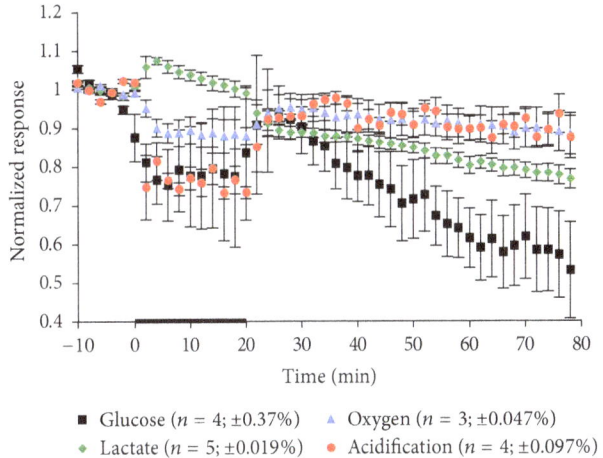

■ Glucose ($n = 4$; ±0.37%) ▲ Oxygen ($n = 3$; ±0.047%)
♦ Lactate ($n = 5$; ±0.019%) ● Acidification ($n = 4$; ±0.097%)

FIGURE 6: Average metabolic response of RAW 264.7 cells to a 20 min exposure of 25 μg/mL mouse LOX-1/OLR1 antibody. The black bar indicates the exposure time. The number of experiments and standard error of the mean for the basal metabolic rate 10 min before exposure to the antibody are given in parentheses.

metabolic effects from the introduction of the antibody into the extracellular solution, we exposed macrophages to 25 μg/mL of mouse LOX-1/OLR1 antibody for 20 min. The dynamic trace of the exposure and recovery is seen in Figure 6 and Table 6. Results indicated that, upon exposure, there were statistically significant decreases in glucose and oxygen consumption as well as extracellular acidification, while lactate production had a significant increase. The only significant decrease reported during recovery was with glucose consumption; however this decrease over time is often found in glucose consumption due to enzyme limitations of glucose oxidase. The rapid changes that occur during exposure indicate a decrease in aerobic respiration, suggesting an inhibition of oxidative burst. The inhibiting nature of the mouse LOX-1/OLR1 antibody toward oxidative burst may aid in oxidative burst prevention during oxLDL exposure.

To ensure that the aforementioned change during mouse LOX-1/OLR1 antibody exposure was not due to trehalose, a concentration of 50 mg/mL was introduced to the cells for an hour to see if there was any significant effect. During the course of exposure, there were significant decreases seen in glucose consumption, lactate production, and extracellular acidification as can be seen in Figure 7 and Table 7. Compared with the changes seen during mouse LOX-1/OLR1 antibody exposure, there is not a recovery associated with trehalose exposure. Additionally, during trehalose exposure lactate production significantly decreases while oxygen consumption has an increasing trend, which is inverse during the mouse LOX-1/OLR1 antibody exposure. Therefore, this data suggests there is no lasting effect caused by trehalose presence within the mouse LOX-1/OLR1 antibody solution.

TABLE 8: Metabolic effects of 25 μg/mL Rat IgG2A Isotype Control exposure on macrophages. Mean peak MAMP height changes and standard errors for each experimental condition during exposure and recovery are shown. All p values are calculated based on peak height versus 10 min average basal value prior to exposure.

| | Rat IgG2A isotype control exposure | | | |
	Glucose (%)	Lactate (%)	Oxygen (%)	Acidification (%)
During exposure	39.2 ± 11.7	16.6 ± 3.1	38.4 ± 28.7	28.2 ± 7.9
	$p = 0.4268$	$p = 0.3539$	$p = 0.4283$	$p = 0.2637$

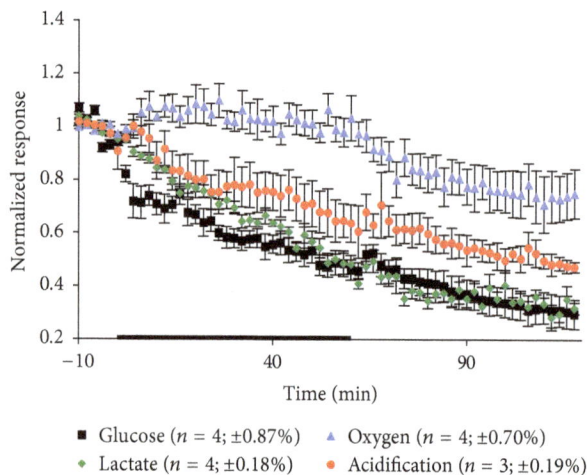

FIGURE 7: Average metabolic response of RAW 264.7 cells to a 60 min exposure of 50 mg/mL trehalose. The black bar indicates the exposure time. The number of experiments and standard error of the mean for the basal metabolic rate 10 min before exposure to the antibody are given in parentheses.

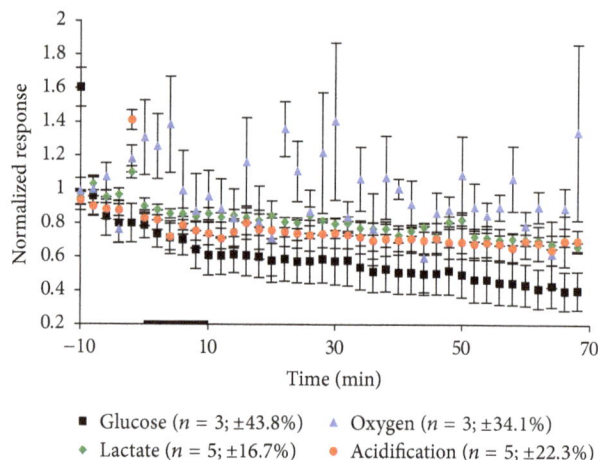

FIGURE 8: Average metabolic response of RAW 264.7 cells to a 10 min exposure of 25 μg/mL rat IgG2A isotype control. The black bar indicates the exposure time. The number of experiments and standard error of the mean for the basal metabolic rate 10 min before exposure to the antibody are given in parentheses.

As a control, rat IgG2A isotype control was exposed to the cells at the same concentration as mouse LOX-1/OLR1 antibody, to ensure there were no significant changes in metabolism due to additive stability compounds. This data can be seen in Figure 8 and Table 8. There were no statistically significant metabolic changes during the 10 min exposure, suggesting the dramatic changes seen during mouse LOX-1/OLR1 antibody exposure are due solely to the antibody interaction with the macrophage.

Due to the ability of highly oxidized LDL to induce oxidative burst, Kalen high oxLDL was exposed to the macrophages after a 10 min exposure of mouse LOX-1/OLR1 antibody. This data can be seen in Table 9. Statistically significant lactate production changes from the basal metabolic rates that occurred during antibody exposure in the lower two dosages and oxygen consumption for the 100 μg/mL dosage. These findings suggest that the antibody may not have significant deleterious effects on the macrophage metabolism when dosed in shorter time frames. Statistically significant changes from the basal rate were also seen during oxLDL exposure, but this was limited to two instances: lactate production at 10 μg/mL and glucose consumption at 100 μg/mL. The lack of statistically significant metabolic changes from the basal rate during antibody or oxLDL exposure, as well as during recovery, suggests that the antibody inhibited oxLDL to macrophage binding, thus prohibiting engulfment and oxidative burst.

4. Discussion

In the present study, we were able to identify peak height increases in metabolism during LDL exposure. Some anaerobic increases were statistically significant; however they were not seen in each analyte or to a high degree, indicating a lack of oxidative burst incitement. The metabolic hallmarks of oxidative burst are increases in all three metabolic analytes coupled with increases in the extracellular acidification. During LDL exposure, the most dramatic observable increase was in lactate production alone. This indicates that anaerobic respiration is upregulated while aerobic respiration remains fairly stable, which is not related to oxidative burst. Instead these changes indicate that lipoprotein exposure causes an increase in cellular uptake to promote LDL engulfment and subsequent breakdown for removal from the system.

In our previous study with oxLDL, significant responses were seen during or immediately following exposure. The oxLDL used in those studies was purchased through Intracel (Frederick, MD) who prepares each solution using human LDL. The differences in responses seen in this study could be caused from the variations in preparation techniques between Intracel and Kalen Biomedical, who report using varying oxidation conditions of human LDL for their oxLDL kit. However, there are responses, both statistically significant and nonsignificant, seen in the medium and high oxLDL

TABLE 9: Metabolic effects of mouse LOX-1/OLR1 antibody 10 min exposure followed by 6 min exposure to Kalen high oxLDL on macrophages. Mean peak MAMP height changes and standard errors for each experimental condition during antibody exposure, oxLDL exposure, and recovery are shown. All p values are calculated based on peak height versus 10 min average basal value prior to exposure. Terms in bold are representative of $p < 0.05$, indicating statistical significance.

| | | Mouse LOX-1/OLR1 antibody and oxLDL exposure | | | |
		Glucose (%)	Lactate (%)	Oxygen (%)	Acidification (%)
		$n = 3$	$n = 3$	$n = 3$	$n = 4$
	Antibody exposure	20.1 ± 17.8	**3.15 ± 0.6**	43.2 ± 39.9	7.9 ± 7.7
		$p = 0.9699$	**$p = 0.0046$**	$p = 0.3284$	$p = 0.4077$
$1 \mu g/mL$	oxLDL exposure	32.9 ± 34.3	5.1 ± 2.1	20.4 ± 13.3	5.27 ± 6.5
		$p = 0.446$	$p = 0.0606$	$p = 0.1868$	$p = 0.5293$
	Recovery	27.3 ± 44.0	**6.61 ± 1.9**	92.7 ± 56.1	29.7 ± 34.5
		$p = 0.5972$	**$p = 0.0192$**	$p = 0.1594$	$p = 0.4217$
		$n = 3$	$n = 4$	$n = 4$	$n = 2$
	Antibody exposure	1.92 ± 11.5	**3.55 ± 0.91**	17.34 ± 9.8	1.72 ± 0.97
		$p = 0.9584$	**$p = 0.0109$**	$p = 0.1785$	$p = 0.1754$
$10 \mu g/mL$	oxLDL exposure	14.8 ± 27.1	**6.4 ± 0.99**	10.8 ± 2.6	11.2 ± 4.2
		$p = 0.7436$	**$p = 0.0007$**	$p = 0.1529$	$p = 0.0759$
	Recovery	12.3 ± 38.1	**9.9 ± 1.1**	0.2 ± 1.4	9.64 ± 6.9
		$p = 0.8174$	**$p < 0.0001$**	$p = 0.9758$	$p = 0.2568$
		$n = 4$	$n = 3$	$n = 2$	$n = 2$
	Antibody exposure	15.5 ± 20.8	5.7 ± 9.5	124.21 ± 124.2	4.7 ± 1.9
		$p = 0.781$	$p = 0.5755$	$p = 0.391$	$p = 0.0995$
$50 \mu g/mL$	oxLDL exposure	14.0 ± 15.9	11.8 ± 8.9	141.9 ± 173.6	20.6 ± 10.9
		$p = 0.7947$	$p = 0.2425$	$p = 0.4737$	$p = 0.1557$
	Recovery	138.7 ± 104.2	1.1 ± 5.6	135.8 ± 166.9	**40.2 ± 6.2**
		$p = 0.2679$	$p = 0.8496$	$p = 0.4754$	**$p = 0.0076$**
		$n = 3$	$n = 3$	$n = 3$	$n = 2$
	Antibody exposure	1.85 ± 6.7	0.67 ± 0.25	**34.4 ± 10.2**	1.05 ± 1.51
		$p = 0.7939$	$p = 0.2218$	**$p = 0.0203$**	$p = 0.5369$
$100 \mu g/mL$	oxLDL exposure	**8.56 ± 1.8**	4.34 ± 2.4	30.4 ± 12.1	34.6 ± 30.1
		$p = 0.0057$	$p = 0.1348$	$p = 0.0544$	$p = 0.3337$
	Recovery	**18.5 ± 1.9**	2.06 ± 1.9	10.3 ± 8.1	57.2 ± 48.6
		$p = 0.0002$	$p = 0.3377$	$p = 0.2633$	$p = 0.324$

types, suggesting an altering of the macrophage metabolism, likely due to the oxidative burst. Intracel oxLDL was also exposed to the macrophages at all four dosages and compared with the high oxLDL from Kalen. The responses were similar and indicated oxidative burst occurrence. Most notably, in both species of oxLDL the lower dosages exhibited greater changes in metabolism than the highest dosage. This data suggests the macrophage oxLDL receptors are saturated when exposed to higher concentrations of oxLDL, prohibiting oxidative burst from occurring.

The slight significant responses seen during CpG-DNA exposure do not follow expected oxidative burst metabolism. Current literature suggests that there is an acute inflammatory response to CpG-DNA due to its ability to overexpress LOX-1 receptors, but unfortunately reported exposure times vary [9]. Further experiments should examine longer exposure times and basal metabolic rates of macrophages that have been preincubated with CpG-DNA. Alternate concentrations of CpG-DNA should also be used to ensure maximal inflammatory response is gained.

Significant and rapid responses were seen upon mouse LOX-1/OLR1 antibody exposure. The responses seen indicate

a rapid decrease in aerobic respiration, inhibiting oxidative burst. Control studies suggested that the response seen during mouse LOX-1/OLR1 antibody exposure is due to the antibody and not due to stability additives. This immediate response is likely due to the mouse LOX-1/OLR1 antibody interaction with the LOX-1 receptor, which typically interacts with oxLDL to promote cellular uptake and oxidative burst. Interestingly, the inhibition of aerobic respiration shows that the antibody/LOX-1 receptor interaction is actively altering cellular metabolism, suggesting LOX-1 receptor availability is vital to basal macrophage respiration. Further studies should be undertaken to assess the importance of LOX-1 to basal metabolism. Other studies should investigate various dosage concentrations of mouse LOX-1/OLR1 antibody to avoid oversaturation of the receptors.

Finally, preexposure of the macrophages with the mouse LOX-1/OLR1 antibody for a shorter time frame (10 min) showed limited metabolic responses during the mouse LOX-1/OLR1 antibody exposure. A subsequent high oxLDL dosage also showed limited changes of metabolic responses from the baseline indicating the mouse LOX-1/OLR1 antibody successfully prohibited significantly high oxLDL uptake that would lead to oxidative burst and foam cell formation.

References

[1] H. Yoshida, N. Kondratenko, S. Green, D. Steinberg, and O. Quehenberger, "Identification of the lectin-like receptor for oxidized low-density lipoprotein in human macrophages and its potential role as a scavenger receptor," *Biochemical Journal*, vol. 334, no. 1, pp. 9–13, 1998.

[2] T. Navarra, S. Del Turco, S. Berti, and G. Basta, "The lectin-like oxidized low-density lipoprotein receptor-1 and its soluble form: Cardiovascular implications," *Journal of Atherosclerosis and Thrombosis*, vol. 17, no. 4, pp. 317–331, 2010.

[3] H. Moriwaki, N. Kume, H. Kataoka et al., "Expression of lectin-like oxidized low density lipoprotein receptor-1 in human and murine macrophages: upregulated expression by TNF-α," *FEBS Letters*, vol. 440, no. 1-2, pp. 29–32, 1998.

[4] N. Kume, H. Moriwaki, H. Kataoka et al., "Inducible expression of LOX-1, a novel receptor for oxidized ldl, in macrophages and vascular smooth muscle cells," in *Atherosclerosis V: The Fifth Saratoga Conference*, F. Numano and M. A. Gimbrone, Eds., vol. 902, pp. 323–327, New York Acad Sciences, New York, NY, USA, 2000.

[5] D. W. Kimmel, W. P. Dole, and D. E. Cliffel, "In preparation for tox let," 2012.

[6] S. Ogura, A. Kakino, Y. Sato et al., "LOX-1: The multifunctional receptor underlying cardiovascular dysfunction," *Circulation Journal*, vol. 73, no. 11, pp. 1993–1999, 2009.

[7] D. P. Via, A. L. Plant, I. F. Craig, A. M. Gotto Jr., and L. C. Smith, "Metabolism of normal and modified low-density lipoproteins by macrophage cell lines of murine and human origin," *Biochimica et Biophysica Acta (BBA)—Lipids and Lipid Metabolism*, vol. 833, no. 3, pp. 417–428, 1985.

[8] A. Tanimoto, Y. Murata, M. Nomaguchi et al., "Histamine increases the expression of LOX-1 via H2 receptor in human monocytic THP-1 cells," *FEBS Letters*, vol. 508, no. 3, pp. 345–349, 2001.

[9] H. Hemmi, O. Takeuchi, T. Kawai et al., "A Toll-like receptor recognizes bacterial DNA," *Nature*, vol. 408, pp. 740–745, 2000.

22

Normal Non-HDL Cholesterol, Low Total Cholesterol, and HDL Cholesterol Levels in Sickle Cell Disease Patients in the Steady State

Richard K. D. Ephraim,[1] **Patrick Adu,**[1] **Edem Ake,**[1] **Hope Agbodzakey,**[1] **Prince Adoba,**[2] **Obed Cudjoe,**[3] **and Clement Agoni**[1]

[1]*Department of Medical Laboratory Science, School of Allied Health Sciences, College of Health and Allied Sciences, University of Cape Coast, Cape Coast, Ghana*
[2]*Department of Molecular Medicine, School of Medical Sciences, College of Health, KNUST, Kumasi, Ghana*
[3]*Department of Microbiology, School of Medical Sciences, College of Health and Allied Sciences, University of Cape Coast, Cape Coast, Ghana*

Correspondence should be addressed to Richard K. D. Ephraim; rephraim@ucc.edu.gh

Academic Editor: Maurizio Averna

Background. Abnormal lipid homeostasis in sickle cell disease (SCD) is characterized by defects in plasma and erythrocyte lipids and may increase the risk of cardiovascular disease. This study assessed the lipid profile and non-HDL cholesterol level of SCD patients. *Methods.* A hospital-based cross-sectional study was conducted in 50 SCD patients, in the steady state, aged 8–28 years, attending the SCD clinic, and 50 healthy volunteers between the ages of 8–38 years. Serum lipids were determined by enzymatic methods and non-HDL cholesterol calculated by this formula: non-HDL-C = TC-HDL-C. *Results.* Total cholesterol (TC) ($p = 0.001$) and high-density lipoprotein cholesterol (HDL-C) ($p < 0.0001$) were significantly decreased in cases compared to controls. The levels of non-HDL-C, low-density lipoprotein cholesterol (LDL-C), and triglyceride (TG) were similar among the participants. The levels of decrease in TC and HDL were associated with whether a patient was SCD-SS or SCD-SC. Systolic blood pressure and diastolic blood pressure were each significantly associated with increased VLDL [SBP, $p = 0.01$, OR: 0.74 (CI: 0.6–0.93); DBP, $p = 0.023$, OR: 1.45 (CI: 1.05–2.0)]. *Conclusion.* Dyslipidemia is common among participants in this study. It was more pronounced in the SCD-SS than in SCD-SC. This dyslipidemia was associated with high VLDL as well as increased SBP and DBP.

1. Background

Sickle cell disease (SCD) is a genetic disorder caused by the substitution of valine for glutamic acid at the sixth position of the amino acid β-chain of the haem molecule [1, 2] and is characterized by the possession of sickle haemoglobin [3]. It is a significant cause of morbidity and mortality among black individuals and descendants of Negroid race [4]. Life expectancy is shortened with studies reporting average life expectancy of 42 and 48 years for males and females, respectively [5].

SCD is characterized by defect in plasma and erythrocyte lipids associated with chronic oxidative stress [6]. These two morbid processes disturb lipid homeostasis which in turn may lead to atherosclerosis in these patients [7, 8]. Abnormal lipid homeostasis, as well as other haematological disorders, has been reported in SCA and this has been suggested to have the potential to alter membrane fluidity and functions of red blood cells (RBC) in individuals with SCD [9–11].

Mostly, a standard lipid profile (triglyceride, total cholesterol, high-density lipoprotein cholesterol, and low-density lipoprotein cholesterol) is used to assess the risk of coronary artery disease (CAD). Earlier studies in patients with SCD recorded a significant increase in triglyceride (TG) levels and decreased levels of total cholesterol (TC), high-density lipoprotein cholesterol (HDL-C), and low-density lipoprotein cholesterol (LDC-C) [9]. This remains a cause for concern as it is associated with increased mortality [10].

Atherosclerosis, often associated with CAD, is characterized by elevated levels of cholesterol and LDL-C [12]. Pulmonary hypertension, the main form of cardiovascular dysfunction in SCD, is often characterized by low levels of TC and LDL-C. Relying on LDL-C levels to assess cardiovascular disease may be misleading as reported in a recent meta-analysis [11]. That study recorded low levels of LDL-C but high levels of non-HDL-C (TC-HDL-C) in people with cardiovascular disease [11, 13]. It is imperative that these reported dyslipidemia cases in SCD should be interrogated in light of the new findings. In addition, there is a paucity of data on lipid profile in SCD patients in Ghana. In light of the above, we sought to determine the lipid profile and non-HDL-C levels in SCD patients in the steady state.

2. Methods

2.1. Setting/Design. This hospital-based cross-sectional study was carried out at the sickle cell unit of Tema General Hospital in the Greater Accra region of Ghana.

2.2. Participants. A total of 50 SCD patients (12 with HbS and 38 with HbSC haemoglobin variants) were recruited for the study. A total number of 50 healthy, age- and sex-matched controls also participated in the study.

2.3. Exclusion/Inclusion Criteria. Patients who have been diagnosed with the sickle cell disease, those with the genotype HbSS and HbSC who are in the steady state, and those who allowed parental consent were included in the study. Exclusion criteria included patients with inflammatory episodes, patients with sickle cell trait, patients on medications that affect lipid metabolism, and those who have had blood transfusion four (4) months prior to the study.

2.4. Ethical Consideration. All protocols for the study were approved by the Institutional Review Board (IRB) of the University of Cape Coast as well as the sickle cell clinic of the Tema General Hospital. Participation was voluntary and involved only Ghanaians. Written informed consent was obtained from each participant. All data was deidentified before analysis.

2.5. Sampling. In all subjects, 5 mL of overnight fasting venous samples was collected from all eligible subjects: 3 mL was put in plain tube, allowed to clot, and centrifuged at 2500 rpm for 5 minutes and the serum was used for estimation of lipid profile; 2 mL was put in EDTA tube and used for confirmation of their haemoglobin phenotype by cellulose acetate electrophoresis.

2.6. Testing. TC and TG concentrations were analyzed by enzymatic assay, whereas HDL-C was estimated calorimetrically. The calculation of VLDL-C was done by VLDL-C = triglyceride/2.2 and LDL-C calculation was done by the following Friedewald equation: LDL-C = TC-HDL-C –

TABLE 1: Demographics and clinical characteristics of participants.

Variable	Cases (n = 50)	Controls (n = 50)	p value
Age (years)	18.14 ± 4.63	21.42 ± 7.76	**0.060**
Gender			0.488
Male	11 (22.0)	14 (28.0)	
Female	39 (78.0)	36 (72.0)	
Age group, n (%)			0.056
<10	2 (4.0)	2 (4.0)	
10–19	33 (66.0)	20 (20.0)	
20–29	15 (30.0)	19 (38.0)	
30–39	0 (0.0)	9 (18.0)	
BMI (kg/m²)	20.67 ± 3.46	23.22 ± 4.50	**0.002**
BMI, n (%)			0.077
Underweight	15 (30.0)	9 (18.0)	
Normal	30 (60.0)	26 (52.0)	
Overweight	4 (8.0)	11 (222.0)	
Obese	1 (2.0)	4 (8.0)	
Blood pressure (mmHg)			
SBP	108.90 ± 8.13	115.52 ± 5.17	**<0.0001**
DBP	71.92 ± 6.88	76.98 ± 5.98	**<0.0001**
Lipid profile			
TC (mmol/L)	3.62 ± 0.78	4.20 ± 0.98	**0.001**
TG (mmol/L)	0.98 ± 0.67	0.97 ± 0.41	0.967
LDL-C (mmol/L)	2.15 ± 0.71	4.27 ± 2.02	0.299
VLDL-C (mmol/L)	0.58 ± 0.08	0.44 ± 0.03	0.108
HDL-C (mmol/L)	1.03 ± 0.33	1.50 ± 0.47	**<0.0001**
Non-HDL-C (mmol/L)	2.62 ± 0.68	2.68 ± 1.06	0.738
TG/HDL ratio	1.48 ± 0.47	0.71 ± 0.36	0.109

SBP: systolic blood pressure; DBP: diastolic blood pressure; TC: total cholesterol; TG: triglyceride; LDL-C: low-density lipoprotein cholesterol; VLDL-C: very-low-density lipoprotein cholesterol; HDL-C: high-density lipoprotein cholesterol.

(TG/2.2) [14]. Non-HDL-C was calculated by TC-HDL-C [15].

2.7. Statistical Analysis. Data was entered into Microsoft Excel (Microsoft, Redmond, WA, USA) and analyzed with SPSS version 16.0 (SPSS Inc., Chicago, IL, USA). The results were expressed as mean ± standard deviation and *t*-test was used to calculate the level of significance. A *p* value of ≤ 0.05 was considered statistically significant. Multivariate logistic regression was done to determine the independent factors of dyslipidemia in SCD.

3. Results

Table 1 presents the baseline characteristics of the participants. The mean ages of the sickle cell disease (SCD) patients and those without SCD were 18.14 ± 4.63 years and 21.42 ± 7.76 years, respectively. SCD was found to be more prevalent in females (78.0%) than in males (22.0%), with majority of the SCD patients, 33 (66.0%), aged between 10 and 19 years. Assessments of obesity using BMI was significantly

TABLE 2: BMI and lipid profiles in participants stratified by the type of haemoglobin variants.

Parameter	AA (n = 50)	SC (n = 12)	SS (n = 38)	p value
		Sickling phenotype		
BMI (kg/m²)	23.21 ± 4.50	20.48 ± 4.18*	20.73 ± 3.26*	**0.009**
BMI, n (%)				0.181
Underweight	9 (18.0)	4 (33.3)	11 (28.9)	
Normal	26 (52.0)	7 (58.3)	23 (60.5)	
Overweight	11 (22.0)	0 (0.0)	4 (10.5)	
Obese	4 (8.0)	1 (8.3)	0 (0.0)	
Blood pressure (mmHg)				
SBP	115.52 ± 5.17	107.92 ± 7.22*	109.21 ± 8.47*	**<0.0001**
DBP	76.98 ± 5.98	70.42 ± 6.86*	72.39 ± 6.91*	**0.001**
Lipid profile				
TC (mmol/L)	4.20 ± 0.98	3.23 ± 0.81*	3.74 ± 0.75*	**0.001**
TG (mmol/L)	0.97 ± 0.41	0.99 ± 0.81	0.97 ± 0.63	0.989
LDL-C (mmol/L)	4.27 ± 2.02	1.86 ± 0.76	2.25 ± 0.68	0.581
VLDL-C (mmol/L)	0.44 ± 0.03	0.64 ± 0.20	0.57 ± 0.09	0.241
HDL-C (mmol/L)	1.50 ± 0.47	0.93 ± 0.31*	1.06 ± 0.33*	**<0.0001**
Non-HDL-C (mmol/L)	2.68 ± 1.06	2.38 ± 0.78	2.70 ± 0.65	0.533
TG/HDL ratio	0.71 ± 0.36	2.71 ± 1.91*	1.10 ± 0.19+	0.035

*Significantly different from the control group (AA) (p < 0.05). +Significantly different from the group (SC) (p < 0.05).

TABLE 3: Factors associated with high TC, TG, and VLDL and low HDL in sickle cell disease patients.

Parameters	TC ≥ 5.0 mmol/L OR (95% CI)	p value	TG ≥ 1.70 mmol/L OR (95% CI)	p value	VLDL ≥ 1.04 OR (95% CI)	p value	HDL < 1.10 mmol/L OR (95% CI)	p value
Age (years)	1.12 (0.62–1.67)	0.943	1.03 (0.77–1.38)	0.837	1.21 (0.95–1.52)	0.119	0.97 (0.85–1.12)	0.696
BMI (kg/m²)	0.71 (0.30–1.71)	0.444	0.80 (0.52–1.23)	0.309	0.75 (0.50–1.12)	0.154	0.89 (0.73–1.07)	0.200
SBP	95.46 (—)	0.993	0.85 (0.69–1.06)	0.154	0.74 (0.60–0.93)	**0.010**	0.87 (0.76–0.98)	0.260
DBP	0.68 (0.14–3.19)	0.621	1.26 (0.91–1.74)	0.166	1.45 (1.05–2.00)	**0.023**	1.12 (0.97–1.29)	0.130

(= 0.002) lower in the SCD patients than in those with no SCD (Table 1). Both systolic and diastolic blood pressure parameters of the SCD patients were significantly lower (<0.0001) compared to healthy subjects. Serum lipid profile showed no statistically significant difference between the two groups as TG, LDL, VLDL, and non-HDL were compared (p > 0.05) except for TC and HDL (p = 0.001, <0.0001, resp.) (Table 1).

The plasma lipid concentrations, BMI, and blood pressure of the three groups are shown in Table 2. Mean BMI, SBP, and DBP were significantly lower in SC and SS patients (p < 0.05) compared to the healthy controls (AA).

TC and HDL cholesterol were significantly lower in SC and SS patients compared to the control groups (p < 0.05), despite being higher in the sickle cell patients with SS genotype than in those with SC. In addition, TG/HDL ratio was significantly higher in SC patients than in the healthy controls and SS patients.

Table 3 shows the factors associated with dyslipidemia in SCD patients. Both SBP and DBP were each significantly associated with VLDL [SBP, p = 0.01, OR: 0.74 (CI: 0.6–0.93); DBP, p = 0.023, OR: 1.45 (CI 1.05–2.0)].

Our study also investigated the relationship of BMI and blood pressure variables with serum lipid profile in SCD patients and healthy individuals. Body mass index showed a nonsignificant inverse correlation with TC, TG, and VLDL in both SCD and healthy patients (see Supplementary Data S1 in Supplementary Material available online at http://dx.doi.org/10.1155/2016/7650530). However, BMI correlation with LDL was negative for SCD patients and positive for healthy controls and vice versa for HDL despite being significant for SCD patients (Supplementary Data S1). SBP was directly related to all lipid profile parameters in the controls with the exception of LDL whereas TG and VLDL showed inverse correlation with SBP in the cases. On the other hand, with the exception of HDL, DBP was positive but not significantly related to TC, TG, LDL, and VLDL in cases whereas in the controls all the lipid profile parameters with the exception of VLDL and non-HDL-C showed a nonsignificant positive correlation with DBP.

4. Discussion

This study sought to assess the lipid profile and non-HDL-C levels of sickle cell disease patients in the steady state compared to healthy controls. Our findings showed that cholesterol (total cholesterol, HDL) levels decreased in SCD patients and were dependent on whether the patient has haemoglobin SS or SC; non-HDL remained unchanged between the two groups [16].

Non-HDL-C is a significant predictor of cardiovascular disease among diabetes patients [17]. We observed no significant difference in non-HDL-C levels among our participants. This to our knowledge is the first report on non-HDL-C levels in SCD and thus gives credence to the established evidence that cardiovascular disease in sickle cell disease is mostly due to pulmonary hypertension and not atherosclerosis associated with elevated levels of TC, HDL, and LDL [9, 18, 19].

The decreased TC and HDL in SCD are well documented in almost all the studies that have examined lipids in patients with SCD [20–22]. Hypocholesterolemia in SCD has been attributed to increased erythropoiesis in response to the anaemia associated with SCD as stated in earlier studies [10, 23]. Hypocholesterolemia has been identified as a potential biomarker of the clinical severity of SCD [24].

Consistent with some but not all studies, we recorded low HDL in SCD patients compared to controls [25]. Several reasons including small sample sizes, differences in gender, age, and weight, and variations in disease severity have been ascribed for these inconsistencies [26]. In the general population, low HDL is a recognized risk factor of cardiovascular disease but low HDL also remains a common feature of pulmonary hypertension which is the main cardiovascular disorder associated with SCD [27]. SCD patients with low HDL are more likely to have received more blood transfusions, an indication of the severity of the patient's condition [28].

TG and LDL levels were not significantly different among our participants. The observed TG level is consistent with the findings of Reaven [29] in SCD patients in Nigeria. However, we cannot proffer any reasons for the observed levels of LDL which seems to be at variance with what is recorded in most studies [24, 29] except to say that the use of steady state patients could account for this observation.

The TG/HDL-C ratio also known as the atherogenic index has been implicated in endothelial dysfunction associated with insulin resistance [30]. The increase in this index as observed in this study suggests an increased risk of pulmonary hypertension among our participants [31]. Also, we noted that the TG/HDL-C ratio was higher in SC-SCD patients than in SS-SCD patients, further providing evidence of the high level of TG in the SCD participants. Multivariate analysis showed an association between elevated VLDL levels and blood pressure (BP) with SCD patients being more liable to developing diastolic dysfunction (depicted here by elevated DBP) which ultimately increases mortality [31]. The observed VLDL level is strengthened by the earlier observation of the positive correlation between TG and VLDL. The role of TG in the development of pulmonary hypertension in SCD is well elucidated with high VLDL levels similar to the one recorded further buttressing the TG levels recorded in this study.

5. Conclusion

It was evident from this study that dyslipidemia characterized by low HDL and TC was present among the SCD patients who participated this study. It was more pronounced in the SCD-SS than in SCD-SC. Non-HDL levels were unchanged in cases compared to controls. Dyslipidemia especially high VLDL is associated with increased SBP and decreased DBP.

Competing Interests

The authors declare that there are no competing interests.

Acknowledgments

The authors are thankful to the staff and patients of the sickle cell clinic and also to the Laboratory Department of the Tema General Hospital.

References

[1] A. V. Hoffbrand, P. Moss, and J. Pettit, Eds., *Genetic Disorder of Haematology: Essential Haematology*, Blackwell, 5th edition, 2006.

[2] J. A. Switzer, D. C. Hess, F. T. Nichols, and R. J. Adams, "Pathophysiology and treatment of stroke in sickle-cell disease: present and future," *Lancet Neurology*, vol. 5, no. 6, pp. 501–512, 2006.

[3] A. A. Uwakwe, C. Onwuegbuke, and N. M. Nwinuka, "Effect of caffeine on the polymerization of HbS and sickling rate osmotic fragility of HbS erythrocytes," *Journal of Applied Sciences and Environmental Management*, vol. 6, no. 1, pp. 69–72, 2002.

[4] R. J. Dunlop and K. C. Benneth, "Pain management for sickle cell disease," *Cochrane Database of Systematic Reviews*, vol. 19, pp. 246–248, 2006.

[5] J. K. Nnodim, A. U. Opara, H. U. Nwanjo, and O. A. Ibeaja, "Plasma lipid profile in sickle cell disease patients in Owerri, Nigeria," *Pakistan Journal of Nutrition*, vol. 11, no. 1, pp. 64–65, 2012.

[6] O. E. Yesim, S. Suna, U. Selma, O. Hilal, and O. Nuriman, "Hypocholesterolemia is associated negatively with hemolysate lipid peroxidation in sickle cell anemia patients," *Clinical and Experimental Medicine*, vol. 11, no. 3, pp. 195–198, 2011.

[7] A. Diatta, N. Sall, N. Sarr, and F. Diallo, "Évaluation du stress oxydatifdans la maladiedrepanocytaire," *L'Eurobiologiste*, vol. 33, no. 241, pp. 57–60, 1999.

[8] C. Rice-Evans, S. C. Omorphos, and E. Baysal, "Sickle cell membranes and oxidative damage," *Biochemical Journal*, vol. 237, no. 1, pp. 265–269, 1986.

[9] M. S. Buchowski, L. L. Swift, S. A. Akohoue, S. M. Shankar, P. J. Flakoll, and N. Abumrad, "Defects in postabsorptive plasma homeostasis of fatty acids in sickle cell disease," *Journal of Parenteral and Enteral Nutrition*, vol. 31, no. 4, pp. 263–268, 2007.

[10] S. Zorca, L. Freeman, M. Hildesheim et al., "Lipid levels in sickle-cell disease associated with haemolytic severity, vascular dysfunction and pulmonary hypertension," *British Journal of Haematology*, vol. 149, no. 3, pp. 436–445, 2010.

[11] K. Bergmann, "Non-HDL cholesterol and evaluation of cardiovascular disease risk," *The New England Journal of Medicine*, vol. 364, pp. 127–135, 2011.

[12] M. Rumińska, A. Czerwonogrodzka, B. Pyrzak, A. Majcher, and D. Janczarska, "Utility of non-HDL in abdominal obesity in children and adolescents," *Pediatria Polska*, vol. 85, no. 1, pp. 35–40, 2010.

[13] S. S. Martin, M. J. Blaha, M. B. Elshazly et al., "Comparison of a novel method vs the Friedewald equation for estimating low-density lipoprotein cholesterol levels from the standard lipid profile," *The Journal of the American Medical Association*, vol. 310, no. 19, pp. 2061–2068, 2013.

[14] S. M. Boekholdt, B. J. Arsenault, S. Mora et al., "Association of LDL cholesterol, non-HDL cholesterol, and apolipoprotein B levels with risk of cardiovascular events among patients treated with statins: a meta-analysis," *JAMA*, vol. 307, no. 12, pp. 1302–1309, 2012.

[15] Z. Rahimi, A. Merat, M. Haghshenass, H. Madani, M. Rezaei, and R. L. Nagel, "Plasma lipids in Iranians with sickle cell disease: hypocholesterolemia in sickle cell anemia and increase of HDL-cholesterol in sickle cell trait," *Clinica Chimica Acta*, vol. 365, no. 1-2, pp. 217–220, 2006.

[16] J. Liu, C. Sempos, R. P. Donahue, J. Dorn, M. Trevisan, and S. M. Grundy, "Joint distribution of non-HDL and LDL cholesterol and coronary heart disease risk prediction among individuals with and without diabetes," *Diabetes Care*, vol. 28, no. 8, pp. 1916–1921, 2005.

[17] Z. M. Marzouki and S. M. Khoja, "Plasma and red blood cells membrane lipid concentration of sickle cell disease patients," *Saudi Medical Journal*, vol. 24, no. 4, pp. 376–379, 2003.

[18] M. A. F. El-Hazmi, A. S. Warsy, A. Al-Swailem, A. Al-Swailem, and H. Bahakim, "Red cell genetic disorders and plasma lipids," *Journal of Tropical Pediatrics*, vol. 41, no. 4, pp. 202–205, 1995.

[19] D. J. VanderJagt, J. Shores, A. Okorodudu, S. N. Okolo, and R. H. Glew, "Hypocholesterolemia in Nigerian children with sickle cell disease," *Journal of Tropical Pediatrics*, vol. 48, no. 3, pp. 156–161, 2002.

[20] J. Shores, J. Peterson, D. VanderJagt, and R. H. Glew, "Reduced cholesterol levels in African-American adults with sickle cell disease," *Journal of the National Medical Association*, vol. 95, no. 9, pp. 813–817, 2003.

[21] J. Sasaki, M. R. Waterman, G. R. Buchanan, and G. L. Cottam, "Plasma and erythrocyte lipids in sickle cell anaemia," *Clinical & Laboratory Haematology*, vol. 5, no. 1, pp. 35–44, 1983.

[22] M. P. Westerman, "Hypocholesterolaemia and anaemia," *British Journal of Haematology*, vol. 31, no. 1, pp. 87–94, 1975.

[23] M. O. Seixas, L. Rocha, M. Carvalho et al., "Lipoprotein cholesterol and triglyceride in children with steady-state sickle cell disease," *Blood*, vol. 114, no. 22, pp. 1547–1547, 2009.

[24] W. L. Stone, P. H. Payne, and F. O. Adebonojo, "Plasma-vitamin E and low plasma lipoprotein levels in sickle cell anemia patients," *Journal of the Association for Academic Minority Physicians*, vol. 1, no. 2, pp. 12–16, 1990.

[25] E. Choy and N. Sattar, "Interpreting lipid levels in the context of high-grade inflammatory states with a focus on rheumatoid arthritis: a challenge to conventional cardiovascular risk actions," *Annals of the Rheumatic Diseases*, vol. 68, no. 4, pp. 460–469, 2009.

[26] S. Yuditskaya, A. Tumblin, G. T. Hoehn et al., "Proteomic identification of altered apolipoprotein patterns in pulmonary hypertension and vasculopathy of sickle cell disease," *Blood*, vol. 113, no. 5, pp. 1122–1128, 2009.

[27] M. O. Seixas, L. C. Rocha, M. B. Carvalho et al., "Levels of high-density lipoprotein cholesterol (HDL-C) among children with steady-state sickle cell disease," *Lipids in Health and Disease*, vol. 9, article 91, 2010.

[28] K. Akinlade, C. Adewale, S. Rahamon, F. Fasola, J. Olaniyi, and A. Atere, "Defective lipid metabolism in sickle cell anaemia subjects in vaso-occlusive crisis," *Nigerian Medical Journal*, vol. 55, no. 5, pp. 428–431, 2014.

[29] G. Reaven, "Metabolic syndrome: pathophysiology and implications for management of cardiovascular disease," *Circulation*, vol. 106, no. 3, pp. 286–288, 2002.

[30] P. L. Da Luz, D. Favarato, J. R. Faria-Neto Jr., P. Lemos, and A. C. P. Chagas, "High ratio of triglycerides to HDL-cholesterol predicts extensive coronary disease," *Clinics*, vol. 63, no. 4, pp. 427–432, 2008.

[31] V. Sachdev, R. F. Machado, Y. Shizukuda et al., "Diastolic dysfunction is an independent risk factor for death in patients with sickle cell disease," *Journal of the American College of Cardiology*, vol. 49, no. 4, pp. 472–479, 2007.

23

The Effect of Proprotein Convertase Subtilisin/Kexin Type 9 Inhibitors on Nonfasting Remnant Cholesterol in a Real World Population

Anthony P. Morise ⓘ,[1] Jennifer Tennant ⓘ,[1] Sari D. Holmes ⓘ,[1] and Danyel H. Tacker ⓘ[2]

[1]*Section of Cardiology, West Virginia University Heart and Vascular Institute, Morgantown, WV, USA*
[2]*Department of Pathology, Anatomy, and Laboratory Medicine, West Virginia University School of Medicine, Morgantown, WV, USA*

Correspondence should be addressed to Anthony P. Morise; tmorise@hsc.wvu.edu

Academic Editor: Akihiro Inazu

Background. Proprotein convertase subtilisin/kexin type 9 (PCSK9) inhibitors have demonstrated significant effects on low-density lipoprotein (LDL) cholesterol and nonhigh density lipoprotein (HDL) cholesterol. To date, there have been limited reports on the effect of PCSK9 inhibitors on remnant cholesterol. *Objectives.* Assess the effect of PCSK9 inhibitors on nonfasting remnant cholesterol in a real world population. Identify whether pretreatment triglyceride levels are associated with PCSK9 inhibition success as indicated by changes in remnant cholesterol levels. *Methods.* Patients in our adult lipid clinic (*n* = 109) receiving PCSK9 inhibition for atherosclerotic cardiovascular disease or familial hypercholesterolemia who had available pre- and post-PCSK9 inhibition standard nonfasting lipid data were, retrospectively, selected for data analysis. Remnant cholesterol was the difference between non-HDL and LDL cholesterol. LDL cholesterol was measured directly and calculated from Friedewald and Martin/Hopkins methods. Data were analyzed using repeated measures ANOVA and multivariable linear regression for differential effects on remnant and LDL cholesterol based upon pretreatment nonfasting triglyceride levels. *Results.* Remnant cholesterol as well as total, LDL, non-HDL cholesterol, and triglycerides decreased significantly (*P*<0.001) after PCSK9 inhibition. Patients with higher pretreatment triglyceride levels showed greater decrease in remnant cholesterol after PCSK9 inhibition (*P*<0.001) than those with lower pretreatment triglycerides. *Conclusions.* In patients receiving PCSK9 inhibitors, remnant cholesterol as determined from nonfasting blood was reduced in proportion to pretreatment triglycerides.

1. Introduction

Proprotein convertase subtilisin/kexin type 9 (PCSK9) inhibitors have demonstrated significant effects on most lipoprotein particles and their respective cholesterol content. The recently completed Fourier outcomes trial demonstrated significant reductions in low-density lipoprotein (LDL) cholesterol, nonhigh density lipoprotein (HDL) cholesterol, total cholesterol, Apolipoproteins B (Apo B) and A1, triglycerides, and lipoprotein a in a large population of patients with stable coronary disease using evolocumab [1]. In earlier efficacy trials, both alirocumab and evolocumab had been demonstrated to lower concentrations of intermediate density cholesterol (IDL) [2, 3]. The pharmacologic reduction

of most particles that include Apo B has shown prognostic benefit [4]. Very low-density lipoprotein (VLDL) remnants, IDL, and chylomicron remnants contain Apo B (either Apo B100 or 48) and are often collectively referred to as remnant cholesterol or triglyceride-rich lipoproteins (TRL). In fasting specimens, directly measured remnant lipoprotein cholesterol has been demonstrated to be associated with increased risk of coronary heart disease and large artery atherosclerotic stroke [5, 6]. In nonfasting specimens, a simpler method to estimate remnant cholesterol content has also been suggested to be an independent causal risk factor for ischemic heart disease [7]. While all patients prescribed PCSK9 inhibitors will have unacceptably high LDL cholesterol, all patients prescribed PCSK9 inhibitors will not necessarily have elevated remnant cholesterol or triglycerides.

To date, there are limited reports on the effect of PCSK9 inhibitors on remnant cholesterol [2, 3]. The purpose of this study was to evaluate the effect of these medications on nonfasting remnant cholesterol in a real world population of patients with FDA- and payer-approved clinical indications for the medication and to confirm that the effect of PCSK9 inhibition on remnant cholesterol is dependent upon the baseline triglyceride level.

2. Materials and Methods

2.1. Patient Population. All patients were >18 years of age and referred to our adult lipid clinic at the West Virginia University Heart and Vascular Institute because of an unacceptably high LDL cholesterol level. All patients had either atherosclerotic cardiovascular disease (ASCVD) or heterozygous familial hypercholesterolemia (FH) without clinical ASCVD as qualifying diagnoses for approval of PCSK9 inhibitors. The vast majority of patients had some adjustment made to their lipid-lowering therapy upon arrival to the lipid clinic. However, all patients were on a stable regimen of lipid-lowering therapy at the time of enrollment and use of the PCSK9 inhibitor. No changes in lipid-lowering therapy were made during the initiation and follow-up of the PCSK9 inhibitor. All patients received approval from their respective payers for the administration of the particular PCSK9 inhibitor that was acceptable to the payer. Criteria used were the patient's qualifying diagnosis, documented record of lipid-lowering therapy, and current lipid levels on that therapy. All patients had unacceptably high LDL cholesterol levels on the maximal medical therapy they were able to tolerate, meaning that their LDL cholesterol levels were >70 mg/dl for ASCVD and >100 mg/dl for FH. Patients could be on moderate to high-intensity statin therapy, completely statin intolerant on 0 mg of statin, tolerant of small doses of statin, on other nonstatin therapies, or some combination of these. All patients included in this analysis had standard nonfasting lipid laboratory data available just before administration and after the administration of at least 3 doses of the PCSK9 inhibitor. Dosing interval for PCSK9 inhibitors was biweekly. All patients received either evolocumab or alirocumab. Approval to use patient data was obtained from the institutional review board and a waiver of consent was granted.

2.2. Laboratory Testing. All patients had nonfasting lipid testing. Standard lipid measurements were performed on ARCHITECT c-analyzers using ARCHITECT reagents (Abbott Diagnostics, Abbott Park, IL, USA) including measured lipid profile components (total cholesterol, HDL cholesterol, and triglycerides) and direct LDL measurements at baseline (pretreatment) and minimally after the third dose of PCSK9 inhibitor. An attempt was also made to obtain Apo B measurements before and after therapy. Total cholesterol measurements were performed using standard enzymatic assay employing coupled cholesterol esterase/cholesterol oxidase reactions and detection of quinoneimine product at 500 nm. HDL cholesterol measurements were performed employing selective dissolution of HDL with proprietary detergent and subsequent coupled cholesterol esterase/cholesterol oxidase reactions as for total cholesterol measurement. Triglycerides measurements were performed using standard enzymatic assay employing coupled lipase/glycerol kinase/glycerol phosphate oxidase reactions and detection of quinoneimine product at 500 nm. Direct LDL cholesterol measurements were performed employing selective dissolution of LDL with proprietary detergent and subsequent coupled cholesterol esterase/cholesterol oxidase reactions as for total cholesterol measurement. Apo B measurements were performed at a reference laboratory using the Tina-quant Apolipoprotein B immunoturbidimetric assay (Roche Diagnostics, Indianapolis, IN, USA).

2.3. Calculations. Remnant cholesterol was determined by calculating the difference between non-HDL cholesterol and LDL cholesterol as previously described [7]. Non-HDL cholesterol was determined by calculating the difference between total cholesterol and HDL cholesterol. In addition to the direct measurement, LDL cholesterol was calculated using 2 methods: Friedewald equation (LDL = total cholesterol – HDL cholesterol – (triglycerides/5)) [8] and the recently described Martin/Hopkins method [8]. Because the Friedewald equation and Martin/Hopkins method both fail when triglycerides are > 400 mg/dL, the direct LDL measurement was substituted.

2.4. Statistical Analysis. All analyses were performed with SPSS Statistics Version 24.0 (IBM Corp., Armonk, NY). P value < 0.05 was considered statistically significant. Continuous variables were presented as mean ± standard deviation (SD) and categorical variables were presented as frequency (percent). Pretreatment to posttreatment lipid measures were compared using paired-samples t tests. Pearson correlations were conducted to compare changes in remnant cholesterol with changes in other lipid measures. To examine the impact of pretreatment triglyceride levels on change in remnant cholesterol measures, groups were created above and below the median value of pretreatment triglycerides in this sample (>223 versus ≤223). Changes in LDL and remnant cholesterol by pretreatment triglyceride groups were examined using repeated measures ANOVA. Multivariable linear regression analyses were used to examine the impact of pretreatment triglyceride groups on change in remnant cholesterol measures after adjustment for age, gender, diabetes, ASCVD, any statin use, and type/dose of PCSK9 inhibitor (evolocumab 140 mg, alirocumab 75 mg, and alirocumab 150 mg). Beta and B coefficient values are presented from these models to show the standardized and unstandardized estimate of effect, respectively, in remnant cholesterol change for each factor in the model, although only factors significantly associated with remnant cholesterol change were interpreted. The B coefficients can be used to form the regression equation for each model and represent the estimate of effect for each variable using the original units of measurement, whereas the beta coefficient is standardized and can demonstrate the

TABLE 1: Baseline characteristics of patient sample.

	Statin Intolerance	Total Sample $N = 109$	Any LLRx
Age (years)		64.1 ± 11.7	
Female		65 (60)	
PCSK9 Inhibitor			
Evolocumab 140mg		62 (57)	
Alirocumab 75mg		23 (21)	
Alirocumab 150 mg		24 (22)	
ASCVD		98 (90)	
Coronary		87	
Peripheral		2	
Cerebral		1	
Polyvascular		2	
Coronary calcium		6	
FH without ASCVD		11 (10)	
Diabetes		37 (34)	
Lipid diagnosis ASCVD			
Hypercholesterolemia alone		61 (56)	
Combined hyperlipidemia		37 (34)	
Therapy	Statin Intolerance	Any Statin	Any LLRx
ASCVD	78 (80)	20 (20)	44 (45)
FH	9 (82)	2 (18)	2 (18)

ASCVD, atherosclerotic cardiovascular disease; FH, familial hypercholesterolemia; LLRx, lipid lowering therapy; PCSK9, proprotein convertase subtilisin kexin 9.
Data presented as frequency (%) or mean ± SD.

relative strength of effect amongst all factors in the model regardless of differing units of measurement.

It is possible that the phenomenon of regression to the mean could play a role in the results of this study given that there was only one pretreatment measurement and one posttreatment measurement. Therefore, for the purpose of sensitivity analyses, an equation was used to estimate the regression to the mean effect [9] and doubly robust ANCOVA analyses of the triglyceride group effect on change in remnant cholesterol levels, adjusted again by baseline remnant cholesterol levels, were conducted.

3. Results

3.1. Patient Population. Between September 2015 and January 2018, 122 patients were treated with PCSK9 inhibitors. At the time of this analysis, 10 patients had not received 3 doses of drug and had not obtained follow-up lipid profiles. Three additional patients failed to get adequate follow-up lipid profiles after beginning medication. Therefore, 109 patients had lipid profile data before and after PCSK9 inhibition, including 62 patients with evolocumab and 47 patients with alirocumab (Table 1). Most patients were treated because of ASCVD (mean age = 65.4 ± 10.4 years). Most with ASCVD had statin intolerance, hypercholesterolemia alone, and coronary artery disease as the principle manifestation and were not on any lipid-lowering therapy. Of the 10%

of patients who had familial hypercholesterolemia without ASCVD (mean age = 52.0 ± 15.6 years), most had statin intolerance, pure hypercholesterolemia ($n = 1$ with combined hyperlipidemia) and were not on any lipid-lowering therapy. Diabetes of any type was present in 34% of patients, all with ASCVD.

3.2. Basic Lipid Data. With the exception of HDL cholesterol, all lipid measures decreased significantly after PCSK9 inhibition (Table 2). Examining continuous variable change scores, greater absolute decrease in triglycerides after PCSK9 inhibition was significantly correlated with greater absolute decrease in remnant cholesterol when LDL was derived using any of the 3 methods (Friedewald equation [$r = 0.87$, $P < 0.001$], Martin/Hopkins method [$r = 0.77$, $P < 0.001$], and direct measurement [$r = 0.71$, $P < 0.001$]). The absolute decrease in remnant cholesterol was significantly greater when LDL was derived by Martin/Hopkins method than when derived by Friedewald equation (-14.3 versus -11.7, $P = 0.002$) or measured directly (-14.3 versus -10.99, $P = 0.022$).

3.3. Impact of Pretreatment Triglyceride Levels. Repeated measures ANOVA revealed that patients in the high pretreatment triglyceride group ($n = 54$) had greater reductions in remnant cholesterol than patients in the low pretreatment triglyceride group ($n = 55$) regardless of the way LDL was determined (Friedewald equation [$F = 20.0$, $P < 0.001$], direct

TABLE 2: Lipid levels before and after treatment.

	Pre-treatment	Post-treatment	% Change	P value
Total cholesterol	259.6 ± 70.1	151.5 ± 52.7	−41%	<0.001
HDLc	45.1 ± 11.3	46.0 ± 11.8	2%	0.228
NonHDLc	215.1 ± 68.6	106.5 ± 52.4	−50%	<0.001
Triglycerides	255.3 ± 161.7	191.9 ± 99.8	−24%	<0.001
LDLc				
Friedewald equation	167.0 ± 57.2	70.11 ± 47.2	−58%	<0.001
Direct measurement	174.5 ± 57.2	76.9 ± 46.4	−56%	<0.001
Martin/Hopkins method	171.3 ± 56.4	76.9 ± 46.0	−55%	<0.001
Remnant cholesterol				
Friedewald equation LDLc	48.1 ± 28.2	36.4 ± 16.5	−24%	<0.001
Direct measurement LDLc	40.6 ± 29.9	29.6 ± 15.4	−27%	<0.001
Martin/Hopkins method LDLc	43.8 ± 27.2	29.6 ± 13.1	−32%	<0.001
Apolipoprotein B ($n = 98$)	141.4 ± 40.8	74.2 ± 30.5	−48%	<0.001

HDLc = high density lipoprotein cholesterol, LDLc = low density lipoprotein cholesterol.

measurement [$F = 11.4$, $P = 0.001$], and Martin/Hopkins method [$F = 16.9$, $P < 0.001$]; Figure 1). Repeated measures ANOVA also revealed that patients in the high pretreatment triglyceride group had similar reductions in Friedewald derived LDL ($F = 0.3$, $P = 0.591$), direct measurement LDL ($F = 1.2$, $P = 0.275$), and Martin/Hopkins derived LDL ($F = 1.0$, $P = 0.315$) to patients in the low pretreatment triglyceride group (Figure 2).

In multivariable analyses, pretreatment triglycerides group (low versus high) was a significant independent factor associated with changes in remnant cholesterol after PCSK9 inhibition (Table 3). Specifically, patients in the high pretreatment triglyceride group had greater reductions in remnant cholesterol regardless of the way LDL was determined (Friedewald equation [$B = -20.1$, $P < 0.001$], Martin/Hopkins method [$B = -16.8$, $P < 0.001$], and direct measurement [$B = -16.3$, $P = 0.001$]). Within these multivariable models, the high pretreatment triglyceride group was associated with a 20.1 point, 16.8 point, and 16.3 point greater reduction in remnant cholesterol, respectively, when LDL was determined by Friedewald equation, Martin/Hopkins method, and direct measurement. Multivariable analyses also demonstrated that the type of drug used was significantly associated with changes in remnant cholesterol when LDL was derived with the Friedewald equation and Martin/Hopkins method, but not when direct LDL measurement was utilized. Specifically, patients who received alirocumab 150 mg had greater reductions in remnant cholesterol than patients who received evolocumab 140 mg when LDL was derived by the Friedewald equation ($B = -12.1$, $P = 0.036$) and Martin/Hopkins method ($B = -12.8$, $P = 0.015$). Within these multivariable models, alirocumab 150 mg was associated with a 12.1-point and 12.8-point greater reduction in remnant cholesterol compared to evolocumab 140 mg when LDL was derived by Friedewald equation and Martin/Hopkins method, respectively.

The results of sensitivity analysis estimates found that it is possible that the change in remnant cholesterol levels for the low triglyceride group was mostly due to regression to the mean, but the change in remnant cholesterol levels

for the high triglyceride group had an effect over and above the effect for regression to the mean. In addition, doubly robust ANCOVA analyses found similar effects for triglyceride group as the primary univariate analyses in the study (Friedewald equation [$F = 3.8$, $P = 0.055$], direct measurement [$F = 4.9$, $P = 0.029$], and Martin/Hopkins method [F = 5.3, $P = 0.023$]).

4. Discussion

This study demonstrates that remnant cholesterol, like all other Apo B containing lipid fractions, is effectively lowered by PCSK9 inhibition. Importantly, these observations were made in a real world population outside of a clinical trial. In addition, as expected, our findings demonstrate that the degree of remnant cholesterol lowering by PCSK9 inhibition is associated with pretreatment triglyceride concentration, although the same is not true for LDL cholesterol lowering. This finding is consistent with the concept that remnant cholesterol containing larger TRL particles is predicted to be elevated when the serum triglycerides are elevated.

While all of the study patients had approved indications for PCSK9 inhibition, most were not treated with high-intensity statins and thus were considered to be statin intolerant. As a result, pretreatment LDL cholesterol was significantly higher in our study patients than those seen in clinical trials [1]. When compared with the results of the Fourier study [1], the percent decrease in nonHDL, Apo B, and LDL cholesterol in our study was comparable. However, the percent decrease in triglycerides was greater in our study, likely reflecting the nonfasting lipid testing and higher pretreatment triglyceride levels.

We used 3 different methods to generate LDL cholesterol results and then subsequent remnant cholesterol results. Each of the 3 methodologies has strengths and weaknesses. The commonly calculated and nearly universal Friedewald method should not be used when triglycerides are > 400 mg/dL; in such instances, direct LDL measurement is preferred. The recently described Martin/Hopkins method

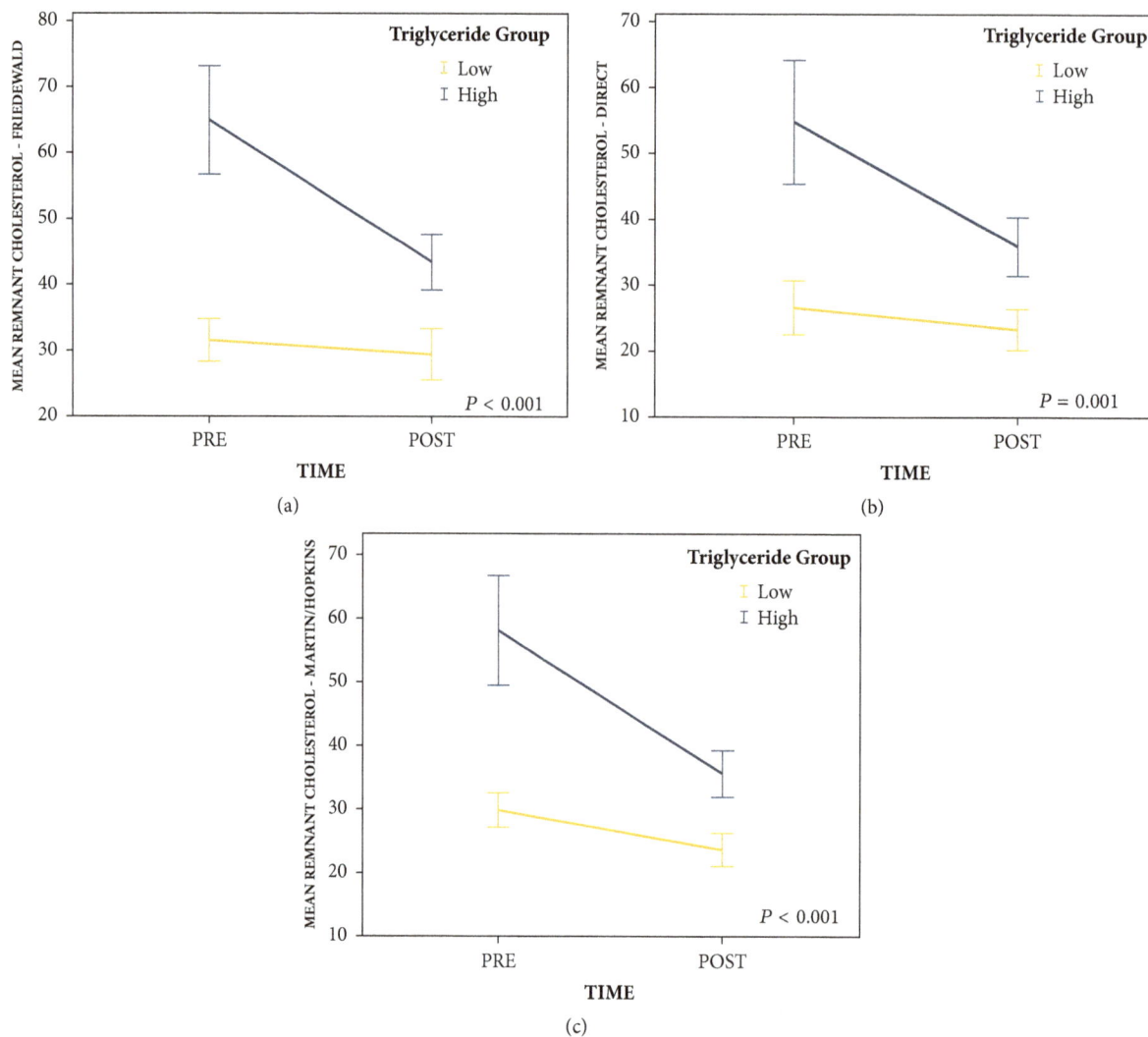

FIGURE 1: Change in remnant cholesterol levels from pretreatment to posttreatment by pretreatment triglyceride groups (error bars: 95% CI).

[8] utilizes the same measured variables as the Friedewald method. However, when rapid ultracentrifugation is used as the standard, the Martin/Hopkins method is more accurate than the Friedewald method in estimate of LDL cholesterol especially when the LDL levels are low [10]. Nevertheless, the Martin/Hopkins method still has the limitation when triglycerides are > 400 mg/dL. Direct LDL measurement is not dependent on the measurement of triglycerides for its determination and should not be effected by elevated triglycerides as are the other 2 methods. In fact in the current study, the pretreatment direct LDL was higher than the LDLs estimated by the other 2 methods perhaps reflecting the absence of the triglyceride influence. While directly measured LDL has none of the aforementioned limitations due to triglycerides, it is not as standardized as are the measurements in the standard lipid profile (total cholesterol, HDL, and triglycerides) used to calculate LDL cholesterol. The effect of PCSK9 inhibition on LDL and remnant cholesterol is qualitatively similar using each of the 3 methods. The quantitative

differences we may have found in this study will need to be reevaluated using a larger sample size.

Current knowledge concerning how PCSK9 inhibitors work indicates that the interaction between LDL particles and the LDL receptor is effected by PCSK9 such that the LDL receptor (as well as the LDL particle) is ultimately destroyed and never recycled back to the hepatic cellular surface. Remnant particles have a somewhat different pathway for clearance [11]; as the larger TRL particles such as VLDL have their triglyceride content reduced by lipase hydrolysis, they eventually become IDL or smaller VLDL particles. Given sufficient time, they can eventually become LDL particles. Most TRL particles have Apolipoprotein E (Apo E) as well as Apo B on their surfaces, acquired during the process of VLDL production in the liver. Apo E is a ligand for the LDL receptor in the liver as well as for hepatic heparan sulfate proteoglycan receptors (HSPG-R). HSPG-R contain hepatic lipase and polypeptide strands used to capture lipoproteins. Due to con-figurational issues, many VLDL remnants cannot bind to LDL

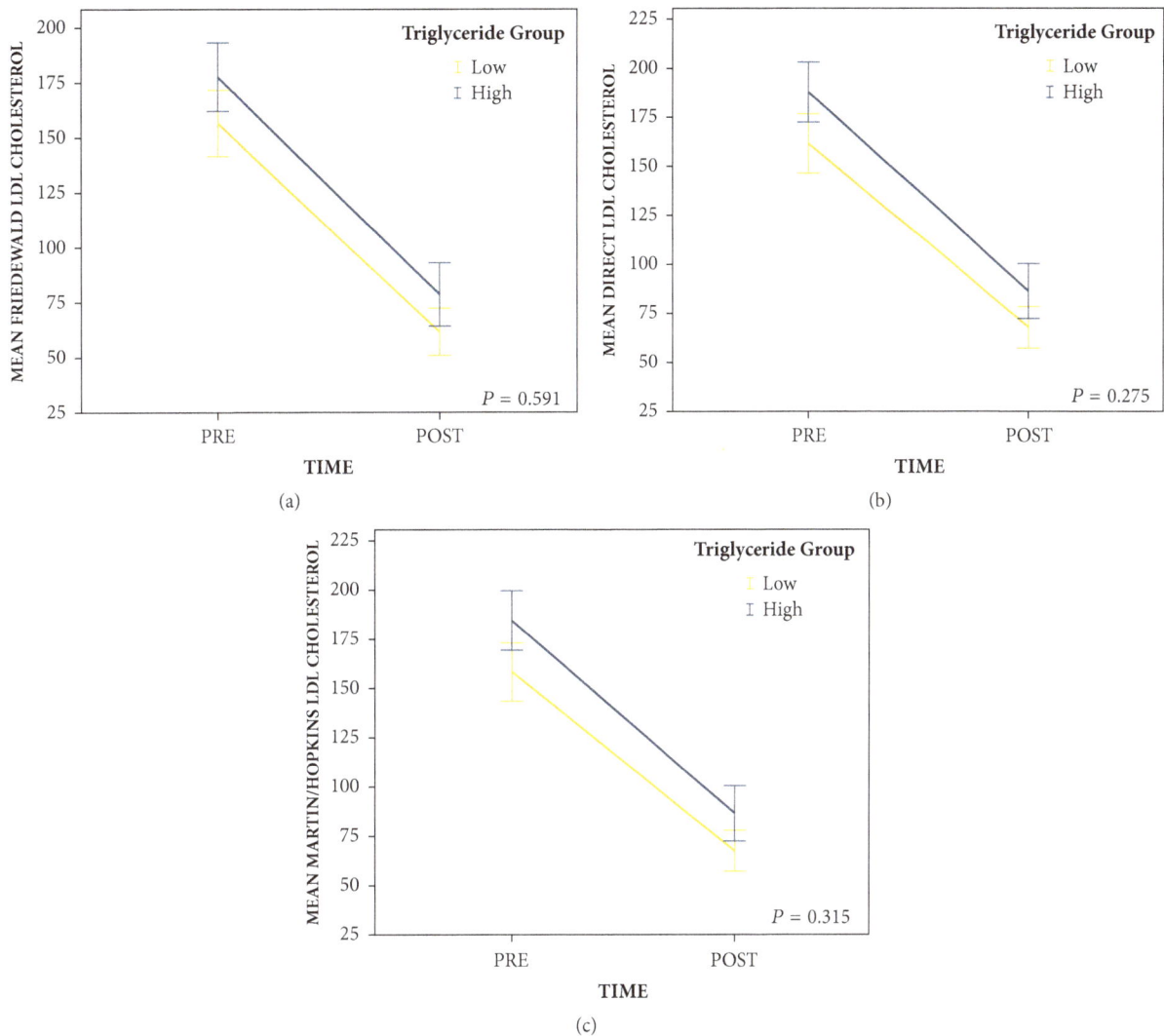

FIGURE 2: Change in LDL cholesterol levels from pretreatment to posttreatment by pretreatment triglyceride groups (error bars: 95% CI).

receptors and must rely on HSPG-R for remnant clearance. Remnant particles become attached to HSPG-R via Apo E binding and undergo further triglyceride lipolysis by hepatic lipase. By this mechanism, many remnant particles diminish in size and become LDL particles suitable for clearance by the LDL receptor. Recent information [12] suggests that, in addition to being involved in remnant metabolism, HSPG-R are PCSK9 receptors that facilitate subsequent PCSK9 and LDL receptor complex formation. Much like the interference with LDL clearance, this PCSK9/LDL receptor complex formation would interfere with the process of remnant clearance as well, due to the reduction in available LDL receptors as VLDL and IDL become LDL particles. Therefore, it would be expected that PCSK9 inhibition by currently available monoclonal antibodies would facilitate the process of remnant as well as LDL clearance.

The differences found by multivariable analysis concerning the effect of equipotent doses of 2 PCSK9 inhibitors on the reduction of remnant cholesterol may be related to sample size or chance alone. Thus, further study in larger groups of patients is needed.

4.1. Study Limitations. This study is limited by its lack of a control group, small sample size, lack of specific remnant particle analysis, and direct measurement of remnant cholesterol content. Previously published data involved 81 healthy, normolipemic, nonobese men for evolocumab [2], and 18 healthy volunteers for alirocumab [3]. Evolocumab significantly reduced remnant pool size and increased the fractional catabolic rate of IDL and VLDL [2]. Alirocumab significantly reduced IDL cholesterol content but not VLDL cholesterol content [3]. Larger pools of more clinically relevant patients with appropriate control groups are available from the Fourier [1] (evolocumab, $N = 27,564$), Spire [13] (bococizumab, $N = 27,438$), and Odyssey [14] (alirocumab, $N = 18,924$) outcomes trials. It is unclear whether these large datasets have detailed remnant particle data or directly measured remnant cholesterol as used in prior studies

TABLE 3: Results of the multivariable linear regression analyses for change in remnant cholesterol when LDL was derived by 3 different methodologies.

Friedewald Equation	B	Beta	P Value
Age	−0.21	−0.10	0.708
Female	8.79	0.18	0.074
Diabetes	−5.25	−0.10	0.278
ASCVD	0.87	0.01	0.913
Statin	−2.45	−0.04	0.669
Alirocumab 75mg vs evolocumab 140mg	−8.39	−0.14	0.156
Alirocumab 150mg vs evolocumab 140mg	−12.09	−0.21	0.036
Triglyceride group (high vs low)	−20.06	−0.41	<0.001
Direct Measurement			
Age	−0.31	−0.14	0.193
Female	8.39	0.16	0.113
Diabetes	−1.61	−0.03	0.757
ASCVD	2.90	0.04	0.734
Statin	−8.60	−0.14	0.165
Alirocumab 75mg vs evolocumab 140mg	−1.56	−0.03	0.806
Alirocumab 150mg vs evolocumab 140mg	−7.29	−0.12	0.236
Triglyceride group (high vs low)	−16.29	−0.32	0.001
Martin/Hopkins Method			
Age	−0.27	−0.14	0.185
Female	8.62	0.19	0.053
Diabetes	−4.16	−0.09	0.342
ASCVD	1.15	0.02	0.872
Statin	−0.87	−0.02	0.866
Alirocumab 75mg vs evolocumab 140mg	−6.04	−0.11	0.259
Alirocumab 150mg vs evolocumab 140mg	−12.76	−0.24	0.015
Triglyceride group (high vs low)	−16.76	−0.38	<0.001

ASCVD = atherosclerotic cardiovascular disease.

[5, 6, 15]. Nevertheless, the analysis of remnant cholesterol and risk as presented by Varbo et al. [7] suggests that the simple calculation of remnant cholesterol from non-HDL cholesterol minus LDL cholesterol could be sufficient to assess clinical impact. These simpler remnant cholesterol data are available in the 3 large randomized trial populations and could be explored to confirm the findings of our study and determine whether these changes in remnant cholesterol have a prognostic impact independent of the effect on LDL.

We limited our evaluation to nonfasting specimens principally for the sake of patient convenience. However, the use of nonfasting lipids is endorsed by numerous societies, guidelines, and statements including the American Heart Association, the European Society of Cardiology, and the American Association of Clinical Endocrinology [16]. In addition, Varbo et al. [7] have demonstrated through both observational hazard assessments and Mendelian randomization studies that nonfasting remnant cholesterol assessments using the simpler calculation have merit as a means to assess for an effect on clinical outcomes.

4.2. Summary. In this real world population receiving PCSK9 inhibition for approved indications and using nonfasting lipid measurements, all Apo B containing particle subgroups had significant reductions in their respective cholesterol concentrations, including remnants. Remnant cholesterol was estimated by the simple subtraction of LDL cholesterol from non-HDL cholesterol. As anticipated, those patients with higher pretreatment triglyceride levels and therefore higher levels of pretreatment remnant cholesterol had a greater decrease in remnant cholesterol after treatment than patients with relatively normal pretreatment triglyceride and remnant levels. Further study in larger populations and with more measurement points will be needed to confirm these findings and determine if these findings will have independent prognostic significance.

References

[1] M. S. Sabatine, R. P. Giugliano, A. C. Keech et al., "Evolocumab and clinical outcomes in patients with cardiovascular disease," *The New England Journal of Medicine*, vol. 376, no. 18, pp. 1713–1722, 2017.

[2] G. F. Watts, D. C. Chan, R. Dent et al., "Factorial effects of evolocumab and atorvastatin on lipoprotein metabolism," *Circulation*, vol. 135, no. 4, pp. 338–351, 2017.

[3] G. Reyes-Soffer, M. Pavlyha, C. Ngai et al., "Effects of PCSK9 inhibition with alirocumab on lipoprotein metabolism in healthy humans," *Circulation*, vol. 135, no. 4, pp. 352–362, 2017.

[4] A. D. Sniderman, K. Williams, J. H. Contois et al., "A meta-analysis of low-density lipoprotein cholesterol, non-high-density lipoprotein cholesterol, and apolipoprotein b as markers of cardiovascular risk," *Circulation: Cardiovascular Quality and Outcomes*, vol. 4, no. 3, pp. 337–345, 2011.

[5] P. H. Joshi, A. A. Khokhar, J. M. Massaro et al., "Remnant Lipoprotein Cholesterol and Incident Coronary Heart Disease: The Jackson Heart and Framingham Offspring Cohort Studies," *Journal of the American Heart Association*, vol. 5, no. 5, p. e002765, 2016.

[6] J. Kim, J. Park, S. Jeong et al., "High Levels of Remnant Lipoprotein Cholesterol Is a Risk Factor for Large Artery Atherosclerotic Stroke," *Journal of Clinical Neurology*, vol. 7, no. 4, p. 203, 2011.

[7] A. Varbo, M. Benn, A. Tybjærg-Hansen, A. B. Jørgensen, R. Frikke-Schmidt, and B. G. Nordestgaard, "Remnant cholesterol as a causal risk factor for ischemic heart disease," *Journal of the American College of Cardiology*, vol. 61, no. 4, pp. 427–436, 2013.

[8] S. S. Martin, M. J. Blaha, M. B. Elshazly et al., "Comparison of a novel method vs the Friedewald equation for estimating low-density lipoprotein cholesterol levels from the standard lipid profile," *The Journal of the American Medical Association*, vol. 310, no. 19, pp. 2061–2068, 2013.

[9] A. G. Barnett, J. C. Van Der Pols, and A. J. Dobson, "Correction to: Regression to the mean: What it is and how to deal with it," *International Journal of Epidemiology*, vol. 34, no. 1, pp. 215–220, 2005.

[10] V. Sathiyakumar, J. Park, A. Golozar et al., "Fasting Versus Non-fasting and Low-Density Lipoprotein Cholesterol Accuracy," *Circulation*, vol. 137, no. 1, pp. 10–19, 2017.

[11] C. Koopal, A. D. Marais, J. Westerink, and F. L. J. Visseren, "Autosomal dominant familial dysbetalipoproteinemia: A pathophysiological framework and practical approach to diagnosis and therapy," *Journal of Clinical Lipidology*, vol. 11, no. 1, pp. 12–23.e1, 2017.

[12] C. Gustafsen, D. Olsen, J. Vilstrup et al., "Heparan sulfate proteoglycans present PCSK9 to the LDL receptor," *Nature Communications*, vol. 8, no. 1, 2017.

[13] P. M. Ridker, J. Revkin, P. Amarenco et al., "Cardiovascular efficacy and safety of bococizumab in high-risk patients," *The New England Journal of Medicine*, vol. 376, no. 16, pp. 1527–1539, 2017.

[14] "Presented by Dr. Philippe Steg at the American College of Cardiology Annual Scientific Session (ACC 2018)," Orlando, FL, March 10, 2018.

[15] M. V. Holmes, I. Y. Millwood, and C. Kartsonaki, "Lipids, lipoproteins, and metabolites and risk of myocardial infarction and stroke," *Journal of the American College of Cardiology*, vol. 71, pp. 630–632, 2018.

[16] B. G. Nordestgaard, "A Test in Context: Lipid Profile, Fasting Versus Nonfasting," *Journal of the American College of Cardiology*, vol. 70, no. 13, pp. 1637–1646, 2017.

24

Apolipoproteins A-I, B, and C-III and Obesity in Young Adult Cherokee

Wenyu Wang,[1] Piers Blackett,[2] Sohail Khan,[3] and Elisa Lee[1]

[1]*Center for American Indian Health Research, College of Public Health, University of Oklahoma Health Sciences Center,
 Oklahoma City, OK 73190, USA*
[2]*Section of Diabetes and Endocrinology, Department of Pediatrics, Harold Hamm Diabetes Center,
 University of Oklahoma Health Sciences Center, Oklahoma City, OK 73104, USA*
[3]*The Cherokee Nation, P.O. Box 948, Tahlequah, OK 74465, USA*

Correspondence should be addressed to Piers Blackett; piersblackett@att.net

Academic Editor: Gerd Schmitz

Since young adult Cherokee are at increased risk for both diabetes and cardiovascular disease, we assessed association of apolipoproteins (A-I, B, and C-III in non-HDL and HDL) with obesity and related risk factors. Obese participants (BMI $\geq$ 30) aged 20–40 years (n = 476) were studied. Metabolically healthy obese (MHO) individuals were defined as not having any of four components of the ATP-III metabolic syndrome after exclusion of waist circumference, and obese participants not being MHO were defined as metabolically abnormal obese (MAO). Associations were evaluated by correlation and regression modeling. Obesity measures, blood pressure, insulin resistance, lipids, and apolipoproteins were significantly different between groups except for total cholesterol, LDL-C, and HDL-apoC-III. Apolipoproteins were not correlated with obesity measures with the exception of apoA-I with waist and the waist : height ratio. In a logistic regression model apoA-I and the apoB : apoA-I ratio were significantly selected for identifying those being MHO, and the result (C-statistic = 0.902) indicated that apoA-I and the apoB : apoA-I ratio can be used to identify a subgroup of obese individuals with a significantly less atherogenic lipid and apolipoprotein profile, particularly in obese Cherokee men in whom MHO is more likely.

1. Introduction

Since obesity predicts atherosclerotic cardiovascular disease (ACVD), it has significant worldwide health and economic implications [1]. This is particularly true in the Cherokee and other American Indian populations [2] in whom obesity is associated with the metabolic syndrome, which often precedes type 2 diabetes (T2D) [3]. Consequently it has become important to study association of obesity with apolipoproteins, since obesity-associated changes in lipid transport precede and predict subsequent insulin resistance and ultimately the development of ACVD and T2D [4, 5]. Therefore, we selected apolipoproteins known to predict atherosclerosis for study. We also proposed that the obese participants could be classified as two distinct groups based on the presence of metabolic complications including

dyslipidemia [6] and that apolipoprotein levels might serve to identify differences between the metabolically healthy obese (MHO) and metabolically abnormal obese (MAO) groups.

Apolipoprotein B (apoB) represents the total number of apoB-containing lipoproteins [7] and is considered to be superior to LDL-C and non-HDL-C in predicting cardiovascular disease [8], whereas apolipoprotein A-I (apoA-I) has a known inverse association and low levels are associated with increased body mass index (BMI) [9]. Furthermore, the ratio of apoB to apoA-I (B : A-I ratio), representing the combination of two atherogenic processes, is an even stronger predictor [10]. Apolipoprotein C-III (apoC-III) is secreted with VLDL and becomes distributed among circulating lipoproteins [11] conferring harmful properties resulting in ACVD [3, 12]. LDL particles containing apoC-III are more atherogenic than particles without apoC-III [13] and apoC-III on non-HDL lipoprotein particles independently predicted recurrent coronary events [14] and progression of carotid intima-media thickness during treatment [15]. Following hepatic secretion of apoC-III as VLDL, its subsequent distribution on HDL particles may also be harmful, since HDL-apoC-III predicted angiographic progression of atherosclerosis in bypass grafts [16] and more recently HDL-apoC-III has been identified as a proatherogenic HDL subtype with loss of its anti-inflammatory properties [16].

Genetic deficiency [17] and targeted gene disruption [18] of apoC-III have been shown to be associated with protection from atherosclerosis [17]. However, the relative role of apoC-III's distribution on lipoproteins remains uncertain [19], and preliminary evidence suggests that obesity may play a central role in determining apoC-III levels, lipoprotein distribution, and clinical outcomes [20]. Consequently this study and analysis were done to examine association of obesity with apoB, apoA-I, and apoC-III content of both non-HDL and HDL.

2. Methods

With collaboration of the Cherokee Nation of Oklahoma, adults aged 20–40 years in the Cherokee Diabetes Study cohort residing in a 5-county area in northeastern Oklahoma participated in the study ($n = 1051$). Of this group 477 (45%) were obese, defined as having a BMI greater than or equal to 30. Nondiabetic participants were excluded according to American Diabetes Association criteria for fasting plasma glucose (FPG) defined as being greater than or equal to 126 mg/dl or being on medications for diabetes. Informed consent was obtained from each subject or his/her legal guardian, following approval of the Institutional Review Boards of the University of Oklahoma Health Sciences Center and the Cherokee Nation.

After obtaining clinical measurements, fasting blood specimens were collected for determining FPG, insulin, lipids, and apolipoproteins.

2.1. Lipids and Apolipoproteins.
An Abbott VP-Super System automatic analyzer and commercial reagents were used to determine levels of glucose, cholesterol (Boehringer, Mannheim, Germany), and triglyceride (Miles Inc., Tarrytown, NJ) by enzymatic methodology. HDL-C was measured using the heparin-manganese precipitation procedure of the Lipid Research Clinics program and LDL-C was calculated by the Friedewald formula. ApoA-I, apoB, and apoC-III were determined by previously validated electroimmunoassays [21–23]. The apoC-III concentrations in whole plasma and heparin-manganese supernatant were determined by separate assays. ApoC-III in the precipitate was calculated by subtracting the supernatant value from the total plasma apoC-III.

2.2. Glucose and Insulin.
Fasting insulin levels were determined in a National Institutes of Health core laboratory at the Endocrinology Department at the University of Chicago.

Insulin was measured in serum samples using a competitive double antibody radioimmunoassay and glucose by an automated method using glucose oxidase (Alfa Wassermann, Inc., West Caldwell, NJ). The homeostasis index (HOMA-IR) was computed from the product of insulin (IU/mL) and glucose (mmol/L) divided by 22.5.

2.3. Blood Pressure, Waist Circumference, and Height Measurements.
Three consecutive measurements of systolic blood pressure (SBP) and diastolic blood pressure (DBP) were performed on the right arm using a Baum mercury sphygmomanometer (W.A. Baum Co., Copiague, NY), and the average of the second and third measurements was recorded. The average of duplicate measures of the waist circumference and height was obtained using a nonstretchable linen tape for the waist circumference and a wall-mounted calibrated stadiometer for the height measurements.

2.4. Statistical Analysis.
Means and standard deviations were estimated on all measurements on obese participants aged 20–40 years. Obesity was defined as having a BMI $\geq$ 30. Participants were separated by subgroups of metabolically healthy obese (MHO) individuals, defined as having none of four National Cholesterol Education Program (NCEP) metabolic syndrome components after excluding waist circumference. Metabolically abnormal obese (MAO) individuals had at least one of the four criteria. The mean difference of a variable between the two subgroups was calculated after adjusting for age and gender. Logarithmic transformation was used if a variable was not symmetrically distributed such as triglyceride. Spearman partial correlation was done between variables after adjusting for age and gender. To explore odds that an obese participant is classified as being MHO and distinguish from being MAO, logistic regression was used to select significant determinants with adjustment for age and gender. Statistical analyses were conducted with SAS (version 9.4), and a $P < 0.05$ was considered as significant.

3. Results

Measures of obesity (BMI, waist circumference, and waist to height ratio), blood pressure (systolic and diastolic), glucose homeostasis (fasting glucose, insulin, and the homeostasis index), lipids, and apolipoproteins were different between individuals being MHO and MAO with the exception of TC, LDL-C, and HDL-apoC-III (Table 1).

Significant correlations were observed between waist and waist to height ratio with apoA-I; diastolic blood pressure with HDL-apoC-III; and insulin and HOMA-IR with apoB, apoA-I, apoB : apoA-I ratio, and non-HDL-apoC-III (Table 2). Lipids showed expected correlations with apolipoproteins with exception of triglyceride with apoA-I.

In the logistic regression model for odds or probability of an obese individual being MHO, apoA-I and apoB : apoA-I ratio were significantly selected among apoB, apoA-I, apoB : apoA-I, LpA-I, LpA-I : A-II, non-HDL-apoC-III, and HDL-apoC-III into the model for identifying those with

TABLE 1: Mean values, standard deviations, and differences ($P < 0.05$, bold type) between metabolically healthy obese (MHO) individuals, defined as those who are obese without any of four abnormal National Cholesterol Education Program (NCEP) metabolic syndrome components after excluding waist circumference, and metabolically abnormal obese (MAO) individuals defined as having one or more of the four abnormal criteria. Obesity was defined as body mass index (BMI) $\geq$ 30 for young adult Cherokee aged 20–40.

	MHO ($n = 41$)		MAO ($n = 435$)		P
	Mean	STD	Mean	STD	
Obesity					
Male	56.1%		41.1%		0.0643
BMI	33.87	3.87	36.81	6.90	**0.0096**
Waist (cms)	102.95	10.78	109.43	13.54	**0.0006**
Waist : height	0.62	0.05	0.66	0.09	**0.0022**
Blood pressure (mm Hg)					
SBP	118.59	7.57	122.88	13.01	**0.0009**
DBP	75.55	6.9676	79.41	9.6968	**0.0007**
Insulin resistance					
FPG (mmol/L)	4.55	0.55	4.85	0.62	**0.0047**
Insulin (μU/ml)	114.60	62.45	181.24	116.0	**0.0002**
log(Insulin)	4.59	0.57	5.03	0.57	**<0.0001**
HOMA-IR	3.88	2.23	6.58	4.51	**0.0001**
Lipids (mmol/L)					
TC	4.37	0.87	4.42	0.89	0.8534
TG	0.87	0.36	1.42	0.80	**<0.0001**
log(TG)	4.25	0.4570	4.69	0.5366	**<0.0001**
LDL-C	2.65	0.81	2.83	0.78	0.2249
HDL-C	1.31	0.19	0.95	0.22	**<0.0001**
Non-HDL-C	3.05	0.88	3.47	0.88	**0.0040**
Apolipoproteins (μmol/L)					
ApoB	1.56	0.37	1.75	0.40	**0.0023**
ApoA-I	2.56	0.31	2.15	0.32	**<0.0001**
ApoB : apoA-I	0.62	0.1542	0.83	0.2046	**<0.0001**
LpA-1	0.57	0.08	0.47	0.09	**<0.0001**
LpA-1 : A-II	1.99	0.24	1.68	0.24	**<0.0001**
Non-HDL-apoC-III	0.06	0.03	0.09	0.05	**0.0001**
HDL-apoC-III	0.09	0.02	0.09	0.02	0.4069

P, P value for difference between the two groups after adjusting for age and gender, and significant values ($P < 0.05$) are in bold font; STD, standard deviation.

MHO among all the obese participants (Table 3). ApoA-I has a positive association while apoB : apoA-I ratio has a strong negative association with the odds of an obese individual being MHO. Assuming that the other variables in the model are the same, males had a more than four times chance of being MHO than females; and for one standard deviation higher apoA-I or apoB : apoA-I, the chance of being MHO is increased by 2.57 times or reduced by 69%, respectively (Table 3).

Figure 1 shows the receiver operating characteristics (ROC) curve derived from the logistic regression. The area under the curve (C-statistic) was 0.902, implying high discrimination ability of the model and consequently of both apoA-I and the apoB : apoA-I ratio in identifying those being MHO among all the obese participants. For a sensitivity of 90% the respective highest specificity is 81% and the corresponding cutoff probability is 0.0855 from the ROC curve (Figure 1).

4. Discussion

Paradoxically low non-HDL-apoC-III with relatively less atherogenic lipids and lipoproteins, resembling "metabolically healthy" obesity [6] with reduced cardiovascular risk [24] was present in a young adult Cherokee population. This group had reduced adiposity and less insulin resistance as has been observed in cross-sectional studies [25, 26]. This distinct entity may represent a transient phase in a sequence of worsening insulin resistance, a concept supported by the Atherosclerosis Risk in Communities (ARIC) study showing that risk factors increase after three years of follow-up when compared to a nonobese group [27]. Therefore, prescription of weight management for all obese patients, including the "metabolically healthy," would be prudent to avoid atherogenic risk factor progression.

Insulin resistance, measured as the homeostasis index (HOMA-IR) and the fasting insulin level, correlated with

TABLE 2: Spearman partial correlation between selected variables after adjusting for age and gender for obese young adult Cherokee aged 20–40 ($N = 476$).

	ApoB		ApoA-I		ApoB : apoA-I		Non-HDL-apoC-III		HDL-apoC-III	
	R	P	R	P	R	P	R	P	R	P
Obesity										
BMI	−0.012	0.803	−0.085	0.074	0.047	0.326	0.052	0.275	−0.058	0.217
Waist	−0.029	0.542	−0.123	**0.009**	0.044	0.349	0.070	0.140	−0.062	0.189
Waist : height	−0.015	0.755	−0.109	**0.021**	0.056	0.241	0.049	0.304	−0.077	0.106
Blood pressure										
SBP	0.037	0.429	0.062	0.192	−0.008	0.863	−0.003	0.956	0.038	0.423
DBP	0.049	0.296	0.011	0.824	0.035	0.464	0.038	0.428	0.099	**0.036**
Insulin resistance										
FPG	0.068	0.154	−0.037	0.441	0.066	0.164	0.027	0.568	0.024	0.610
HOMA-IR	0.112	**0.018**	−0.106	**0.025**	0.173	**0.000**	0.183	**0.000**	0.037	0.441
Insulin	0.107	**0.023**	−0.102	**0.030**	0.172	**0.000**	0.190	**0.000**	0.035	0.463
Lipids										
TC	0.814	**<0.001**	0.340	**<0.001**	0.505	**<0.001**	0.510	**<0.001**	0.369	**<0.001**
LDL-C	0.725	**<0.001**	0.178	**0.000**	0.528	**<0.001**	0.299	**<0.001**	0.163	**0.001**
HDL-C	−0.143	**0.002**	0.759	**<0.001**	−0.562	**<0.001**	−0.299	**<0.001**	0.192	**<0.001**
Non-HDL-C	0.862	**<0.001**	0.132	**0.005**	0.672	**<0.001**	0.588	**<0.001**	0.316	**<0.001**
TG	0.602	**<0.001**	−0.090	0.057	0.574	**<0.001**	0.886	**<0.001**	0.449	**<0.001**

P, *P* value for Spearman partial correlation after adjusting for age and gender, and significant values ($P < 0.05$) are in bold font; *R*, Spearman partial correlation.

TABLE 3: Logistic regression model for odds or probability of an obese participant being metabolically healthy obese (MHO).

Variable	Estimate	SE	P	Unit[*]	OR	95% CI
Intercept	−4.245	1.888	0.0246			
Age	−0.049	0.036	0.1770	5	0.78	0.54, 1.11
Male versus female	1.433	0.401	**0.0004**		4.19	1.94, 9.46
ApoA-I	0.055	0.012	**<0.0001**	17.19	2.57	1.75, 3.90
ApoB : apoA-I	−5.594	1.361	**<0.0001**	0.21	0.31	0.17, 0.53
C-statistic	0.902					

[*]One standard deviation used for apoA-I and apoB : apoA-I.
CI, confidence interval; C-statistic, the area under the receiver operating characteristic curve; Estimate, estimated coefficient; OR, odds ratio; *P*, *P* value; SE, standard error.

apoB, with the apoB : apoA-I ratio, and inversely with apoA-I supporting known association of insulin resistance with lipoprotein transport [12]. Correlations of HOMA-IR and fasting insulin with non-HDL-apoC-III but not with HDL-apoC-III can be accounted for by an effect of insulin resistance on VLDL prior to apoC-III's transfer to HDL during lipolysis [28]. Since apoC-III has a known role in cardiovascular disease [14, 20] it may be a marker for atherosclerosis associated with obesity. Non-HDL-apoC-III is an independent predictor of atherosclerosis [15], and LDL containing both apoB and apoC-III is more atherogenic than LDL containing apoB alone [13], supporting an obesity-associated increase in risk attributable to apoC-III. Its atherogenic properties are further supported by adverse effects on the arterial wall including enhanced LDL binding to biglycan [29] and proinflammatory effects via nuclear factor kB-mediated VCAM-1 expression and monocyte adhesion [30]. We have previously shown that non-HDL-apoC-III is proportionate to the number of metabolic syndrome criteria

[3] supporting contribution to cardiovascular risk in obesity, a proposed contributor to the syndrome [4]. Furthermore, hepatic insulin resistance influences apoC-III transcription [31, 32] beginning in childhood [32]. Also the apoC-III promotor contains a carbohydrate response element suggesting that dietary glucose and possibly saturated fatty acids increase transcription [33, 34] and contribute to increasing levels [35].

We confirmed known association of obesity with triglyceride and inverse association with HDL-C [36]. We observed an inverse correlation of the waist circumference and waist to height ratio, but not BMI, with apoA-I. Although height may influence the association, since it associates negatively with liver fat content but not with visceral fat mass [37], our observation supports the role of visceral fat in HDL metabolism in obesity and the concept that low HDL-C in obesity may be associated with apoA-I degradation and impaired cholesterol efflux from lipoprotein and cellular sources [38]. However, there is considerable heterogeneity

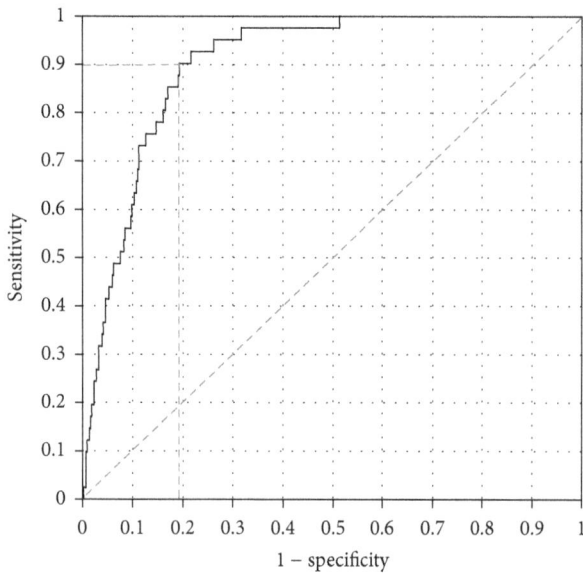

FIGURE 1: The receiver operating characteristic (ROC) curve from logistic regression selecting apoA-I and apoB to apoA-I ratio as identifying those with MHO among all obese individuals. The area under the ROC was 0.902 (that is, C-statistic = 0.902) indicating high discrimination of the model.

in prediabetes phenotypes associated with obesity, and we did not evaluate non-alcoholic fatty liver disease (NAFLD) manifesting as increased hepatic fat. Furthermore, the presence of NAFLD could account for increased carotid intima-media thickness (cIMT) and a more dysregulated lipid and apolipoprotein profile [39]. Also male participants were much more likely to have MHO; men had a more than four times chance of being MHO than women. This finding is unique in obese Cherokee participants and different to sex differences for MHO among other populations that have favored females [40]. This observation can possibly be accounted for by differences in daily exercise or unknown factors [41]; however we have previously observed higher rates for the metabolic syndrome in 20–40-year-old males (30.9%) than in females (22.0%) consistent with higher overall risk for males [3].

Our findings emphasize the atherogenic role of low and dysfunctional HDL-C in obesity since low HDL-C is associated with risk for cardiovascular disease [42] even when LDL-C is lowered [43]. Both apoA-I and HDL-C are known to be low in obesity and the low levels are attributed to HDL's interaction with VLDL mediated by hepatic triglyceride lipase [44]. These metabolic events are in part attributed to insulin resistance [44] and account for our finding that MHO participants have less insulin resistance and higher apoA-I, possibly accounting for a less atherogenic effect, particularly when combined with lower apoB and the apoB : apoA-I ratio [45]. The high C-statistic and retention in the model support prediction of MHO and lower risk attributable to the apoB : apoA-I ratio [45, 46] and support use of the ratio as a measure of cardiovascular risk attributable to obesity.

5. Conclusions

Measurements of apoA-I, apoB, and apoC-III contained in HDL and non-HDL contribute information on risk and provide rationale for cholesterol, triglyceride, and apoC-III lowering strategies in obese individuals. Association of apoC-III in non-HDL, a known correlate of triglyceride, with insulin resistance and lipids supports early risk reduction treatments including lifestyle and lipid-lowering medications. Obese cases may present with relatively normal metabolic syndrome criteria and should be managed with caution since obese individuals may increase their risk for both atherosclerosis and T2D. ApoA-I was positively and apoB : apoA-I was negatively associated with the chance of being MHO, and both were good identifiers for being MHO suggesting that these apolipoprotein measures can be helpful in assessing risk in obese individuals, especially in men who have a much higher chance of being MHO than women.

Abbreviations

Apo:	Apolipoprotein before A-I, B, and C-III
BMI:	Body mass index
FPG:	Fasting blood glucose
TC:	Total cholesterol
TG:	Triglyceride
LDL-C:	LDL-cholesterol
HDL-C:	HDL-cholesterol
apoB : apoA-I ratio:	Apolipoprotein B to A-I ratio
LpA-1 : A-II:	Lipoproteins containing apoA-I and apoA-II
SBP:	Systolic blood pressure
DBP:	Diastolic blood pressure
Log:	Logarithm
HOMA-IR:	Homeostasis index for insulin resistance
WHR:	Waist to height ratio
Waist:	Waist circumference
Non-HDL-apoC-III:	ApoC-III in non-HDL
HDL-apoC-III:	ApoC-III in HDL.

Acknowledgments

The authors wish to express their appreciation to the Cherokee people and to the health care and administrative officials of the Cherokee Nation in Tahlequah, Oklahoma, for their support and assistance. The authors also thank the administrators and staff of the Cherokee Nation clinics at Stilwell, Salina, Sallisaw, and Jay. The Cherokee Diabetes Study is supported by the National Institute of Diabetes and Digestive and Kidney Diseases, Bethesda, MD, by Grant R01 DK47920 awarded to Elisa T. Lee, Ph.D. The late Petar Alaupovic, Ph.D., Head of The Lipid and Lipoprotein Laboratory at the Oklahoma Medical Research Foundation provided laboratory support and initiated the hypothesis and analysis.

References

[1] R. H. Eckel, W. W. Barouch, and A. G. Ershow, "Report of the National Heart, Lung, and Blood Institute-National Institute of Diabetes and Digestive and Kidney Diseases working group on the pathophysiology of obesity-associated cardiovascular disease," *Circulation*, vol. 105, no. 24, pp. 2923–2928, 2002.

[2] B. V. Howard, E. T. Lee, L. D. Cowan et al., "Rising tide of cardiovascular disease in American Indians: The Strong Heart Study," *Circulation*, vol. 99, no. 18, pp. 2389–2395, 1999.

[3] P. Alaupovic, P. Blackett, W. Wang, and E. Lee, "Characterization of the metabolic syndrome by apolipoproteins in the Oklahoma Cherokee," *Journal of the Cardiometabolic Syndrome*, vol. 3, no. 4, pp. 193–199, 2008.

[4] S. M. Haffner, M. P. Stern, H. P. Hazuda, B. D. Mitchell, and J. K. Patterson, "Cardiovascular risk factors in confirmed prediabetic individuals. Does the clock for coronary heart disease start ticking before the onset of clinical diabetes?" *The Journal of the American Medical Association*, vol. 263, no. 21, pp. 2893–2898, 1990.

[5] J. B. Meigs, P. W. Wilson, C. S. Fox et al., "Body mass index, metabolic syndrome, and risk of type 2 diabetes or cardiovascular disease," *The Journal of Clinical Endocrinology & Metabolism*, vol. 91, no. 8, pp. 2906–2912, 2006.

[6] N. Stefan, H. U. Häring, F. B. Hu, and M. B. Schulze, "Metabolically healthy obesity: epidemiology, mechanisms, and clinical implications," *The Lancet Diabetes & Endocrinology*, vol. 1, no. 2, pp. 152–162, 2013.

[7] J. Elovson, J. E. Chatterton, G. T. Bell et al., "Plasma very low density lipoproteins contain a single molecule of apolipoprotein B," *Journal of Lipid Research*, vol. 29, no. 11, pp. 1461–1473, 1988.

[8] A. D. Sniderman, "Applying apoB to the diagnosis and therapy of the atherogenic dyslipoproteinemias: a clinical diagnostic algorithm," *Current Opinion in Lipidology*, vol. 15, no. 4, pp. 433–438, 2004.

[9] P. R. Blackett, K. S. Blevins, M. Stoddart et al., "Body mass index and high-density lipoproteins in Cherokee Indian children and adolescents," *Pediatric Research*, vol. 58, no. 3, pp. 472–477, 2005.

[10] G. Walldus, I. Jungner, A. H. Aastveit, I. Holme, C. D. Furberg, and A. D. Sniderman, "The apoB/apoA-I ratio is better than the cholesterol ratios to estimate the balance between plasma proatherogenic and antiatherogenic lipoproteins and to predict coronary risk," *Clinical Chemistry and Laboratory Medicine*, vol. 42, no. 12, pp. 1355–1363, 2004.

[11] M. Burstein and R. Morlin, "Precipitation of serum lipoproteins by anionic detergents in the presence of bivalent cations," *Revue Européenne D'études Cliniques et Biologiques*, vol. 15, no. 1, pp. 109–113, 1970.

[12] W. Wang, S. Khan, P. Blackett, P. Alaupovic, and E. Lee, "Apolipoproteins A-I, B, and C-III in young adult Cherokee with metabolic syndrome with or without type 2 diabetes," *Journal of Clinical Lipidology*, vol. 7, no. 1, pp. 38–42, 2013.

[13] C. O. Mendivil, E. B. Rimm, J. Furtado, S. E. Chiuve, and F. M. Sacks, "Low-density lipoproteins containing apolipoprotein C-III and the risk of coronary heart disease," *Circulation*, vol. 124, no. 19, pp. 2065–2072, 2011.

[14] F. M. Sacks, P. Alaupovic, L. A. Moye et al., "VLDL, apolipoproteins B, CIII, and E, and risk of recurrent coronary events in the cholesterol and recurrent events (CARE) trial," *Circulation*, vol. 102, no. 16, pp. 1886–1892, 2000.

[15] D. H. Blankenhorn, R. H. Selzer, D. W. Crawford et al., "Beneficial effects of colestipol-niacin therapy on the common carotid artery. Two- and four-year reduction of intima-media thickness measured by ultrasound," *Circulation*, vol. 88, no. 1, pp. 20–28, 1993.

[16] M. K. Jensen, E. B. Rimm, J. D. Furtado, and F. M. Sacks, "Apolipoprotein C-III as a potential modulator of the association between HDL-cholesterol and incident coronary heart disease," *Journal of the American Heart Association*, vol. 1, no. 2, Article ID e000232, 2012.

[17] The TG and HDL Working Group of the Exome Sequencing Project, National Heart, Lung, and Blood Institute, "Loss-of-function mutations in APOC3, triglycerides, and coronary disease," *The New England Journal of Medicine*, vol. 371, no. 1, pp. 22–31, 2014.

[18] M. J. Graham, R. G. Lee, T. A. Bell III et al., "Antisense oligonucleotide inhibition of apolipoprotein C-III reduces plasma triglycerides in rodents, nonhuman primates, and humans," *Circulation Research*, vol. 112, no. 11, pp. 1479–1490, 2013.

[19] M. C. Wyler Von Ballmoos, B. Haring, and F. M. Sacks, "The risk of cardiovascular events with increased apolipoprotein CIII: a systematic review and meta-analysis," *Journal of Clinical Lipidology*, vol. 9, no. 4, pp. 498–510, 2015.

[20] B. Talayero, L. Wang, J. Furtado, V. J. Carey, G. A. Bray, and F. M. Sacks, "Obesity favors apolipoprotein E- and C-III-containing high density lipoprotein subfractions associated with risk of heart disease," *Journal of Lipid Research*, vol. 55, no. 10, pp. 2167–2177, 2014.

[21] M. D. Curry, W. J. McConathy, J. D. Fesmire, and P. Alaupovic, "Quantitative determination of human apolipoprotein Cc-III by electroimmunoassay," *Biochimica et Biophysica Acta (BBA)—Lipids and Lipid Metabolism*, vol. 617, no. 3, pp. 503–513, 1980.

[22] M. D. Curry, P. Alaupovic, and C. A. Suenram, "Determination of apolipoprotein A and its constitutive A-I and A-II polypeptides by separate electroimmunoassays," *Clinical Chemistry*, vol. 22, no. 3, pp. 315–322, 1976.

[23] M. D. Curry, A. Gustafson, P. Alaupovic, and W. J. McConathy, "Electroimmunoassay, radioimmunoassay, and radial immunodiffusion assay evaluated for quantification of human apolipoprotein B," *Clinical Chemistry*, vol. 24, no. 2, pp. 280–286, 1978.

[24] D. Samocha-Bonet, D. Chisholm, K. Tonks, L. Campbell, and J. Greenfield, "Insulin-sensitive obesity in humans—a "favorable fat" phenotype?" *Trends in Endocrinology & Metabolism*, vol. 23, no. 3, pp. 116–124, 2012.

[25] N. Kloting, M. Fasshauer, A. Dietrich et al., "Insulin-sensitive obesity," *American Journal of Physiology. Endocrinology and Metabolism*, vol. 299, no. 3, pp. E506–E515, 2010.

[26] N. Stefan, K. Kantartzis, J. Machann et al., "Identification and characterization of metabolically benign obesity in humans," *Archives of Internal Medicine*, vol. 168, no. 15, pp. 1609–1616, 2008.

[27] Z. Cui, K. P. Truesdale, P. T. Bradshaw, J. Cai, and J. Stevens, "Three-year weight change and cardiometabolic risk factors in obese and normal weight adults who are metabolically healthy: the atherosclerosis risk in communities study," *International Journal of Obesity*, vol. 39, no. 8, pp. 1203–1208, 2015.

[28] S. Eisenberg, J. R. Patsch, J. T. Sparrow, A. M. Gotto, and

T. Olivecrona, "Very low density lipoprotein. Removal of apolipoproteins C-II and C-III-1 during lipolysis in vitro," *The Journal of Biological Chemistry*, vol. 254, no. 24, pp. 12603–12608, 1979.

[29] K. Olin-Lewis, R. M. Krauss, M. La Belle et al., "ApoC-III content of apoB-containing lipoproteins is associated with binding to the vascular proteoglycan biglycan," *Journal of Lipid Research*, vol. 43, no. 11, pp. 1969–1977, 2002.

[30] C. Zheng, V. Azcutia, E. Aikawa et al., "Statins suppress apolipoprotein CIII-induced vascular endothelial cell activation and monocyte adhesion," *European Heart Journal*, vol. 34, no. 8, pp. 615–624, 2013.

[31] J. Altomonte, L. Cong, S. Harbaran et al., "Foxo1 mediates insulin action on apoC-III and triglyceride metabolism," *The Journal of Clinical Investigation*, vol. 114, no. 10, pp. 1493–1503, 2004.

[32] P. R. Blackett, K. S. Blevins, E. Quintana et al., "ApoC-III bound to apoB-containing lipoproteins increase with insulin resistance in Cherokee Indian youth," *Metabolism: Clinical and Experimental*, vol. 54, no. 2, pp. 180–187, 2005.

[33] V. Ceccarelli, G. Nocentini, C. Riccardi et al., "Effect of dietary saturated fatty acids on HNF-4α DNA binding activity and ApoCIII mRNA in sedentary rat liver," *Molecular and Cellular Biochemistry*, vol. 347, no. 1-2, pp. 29–39, 2011.

[34] N. Faghihnia, L. M. Mangravite, S. Chiu, N. Bergeron, and R. M. Krauss, "Effects of dietary saturated fat on LDL subclasses and apolipoprotein CIII in men," *European Journal of Clinical Nutrition*, vol. 66, no. 11, pp. 1229–1233, 2012.

[35] S. Caron, A. Verrijken, I. Mertens et al., "Transcriptional activation of apolipoprotein CIII expression by glucose may contribute to diabetic dyslipidemia," *Arteriosclerosis, Thrombosis, and Vascular Biology*, vol. 31, no. 3, pp. 513–519, 2011.

[36] B. V. Howard, G. Ruotolo, and D. C. Robbins, "Obesity and dyslipidemia," *Endocrinology and Metabolism Clinics of North America*, vol. 32, no. 4, pp. 855–867, 2003.

[37] N. Stefan, H. Häring, F. B. Hu, and M. B. Schulze, "Divergent associations of height with cardiometabolic disease and cancer: epidemiology, pathophysiology, and global implications," *The Lancet Diabetes & Endocrinology*, vol. 4, no. 5, pp. 457–467, 2016.

[38] G. F. Lewis and D. J. Rader, "New insights into the regulation of HDL metabolism and reverse cholesterol transport," *Circulation Research*, vol. 96, no. 12, pp. 1221–1232, 2005.

[39] N. Stefan, A. Fritsche, F. Schick, and H. Häring, "Phenotypes of prediabetes and stratification of cardiometabolic risk," *The Lancet Diabetes & Endocrinology*, vol. 4, no. 9, pp. 789–798, 2016.

[40] C. M. Phillips, C. Dillon, J. M. Harrington et al., "Defining metabolically healthy obesity: role of dietary and lifestyle factors," *PLoS ONE*, vol. 8, no. 10, Article ID e76188, 2013.

[41] S. Velho, F. Paccaud, G. Waeber, P. Vollenweider, and P. Marques-Vidal, "Metabolically healthy obesity: different prevalences using different criteria," *European Journal of Clinical Nutrition*, vol. 64, no. 10, pp. 1043–1051, 2010.

[42] N. E. Miller, D. S. Thelle, O. H. Forde, and O. D. Mjos, "The Tromso heart-study. High-density lipoprotein and coronary heart-disease: a prospective case-control study," *The Lancet*, vol. 1, no. 8019, pp. 965–968, 1977.

[43] P. Barter, A. M. Gotto, J. C. LaRosa et al., "HDL cholesterol, very low levels of LDL cholesterol, and cardiovascular events," *The New England Journal of Medicine*, vol. 357, no. 13, pp. 1301–1310, 2007.

[44] B. Vergès, M. Adiels, J. Boren et al., "Interrelationships between the kinetics of VLDL subspecies and hdl catabolism in abdominal obesity: a multicenter tracer kinetic study," *Journal of Clinical Endocrinology and Metabolism*, vol. 99, no. 11, pp. 4281–4290, 2014.

[45] G. Walldius, I. Jungner, I. Holme, A. H. Aastveit, W. Kolar, and E. Steiner, "High apolipoprotein B, low apolipoprotein A-I, and improvement in the prediction of fatal myocardial infarction (AMORIS study): a prospective study," *The Lancet*, vol. 358, no. 9298, pp. 2026–2033, 2001.

[46] W. A. Van Der Steg, S. M. Boekholdt, E. A. Stein et al., "Role of the apolipoprotein B-apolipoprotein A-I ratio in cardiovascular risk assessment: a case-control analysis in EPIC-Norfolk," *Annals of Internal Medicine*, vol. 146, no. 9, pp. 640–648, 2007.

25

Cholesterol Efflux Capacity of Apolipoprotein A-I Varies with the Extent of Differentiation and Foam Cell Formation of THP-1 Cells

Kouji Yano,[1] Ryunosuke Ohkawa,[1] Megumi Sato,[1] Akira Yoshimoto,[1] Naoya Ichimura,[1] Takahiro Kameda,[2] Tetsuo Kubota,[3] and Minoru Tozuka[1]

[1]Analytical Laboratory Chemistry, Field of Applied Laboratory Science, Graduate School of Health Care Sciences, Tokyo Medical and Dental University, 1-5-45 Yushima, Bunkyo-ku, Tokyo 113-8519, Japan
[2]Department of Medical Technology, School of Health Sciences, Tokyo University of Technology, 5-23-22 Nishi-Kamata, Ohta-ku, Tokyo 144-8535, Japan
[3]Microbiology and Immunology, Field of Applied Laboratory Science, Graduate School of Health Care Sciences, Tokyo Medical and Dental University, 1-5-45 Yushima, Bunkyo-ku, Tokyo 113-8519, Japan

Correspondence should be addressed to Ryunosuke Ohkawa; ryu-th@umin.ac.jp

Academic Editor: Gerd Schmitz

Apolipoprotein A-I (apoA-I), the main protein component of high-density lipoprotein (HDL), has many protective functions against atherosclerosis, one of them being cholesterol efflux capacity. Although cholesterol efflux capacity measurement is suggested to be a key biomarker for evaluating the risk of development of atherosclerosis, the assay has not been optimized till date. This study aims at investigating the effect of different states of cells on the cholesterol efflux capacity. We also studied the effect of apoA-I modification by homocysteine, a risk factor for atherosclerosis, on cholesterol efflux capacity in different states of cells. The cholesterol efflux capacity of apoA-I was greatly influenced by the extent of differentiation of THP-1 cells and attenuated by excessive foam cell formation. N-Homocysteinylated apoA-I indicated a lower cholesterol efflux capacity than normal apoA-I in the optimized condition, whereas no significant difference was observed in the cholesterol efflux capacity between apoA-I in the excessive cell differentiation or foam cell formation states. These results suggest that cholesterol efflux capacity of apoA-I varies depending on the state of cells. Therefore, the cholesterol efflux assay should be performed using protocols optimized according to the objective of the experiment.

1. Introduction

Atherosclerosis is a major trigger for cardiovascular and cerebrovascular diseases. Since these are the leading causes of death in the developed countries [1], a prevention of atherosclerosis is very crucial. In humans, many defense mechanisms exist against the development of atherosclerosis, and apolipoprotein A-I (apoA-I), the major protein component of high-density lipoprotein (HDL), is associated with some of these mechanisms. Antioxidant [2, 3] and anti-inflammatory abilities [2, 4, 5] and cholesterol efflux capacity [6, 7] are widely recognized as antiatherosclerotic functions of apoA-I. Cholesterol efflux plays a particularly important role in the prevention of progression of atherosclerotic lesions since excessive lipids, such as cholesterol, are removed from the atherosclerotic lesions by apoA-I via the ATP-binding cassette transporter A1 (ABCA1). Recently, studies have reported that the cholesterol efflux capacity is inversely associated with the occurrence of cardiovascular events [8–10].

It is known that apoA-I, which plays a central role in cholesterol efflux capacity, is modified by various reactions such as oxidation, glycation, and homocysteinylation [11–15]. These modifications of apoA-I, presumed to lead to the development of atherosclerosis, have been the focal point of many studies investigating dysfunctionalities in cholesterol efflux. Actually, there are various studies on the relationship between modification of apoA-I and cholesterol efflux capacity that have been reported [16–18]. However, the assay for measurement of cholesterol efflux capacity has not been optimized even by using the same cell line and it is not known whether cellular conditions could lead to a difference in cholesterol efflux capacity. Therefore, in this study, we assessed the cholesterol efflux capacity of apoA-I by using various states of cells.

Moreover, we focused on the role of homocysteine (Hcy), one of the risk factors for atherosclerosis, in cholesterol efflux [19, 20]. Hcy is a metabolite of methionine, an essential amino acid, and is converted to methionine and cysteine [21]. A part of Hcy is converted to homocysteine thiolactone (HcyT) by methionyl-tRNA synthetase [22]. HcyT is rich in reactivity and is known to bind to lysine residues (N-homocysteinylation) of many proteins in normal human plasma such as albumin, hemoglobin, and fibrinogen [23]. We recently reported that N-homocysteinylated apoA-I (N-Hcy apoA-I) exists in normal human serum in the ratio of 1.0–7.4% to total apoA-I [24]. However, the effect of N-homocysteinylation of apoA-I on the cholesterol efflux capacity is not yet confirmed. We also evaluated the cholesterol efflux capacity of N-Hcy apoA-I using various states of cells and compared the results to that obtained by using normal apoA-I.

2. Materials and Methods

2.1. Chemicals and Blood Samples. Unless otherwise stated, all reagents were purchased from Wako Pure Chemical. Blood samples were obtained from healthy volunteers after their informed consent, and the study was approved by the ethics committee of Tokyo Medical and Dental University (number 1441).

2.2. Cell Culture. THP-1, a human monocyte cell line, was purchased from ATCC and maintained in RPMI-1640 (Sigma-Aldrich) containing 10% fetal bovine serum (FBS), 0.1% penicillin/streptomycin, and 0.1% nonessential amino acids at 37°C under 5% CO_2.

2.3. Cholesterol Efflux Capacity. THP-1 cells were incubated with RPMI-1640 medium containing 100 ng/mL of phorbol 12-myristate 13-acetate (PMA; Sigma-Aldrich) supplemented with 0.2% bovine serum albumin (BSA) in 24-well cell culture plates at a density of 2.5×10^5 cells/well. For preparation of various states of differentiated cells, the incubation time was varied from 1 to 5 days as indicated in Figure 1(a). After removing the supernatant, THP-1 cells were loaded with RPMI-1640 containing acetylated low-density lipoprotein (acLDL) (50 μg protein/mL), T0901317 (1 mmol/L; Enzo Life Sciences), a Liver X receptor (LXR) agonist for promoting expression of ABCA1, ^{3}H-cholesterol (1 μCi/mL; PerkinElmer), and BSA (0.2%). The loading time was also varied over 1 to 5 days to generate different stages of foam cells. THP-1 cells were washed thrice with RPMI-1640 medium containing 0.2% BSA before equilibration with RPMI-1640 supplemented with T0901317 (1 mmol/L) and BSA (0.2%) for 18 hours. After removing the supernatant, cells were incubated

with apoA-I (10 μg/mL) and T0901317 (1 mmol/L) in RPMI-1640 for additional 4 hours. Efflux medium was separated from the cells and NaOH (0.1 mol/L) was added directly to the cells for lysis. Both the media and cell lysates were counted for radioactivity using the liquid scintillation counter. Cholesterol efflux capacity was calculated as a percentage of radioactivity in the medium as per the given formula: {^{3}H-cholesterol in medium/(^{3}H-cholesterol in medium + ^{3}H-cholesterol in cells)} × 100 – percentage of passive diffusion in the case of no acceptor (apoA-I).

2.4. Morphological Analysis. For microscopic observation, THP-1 cells were differentiated and loaded at a density of 2.5×10^5 cells/well for different time periods on cover glass (Thermo Fisher Scientific) in 24-well cell culture plates. After incubation, adherent cells were washed with phosphate buffered saline (PBS) and then fixed with 3.7% formaldehyde solution. After washing with PBS, differentiated cells were stained with hematoxylin solution, and loaded cells after differentiation were stained with hematoxylin solution followed by working solution of oil red O for assessing the extent of foam cell formation. The plates were washed with distilled water and observed under the microscope (BX53-34-FL, Olympus). A semiquantification of lipid droplets stained by oil red O was performed with ImageJ (National Institutes of Health).

2.5. Flow Cytometric Analysis. THP-1 cells were treated with PMA in 6-well culture plates (1.0×10^6 cells/well) for various time periods as described above, and adherent cells were then washed with PBS. After washing, cells were detached gently with a cell scraper and stained with fluorescein isothiocyanate- (FITC-) conjugated anti-CD11b mouse IgG (Abcam, ab25727, monoclonal) (dilution 1 : 100) for 30 min in a dark place at 4°C. After washing, cells were fixed with 3% paraformaldehyde/PBS. To remove aggregates, cells were filtered using a nylon filter and the concentration was adjusted to approximately 1.0×10^5 cells/mL. CD11b expression was analyzed by flow cytometry (Epics XL, Beckman Coulter) with excitation at 505 nm and emission at 545 nm. Samples were controlled by using FITC-conjugated mouse IgG as an isotype control (Abcam, ab91365, monoclonal) in each day and the percentage of cells which exceeded the maximum fluorescent intensity of isotype control was defined as certain CD11b positive cells.

2.6. Preparation of LDL, HDL, and apoA-I. LDL (1.019–1.063 g/mL) and HDL (1.063–1.21 g/mL) were isolated from serum obtained from healthy subjects by ultracentrifugation, as described previously [25]. Next, LDL and HDL fractions were dialyzed against PBS. To purify apoA-I, HDL fraction was delipidated with ethanol/ether 3 : 2 (v/v) and applied to S-200-HR (Sigma-Aldrich) column (2.5 × 90 cm), equilibrated with 20 mmol/L Tris-HCl buffer (pH 7.4) containing 6.8 mol/L urea. The isolated apoA-I fraction was dialyzed against PBS and stored at −20°C until use (Supplemental Figure 1 in Supplementary Material available online at http://dx.doi.org/10.1155/2016/9891316). LDL fraction was acetylated as reported previously [26].

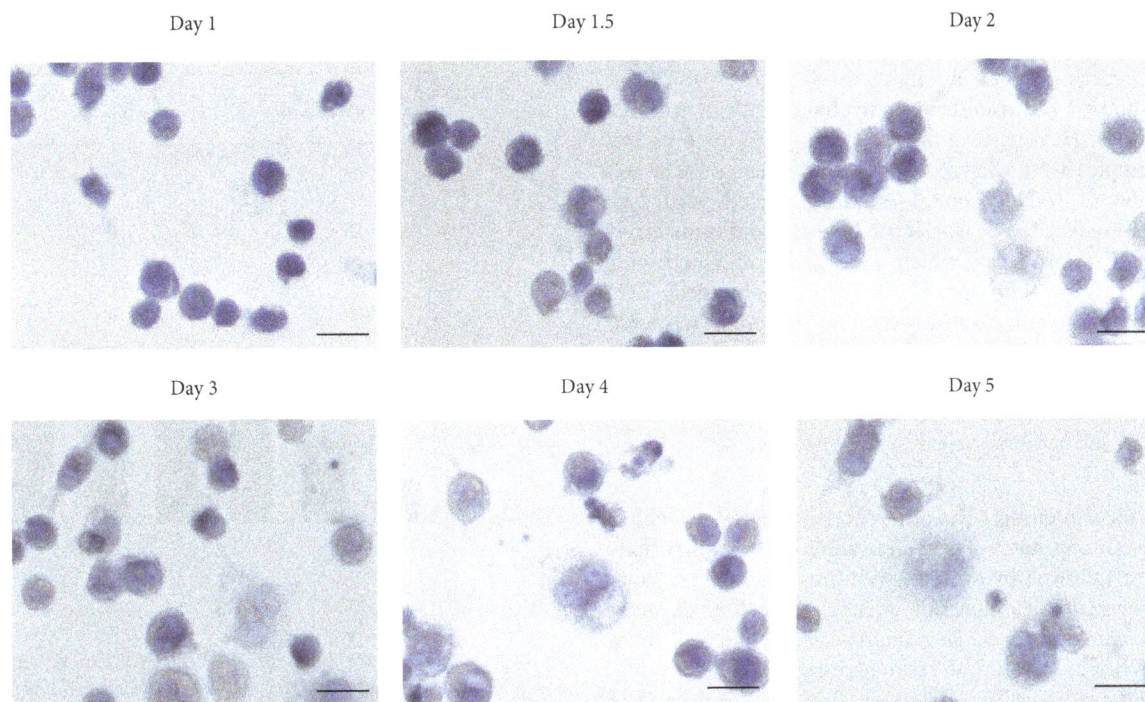

(a)

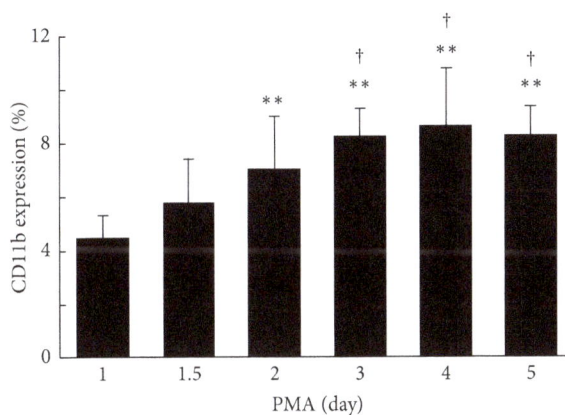

(b)

FIGURE 1: The morphological changes in THP-1 cells. THP-1 cells were treated with 100 ng/mL PMA for different duration (1 to 5 days). After treatment, (a) THP-1 cells were fixed with 3.7% formaldehyde and stained with hematoxylin (scale bars, 20 μm). (b) THP-1 cells were incubated with FITC-conjugated anti-CD11b antibody, and CD11b expression was analyzed using flow cytometry. The percentage of CD11b positive cells which exceeded the maximum fluorescent intensity of the isotype control in each day was indicated. The value at day 1 was defined as 1. The values were indicated by mean + SD ($n = 5$, $^{**}P < 0.01$ versus day 1, $^{\dagger}P < 0.05$ versus day 1.5).

2.7. Western Blotting. The expression of CD11b after PMA treatment for different time periods was evaluated by western blotting as described previously [24]. THP-1 cells were lysed with radioimmunoprecipitation buffer (RIPA buffer; 50 mmol/L Tris-HCl pH 8.0 including 150 mmol/L sodium chloride, 0.5% sodium deoxycholate, and 1.0% nonidet P-40 (MP Biomedicals)) containing protease inhibitor cocktails (Roche) at 4°C. The amount of total protein in cell lysates was measured by Folin-Lowry method. Cell lysates were separated by sodium dodecyl sulfate-polyacrylamide gel electrophoresis (SDS-PAGE) and transferred onto PVDF membranes (Millipore), which were then incubated with 5% skim milk in Tris-buffered saline containing 0.05% Tween 20 (washing buffer). PVDF membranes were incubated with primary antibodies, anti-CD11b (Abcam, ab52478, monoclonal) and anti-β-actin (Santa Cruz, sc-13-656, polyclonal). After three washes with washing buffer, the membranes were incubated with peroxidase-labeled secondary antibody (Beckman Coulter, IM0831, polyclonal). The bands of CD11b and β-actin were visualized by ECL Prime western blotting detection reagent (GE Healthcare). Semiquantification was performed by densitometry (CS Analyzer, ATTO).

2.8. N-Homocysteinylation of apoA-I. *N*-Homocysteinylation of apoA-I was performed according to the previously reported method with a slight modification [15]. The purified apoA-I (70 μmol/L) was incubated with or without 10 mmol/L of HcyT (MP Biomedicals) at 37°C for 6 hours. Then, samples were dialyzed against PBS. To confirm the production of *N*-Hcy apoA-I, samples treated with cysteamine were isolated by isoelectric focusing and transferred onto PVDF membranes, which were then incubated with primary antibody, anti-apoA-I (Academy Biomedical, 11A-G2b, polyclonal). After washing thrice, the membranes were incubated with peroxidase-labeled secondary antibodies (Medical and Biological Laboratories). The bands containing apoA-I were visualized with 3,3'-diaminobenzidine-4HCl and H_2O_2, and semiquantified by CS Analyzer.

2.9. Statistical Analysis. The differences of cholesterol efflux capacities among states of cells were analyzed using Kruskal-Wallis test followed by a Mann–Whitney U test with Bonferroni correction. Comparison of cholesterol efflux capacity between apoA-I and *N-Hcy apoA-I* was evaluated by a Mann–Whitney U test. The results are expressed as mean + standard deviation (SD). A P value < 0.05 was considered as statistically significant.

3. Results

3.1. Effect of Differentiation on Cholesterol Efflux Capacity of apoA-I. THP-1 cells differentiated following treatment with PMA for 1 to 5 days. To confirm differentiation of the cells at each stage, the morphology of THP-1 cells was assessed by staining with hematoxylin (Figure 1(a)). Mild morphological changes were observed at day 1.5, and most of these were adendritic cells with approximately 20 μm size. From day 1.5 to day 2, dynamic changes in cells were observed. Cells grew larger and showed the appearance of dendrite formation corresponding to the increase in stimulation. Next, we evaluated the expression of CD11b, an adhesion molecule, on THP-1 cells by flow cytometric analysis (Figure 1(b)). The expression of CD11b was essentially increased in a stimulation with PMA on THP-1 macrophages, while the expression on THP-1 monocytes without PMA treatment was not almost detected (data not shown). Conclusively, expression of CD11b at day 2 was significantly higher than that at day 1 (P < 0.01) and maintained a plateau state (Figure 1(b)). A similar tendency was also observed in the western blotting analysis (Figures 2(a) and 2(b)). The effect of differentiation of THP-1 cells on cholesterol efflux capacity of apoA-I was evaluated on the condition that the period of foam cell formation was fixed at 1 day. The cholesterol efflux capacities were the highest and the second highest at day 2 and day 1 of differentiation, respectively, and gradually deceased after day 3 (Figure 3). Although the period of foam cell formation was fixed for 1 day, total ^{3}H-cholesterol taken up by the cells before efflux was largely different according to the differentiation periods, the lowest level at day 1, a peak at day 2, and gradual decrease after day 3 (Supplementary Figure 2(a)). In addition, ^{3}H-cholesterol mass in the medium after efflux also showed a

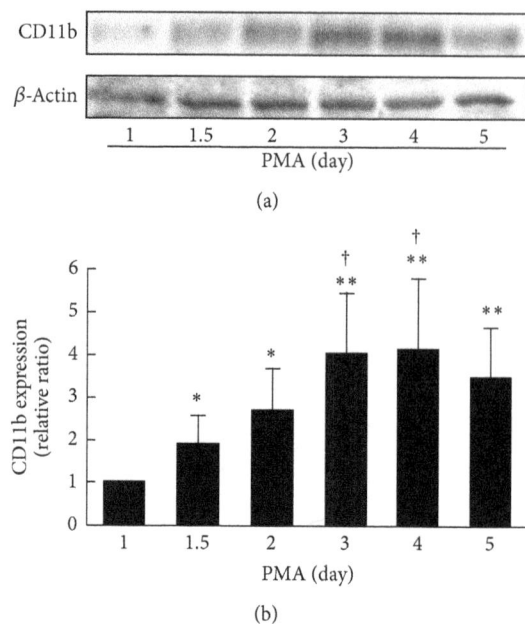

(a)

(b)

FIGURE 2: CD11b expression on THP-1 cells treated with PMA for different periods. (a) After PMA treatment for 1 to 5 days, cell lysates (10 μg of proteins) were separated by SDS-PAGE (7% polyacrylamide gel) and detected with anti-CD11b or anti-β-actin antibodies. (b) The expression of CD11b also shows bar graphs. The value at day 1 was defined as 1. The values were indicated by mean + SD ($n = 4$, $^*P$ < 0.05 versus day 1, $^{**}P$ < 0.01 versus day 1, and $^\dagger P$ < 0.05 versus day 1.5).

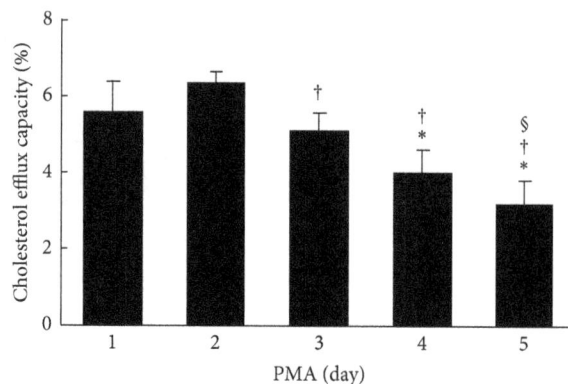

FIGURE 3: The effect of differentiation on cholesterol efflux capacity of apoA-I. THP-1 cells were treated with 100 ng/mL PMA for different periods (1 to 5 days) and loaded with ^{3}H-cholesterol (1 μCi/mL), acLDL (50 μg protein/mL), and T0901317 (1 μmol/L) for 1 day. The cholesterol efflux capacity of apoA-I (10 μg/mL) was determined under these conditions. The values were indicated by mean + SD ($n = 3$, $^*P$ < 0.05 versus day 1, $^\dagger P$ < 0.05 versus day 2, and $^\S P$ < 0.05 versus day 3).

similar tendency to total ^{3}H-cholesterol taken up by the cells (Supplementary Figure 2(b)). Cholesterol efflux capacities expressed in percentage were largely different from the actual efflux of ^{3}H-cholesterol mass. In contrast, ABCA1 expression was the highest at day 1 and gradually reduced in parallel with the duration of PMA treatment (Supplementary Figure 3).

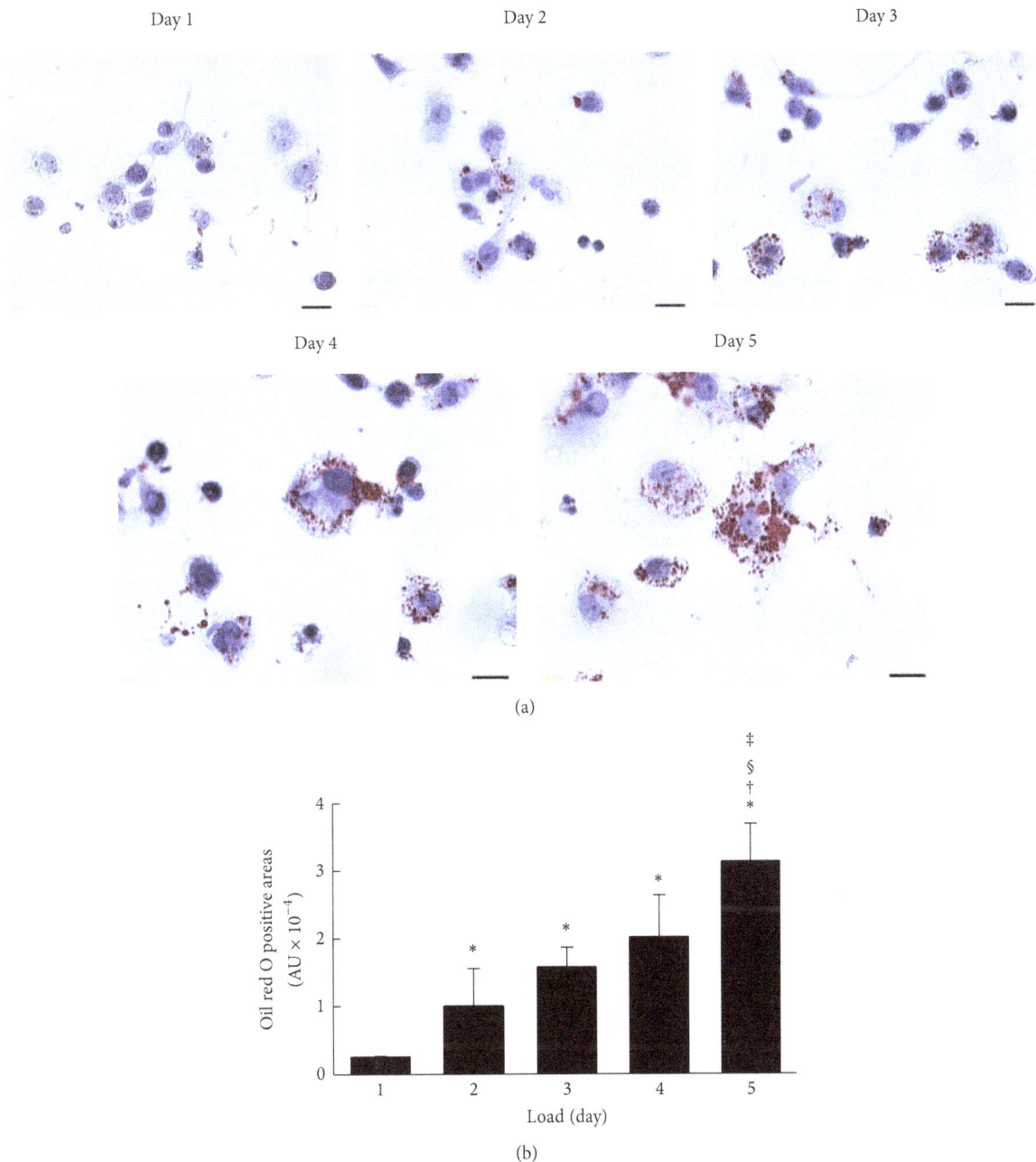

FIGURE 4: Foam cell formation. (a) After PMA treatment for 2 days, cells were loaded with acLDL (50 μg protein/mL) and T0901317 (1 μmol/L) for indicated periods. After equilibration, cells were stained with oil red O and hematoxylin (scale bars, 20 μm). (b) Semiquantitative analysis of oil red O positive cells was performed using the ImageJ (National Institutes of Health). The values were indicated by mean + SD ($n = 3$, [*]$P < 0.05$ versus day 1, [†]$P < 0.05$ versus day 2, [§]$P < 0.05$ versus day 3, and [‡]$P < 0.05$ versus day 4).

3.2. Effect of Foam Cell Formation on Cholesterol Efflux Capacity of apoA-I. In experiments wherein the stage of differentiation of THP-1 cells with PMA treatment was fixed for 2 days, THP-1 macrophages were incubated with acLDL for 1 to 5 days. The accumulations of intracellular lipid droplets were evaluated using oil red O staining (Figure 4(a)). Lipid droplets stained with oil red O were faintly observed from day 1 and dramatically increased depending on the loading period of acLDL (Figure 4(b)). Subsequently, the cholesterol efflux capacities of apoA-I in each stage of cells were determined. Cholesterol efflux capacity was highest for 1 day after loading with acLDL and significantly decreased in a loading time-dependent manner (Figure 5). Cellular uptake of total ^{3}H-cholesterol before efflux showed a gradual increase, whereas medium ^{3}H-cholesterol after efflux was slightly decreased during acLDL loading (Supplementary Figures 2(c) and 2(d)). In addition, no significant change in ABCA1 expression was observed (Supplementary Figure 4).

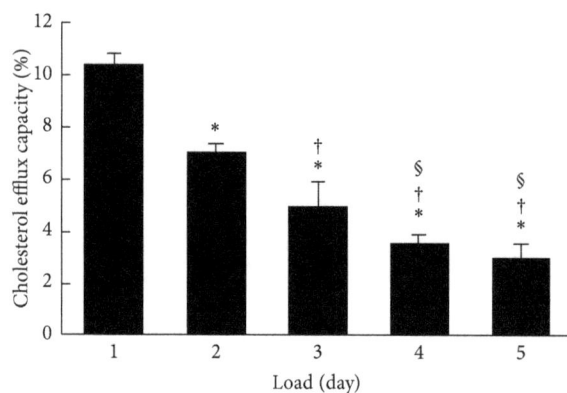

FIGURE 5: The effect of foam cell formation on cholesterol efflux capacity of apoA-I. After PMA treatment for 2 days, cells were loaded with ^{3}H-cholesterol (1 μCi/mL), acLDL (50 μg protein/mL), and T0901317 (1 μmol/L) for 1 to 5 days. Under these conditions, the cholesterol efflux capacity was determined. The values were indicated by mean + SD ($n = 3$, $^*P < 0.05$ versus day 1, $^†P < 0.05$ versus day 2, $^§P < 0.05$ versus day 3, and $^§P < 0.05$ versus day 3).

3.3. Cholesterol Efflux Capacity of N-Hcy apoA-I.

Using various stages of cells regulated differentiation and foam cell formation, the effect of N-homocysteinylation on cholesterol efflux capacity of apoA-I was evaluated. To confirm the extent of N-homocysteinylation of apoA-I, apoA-I incubated with HcyT was tested using an isoelectric focusing. Compared to normal apoA-I, N-Hcy apoA-I indicates higher isoelectric point (pI) depending on the number of attached homocysteine compounds (Figure 6(a)). The ratio of N-Hcy apoA-I to total apoA-I on the isoelectric focusing pattern was 51.9 ± 5.9% (Figure 6(b)). Cholesterol efflux capacities of apoA-I and N-Hcy apoA-I reached those peaks at day 2 and gradually decreased depending on the periods of the treatment with PMA (Figure 7(a)). In case of foam cell formation, a significant decrease was observed in cholesterol efflux capacities of apoA-I and N-Hcy apoA-I according to the acLDL loading (Figure 7(b)). In addition, the cholesterol efflux capacity of N-Hcy apoA-I was significantly lower than that of intact apoA-I by differentiation and foam cell formation at day 1 and day 2, respectively; however no significant difference was observed after day 3 (Figure 7).

4. Discussion

Cholesterol efflux is essential for the antiatherosclerotic function of apoA-I, and the evaluation of its capacity would provide important information in terms of prevention and diagnosis of atherosclerosis. However, in previous reports assays for measurement of cholesterol efflux capacity of apoA-I and modified apoA-I have been performed using different methods. Therefore, the effect of modification of apoA-I on cholesterol efflux capacity is often controversial [16–18]. Our aim was to evaluate the variation in cholesterol efflux capacity of apoA-I under various cellular conditions such as differentiation and foam cell formation and to investigate the effect of N-homocysteinylation of apoA-I on cholesterol efflux capacity in these conditions. First, we confirmed the

morphological features and expression of CD11b, which is an established differentiation marker of macrophages of THP-1 cells treated with PMA [27]. THP-1 cells treated with PMA for 2 days showed a behavior similar to macrophages morphologically, including increase in cell volume [28]. It is known that PMA induces CD11b expression via protein kinase C delta (PKC-δ) [27]. An increase in the expression of CD11b on THP-1 cells treated with PMA for 2 days was observed by flow cytometric and western blotting analysis while the expression of ABCA1 decreased from day 1 to day 5 (Supplementary Figure 3). PMA is presumed to have reduced the ABCA1 gene expression via PKC activation [29], since PKC was activated with PMA as stated above. Taken together, these results suggest that variously differentiated cells were prepared. Second, we evaluated the effect of differentiation of THP-1 cells on cholesterol efflux capacity. Cholesterol efflux capacity reached a peak at day 2 and decreased after day 3 (Figure 3). At day 1, the uptake of ^{3}H-cholesterol into cells and levels detected in the medium were low (Supplementary Figures 2(a) and 2(b)). At day 2, the uptake of intracellular ^{3}H-cholesterol was about three times as high as at day 1. On the contrary, after day 3, both intracellular and medium ^{3}H-cholesterol were decreased, and the decreased ratios of medium ^{3}H-cholesterol were slightly larger than those of intracellular ^{3}H-cholesterol. These results suggest that the increase of the cholesterol efflux capacity indicated as a percentage at day 2 was caused by the differentiation of THP-1 macrophages, and a decrease after day 3 would be determined by the reduced amount of the expression of ABCA1, a pivotal transporter in cholesterol efflux to apoA-I. In many studies, the duration of PMA stimulation has been inconsistent for the evaluation of cholesterol efflux capacity [16–18]. Our results indicate that the optimum period of PMA treatment is 2 days if the cholesterol efflux capacity is required to be assessed at the maximum stage.

Next, we examined the effect of foam cell formation on cholesterol efflux capacity. Oil red O positive cells, which indicate the formation of lipid droplets, were increased in proportion to days of loading acLDL (Figure 4). Meanwhile, cholesterol efflux capacity was decreased in a loading time-dependent manner (Figure 5). Although there was an increase in the intracellular ^{3}H-cholesterol over the passage of days, the levels of secreted ^{3}H-cholesterol in the medium by apoA-I were decreased gradually (Supplementary Figures 2(c) and 2(d)). Moreover, no significant change was observed in ABCA1 expression (Supplementary Figure 4). The excessive accumulation of cholesterol ester in cells is reported to inhibit lysosomal acid lipase (LAL) activity, an enzyme related to the cholesterol metabolism [30]. Free cholesterol taken up into cells is esterified by acyl coenzyme-A cholesterol acyltransferase (ACAT) and stored as lipid droplets [31]. Since cholesterol ester is hydrolyzed by LAL and removed as free cholesterol through ABCA1, the inhibition of LAL may lead to a reduction in the cholesterol efflux regardless of the degree of ABCA1 expression on the cells surface. Hence, decrease in cholesterol efflux without reduction of ABCA1 expression may be related to the loss of LAL activity caused by the excessive uptake of acLDL, including

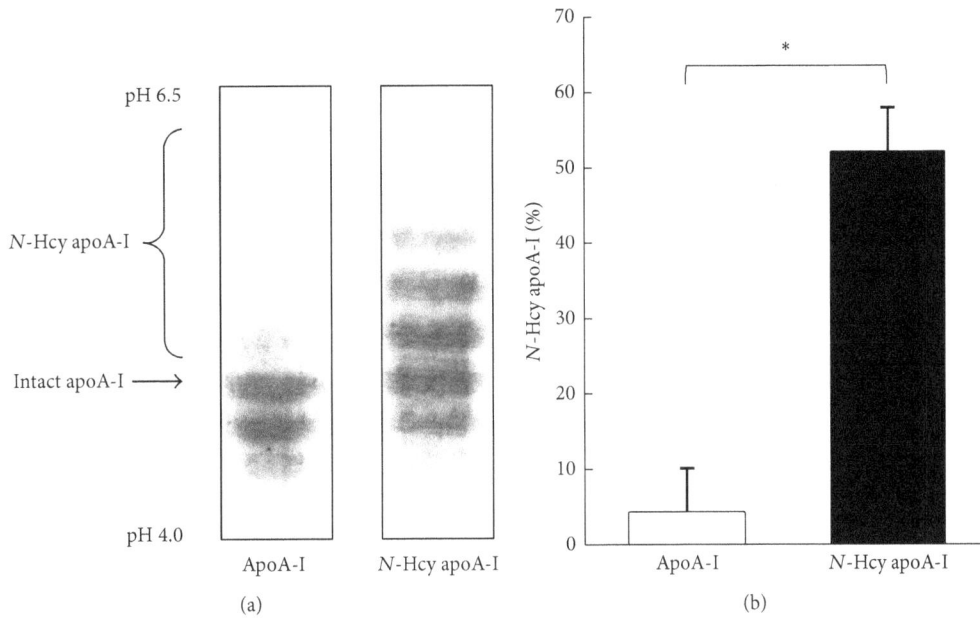

FIGURE 6: Isoelectric focusing for apoA-I and apoA-I treated with homocysteine thiolactone. (a) Representative isoelectric focusing patterns of apoA-I and N-Hcy apoA-I. Purified apoA-I was incubated with or without 10 mmol/L of homocysteine thiolactone at 37°C for 6 hours. After dialysis against PBS, the samples were analyzed by isoelectric focusing followed by western blotting using anti-apoA-I antibody. (b) Histogram represents the percentage of N-Hcy apoA-I to total apoA-I using CS analyzer. The values were indicated by mean + SD ($n = 3$, $^{*}P < 0.05$).

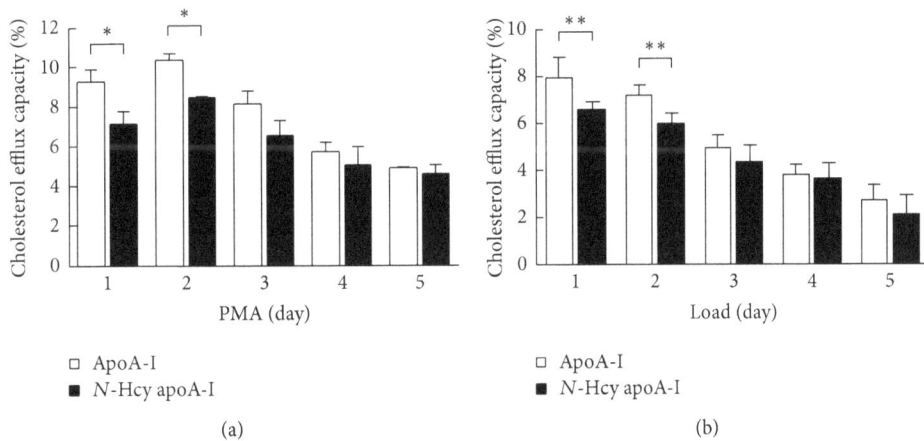

FIGURE 7: The effect of differentiation and foam cell formation on cholesterol efflux capacity of apoA-I and N-Hcy apoA-I. (a) After THP-1 cells were differentiated with PMA for different periods (1 to 5 days) and were loaded with ^{3}H-cholesterol (1 μCi/mL), acLDL (50 μg/mL), and T0901317 (1 μmol/L) for 1 day, cholesterol efflux capacity was assessed by using apoA-I (10 μg/mL) or N-Hcy apoA-I (10 μg/mL). The values were indicated by mean + SD ($n = 3$, $^{*}P < 0.05$). (b) After PMA treatment for 2 days and loading with ^{3}H-cholesterol (1 μCi/mL), acLDL (50 μg/mL), and T0901317 (1 μmol/L) for different periods (1 to 5 days), cholesterol efflux capacity was determined by using apoA-I (10 μg/mL) or N-Hcy apoA-I (10 μg/mL). The values were indicated by mean + SD ($n = 6$, $^{**}P < 0.01$).

^{3}H-cholesterol, and the effect might be reversed in the presence of an ACAT inhibitor. These results indicate that the optimum period of loading cholesterol is 1 day if the cholesterol efflux capacity has to be assessed at the maximum stage. Longer durations of loading cholesterol decrease cholesterol efflux capacity because of excessive cholesterol accumulation.

To the best of our knowledge, this study is the first to indicate the relationship between cholesterol efflux capacity and

condition of THP-1 cells such as differentiation and foam cell formation, in depth. Using these different states of cells, we examined the effect of N-homocysteinylation on cholesterol efflux capacity of apoA-I. N-Hcy apoA-I was detected as a more positively charged band than intact apoA-I (Figure 6) and was defined as a percentage to total apoA-I, as described previously [15]. The variation in cholesterol efflux capacities of both apoA-I and N-Hcy apoA-I using THP-1 cells treated with PMA in different time intervals indicates similar trends, as compared to the values that reached the peaks at day 2 and then gradually decreased (Figure 7(a)). However, significant differences were observed in the cholesterol efflux capacities between apoA-I and N-Hcy apoA-I at day 1 and day 2. On the contrary, no significant difference was observed in those capacities at day 3 and after. Furthermore, cholesterol efflux capacities of both apoA-I and N-Hcy apoA-I were also evaluated using THP-1 cells treated with cholesterol loading at different time intervals. Similarly, the cholesterol efflux capacities of these two types of apoA-I showed the same behavior, and N-Hcy apoA-I indicated significantly lower capacities than apoA-I at day 1 and day 2 after loading. Only a few studies have been reported about the relationship between Hcy and cholesterol efflux capacity [32, 33]. We previously reported that the cholesterol efflux capacity of apoA-I was not significantly reduced with N-homocysteinylation [15]. In the previous study the period of PMA stimulation was 3 days, indicating that the cells might be in the state of excessive differentiation to consider the possible effect of N-homocysteinylation. To confirm the effect of N-homocysteinylation on cholesterol efflux capacity of apoA-I, the influence of modification site of apoA-I and its dose response should be investigated. However, our results in this study reiterate the importance of the cell conditions such as the degree of differentiation and foam cell formation when the effects of any modifications of apoA-I, not limited to N-homocysteinylation, on cholesterol efflux capacity are evaluated. Optimization of experimental conditions will be required depending on the purpose of the measurement. However, we have focused on THP-1 cells in this study. Further investigations are needed to assess the cholesterol efflux capacity using other monocytes or monocytic cell lines [34–40]. Although this study has only examined apoA-I, studies using HDL could be necessary to confirm a precise mechanism. Further experiments will provide us a hint for evaluation of various stages of atherosclerosis.

5. Conclusions

In conclusion, the cholesterol efflux capacities of apoA-I and N-Hcy apoA-I vary depending on the state of the cells; hence, it is important to evaluate the cholesterol efflux capacity in several states of the cells adapted for a purpose.

Competing Interests

The authors declare that there is no conflict of interests regarding the publication of this paper.

Acknowledgments

The authors thank Associate Professor Masayuki Hara (General Isotope Research Division, Research Center for Medical and Dental Sciences) for generously providing technical advice especially, regarding the handling of isotopes. This work was supported by a Grant-in-Aid for Scientific Research (KAKENHI) from Japan Society for the Promotion of Science (JSPS) Grants nos. 15K19174 (Ryunosuke Ohkawa) and 26460642 (Minoru Tozuka).

References

[1] C. J. L. Murray and A. D. Lopez, "Alternative projections of mortality and disability by cause 1990–2020: Global Burden of Disease Study," *The Lancet*, vol. 349, no. 9064, pp. 1498–1504, 1997.

[2] P. J. Barter, S. Nicholls, K.-A. Rye, G. M. Anantharamaiah, M. Navab, and A. M. Fogelman, "Antiinflammatory properties of HDL," *Circulation Research*, vol. 95, no. 8, pp. 764–772, 2004.

[3] M. Navab, S. Y. Hama, C. J. Cooke et al., "Normal high density lipoprotein inhibits three steps in the formation of mildly oxidized low density lipoprotein: step 1," *Journal of Lipid Research*, vol. 41, no. 9, pp. 1481–1494, 2000.

[4] A. J. Murphy, K. J. Woollard, A. Hoang et al., "High-density lipoprotein reduces the human monocyte inflammatory response," *Arteriosclerosis, Thrombosis, and Vascular Biology*, vol. 28, no. 11, pp. 2071–2077, 2008.

[5] A. Urundhati, Y. Huang, J. A. Lupica, J. D. Smith, J. A. DiDonato, and S. L. Hazen, "Modification of high density lipoprotein by myeloperoxidase generates a pro-inflammatory particle," *The Journal of Biological Chemistry*, vol. 284, no. 45, pp. 30825–30835, 2009.

[6] R. Ohashi, H. Mu, H. Wang, Q. Yao, and C. Chen, "Reverse cholesterol transport and cholesterol efflux in atherosclerosis," *Quarterly Journal of Mathematics*, vol. 98, no. 12, pp. 845–856, 2005.

[7] E. Favari, A. Chroni, U. J. Tietqe et al., "Cholesterol efflux and reverse cholesterol transport," in *High Density Lipoproteins: From Biological Understanding to Clinical Exploitation*, vol. 224 of *Handbook of Experimental Pharmacology*, pp. 181–206, Springer, Berlin, Germany, 2015.

[8] A. Rohatgi, A. Khera, J. D. Berry et al., "HDL cholesterol efflux capacity and incident cardiovascular events," *New England Journal of Medicine*, vol. 371, no. 25, pp. 2383–2393, 2014.

[9] D. Saleheen, R. Scott, S. Javad et al., "Association of HDL cholesterol efflux capacity with incident coronary heart disease events: a prospective case-control study," *The Lancet Diabetes and Endocrinology*, vol. 3, no. 7, pp. 507–513, 2015.

[10] A. Ritsch, H. Scharnagl, and W. März, "HDL cholesterol efflux capacity and cardiovascular events," *The New England Journal of Medicine*, vol. 372, no. 19, pp. 1870–1871, 2015.

[11] B. Shao, C. Tang, J. W. Heinecke, and J. F. Oram, "Oxidation of apolipoprotein A-I by myeloperoxidase impairs the initial interactions with ABCA1 required for signaling and cholesterol export," *Journal of Lipid Research*, vol. 51, no. 7, pp. 1849–1858, 2010.

[12] B. Shao, S. Pennathur, I. Pagani et al., "Modifying apolipoprotein A-I by malondialdehyde, but not by an array of other reactive carbonyls, blocks cholesterol efflux by the ABCA1 pathway," *Journal of Biological Chemistry*, vol. 285, no. 24, pp. 18473–18484, 2010.

[13] B. E. Brown, J. Nobecourt, J. Zeng, A. J. Jenkins, K.-A. Rye, and M. J. Davies, "Apolipoprotein A-I glycation by glucose and reactive aldehydes alters phospholipid affinity but not cholesterol export from lipid-laden macrophages," *PLoS ONE*, vol. 8, no. 5, Article ID e65430, 2013.

[14] A. Hoang, A. J. Murphy, M. T. Coughlan et al., "Advanced glycation of apolipoprotein A-I impairs its anti-atherogenic properties," *Diabetologia*, vol. 50, no. 8, pp. 1770–1779, 2007.

[15] A. Miyazaki, N. Sagae, Y. Usami et al., "N-homocysteinylation of apolipoprotein A-I impairs the protein's antioxidant ability but not its cholesterol efflux capacity," *Biological Chemistry*, vol. 395, no. 6, pp. 641–648, 2014.

[16] Y. Haraguchi, R. Toh, M. Hasokawa et al., "Serum myeloperoxidase/paraoxonase 1 ratio as potential indicator of dysfunctional high-density lipoprotein and risk stratification in coronary artery disease," *Atherosclerosis*, vol. 234, no. 2, pp. 288–294, 2014.

[17] K. A. Hadfield, D. I. Pattison, B. E. Brown et al., "Myeloperoxidase-derived oxidants modify apolipoprotein A-I and generate dysfunctional high-density lipoproteins: comparison of hypothiocyanous acid (HOSCN) with hypochlorous acid (HOCl)," *The Biochemical Journal*, vol. 449, no. 2, pp. 531–542, 2013.

[18] C. R. White, G. Datta, A. K. W. Buck et al., "Preservation of biological function despite oxidative modification of the apolipoprotein A-I mimetic peptide 4F," *Journal of Lipid Research*, vol. 53, no. 8, pp. 1576–1587, 2012.

[19] W. G. Christen, U. A. Ajani, R. J. Glynn, and C. H. Hennekens, "Blood levels of homocysteine and increased risks of cardiovascular disease: causal or casual?" *Archives of Internal Medicine*, vol. 160, no. 4, pp. 422–434, 2000.

[20] D. S. Wald, M. Law, and J. K. Morris, "Homocysteine and cardiovascular disease: evidence on causality from a meta-analysis," *British Medical Journal*, vol. 325, no. 7374, pp. 1202–1206, 2002.

[21] J. J. Strain, L. Dowey, M. Ward, K. Pentieva, and H. McNulty, "B-vitamins, homocysteine metabolism and CVD," *Proceedings of the Nutrition Society*, vol. 63, no. 4, pp. 597–603, 2004.

[22] H. Jakubowski, "The pathophysiological hypothesis of homocysteine thiolactone—mediated vascular disease," *Journal of Physiology and Pharmacology*, vol. 59, supplement 9, pp. 155–167, 2008.

[23] H. Jakubowski, "Homocysteine is a protein amino acid in humans: implications for homocysteine-linked disease," *Journal of Biological Chemistry*, vol. 277, no. 34, pp. 30425–30428, 2002.

[24] N. Ishimine, Y. Usami, S. Nogi et al., "Identification of N-homocysteinylated apolipoprotein AI in normal human serum," *Annals of Clinical Biochemistry*, vol. 47, part 5, pp. 453–459, 2010.

[25] R. J. Havel, H. A. Eder, and J. H. Bragdon, "The distribution and chemical composition of ultracentrifugally separated lipoproteins in human serum," *The Journal of Clinical Investigation*, vol. 34, no. 9, pp. 1345–1353, 1955.

[26] C. J. Andersen, C. N. Blesso, J. Lee et al., "Egg consumption modulates HDL lipid composition and increases the cholesterol-accepting capacity of serum in metabolic syndrome," *Lipids*, vol. 48, no. 6, pp. 557–567, 2013.

[27] H. Schwende, E. Fitzke, P. Ambs, and P. Dieter, "Differences in the state of differentiation of THP-1 cells induced by phorbol ester and 1,25-dihydroxyvitamin D3," *Journal of Leukocyte Biology*, vol. 59, no. 4, pp. 555–561, 1996.

[28] D. G. Hassall, "Three probe flow cytometry of a human foam-cell forming macrophage," *Cytometry*, vol. 13, no. 4, pp. 381–388, 1992.

[29] C. J. Delvecchio and J. P. Capone, "Protein kinase C α modulates liver X receptor α transactivation," *The Journal of Endocrinology*, vol. 197, no. 1, pp. 121–130, 2008.

[30] B. E. Cox, E. E. Griffin, J. C. Ullery, and W. G. Jerome, "Effects of cellular cholesterol loading on macrophage foam cell lysosome acidification," *Journal of Lipid Research*, vol. 48, no. 5, pp. 1012–1021, 2007.

[31] L. L. Rudel, R. G. Lee, and T. L. Cockman, "Acyl coenzyme A: cholesterol acyltransferase types 1 and 2: structure and function in atherosclerosis," *Current Opinion in Lipidology*, vol. 12, no. 2, pp. 121–127, 2001.

[32] J. Julve, J. C. Escolà-Gil, E. Rodríguez-Millán et al., "Methionine-induced hyperhomocysteinemia impairs the antioxidant ability of high-density lipoproteins without reducing in vivo macrophage-specific reverse cholesterol transport," *Molecular Nutrition and Food Research*, vol. 57, no. 10, pp. 1814–1824, 2013.

[33] M. Maranghi, A. Hiukka, R. Badeau, J. Sundvall, M. Jauhiainen, and M.-R. Taskinen, "Macrophage cholesterol efflux to plasma and HDL in subjects with low and high homocysteine levels: a FIELD substudy," *Atherosclerosis*, vol. 219, no. 1, pp. 259–265, 2011.

[34] S. Yamamoto, P. G. Yancey, T. A. Ikizler et al., "Dysfunctional high-density lipoprotein in patients on chronic hemodialysis," *Journal of the American College of Cardiology*, vol. 60, no. 23, pp. 2372–2379, 2012.

[35] P. Nestel, A. Hoang, D. Sviridov, and N. Straznicky, "Cholesterol efflux from macrophages is influenced differentially by plasmas from overweight insulin-sensitive and -resistant subjects," *International Journal of Obesity*, vol. 36, no. 3, pp. 407–413, 2012.

[36] A. V. Khera, M. Cuchel, M. De La Llera-Moya et al., "Cholesterol efflux capacity, high-density lipoprotein function, and atherosclerosis," *The New England Journal of Medicine*, vol. 364, no. 2, pp. 127–135, 2011.

[37] R. Shroff, T. Speer, S. Colin et al., "HDL in children with CKD promotes endothelial dysfunction and an abnormal vascular phenotype," *Journal of the American Society of Nephrology*, vol. 25, no. 11, pp. 2658–2668, 2014.

[38] X.-M. Li, W. H. W. Tang, M. K. Mosior et al., "Paradoxical association of enhanced cholesterol efflux with increased incident cardiovascular risks," *Arteriosclerosis, Thrombosis, and Vascular Biology*, vol. 33, no. 7, pp. 1696–1705, 2013.

[39] C. Charles-Schoeman, Y. Y. Lee, V. Grijalva et al., "Cholesterol efflux by high density lipoproteins is impaired in patients with active rheumatoid arthritis," *Annals of the Rheumatic Diseases*, vol. 71, no. 7, pp. 1157–1162, 2012.

[40] P. Linsel-Nitschke, H. Jansen, Z. Aherrarhou et al., "Macrophage cholesterol efflux correlates with lipoprotein subclass distribution and risk of obstructive coronary artery disease in patients undergoing coronary angiography," *Lipids in Health and Disease*, vol. 8, article 14, 2009.

Permissions

The contributors of this book come from diverse backgrounds, making this book a truly international effort. This book will bring forth new frontiers with its revolutionizing research information and detailed analysis of the nascent developments around the world.

We would like to thank all the contributing authors for lending their expertise to make the book truly unique. They have played a crucial role in the development of this book. Without their invaluable contributions this book wouldn't have been possible. They have made vital efforts to compile up to date information on the varied aspects of this subject to make this book a valuable addition to the collection of many professionals and students.

This book was conceptualized with the vision of imparting up-to-date information and advanced data in this field. To ensure the same, a matchless editorial board was set up. Every individual on the board went through rigorous rounds of assessment to prove their worth. After which they invested a large part of their time researching and compiling the most relevant data for our readers.

The editorial board has been involved in producing this book since its inception. They have spent rigorous hours researching and exploring the diverse topics which have resulted in the successful publishing of this book. They have passed on their knowledge of decades through this book. To expedite this challenging task, the publisher supported the team at every step. A small team of assistant editors was also appointed to further simplify the editing procedure and attain best results for the readers.

Apart from the editorial board, the designing team has also invested a significant amount of their time in understanding the subject and creating the most relevant covers. They scrutinized every image to scout for the most suitable representation of the subject and create an appropriate cover for the book.

The publishing team has been an ardent support to the editorial, designing and production team. Their endless efforts to recruit the best for this project, has resulted in the accomplishment of this book. They are a veteran in the field of academics and their pool of knowledge is as vast as their experience in printing. Their expertise and guidance has proved useful at every step. Their uncompromising quality standards have made this book an exceptional effort. Their encouragement from time to time has been an inspiration for everyone.

The publisher and the editorial board hope that this book will prove to be a valuable piece of knowledge for researchers, students, practitioners and scholars across the globe.

List of Contributors

Lena Foseid, Hanne Devle, Yngve Stenstrøm, Carl Fredrik Naess-Andresen and Dag Ekeberg
Faculty of Chemistry, Biotechnology and Food Science, Norwegian University of Life Sciences, 1432 Ås, Norway

Mousa Abujbara, Mohammed El-Khateeb and Kamel Ajlouni
1The National Center (Institute) for Diabetes, Endocrinology and Genetics (NCDEG)/The University of Jordan, Jordan

Anwar Batieha
Department of Community Medicine, Jordan University of Science and Technology, Irbid, Jordan

Yousef Khader
Jordan University of Science and Technology (JUST), Irbid, Jordan

Hashem Jaddou
The Jordan University of Science and Technology (JUST), Irbid, Jordan

Xingxuan He and Edward H. Schuchman
Department of Genetics & Genomic Sciences, Icahn School of Medicine at Mount Sinai, 1425 Madison Avenue, New York, NY 10029, USA

Kristian Laake, Ingebjørg Seljeflot and Harald Arnesen
Center for Clinical Heart Research, Department of Cardiology, Oslo University Hospital, Ullev°al, 0450 Oslo, Norway
Faculty of Medicine, University of Oslo, 0316 Oslo, Norway
Center for Heart Failure Research, University of Oslo, 0316 Oslo, Norway

Svein Solheim
Center for Clinical Heart Research, Department of Cardiology, Oslo University Hospital, Ullev°al, 0450 Oslo, Norway
Center for Heart Failure Research, University of Oslo, 0316 Oslo, Norway

Peder Myhre
Center for Clinical Heart Research, Department of Cardiology, Oslo University Hospital, Ullev°al, 0450 Oslo, Norway
Center for Heart Failure Research, University of Oslo, 0316 Oslo, Norway
Department of Cardiology, Akershus University Hospital HF, 1478 Lørenskog, Norway

Erik B. Schmidt
Department of Cardiology, Aalborg University Hospital, 9000 Aalborg, Denmark

Arnljot Tveit
Department of Medical Research, Vestre Viken Hospital Trust, Bærum Hospital, 1346 Rud, Norway

Ekaterina Fock, Vera Bachteeva, Elena Lavrova and Rimma Parnova
I. M. Sechenov Institute of Evolutionary Physiology and Biochemistry of the Russian Academy of Sciences, Saint-Petersburg, Russia

Victoria Pons and Montserrat Banquells
Departamento de Investigación del CAR, Av. Alcalde Barnils 3, 08173 Sant Cugat del Vallés, Spain

Franchek Drobnic
Departamento de Investigación del CAR, Av. Alcalde Barnils 3, 08173 Sant Cugat del Vallés, Spain
Servicios Médicos del FC Barcelona, Av. del Sol, s/n, Sant Joan Despí, 08970 Barcelona, Spain

Félix Rueda, Begoña Cordobilla and Joan Carles Domingo
Departamento de Bioqu´ımica y Biolog´ıa Molecular, Facultad de Biolog´ıa, Av. Diagonal 643, 08028 Barcelona, Spain

Mohamadreza Haeri
Department of Biochemistry, Qom University of Medical Sciences, Qom, Iran

Mahmoud Parham
Clinical Research Development Center, Qom University of Medical Sciences, Qom, Iran

Neda Habibi
Gastroenterology and Hepatology Disease Research Center, Qom University of Medical Sciences, Qom, Iran

Jamshid Vafaeimanesh
Gastroenterology and Hepatology Disease Research Center, Qom University of Medical Sciences, Qom, Iran
Gastrointestinal and Liver Diseases Research Center, Iran University of Medical Sciences, Tehran, Iran

Moshrik Abd alamir, Adib Chaus, Leslie Tamura and Alan Brown
Advocate Lutheran General Hospital, Parkside Ste B-01, 1775 Dempster St, Park Ridge, IL 60068, USA

Michael Goyfman
Stony Brook University Hospital, 101 Nicolls Rd, Stony Brook, NY 11794, USA

Firas Dabbous
James R. & Helen D. Russell Institute for Research & Innovation, Advocate Lutheran General Hospital, Center for Advanced Care, 1700 Luther Lane, Suite 1410, Park Ridge, IL 60068, USA

Veit Sandfort
National Institute of Health, 30 Convent Dr, Bethesda, MD 20892, USA

Mathew Budoff
Harbor UCLA Cardiology, 1000W. Carson St, Torrance, CA 90509, USA

Richard K. D. Ephraim, Swithin M. Swaray and Patrick Adu
Department of Medical Laboratory Science, School of Allied Health Sciences, University of Cape Coast, Ghana

Hope Agbodzakey, Prince Adoba, Bright Oppong Afranie, Linda Ahenkorah Fondjo, Samuel Asamoah Sakyi and Beatrice Amoah
Department of Molecular Medicine, School of Medical Sciences, Kwame Nkrumah University of Science and Technology, Ghana

Emmanuel Acheampong and Enoch Odame Anto
Department of Molecular Medicine, School of Medical Sciences, Kwame Nkrumah University of Science and Technology, Ghana

School of Medical and Health Sciences, Edith Cowan University, Western Australia, Australia

Emmanuella Nsenbah Batu
Department of Molecular Medicine, School of Medical Sciences, Kwame Nkrumah University of Science and Technology, Ghana
Department of Biochemistry, Dalian Medical University, China

Megumi Sato, Akira Yoshimoto, Kouji Yano, Ryunosuke Ohkawa and Minoru Tozuka
Analytical Laboratory Chemistry, Field of Applied Laboratory Science, Graduate School of Health Care Sciences, Tokyo Medical and Dental University, ·1-5-45 Yushima, Bunkyo-Ku, Tokyo 113-8519, Japan

Naoya Ichimura
Analytical Laboratory Chemistry, Field of Applied Laboratory Science, Graduate School of Health Care Sciences, Tokyo Medical and Dental University, 1-5-45 Yushima, Bunkyo-Ku, Tokyo 113-8519, Japan
Department of Clinical Laboratory, Medical Hospital of Tokyo Medical and Dental University, 1-5-45 Yushima, Bunkyo-Ku, Tokyo 113-8519, Japan

Takeshi Kasama
Instrumental Analysis Research Division, Research Center for Medical and Dental Sciences, Tokyo Medical and Dental University, 1-5-45 Yushima, Bunkyo-Ku, Tokyo 113-8519, Japan

Antonio Laguna-Camacho
Medical Sciences Research Centre, Autonomous University of the State of Mexico, Toluca, MEX, Mexico

Truong Dang Le
Institute of Biotechnology and Food Technology, Industrial University of Ho Chi Minh City, Ho Chi Minh City, Vietnam

Van Thi Ai Nguyen
Institute of Biotechnology and Food Technology, Industrial University of Ho Chi Minh City, Ho Chi Minh City, Vietnam
Department of Food Technology, Faculty of Chemical Engineering, Ho Chi Minh City University of Technology, Ho Chi Minh City, Vietnam

Hoa Ngoc Phan and Lam Bich Tran
Department of Food Technology, Faculty of Chemical Engineering, Ho Chi Minh City University of Technology, Ho Chi Minh City, Vietnam

L. E. Gutiérrez-Pliego, B. E. Martínez-Carrillo, J. A. Escoto-Herrera, C. A. Rosales-Gómez and R. Valdés-Ramos
Laboratorio de Investigación en Nutrición, Facultad de Medicina, Universidad Aut´onoma del Estado de México, Paseo Tollocan y Venustiano Carranza s/n, Col. Universidad, 50180 Toluca, MEX, Mexico

A. A. Reséndiz-Albor and I. M. Arciniega-Martínez
Laboratorio de Inmunidad de Mucosas, Sección de Investigación y Posgrado, Escuela Superior de Medicina, Instituto Politécnico Nacional, Av. Plan de San Luis S/N, Colonia Casco de Santo Tomas, Miguel Hidalgo, 11350 Ciudad de México, Mexico

Naseeha Jamil
Department of Biomedical and Nutritional Sciences, University of Massachusetts, Lowell, MA, USA

Mahdi Garelnabi
Department of Biomedical and Nutritional Sciences, University of Massachusetts, Lowell, MA, USA
Biomedical Engineering and Biotechnology Program, University of Massachusetts, Lowell, MA, USA

Halleh Mahini and Chinedu Ochin
Biomedical Engineering and Biotechnology Program, University of Massachusetts, Lowell, MA, USA

Gregory Ainsworth
Department of Chemistry, University of Massachusetts, Lowell, MA, USA

E. Derbyshire
Nutritional Insight Ltd, Surrey, UK

Dhrubajyoti Bandyopadhyay
Internal Medicine, Mount Sinai St Luke's Roosevelt Hospital Center, New York, NY, USA

Kumar Ashish
University of Texas MD Anderson Cancer Center, Houston, TX, USA

Adrija Hajra
IPGMER, Kolkata, India

Arshna Qureshi
Department of Medicine, Lady Hardinge Medical College, New Delhi, India

Raktim K. Ghosh
Metro Health, CaseWestern Reserve University, Cleveland, OH, USA

Hidetoshi Yamada, Sayaka Kikuchi, Mayuka Hakozaki, Kaori Motodate and Nozomi Nagahora
Iwate Biotechnology Research Center, 22-174-4 Narita, Kitakami, Iwate 024-0003, Japan

Masamichi Hirose
Department of Molecular and Cellular Pharmacology, Iwate Medical University School of Pharmaceutical Sciences, Shiwa, Iwate 028-3694, Japan

Dhrubajyoti Bandyopadhyay
Department of Internal Medicine, Mount Sinai St Luke's Roosevelt, New York, NY, USA

Arshna Qureshi
Department of Medicine, Lady Hardinge Medical College, New Delhi, India

Sudeshna Ghosh
IPGMER, Kolkata, India

Kumar Ashish
The University of Texas MD Anderson Cancer Center, Houston, TX, USA

Lyndsey R. Heise
Department of Internal Medicine, University of Nebraska Medical Center, Omaha, NE, USA

Adrija Hajra
Department of Internal Medicine, IPGMER, Kolkata, India

Raktim K. Ghosh
Division of Cardiovascular Diseases, Metrohealth Medical Center, CaseWestern Reserve University, Cleveland, OH, USA

Abrha Gebreselema and Yohannes Abere
Department of Biochemistry, College of Medicine and Health Sciences, Bahir Dar University, Bahir Dar, Ethiopia

Zewdie Mekonnen
Department of Biochemistry, College of Medicine and Health Sciences, Bahir Dar University, Bahir Dar, Ethiopia
Department of Biomedical Research, Biotechnology Research Institute, Bahir Dar University, Bahir Dar, Ethiopia

Sikandar Hayat Khan and Syed Mohsin Manzoor
Department of Pathology, PNS Hafeez, Islamabad, Pakistan

Nadeem Fazal and Muhammad Yasir
Department of Medicine, PNS Hafeez, Islamabad, Pakistan

Athar Abbas Gilani Shah
Department of Surgery, PNS Hafeez, Islamabad, Pakistan

Naveed Asif and Aamir Ijaz
Department of Chemical Pathology & Clinical Endocrinology, AFIP, Rawalpindi, Pakistan

Najmusaqib Khan Niazi
Administration Department, PNS Hafeez, Islamabad, Pakistan

Danielle W. Kimmel and David E. Cliffel
Department of Chemistry, Vanderbilt University, VU Station B, Nashville, TN 37235-1822, USA
Vanderbilt Institute for Integrative Biosystems Research and Education, Vanderbilt University, Nashville, TN 37235-1809, USA

William P. Dole
Novartis Institutes for Biomedical Research, 220 Massachusetts Ave. 360C, Cambridge, MA 02139, USA

Richard K. D. Ephraim, Patrick Adu, Edem Ake, Hope Agbodzakey and Clement Agoni
Department of Medical Laboratory Science, School of Allied Health Sciences, College of Health and Allied Sciences, University of Cape Coast, Cape Coast, Ghana

Prince Adoba
Department of MolecularMedicine, School of Medical Sciences, College of Health, KNUST, Kumasi, Ghana

Obed Cudjoe
Department of Microbiology, School of Medical Sciences, College of Health and Allied Sciences, University of Cape Coast, Cape Coast, Ghana

Anthony P. Morise, Jennifer Tennant and Sari D. Holmes
Section of Cardiology, West Virginia University Heart and Vascular Institute, Morgantown, WV, USA

Danyel H. Tacker
Department of Pathology, Anatomy, and LaboratoryMedicine,West Virginia University School ofMedicine,Morgantown, WV, USA

Wenyu Wang and Elisa Lee
Center for American Indian Health Research, College of Public Health, University of Oklahoma Health Sciences Center, Oklahoma City, OK 73190, USA

Piers Blackett
Section of Diabetes and Endocrinology, Department of Pediatrics, Harold Hamm Diabetes Center, University of Oklahoma Health Sciences Center, Oklahoma City, OK 73104, USA

Sohail Khan
The Cherokee Nation, Tahlequah, OK 74465, USA

Kouji Yano, Ryunosuke Ohkawa, Megumi Sato, Akira Yoshimoto, Naoya Ichimura and Minoru Tozuka
Analytical Laboratory Chemistry, Field of Applied Laboratory Science, Graduate School of Health Care Sciences, Tokyo Medical and Dental University, 1-5-45 Yushima, Bunkyo-ku, Tokyo 113-8519, Japan

Takahiro Kameda
Department of Medical Technology, School of Health Sciences, Tokyo University of Technology, 5-23-22 Nishi-Kamata, Ohta-ku, Tokyo 144-8535, Japan

Tetsuo Kubota
Microbiology and Immunology, Field of Applied Laboratory Science, Graduate School of Health Care Sciences, Tokyo Medical and Dental University, 1-5-45 Yushima, Bunkyo-ku, Tokyo 113-8519, Japan

Index

www.ingramcontent.com/pod-product-compliance
Lightning Source LLC
Chambersburg PA
CBHW082038190326
41458CB00010B/3397